The Geological Society of America
Memoir 163

The Miocene Ocean:
Paleoceanography and Biogeography

Edited by

James P. Kennett
Graduate School of Oceanography
University of Rhode Island
Narragansett, Rhode Island 02882-1197

1985

Published by The Geological Society of America, Inc.
3300 Penrose Place, P.O. Box 9140, Boulder, Colorado 80301

Printed in U.S.A.

Library of Congress Cataloging-in-Publication Data
Main entry under title:

The Miocene ocean.

   (Memoir / Geological Society of America ; 163)
   Bibliography: p.
   1. Geology, Stratigraphic—Miocene—Addresses,
essays, lectures. 2. Paleoceanography—Addresses,
essays, lectures. 3. Paleontology—Miocene—Addresses,
essays, lectures. I. Kennett, James P. II. Series:
Memoir (Geological Society of America) ; 163.
QE694.M528   1985   551.7'87   85-14851

Contents

Contents

Preface

This volume is about the paleoceanography and paleobiogeography of the world ocean during the Miocene Epoch, that interval of time between about 24 and 5 Ma. The sedimentary sequences upon which these studies were based resulted from the highly successful Deep Sea Drilling Project (DSPD). Since 1968, deep-sea drilled sections have been obtained from all ocean basins except the Arctic using D. V. *Glomar Challenger.* These sections represent an enormously important resource for earth science investigations.

All but one of the 14 contributions in this volume are products of research within the Cenozoic Paleoceanography project (CENOP), a multi-institutional project established to generate paleoceanographic reconstructions of the Miocene ocean using stable isotopes and quantitative micropaleontological data. CENOP investigations have generated data related to a wide range of parameters and processes of the Miocene ocean including paleogeography, stratigraphy, circulation, water-mass distributions, temperature, variability, vertical water-mass structure, biogeography, and biotic evolution.

Ocean evolution of the Miocene was chosen by CENOP investigators because it was an interval of time of major ice accumulation on Antarctica, with related global paleoclimatic and paleoceanography changes. The Miocene was the interval of time when *major* polar ice sheets became a *permanent* feature of the earth's surface; it was a transition interval between the still unfamiliar Oligocene ocean and the more familiar late Cenozoic ocean. The base of the Miocene is marked by the beginning of major evolutionary radiations of oceanic microfossils, especially planktonic foraminifera, which lead to distinctive Neogene lineages; the end of the Miocene is marked by the terminal Miocene Event represented by an extraordinary number of global paleoenvironmental changes such as global cooling and the Mediterranean salinity crisis.

A number of important changes occurred in the ocean during the Miocene that strongly affected oceanic biogeographic patterns and biotic evolution. Among the changes was a major steepening of thermal gradients of surface waters as the circum-Antarctic Current continued to expand and intensify. Part of this increase in pole to equator surface-water temperature gradient resulted from cooling in high latitudes, but also to warming of low-latitude surface waters.

During the middle Miocene, about 14 Ma, a crucial development occurred in global paleoceanography related to the accumulation of much of the Antarctic ice sheet. This event is marked by a sharp increase in the $\delta^{18}O$ values of oceanic calcareous microfossils, in part reflecting a change in oceanic oxygen isotopic values related to polar ice accumulation, but also partly to a drop in surface temperatures.

Biogeographic patterns of planktonic microfossils underwent distinct changes during the Miocene that have assisted in evaluating paleocirculation patterns. Important differences apparent between the early and late Miocene resulted from changes in surface-water circulation in the Pacific and from the closure of the Indonesian Seaway connection between the tropical Pacific and Indian Oceans.

The early, middle, and late Miocene are each marked by major paleoclimatic and paleoceanographic events that fundamentally changed the oceanic regime as expressed by the oxygen and carbon isotopic records and by changing faunal and floral assemblages.

CENOP investigators chose three time-slice intervals for the purposes of mapping the Miocene ocean before and after the middle Miocene accumulation of Antarctic ice. Used in combination with time-series analyses of selected sections, these time-slice maps have been employed to describe the evolution of the Miocene ocean and its marine assemblages. Time-slices chosen are marked by minimal hiatus development in the deep sea, so as to be represented by sediment in as many stratigraphic sequences as possible throughout the oceans.

A time slice was chosen at 22 Ma in the earliest Miocene to characterize the ocean prior to the major evolutionary diversification of the planktonic foraminifera leading to the Neogene lineages. Another time-slice was chosen across the early Miocene/middle Miocene boundary at 16 Ma following this planktonic foraminiferal radiation and before the distinct oxygen isotopic event that is interpreted to represent a major expansion of the Antarctic ice sheet. A time-slice was chosen in the late Miocene at 8 Ma to characterize the late Miocene ocean following the middle Miocene oxygen isotopic shift and prior to the latest Miocene $\delta^{13}C$ shift (carbon shift).

Although the themes of some papers in this volume are broad and overlapping, for the most part the contributions have been organized in the following manner: paleogeography and paleodepth of the Miocene ocean; stratigraphy; stable isotopic history of the Pacific and Indian Oceans; benthic foraminiferal distributions in relation to bottom and intermediate water structure and history; surface-water biogeography of planktonic foraminifera and radiolaria in relation to paleoceanographic history; and finally, paleoceanographic studies of the Atlantic Ocean.

The paleogeography of the Miocene ocean and its paleoceanographic implications are described by Sclater et al. This paper is based upon charts of the ages and paleodepths of the ocean basins and position of the continents at selected times, including the three CENOP time slices. Barron et al. present a multiple microfossil biochronology (175 datums) for the Miocene, summarizing the stratigraphic basis of the three time-slice intervals mapped by CENOP investigators. Ages are estimated by integration with paleomagnetic stratigraphy and result in a stratigraphic resolution approaching 100,000 years. An additional set of paleomagnetically dated diatom datums for the equatorial and north Pacific areas are presented by Burckle and Opdyke.

A major synthesis of oxygen isotopic data by Savin et al., provides a much more detailed picture of the evolution of surface and near-surface waters for the Miocene than previously obtained. Oxygen isotopic mapping of the three time-slices shows the development of the global latitudinal thermal gradient and of east-west gradients across the Pacific Ocean. This investigation is buttressed by that of Barrera et al. who document the thermal evolution of surface water masses in the eastern North Pacific using time-series of stable isotopes and planktonic foraminifera. Until now, there has been little isotopic information documenting the details of the evolution of surface waters at mid-latitudes of the North Pacific. Oxygen and carbon time-series are presented by Vincent et al. for the tropical Indian Ocean. This forms an important stratigraphic basis for future investigations and has revealed the presence of an early-middle Miocene carbon isotopic shift of much potential importance in interpreting biotic productivity of the circum-Pacific region.

The single, but extensive contribution in this volume that addresses the history of deep and intermediate waters during the Miocene, is by Woodruff. Documented are major changes in Pacific benthic foraminiferal distributions that are related to changes in the vertical water-mass structure, biotic productivity, and equatorial circulation.

This volume contains five contributions about planktonic microfossil paleobiogeography of the Pacific in relation to the development paleoceanographic conditions of the Miocene. Keller infers the depth stratification of Miocene planktonic foraminifera based upon oxygen isotopic values, with the expressed purpose of better understanding changes in vertical surficial water-mass structure of the Pacific Ocean including the equatorial region. Kennett et al. map planktonic foraminiferal distributions in the three time-slices. Major biogeographic differences are evident between the early and late Miocene. These faunal changes are interpreted to reflect both the development, during the middle Miocene, of the Equatorial Undercurrent system when the Indonesian Seaway effectively closed (as a result of Australian northward drift) and the general strengthening of the gyral circulation and the Equatorial Countercurrent that resulted from increased Antarctic glaciation and high-latitude cooling during the middle Miocene. Romine maps the distribution of radiolarians at

8 Ma and documents important differences between the Miocene and Modern Pacific Oceans. These differences are related to further changes in vertical water mass structure across the equatorial Pacific related, in part, to the closure during the Pliocene of the Central American Seaway. A radiolarian time series analysis of Site 298 in the western equatorial Pacific is made by Romine and Lombari and relates to changes in surface water circulation and climate. Lombari maps radiolarian biogeographic patterns across the equatorial Pacific during the two Early Miocene time slices. Also traced in this paper are broad evolutionary patterns of radiolaria; in particular the changing relative importance of nassellarians and spumellarians during the Neogene.

Two contributions in this volume deal with the Atlantic Ocean. Stein presents a time-series of stable isotopic and sedimentary data at Site 366 in the eastern equatorial Atlantic (Sierra Leone Rise). This investigation documents intensified meridional wind circulation and a more arid to semi-arid climate in North Africa in the latest Miocene and late Pliocene. Finally, Hodell and Kennett present maps of oxygen isotopic and planktonic foraminiferal patterns for the three time-slices in the South Atlantic, and infer major paleoceanographic changes between the early and late Miocene.

Many individuals have contributed to the production of this volume. First I would like to thank all of the workers involved in the CENOP Project (1977–1982) for their diverse and valuable contributions over the years. In particular I enjoyed the cooperation of T. Moore, Jr. who co-directed the CENOP Project and provided much-valued advice. At the University of Rhode Island I thank N. Meader and K. Nelson for their assistance in the production of this volume; to D. Hodell for his assistance with the reviews and to M. Emery for help with coordination of the project.

The staff at the Deep Sea Drilling Project, Scripps Institution of Oceanography assisted by their prompt and efficient attention to our numerous sample requests. In particular we are indebted to L. Garifal and A. Altman for their kind cooperation.

There were numerous reviewers of the contributions in this volume whom I thank for their prompt and most valuable constructive advice and criticisms. Those that agreed to be identified as referees during the review process are as follows: M. Bender, W. Berger, R. Casey, J. Caulet, P. Ciesielski, R. Fairbanks, B. Haq, D. Hodell, J. Ingle, G. Jones, D. Johnson, L. Keigwin, D. Kellogg, T. Kellogg, P. Kroopnick, R. Larson, D. Lazarus, M. Ledbetter, P. Lohmann, L. Mayer, K. Miller, T. Moore, C. Nigrini, W. Orr, N. Pisias, R. Poore, W. Prell, D. Rea, W. Riedel, S. Savin, D. Schnitker, N. Shackleton, F. Theyer, and R. Thunell.

Funding for the CENOP Project was provided by the National Science Foundation from the Office of Submarine Geology and Geophysics.

Geological Society of America
Memoir 163
1985

The depth of the ocean through the Neogene

John G. Sclater
The Institute for Geophysics
The University of Texas at Austin
4920 North IH 35
Austin, Texas 78751

Linda Meinke
Andrew Bennett
The Department of Earth and Planetary Sciences
The Massachusetts Institute of Technology
Cambridge, Massachusetts 02139

Cynthia Murphy
The Institute for Geophysics
The University of Texas at Austin
4920 North IH 35
Austin, Texas 78751

ABSTRACT

A simple relation exists between depth and age for ocean floor created at a spreading center. The depth for about 60 percent of the deep oceans can be characterized by this relationship; areas whose depth differs significantly are called residual depth anomalies. They can be separated into two types: aseismic ridges, which are excess crustal loads created at or near a spreading center; and mid-ocean swells, which are thought to be surface manifestations of convection cells in the upper mantle beneath the lithospheric plate. The past position and depth of both types of feature are discussed.

A simple method is introduced for computing the past depth of sediment recovered in individual Deep Sea Drilling Project holes. The method can be used for sites on both normal ocean floor and aseismic ridges. It is accurate from the beginning of the Neogene (25 Ma) to better than 100 m for almost all sites. Four specific examples are discussed in the text.

Charts showing the ages of the ocean floor and the positions of the major plate boundaries and continents are constructed by combining a digitized version of the sea-floor magnetic isochrons with published rotation poles and angles. This report includes constructions for the present, the time of anomaly 5 (9 Ma), and the time of anomaly 6 (20 Ma). Additional charts which include predicted ocean depths are constructed for the present and for three selected ages in the Neogene (22, 16, and 8 Ma). The predicted bathymetries are derived by combining the depth-age relation for normal ocean floor, simple models for aseismic ridges, and individual analyses for mid-ocean swells. A comparison between the predicted and observed bathymetry for the present is presented as evidence that the predicted contours lie within ± 400 m of the observed for about 80 percent of the deep oceans. The marginal basins are excluded.

The paleogeographic position of the continents and the paleodepth contours are evidence that:

1) There was a shallow water passage path around Antarctica well prior to the Neogene.

2) Just before or just after the onset of the Neogene a deep water passage may have developed around Antarctica.

3) The closure of the equatorial circulation system in the late Neogene was proba-

bly complex, starting with the closure of northern Australia and South East Asia, then Africa and Eurasia and finally the formation of the isthmus of Panama.

The opening and closing of these seaways are in the correct direction to produce effects which may be responsible for the major climatic events in the Cenozoic. These include the cooling of the high latitudes in the Late Eocene, the formation of an Antarctic ice sheet in the Middle Miocene and, finally, formation of Northern Hemisphere ice sheets in the Pliocene.

INTRODUCTION

The theory of plate tectonics separates the surface of the Earth into a suite of rigid caps. The caps, most of which include both oceans and continents, interact constantly. These interactions give rise to earthquakes which in turn define the plate boundaries. New plate material is created at a spreading center and moves past other plate material at transform faults. Oceanic plate material is destroyed at trenches, while continental plate is not destroyed but is modified greatly by magmatic processes at the trenches or by continent/continent collision. Plate tectonics is a kinematic description of the surface of an Earth which is in constant motion and is continually evolving. With this description, it is possible to understand the present distribution of earthquake epicenters and, in addition, to construct the past history of ocean basins and the continents which surround them.

Ocean crust is created by the intrusion of hot molten material which cools and contracts as it moves away from a spreading center. As a result of this cooling and contraction, the ocean crust subsides as it gets older. This subsidence gives rise to the world-encircling system of mid-ocean ridges which are characterized by active epicenters at their crests. The cooling of the ocean crust as it moves away from a spreading center occurs predominantly in the vertical direction. In general, it is independent of the speed at which the plate is moving. As a consequence, a simple relation is expected between age and subsidence of a mid-ocean ridge. The depth has been shown to increase with the square root of time for the first 65-80 million years and roughly exponentially thereafter (Parsons and Sclater, 1977).

The depth of the mid-ocean ridges and the height of other topographic features, such as aseismic ridges (e.g., the Ninetyeast Ridge) and oceanic swells (e.g., Hawaii) are known to control the circulation of deep water in the oceans. Thus both the present and the past depth are likely to have had a large effect on the distribution of sediments throughout time. A long term record of deposition in the oceans has been provided by the sediments recovered by the Deep Sea Drilling Project (DSDP). Because of the subsidence of the ocean floor through time not all of the sediment has been deposited at the depth it now occupies in a Deep Sea Drilling Project hole. For example, a site currently at 4600 m in depth on 36 million-year-old crust was created at the crest of a spreading center at an original depth of 2500 m. This site has subsided 2100 m during the 36 million years that sediment has been deposited. The sediment recovered at this site will reflect the history of this subsidence.

Sclater et al. (1980) use the lineation of identified magnetic anomalies, the theory of plate tectonics and a time scale relating anomaly to age to determine the present position of various isochrons on the sea floor. The past position of these isochrons and the continents which surround them can be determined for any selected time by use of the theory of plate tectonics and the rotation parameters which represent the finite motions of the plates between the present and the time in question. Adding paleomagnetic data to obtain the appropriate latitudinal position of the continents and the relation between depth and age to give depth, paleogeographic and paleodepth charts of the oceans can be constructed at these selected times. Such paleodepth charts represent only about 60 percent of the sea floor if the relation between depth and age alone is used and hence are of qualitative rather than quantitative value. However, their usefulness can be substantially improved if the past depth of anomalies such as aseismic ridges and mid-ocean swells are added.

The major goals of this paper are to discuss: a) paleodepth charts and the implications of recent improvements in the understanding of depth anomalies on the ocean floor for these charts; b) a simple method for determining the past depth of sediment recovered in Deep Sea Drilling Project holes; c) preliminary paleodepth charts at selected times in the Neogene (22, 16 and 8 Ma) which include the major aseismic ridges and oceanic swells and finally; (d) the influence of changes in paleogeography and paleodepth on sedimentation in the deep ocean and on climate.

THE RELATION BETWEEN DEPTH AND AGE AND THE CONSTRUCTION OF PALEODEPTH CHARTS

The morphology of the ocean floor is dominated by a world encircling system of mid-ocean ridges (Figure 1 in pocket inside back cover). This ridge system bisects the North and South Atlantic, lies on the eastern side of the Pacific and separates into three quite distinct ridge sequences in the Indian Ocean. Identified magnetic anomaly lineations are observed over most of the ocean floor. Deep Sea Drilling Project sites have been occupied in all the major ocean basins. The age of the ocean floor is determined from the lineations by use of a time scale which relates magnetic anomaly to age. Where these anomalies are absent the age of the oldest sediment recovered in Deep Sea Drilling Project holes is used.

Various authors (Menard, 1969; Sclater et al., 1971; Parsons and Sclater, 1977) show that there is a simple relation between depth and age for mid-ocean ridges. For the first 70 million years the ocean floor subsides as the square root of time (Davis and Lister, 1974), and thereafter the depth increases exponentially to

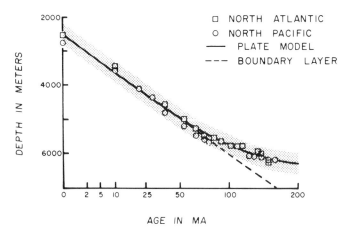

Figure 2. Plots of individual depth measurements used in calculating mean depths and the depth of DSDP Sites for (a) the North Pacific and (b) the North Atlantic. The values illustrate the scatter in depths about the mean value. For the purpose of ease of presentation, the means and standard deviation from the individual profile measurements are offset 2 m.y. The correct age is represented by the first column of the individual points, the others are offset slightly for clarity. The solid circles in Figure 2(b) are points from profiles north of 31°N, including the Azores and Iceland Highs from Sclater et al., (1975).

plate by thermal convection, perturbing this simple cooling relationship (Parsons and McKenzie, 1978).

A compilation of sediment-corrected basement depths from deeper portions of the Atlantic and Pacific oceans is presented by Parsons and Sclater, (1977). Though there is a significant scatter in the data, both sets of observations can be fit to the same empirical curve (Figure 2). The theoretical curve for the creation and cooling of a 125 km thick plate with a bottom boundary temperature of 1330°C best fits the observations. In this concept, a plate of constant thickness is a simple approximation for a more complex model in which the oceanic lithosphere is thought to consist of a rigid upper layer, which is mechanically coherent, and a lower layer, which can convect (Figure 3, Parsons and McKenzie, 1978).

Sclater et al. (1980) compile the world-wide magnetic lineations from the sea floor. They relate these to age using different geomagnetic reversal time scales for the Cenozoic (LaBrecque et al., 1977), Cretaceous (van Hinte, 1976a) and Jurassic (van Hinte, 1976b). From the age of the ocean floor and the tectonic history of the mid-ocean spreading centers, Sclater et al. (1980) construct isochrons of the sea floor for 20, 35, 53, 65, 80, 95, 110, 125, 140 and 165 Ma. A plot of a digitized version of these isochrons and the major plates, used later in this manuscript for the purpose of the reconstructions, is presented as Figure 4a. In addition, the continents and major plate boundaries including spreading centers, transform faults and trenches are shown.

A spreading center has a depth of 2500 m and the 2, 20, 50 and 130 Ma isochrons have depths of 3000, 4000, 5000 and 6000 m respectively (Parsons and Sclater, 1977). The theoretical depth is determined by interpolating the position of these iso-

a constant value (Parsons and Sclater, 1977). This behavior is accounted for by the cooling and contraction of the oceanic plate after it is created at a spreading center. The square root portion of the curve is the region where the plate is cooling from above and where the effect of thermal perturbations from beneath the plate can be ignored. The later exponential portion of the subsidence is thought to result from extra heat brought to the bottom of the

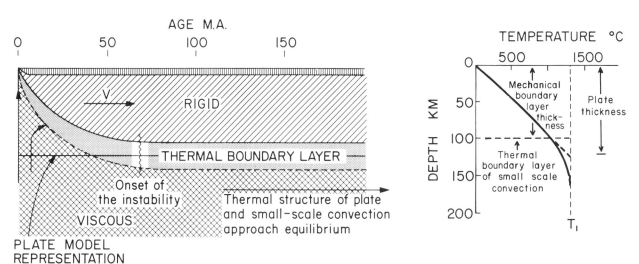

Figure 3. A schematic diagram showing the division of the plate into rigid and viscous regions and the occurrence of an instability in the bottom viscous part (after Parsons and McKenzie, 1978). The temperature structure of the oceanic mechanical and thermal boundary layers, when in equilibrium, is presented to the right of the proposed plate structure.

PRESENT

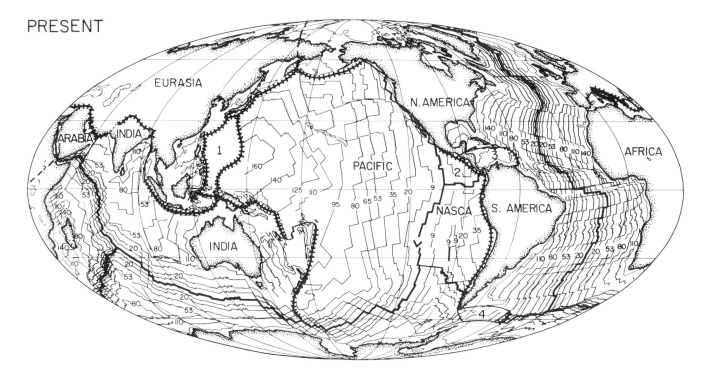

Figure 4a. A Mollweide projection of the Earth showing the major plates, the main plate boundaries, the continents and the 9, 20, 35, 53, 65, 80, 95, 110, 125, 140 and 160 Ma isochrons from Sclater et al. (1980). The minor plates are labeled with the numerals 1, 2, 3 and 4 representing respectively the Philippine, Cocos, Caribbean and Scotia plates. Spreading centers are represented by a heavy line, transform faults by a light line and trenches by cross-hatching.

ANOMALY 5 (9 MA)

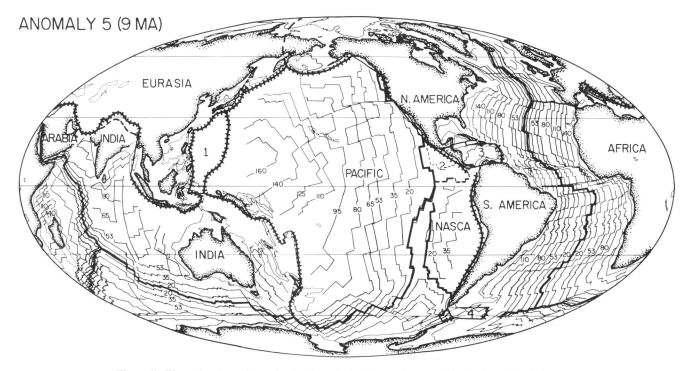

Figure 4b. The major plates, the main plate boundaries, the continents and the isochrons listed above at 9 Ma (anomaly 5). South America is fixed (see text for justification). The paths used for the plate rotations are given in Table 3.

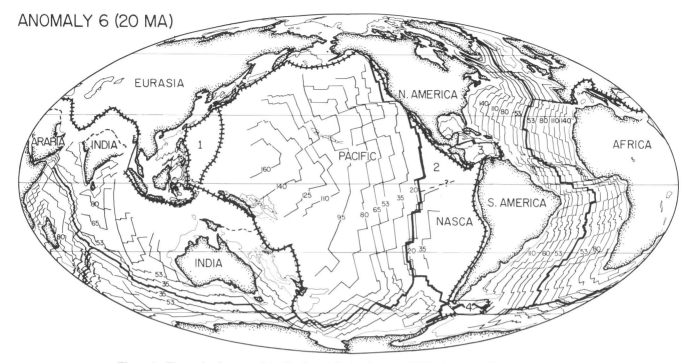

Figure 4c. The major features of the Earth and the isochrons at 20 Ma (anomaly 6). South America is fixed.

chrons on a world-wide chart (Figure 4a). These depths are presented as continuous lines on chart of the theoretical present-day bathymetry (Figure 5). The long-dashed and short-dashed lines are major areas of the sea floor which lie more than ± 400 m from the simple relation between depth and age. These areas are called residual depth anomalies. Parsons and Sclater (1977) deliberately avoid areas of positive depth anomaly and overemphasize the deeper areas of the ocean floor. In general, use of the depth versus age relation of Parsons and Sclater (1977) will lead to slightly too-deep depths and to residual depth anomalies that are almost always positive. The figure of ±400 m for estimating residual depth anomalies is chosen because this figure represents approximately 10 percent of the total subsidence of a mid-ocean ridge from 2500 m at a spreading center to 6400 m when at thermal equilibrium on very old crust.

The match between the observed and the predicted depths is within ± 400 meters throughout most of the Atlantic, except around specific features such as Iceland, the Azores, the Walvis Ridge and the Rio Grande Rise; all of these show up as striking positive depth anomalies. In the Pacific Ocean, the agreement between the observed and predicted depths is satisfactory in the eastern portion of the ocean, except around the Galapagos Triple Junction where there is a broad negative depth anomaly. In the central and western Pacific Ocean, the agreement is much poorer; at least eight major swells and seamount chains have depths much shallower than those predicted by the depth versus age relation of Parsons and Sclater (1977). For the Indian Ocean, the match between predicted and observed depths is reasonable but again there are specific areas which are anomalous; these include the

Mascarene Plateau, the Chagos-Laccadive Ridge, and the Ninetyeast Ridge where the anomalies are positive and one area between Australia and Antarctica where the residual depth anomaly is negative. For the current study, no attempt is made to analyze the depth of the marginal basins. The relation between depth and age in these areas is still not understood. These areas are in need of further study and an analysis of their paleodepth is beyond the scope of this paper.

Residual depth anomalies are substantial areas of the ocean floor which lie more than 400 m either above or below the depth predicted by the relation between depth and age. Contouring these areas with a planimeter and comparing their area to the total area of the ocean shows that they represent 40 percent of the total ocean floor. Regions of thick sediment cover such as the Gulf of Mexico or the Argentine Basin were specifically excluded in this calculation. Forty percent is a large figure and demonstrates that charts constructed using only the depth versus age relation are of qualitative value. Recent advances in the understanding of aseismic ridges and mid-ocean swells may permit the determination of the past depth of these features. Inclusion of these areas which represent the bulk of the residual depth anomalies reduces significantly the errors in the paleodepth charts and makes the charts much more useful.

RESIDUAL DEPTH ANOMALIES

Residual depth anomalies can be separated into two types on the basis of their morphology and geoid anomaly.

The first type exhibits a rapid, short wave-length shallowing

PREDICTED PRESENT

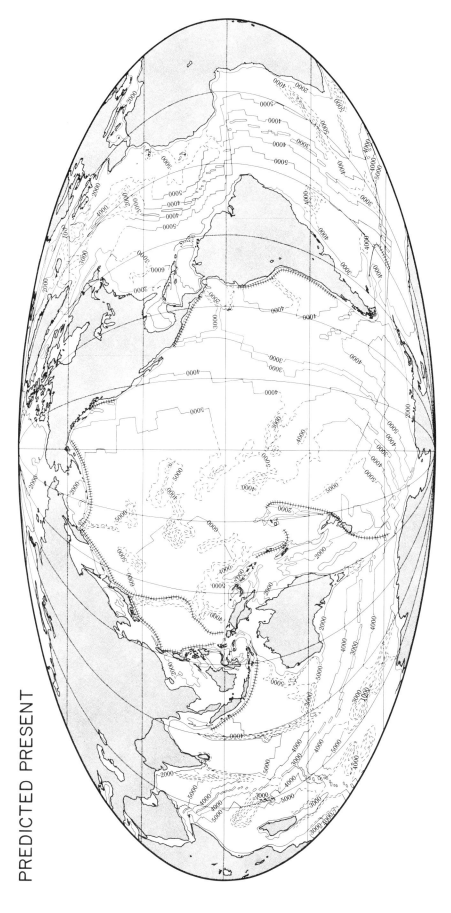

Figure 5. A Mollweide projection of the Earth presenting the continents and the predicted depth of the oceans assuming the depth age relations of Parsons and Sclater (1977) and using the isochrons from Figure 4a to position the 3, 4, 5, and 6 km contours (continuous lines). Areas of positive residual depth anomalies attached to the plate (heavy dashed line) and those resulting from uplift from below the plate (light dashed line) are also shown. The residual depth anomalies are tabulated (Table 1).

TABLE 1. A TABULATION OF RESIDUAL DEPTH ANOMALIES

Residual Depth Anomalies	Correlation with Geoid Anomaly +	Type of anomaly
North Atlantic Ocean		
Iceland	No	probably crustal
Azores/Great Meteor Rise	+	swell
Bermuda Rise	+	swell
Cape Verdes Rise	+	swell
Sierra Leone Rise	?	?
South Atlantic Ocean		
Rio Grande Rise	+	crustal load/swell
Walvis Ridge	No	crustal load
Bouvet Triple Junction Rise	?	?
South Georgia Rise	?	?
North Pacific		
Hawaiian Islands	+	swell
Emperor Seamount Chain	No	crustal load
Mid-Pac Mountains	No	crustal load
Hess Rise	?	?
Shatsky Rise	No	excess sediment
Eauripic Rise	+	crustal load/swell
South Pacific		
Galapagos Triple Junction	--	deep
Cocos Ridge	No	crustal load
Galapagos Ridge	No	crustal load
Nasca Ridge	No	crustal load
Tuamotu and Austral Rise	?	swell
Line Islands Swell	?	crustal load/swell
Manihiki Plateau	No?	crustal load
Ontong Java Plateau	No?	crustal load
Louisville Ridge	No	crustal load
Indian Ocean		
Chagos-Laccadive Ridge	No	crustal load
Ninetyeast Ridge	No	crustal load
Mascarene Plateau	No	crustal load
Madagascar Ridge	+	crustal load/swell
Crozet Swell	+	swell
Crozet Plateau	+	swell
Ob and Lena Swell	+	swell
Mozambique Ridge	No	crustal load
Agulhas Plateau	No	crustal load
Maud Rise	?	?
Kerguelen Plateau	+	swell
Australia Antarctic Discordance	--	deep
Tasman Rise	No	crustal load
Broken Ridge/Wallaby Plateau	+	?

+ From unpublished diagrams of residual depth and geoid anomalies to be published by McKenzie, Watts, Parsons, and Roufouse.

in depth and has a small geoid anomaly. In cross section it is rectangular. In plan view it is generally long and linear and is associated with a fracture zone or a pronounced offset in the magnetic lineations on the sea floor. The Ninetyeast Ridge is the type example. This type of depth anomaly is thought to be excess basalt piled on the ocean floor, either near the ridge axis during the creation of oceanic plate or as off-ridge axis volcanism. As time increases, the excess material moves away from the ridge axis and travels passively on the plate to which it is attached.

Once the material sinks below the sea floor and erosion stops, it subsides at the same rate as the ocean crust on which it rests (Figure 6a, Detrick et al., 1977).

The second type of residual depth anomaly is characterized by a gentle, long wavelength shallowing in depth and a large positive geoid anomaly. In cross section it has the appearance of a swell; in plan view it can be lineated or circular. It has no obvious relation to either fracture zones or offsets in magnetic lineations. In a few cases a linear chain of seamounts lies at the apex of the

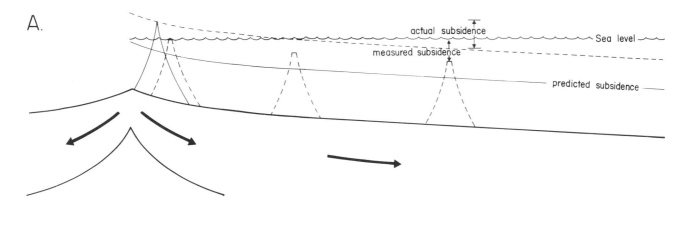

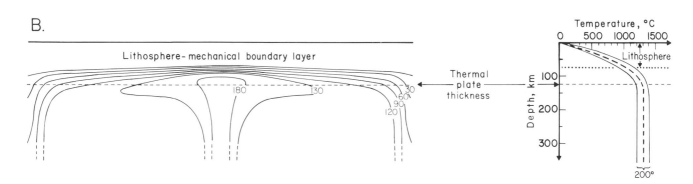

Figure 6. (a) A simple model which can account for the subsidence of many aseismic ridges. These are residual depth anomalies created by an excess crustal load. The measured subsidence is obtained by taking the present isostatically-corrected depth of volcanic basement beneath sea level. The predicted subsidence is indicated by the solid line; actual subsidence is indicated by the dashed line. The measured subsidence will generally be less than the predicted subsidence of the ridge because the ridge or seamount is initially built above sea level and then eroded (from Detrick et al., 1977). (b) A sketch illustrating the temperature variations within the thermally-defined plate produced by convection beneath the lithosphere. These variations will produce a topographic swell and positive geoid anomaly. On the right, a hypothetical temperature profile through such a swell (continuous line) is compared with the horizontally averaged temperature structure (dashed line) (after Parsons and Daly, 1983).

swell. The type example is the Hawaiian Swell. In this case, the residual depth anomaly is largest near the big island of Hawaii and decreases in the direction of Midway (Detrick and Crough, 1978). It has a positive geoid anomaly (McKenzie et al., 1980) and may have a positive residual heat flow anomaly in the vicinity of Midway (Von Herzen et al., 1982). These types of anomalies are probably related to upwelling thermal convection cells in the upper mantle. Such convection cells reduce the effective thermal thickness of the plate (Crough, 1978; Parsons and Daly, 1983), increase the likelihood of volcanism and possibly uplift the plate dynamically (Figure 6b).

Detrick and Crough (1978) show that resetting the isotherms to values equivalent to those for 25 million-year-old lithosphere will account for the subsidence of the Hawaiian Swell. In their concept the plate moves over a hot spot which resets the isotherms and is fixed relative to the plate. This hot spot became

active about 40 million years ago and is currently located under the big island of Hawaii. Parsons and Daly (1983) demonstrate, using numerical experiments on liquids which model the upper mantle, that most of the temperature changes in a convection system occur just beneath the lithosphere. If the temperature changes associated with an upwelling convection cell were to remove the lower thermal boundary layer and thin the upper mechanical layer of the plate, then simple convection explains the observations of Detrick and Crough (1978). It is now generally accepted that some form of convection is the explanation of the Hawaiian Swell. However, many features of this convection are still uncertain. It is not known how much of the topographic anomaly is due to dynamic forces and how much to thermal effects. It is not clear when convection started and over what time span the upwelling cell can be considered fixed. These uncertainties make the past depth of this swell difficult to determine.

Another problem, raised by Menard and McNutt (1982), is that most swells are associated at their apex with massive mid-plate volcanism. If the direction of motion of plate changes abruptly and this volcanism moves off the swell, then the ocean floor subsides at a much faster rate than that predicted by the simple relation between depth and age. The Emperor Seamount Chain may be an example of such a feature. Additional examples may include most aseismic ridges and seamount chains that are not clearly associated with either near ridge axis volcanism at a transform fault or a triple junction. The exact timing of the swell beneath these features is unknown; consequently, it is difficult to compute their paleodepths. Direct recovery of the sedimentary apron by drilling appears to be the only manner in which the past depth can be determined.

Residual depth anomalies are separated by type before being added to the paleodepth charts. The first type of depth anomaly consists of aseismic ridges which are clearly associated with excess volcanism at a ridge axis. It is assumed that these ridges are created at sea level and subside at the same rate as normal ocean floor. The broader residual depth anomalies of the second type are more problematic and major assumptions must be made before they are added to the charts. Two methods are used to treat this anomaly type. The first method assumes that the features are attached to the plate on which they currently lie and that any movement is directly associated with movement of that plate; there is no change in depth. Features which are treated in this manner are the Azores and Iceland highs, the Rio Grande Rise, the Crozet Plateau and the northwest Pacific Island chains on crust 50 Ma or older. In contrast, the Hawaiian, Tuamotu, and Austral swells are assumed to remain fixed in the South American coordinate system. This assumption is approximately equivalent to fixing these three swells with respect to the frame of reference given by the Pacific hot spots.

It is also difficult to model the subsidence of volcanic arcs located landward of their associated trenches and the marginal basins that lie between these arcs and the continents. No attempt is made to add the subsidence of these regions to the current paleodepth charts. However, the positions of the trenches are kept attached to their respective continent or, in the case of the western Pacific, to the Philippine and Indian plates. The predicted position of these trenches are shown on the 9 Ma (anomaly 5) and 20 Ma (anomaly 6) reconstructions (Figures 4b and c).

THE PALEODEPTH OF SEDIMENT RECOVERED AT DEEP SEA DRILLING PROJECT SITES

Depth below sea level has a significant effect upon the type and amount of sediment deposited on the ocean floor. In this chapter a simple method is described for determining the past depth of the sediment recovered in individual DSDP holes.

The method rests on the basic assumption that all DSDP sites are on crust which subsides at the same rate as normal ocean floor. For each site in question the age of the basement, the history of sediment accumulation, the total thickness of sediment

TABLE 2. THE PAST WATER DEPTH FOR THE PRESENT AND THREE DIFFERENT TIMES IN THE NEOGENE (8, 16, AND 22 MA) AT FOUR DSDP SITES

Yr (Ma)	Water Depth	Sediment Thickness	Depth to Basement	Unsedimented Water Depth
Site 14, Age 39. My				
0.	4346.	108.0	4454.0	4413.
8.	4109.	108.0	4212.0	4176.
16.	3839.	108.0	3946.8	3906.
22.	3610.	98.0	3707.6	3670.
End of Site				
Site 400A, Age 115. My				
0.	4399.	3000.0	6399.0	5642.
8.	4498.	1729.5	6227.0	5573.
16.	4513.	1576.9	6090.2	5494.
22.	4477.	1528.0	6005.4	5427.
End of Site				
Site 254, Age 45. My				
0.	1253.	301.0	1554.0	1440.
8.	1054.	369.0	1323.0	1221.
16.	878.	159.7	1037.5	977.
22.	688.	134.0	821.5	771.
End of Site				
Site 317B, Age 115. My				
0.	2598.	910.0	3508.0	3164.
8.	2580.	827.5	3407.1	3094.
16.	2551.	745.6	3297.0	3015.
22.	2520.	689.0	3209.3	2949.

and the water depth are taken from the relevant DSDP report. The depth to basement is computed from the known water depth after isostatically unloading the total sediment thickness (Figure 7). The unloaded depth is compared with that predicted by the empirical relations given by Parsons and Sclater (1977) and the offset is computed. The paleodepth of the site is determined by moving the individual sites back along the subsidence curve for normal ocean floor and then adding the offset. The actual depth to the top of an individual layer at a given time is computed by isostatically adding the total sediment that has accumulated. The details of the method of computation and a worked example for site 14 can be found in Appendix 1. The computer program used to produce the plots and tabulate the output is available upon request. Appendix II presents a compilation of the paleodepths for all the sites considered in this paper in tabular form. As most sites are on either normal ocean floor or aseismic ridges, rather than on swells, the paleodepth plots for the last 22 million years are probably very reliable. This reliability and the fact that the age, depth to the individual sediment layers and depth to basement are determined by direct observation (i.e. drilling) gives these plots an accuracy of ±100 meters. They are much more reliable than the paleodepth charts.

An example of these paleodepth compilations is presented graphically at 8, 16 and 22 Ma (Figure 8); included are sites 14 and 400A on normal ocean crust, site 254 on the aseismic Nine-tyeast Ridge and site 317B on the Manihiki Plateau. Sites 14 and 254, which are underlain by younger crust, shallow much more rapidly going backward in time through the Neogene than do the two older sites. This is to be expected, as young ocean floor

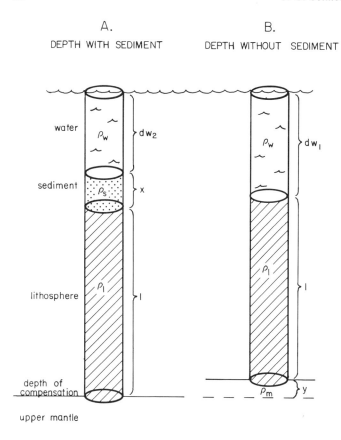

A.
DEPTH WITH SEDIMENT

B.
DEPTH WITHOUT SEDIMENT

Figure 7. A comparison of basement depths (A) loaded with sediment and (B) removed of its sediment load: dw_2 is the water depth when the sediment load is considered; dw_1 is the water depth when this load is removed. Note the isostatic rise of the lithosphere by the amount y when the sediment of thickness x is removed. The equations derived from this conceptual model to compute dw_1 knowing dw_2, x, and the relevant densities are discussed in Appendix 1.

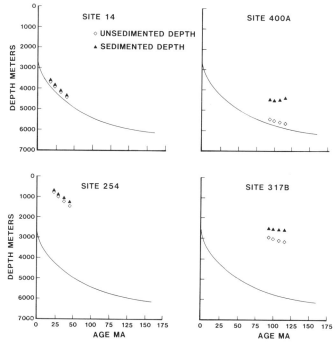

Figure 8. The water depth and unloaded basement depth for four DSDP sites for the present, 8, 16, and 22 Ma. The symbols representing the present are to the right at the maximum age; the sites in general shallow going backwards in time. The continuous line is the theoretical curve for depth versus age given by Parsons and Sclater (1977). (A) Site 14 is on young, normal depth ocean floor in the central South Atlantic; (B) Site 400A is on old, normal ocean floor in the North Atlantic; (C) Site 254 is on the aseismic Ninetyeast Ridge; and (D) Site 317B is on the aseismic Manihiki Plateau in the South Central Pacific.

subsides much more quickly than old ocean floor. In general, the sites shallow going backwards in time. It is possible for them to deepen; an example of this occurs at site 400A between the present and 8 Ma, and as a result of increased sedimentation in the Pliocene.

THE DEPTH OF THE OCEANS AT 22, 16 AND 8 MILLION YEARS AGO

The first step in the construction of the charts of the past depth of the oceans is to digitize the position of the isochrons presented by Sclater et al. (1980). These isochrons, the depth-age relation for normal ocean floor, and the analysis of residual depth anomalies are used to construct a chart of predicted depths for the present oceans (Figure 5). A comparison of this chart with the present bathymetry (Figure 1) permits a qualitative estimate of the reliability of the predicted charts. The isochrons, the residual depth anomalies where appropriate, and the continents are separated by plate; the plates are rotated back to South America. South America is chosen to be fixed in the present coordinate

system as it is the continent that has moved the least in paleolatitude since the Cretaceous (Smith, 1982). Next a two stage process is followed to complete the reconstructions. First the poles and angles used in the reconstructions are checked by rotating the plates, including continent and isochrons, back to their appropriate position at the time of anomalies 5 (9 Ma) and 6 (20 Ma). These charts are used for the tests because the isochrons are picked from the identification of these prominent anomalies. Then the correct poles and interpolated rotation angles, the isochrons and the relation between depth and age are used to construct paleodepth charts for 22, 16, and 8 million years ago. These charts include the residual depth anomalies and the continents.

Selection of the Major Plates

The world is separated, for the purposes of this study, into twelve major plates: the Pacific, Cocos, Nasca, Scotia Sea, Antarctica, Africa, South America, North America, Caribbean, Eurasia (including Southeast Asia and China), Arabia and India/Australia plates. To reduce the number of rotations, China and the marginal basin of the northwestern Pacific Ocean were attached to Eurasia; marginal basins of the southwestern Pacific

Oceans were attached to India. This assumption is not correct for the northwestern Pacific Ocean, as the Philippine Sea is being subducted under the Eurasian plate. However, the subduction rate is probably small and the Philippine plate is thought to be composed principally of back arc basins which are not covered in the present analysis, hence no additional error is introduced by this assumption.

Plate Rotations

The paleodepth charts are constructed for 22, 16 and 8 million years ago. These ages are selected for paleontological reasons. They are not convenient times for paleogeographic reconstructions, which involve the determination and estimation of the accuracy of poles and rotation angles by overlapping isochrons. In order to check the rotation parameters and to find out how well the isochrons do overlap when the plates are moved back toward each other, anomaly 5 (9 Ma) and anomaly 6 (20 Ma) reconstructions are produced of the digitized world-wide set of isochrons (Figures 4b and 4c). The individual plates are rotated back toward each other to close up the present spreading centers (ridges) and then rotated by various paths back to South America. The paths followed for each plate rotation and the poles and angles used for these reconstructions are presented in Tables 3 and 4. The degree of overlap of the isochrons on the 9 and 20

TABLE 3. THE SEQUENCE OF ROTATIONS USED
FOR THE INDIVIDUAL PLATES

```
Eurasia→N. America  Africa→S. America.
Africa→S. America.
Iberia→N. America→Africa→S. America.
N. America→Africa→S. America.
Australia/India→Antarctica→Africa→S. America.
Pacific→Antarctica→Africa→S. America.
Caribbean→N. America→Africa→S. America.
Arabia→Africa→S. America.
India→Africa→S. America.
Antarctica→Africa→S. America.
Nazca→Pacific→Antarctica→Africa→S. America.
Cocos→Pacific→Antarctica→Africa→S. America.
```

Ma reconstructions (Figures 4b and 4c) gives a graphical estimate of the self-consistency of the rotation parameters.

The reconstructions are illuminating. The 9 Ma isochron (anomaly 5) overlaps everywhere (Figure 4b); the only problems for this reconstruction occur in the tectonic development of the Galapagos Spreading Center and the Panama Basin, and in trying to determine the exact development of the Caribbean, Central America and the marginal basins of the northwestern Pacific. This isochron is constructed (Sclater et al., 1980) by rotating a digitized version of the present ridge axis to either side; therefore, the good overlap of this isochron on the 9 Ma reconstruction is not unexpected.

The 20 Ma (anomaly 6) reconstruction is less satisfactory (Figure 4c) because of increased uncertainty as to how to treat the

TABLE 4. THE ROTATIONAL PARAMETERS FOR THE ANOMALY 5 (9 MA)
AND ANOMALY 6 (20 MA) RECONSTRUCTIONS

Total Poles of Rotation to North America Plate

Age Span	Eurasia-N. America	Iberia-N. America	Carib.-N. America
0-9	63.1 142.1 -1.81	63.1 142.1 -1.81	50 116.0 -1.8
0-20	63.1 142.1 -4.02	63.1 142.1 -4.02	50 116.0 -4.0

Total Poles of Rotation to the Pacific Plate

Age Span	Cocos-Pacific[1]	Nazca-Pacific[1]
0-9	-41.3 -108.1 -18.18	56.6 -85.6 -14.76
0-20		64.8 -91.0 -31.4*

Total Poles of Rotation to the Antarctic Plate

Age Span	Australia-Antarctica[2]	Pacific-Antarctica[3]
0-9	9.7 36.5 -6.75	72.0 -70.0 9.75
0-20	13.8 34.6 -12.0	71.25 -73.19 15.41

Total Poles of Rotation to the African Plate

Age Span	N. America-Africa	Antarctica-Africa[4]	India-Africa[4]	A. Arabia-Africa[5]
0-9	70.5 -18.7 2.60	8.4 -42.4 1.4	16.0 48.3 -4.18	26.5 21.5 -1.75
0-20	70.5 -18.7 5.78	8.4 -42.4 3.2	16.0 48.3 -11.5	26.5 21.5 -3.9

Total Poles of Rotation to the South American Plate

Age Span	Africa-South America
0-9	67.3 -39.5 -3.69
0-20	67.3 -39.5 -8.2

All poles are summarized in Sclater et al., 1977, except:
[1]Minster et al., 1974.
[2]Weissel et al., 1977.
[3]Stock and Molnar, 1982.
[4]Fisher and Sclater, 1983.
[5]Norton and Sclater, 1979
*Stock (1981) through 9 Ma then interpolated from 25 Ma pole from Stock and Stork (pers. commun).

Panama Basin, the Caribbean Sea, the Western Pacific marginal basins and the East Pacific Rise north of the junction with the Chile Rise. The difficulty with the East Pacific Rise is caused by the uncertainty of the position of the ridge axis between 28 Ma (anomaly 7) and 9 Ma (anomaly 5). There was a major jump of this axis to the west (Herron, 1972) between 28 Ma and 9 Ma and it is not obvious from the magnetic anomalies (Handschumacher, 1976) whether this jump occurred as a single event or as discrete steps. To simplify our analysis a discrete jump about 15 Ma was assumed. This assumption is an oversimplification and leads to the poor match of anomaly 6 (20 Ma isochron) on either side of the East Pacific Rise in the 20 Ma reconstruction. For the rest of the oceans the 20 Ma isochron overlaps satisfactorily. This chart clearly presents a reasonable base from which to construct preliminary world-wide paleodepth charts.

It is necessary to point out three important features on the 20 Ma reconstruction. First, the Pacific plate is at a considerable distance from the northwestern Pacific marginal basins. Second, a wide gap is open between the northeastern edge of the Indian/ Australian plate and Sumatra/Java Trench. Finally, Drake Passage was not substantially smaller twenty million years ago than it is today. The implications of these features for worldwide deep water circulation and climate are discussed at some length in a later section.

The Paleodepth of the Neogene Ocean at 22, 16 and 8 Million Years Ago

The major objective of this paper is to construct charts of the past depths of the oceans which can be used as a framework for both general and detailed studies of deep sea sedimentation. Three specific dates, 22 Ma (Early Miocene), 16 Ma (Middle Miocene) and 8 Ma (Late Miocene), are selected for paleontological reasons as the times of interest. Paleodepth charts at these dates are constructed in the same manner as for the chart of the present predicted depths (Figure 5). First, the paleogeographic positions of the continents, plate boundaries and isochrons are determined. For the 22 Ma chart, extrapolated versions of the poles and angles used in the anomaly 6 reconstruction are chosen. For the 16 and 8 Ma charts, the poles and angles are chosen by interpolating between those used to construct the anomaly 6 (20 Ma) and anomaly 5 (9 Ma) chart and the present chart. These poles and rotation angles are easily derived from those for the anomaly 5 (9 Ma) and anomaly 6 (20 Ma) charts (Table 4) and are not presented separately. Paleolatitude was included by rotating the plates back to South America, which is assumed to have remained fixed throughout the Cenozoic (Smith, 1982). The depth is added to these charts using the position of the isochrons and the relation between depth and age. The position of the 2, 20, 50 and 130 Ma isochrons for the specific chart in question gives, respectively, the 3000, 4000, 5000 and 6000 m contours. All depths presented are depth to basement. They are not adjusted for sediment load.

The relation between depth and age holds for a significant part but not all of the ocean floor. To obtain a more useful paleodepth chart it is necessary to include the residual depth anomalies. They are separated by type (Table 1) before being added to the paleodepth charts in the manner described in a previous section. For the first type, the aseismic ridges, no attempt is made to remove the sediment cover or to allow them to shallow as they become younger in age. For the second type, the mid-ocean swells, only those associated with the Hawaiian, Austral and Tuamotu Islands have a documented subsidence history (Detrick and Crough, 1978; Crough and Jarrard, 1981). They are fixed in the same frame as South America. The rest are moved with the plate on which they currently reside.

There are problems in constructing the past depth of both types of residual depth anomaly. The effects of sediment cover and shallowing with age have been ignored for aseismic ridges. Both effects can be significant. For the mid-ocean swells the height and width are presumed to remain constant through time. This model can be justified for the three swells discussed above. However, it does not work for the old seamount chains in the Pacific such as the Emperor and Line Island Chains, nor does it work for the swells in other oceans such as the Bermuda and Cape Verde Rises in the Atlantic Ocean and the Crozet Swell in the southwest Indian Ocean. Currently, there is no adequate model for these features. For simplification, these are attached to the plate on which they now lie and are moved with the plate without changing the depth. This technique may also work for the seamount chains in the northwestern Pacific during the short time span involved in these charts. Whether it is valid for the other swells is not known.

Depth charts of the oceans are presented for 22, 16 and 8 Ma (Figures 9, 10 and 11 in pocket inside back cover). They include normal ocean floor, aseismic ridges, and mid-ocean swells. No attempt is made to consider subducted ocean floor, the marginal basins of the western Pacific, the Caribbean Sea or the Mediterranean Sea. In addition, the regions which open between Arabia and Eurasia and between the India plate and Sumatra and Java are left as gaps.

DISCUSSION OF THE PALEODEPTH CHARTS

The oldest chart, the 22 Ma reconstruction, shows some surprising differences from the chart of the present bathymetry (Figure 1). Relative to the present, Arabia and India at 22 Ma are separated from Eurasia and the Tethys sea is open. The Indian plate, which contains Australia and the southern part of New Guinea, is substantially south of the Eurasian Plate. There is a major gap between the northern edge of the Australian continental margin and the marginal basins of the northwestern Pacific. On the floor of the Indian Ocean, the Chagos-Laccadive Ridge and the Mascarene Plateau are nearly coincident, and the Madagascar Ridge is close to the Crozet Plateau. In addition, the Ninetyeast Ridge and Broken Ridge, are adjacent to the Kerguelan Plateau and the Tasman Rise is close to both the Australian-Antarctica Ridge and Antarctica.

Distinct differences between the present and 22 Ma are also observed in the Pacific Ocean. The East Pacific Rise is more centrally located within the ocean basin and has a more north-south orientation. The San Andreas fault is shorter and the Baja Peninsula is attached to the coast of Mexico. Except for the deep, old Philippine Sea, the marginal basins in the Pacific Ocean have disappeared and a substantial amount of newly created sea floor lies to the east of the trench system in the western Pacific and to the west of the Chile trench. Aseismic ridges and swells of the western Pacific have assumed a more central location because of the increase in area to the west of the seamount chains. The Chile Ridge is located more to the south and most of it is separate from the southern tip of South America and the northern tip of Antarctica. The Caribbean Sea is separate from North and South America and a seaway is open between Central and South America. The presence of a deep water passage at 22 Ma, in addition to the shallow water passage implied by this reconstruction, is dependent upon the depth of the younger marginal basins of the eastern Caribbean at this time.

In the North Atlantic, Greenland and Eurasia are closer together. Iceland is nonexistent, but most of the ridge system between Greenland and the Faroe Islands is probably close to or above sea level. This chart cannot verify the presence or absence of a deep water passage between the Faroes and the Wyville-Thompson Ridge at 22 Ma because it is too small a size to measure here, but the question is the subject of active debate (Miller, 1982). To the south, the rest of the Iceland High is closer to Greenland and the Rockall Bank. In the central North Atlantic, The Azores High is smaller in size; the Bermuda Rise, the Cape Verde Plateau and the Meteor Seamount province are closer together. In the South Atlantic, South America and Africa are closer together and the distance between the Rio Grande Rise and Walvis Ridge is reduced. Relative to the other oceans, the Atlantic between 60°N and 50°S shows the fewest changes from the present.

The North Atlantic Ocean does not play a major role in worldwide sedimentation patterns at 22 Ma, with the possible exception of the deep-water gap between the Faroe Islands and the Wyville-Thompson Ridge. In contrast, the southern polar region at 22 Ma is significantly different from the present. The South America-Antarctic ridge system is expanded in size, the South Sandwich trench is located to the west of its present location, and the Scotia Sea is considerably reduced. The control these features place on a deep water passage in the southern oceans has major implications for patterns of oceanic sedimentation on a worldwide scale. Though not totally different from the present, the morphology 22 million years ago is substantially altered. It is easy to imagine that by the Eocene the morphology would be very different than that which is observed today. For example, South America and Antarctica may have coalesced in which case the Tasman Rise would abut the coast of east Antarctica.

The general development of the oceans through the Neogene is illustrated by the 16 Ma and 8 Ma reconstructions (Figures 10 and 11). The regions of elevated depth in the Indian Ocean separate and the Tethys Sea closes. Australia moves further from Antarctica and Timor, and New Guinea collides with the island-arc systems of the western Pacific and southeast Asia. In the Pacific Ocean, the East Pacific Rise moves to the east and then jumps back to the west about the time the Chile Rise collides with South America. The Scotia Sea expands in size and the seaway between Central America and South America closes. The Atlantic Ocean shows a slow but unremarkable progression towards the present day morphology.

The paleodepth charts generated by this study can be directly applied to the problem of the time of onset of both deep and shallow water circum-Antarctic circulation. Though all the continents in the southern ocean close onto Antarctica as one goes backwards in time, it is likely that the final control of both deep and shallow water circulation is constrained by the motion of South America with respect to Antarctica, by the closure of the Scotia Sea and by the motion between South America and Antarctica across Drake Passage. The depth contours on the 22 Ma chart provide evidence that it is likely that neither the Gausberg Ridge, southwest of Kerguelan, nor the Tasman Rise act as a barrier to the flow of either deep or shallow water at this time. However, there is sufficient ambiguity in the depths and positions of these features to suggest that it is possible that the creation of the deep and shallow water circulation patterns are neither concurrent nor due to the opening of the same morphological boundary. Kennett (1982) and Barker and Burrell (1977) calculate the effect of shallowing the Shackleton Fracture Zone as the ocean floor in the vicinity gets progressively younger. They suggest that this was the major barrier to deep water flow and probably occurred in the late Oligocene. Shallow water flow probably did not terminate until South America and Antarctica became one continuous mountain sequence; the timing of this event is not clear. It is suggested by the magnetic anomalies in the Scotia Sea that this occurred during Late Oligocene; however, Norton and Sclater (1979) using final plate rotation, suggest that this passage may have opened as a shallow water connection as early as the Early Eocene.

THE INFLUENCE OF CHANGES IN PALEODEPTH ON SEDIMENTARY DISTRIBUTION AND CLIMATE

During the past twenty years there has been a revolution in the understanding of the ocean floor and the deep water sedimentary record. This revolution is a result of the general acceptance of the theory of plate tectonics and the very successful Deep Sea Drilling Program. Though the sedimentary record goes back much further, it is the Cenozoic and more specifically, the Neogene (the last twenty-five million years) for which there is the most quantitative information. This chapter investigates the possible influence of variations in paleogeography and paleodepth of the oceans on the major sedimentary and climatic changes during this epoch.

Cenozoic paleoclimatic evolution is distinguished by three

major characteristics. First, there was a transition from warm to colder oceans. This is mainly a high latitude effect; the tropics are only slightly affected. Second, there was a sharp drop in temperature at the Eocene/Oligocene boundary that shows up both in paleotemperature analyses of planktonic and benthonic Foraminifera from the Subantarctic Pacific (Shackelton and Kennett, 1975) and also in analyses of changes in the calcium compensation depth (CCD) in the equatorial Pacific (Van Andel et al., 1975). Last, there was a build-up of extensive ice sheets and the onset of the glacial mode during the Neogene.

The paleoclimatic evolution of the Neogene can be subdivided into five general phases. The first phase occurred in the Early Miocene about 22 Ma. During this phase (Figure 9), the ocean basins assumed their modern shape and silicious biogenic productivity began to become prominent at high southern latitudes. The second phase occurred in the Middle Miocene (14 Ma), at which time the Antarctic ice sheet formed, planetary temperature gradients continued to steepen, and temperatures at high and low latitudes became much less closely coupled. In addition, during this phase there was a distinct increase in the number of hiatuses in the equatorial Pacific (Van Andel et al., 1975). The third phase started in the latest Miocene when very cold water expanded to include New Zealand and California and the climate became markedly cooler. This change was associated with an intensification in oceanic circulation, a sea level regression, a carbon shift associated with the cooling, and the isolation and dessication of the Mediterranean, referred to as the Messinian salinity crisis. During the Late Pliocene, the fourth phase became dominant; the Northern Hemisphere ice sheet formed about 2.5 to 3.0 Ma. The formation of the Arctic ice sheet lead directly to the onset of the classic glacial/interglacial cycles. The final phase is the Quaternary between 1.6 Ma and the present. This phase was dominated by a large number of glacial events, the strong latitudinal displacement of climatic zones, large scale fluctuations in ocean circulation, sea level oscillations in the range of 100 m and much erosion.

The major questions for paleosedimentology and paleoceanography center around the cause or causes of all the changes described above. Part of the difficulty in dealing with these questions is that it is unknown to what extent these changes themselves forced the system and to what extent they were forced by the system. The temptation is to assign an external forcing event to such changes; for example, a catastrophe for the Cretaceous extinctions, the opening of Drake Passage or some other strait for the Eocene or Miocene events and finally, the closure of the isthmus of Panama for the mid-Pliocene onset of the northern ice ages.

The closing and opening of straits and the changes in paleodepth are no doubt important (Berggeren and Hollister, 1977) but, as pointed out by Berger et al. (1981), they occur over a long time span. It is difficult to explain features of the sedimentary and climatic record that are abrupt by a cause that has taken so long to develop. Rather than trying to match a stratigraphic or climatic event that is well defined in time with a poorly defined geophysi-

cal cause, Berger et al. (1981) proposed that the question of cause and effect be rephrased in the following manner. Given that it is known that certain morphological changes do take place, are they likely to increase the tendency for sedimentary deposition or climate to move in one or another direction? This approach is followed for the conclusion of this chapter.

Kennett (1982) summarizes the major paleogeographic and paleodepth changes in the Cenozoic that create the modern ocean conditions: 1) the opening of Drake Passage to shallow water flow; 2) the opening of Drake Passage to deep water flow; 3) the subsidence of the Greenland-Iceland-Faroe Ridge; and 4) the closure of Tethys between Asia and Africa and the coalescence of North and South America in the isthmus of Panama.

The paleogeographic and paleodepth charts are examined to estimate the timing of these four changes in an attempt to investigate their effect on paleocirculation, the sedimentary record and climate.

Quantitative information from the reconstructions first becomes available in the early Neogene (22 Ma, Figure 9). At that time, the shallow water passage around Antarctica was well developed. Central and South America were separate and there still existed the possibility of a full Tethyan connection between the Atlantic and Indian Oceans. Clearly an efficient shallow water circulation system around Antarctica was in effect. In addition, it is possible that a limited equatorial system was still active. Owing to the lack of quantitative information on width of the passage in the equatorial region it is not possible to say more about this system. The wide separation of Australia and Antarctica and South America and Antarctica shown on the early Neogene (22 Ma, Figure 9) chart is evidence that the onset of the deep water flow around Antarctica may have occurred about this time. Exactly how water at depths greater than 2000 m flowed through the Kerguelan/Gaussberg Ridge and between the Tasman Rise and the coast of Antarctica is not known. However, it is probable that in both areas there were passages that were sufficiently deep. Thus, it is the region of Drake Passage between South America and Antarctica and the height of the Shackleton Ridge that controlled the onset of deep water circulation. Barker and Burrell (1977) on the basis of a simple subsidence analysis proposed that this fracture zone dropped below 2000 m in depth in the early Miocene and that this was the timing of the onset of deep water flow.

Between 16 Ma, 8 Ma and present (Figures 10, 11 and 1) the justification of a substantial deep water passageway around Antarctica improved and the Tethys completely closed. The straits of Gibralter became a barrier to deep water flow between 8 Ma and the present (Figures 11 and 1) and the isthmus of Panama also closed during the same time period. Australia collided with South East Asia at some time between 16 Ma and present. It is not clear from these charts exactly when this occurred or what effect this would have on shallow and deep water equatorial circulation. The Iceland Faroes Ridge is important but it is difficult to know in detail how it subsided. Shallow water and possibly deep water passage of North Atlantic deep water occurred between the Faroes and the Wyville-Thompson Ridge. Though some scientists argue that these passages have only recently subsided to their present depth, Miller (1982) proposed that they have been open since the Eocene. Multichannel seismic evidence tends to support Miller (1982) by suggesting that at least the Wyville-Thompson Passage is on stretched continental crust. This stretching probably occurred in the early Eocene and the passage could have been deep since then.

Kennett (1977) argues convincingly that the development of circum-Antarctic shallow and deep water circulation will separate ocean temperatures and hence climate into latitudinal zones. Thus, the onset of these circulation patterns would act to decrease temperature in high latitudes, change the sedimentary record and severely effect climatic patterns.

The separation of Australia and Antarctica commenced in the late Cretaceous (Cande and Mutter, 1982) and that between South America and Antarctica in the Eocene (Norton and Sclater, 1979). Extrapolating the early Neogene (22 Ma, Figure 9) chart back in time for the Southern Ocean supports the suggestion of Kennett (1982) that a shallow water circulation system developed around Antarctica sometime in the Eocene. This system separates the oceans latitudinally and it may have been the cause of the sharp drop in temperature and the change in the carbonate compensation level at the Eocene/Oligocene boundary. The onset of a deep water circulation system by the early Miocene reinforced the latitudinality of the temperatures. Again such a system occurs at the right time and acts in the right direction to inverse Antarctic glaciation during the late Oligocene-early Miocene and the buildup of a Southern Hemisphere ice sheet somewhat later (~15 Ma). A similar argument can be made to relate the closing of the isthmus of Panama to the formation of ice sheets 2.5 to 3 million years ago in the Northern Hemisphere.

This study presents evidence that the opening of shallow and deep water seaways probably had a significant effect upon three of the major sedimentary and climatic changes in the Cenozoic. It is not possible with presently available data to state conclusively that the opening of these seaways are the direct cause of these changes. However, they seemed to open at the right time and this alone is suggestive. The opening of the seaways and the onset of major circulation patterns may have an important part to play in the understanding of the sedimentary and climatic record. They are certainly worthy of further detailed study.

ACKNOWLEDGMENTS

We would like to thank the other participants in the CENOP program for the interest and encouragement that enabled us to complete this paper. Fortunately, during the project substantial advances have occurred both in sedimentary analysis and in the understanding of depth anomalies on the deep sea floor. As a result, what started as the routine presentation of some preliminary paleodepth charts became a more ambitious project to examine, in addition, both residual depth anomalies and deep and shallow water seaways from a worldwide perspective.

Support for this work came from the National Science Foundation Oceanography Division, through the CENOP project, and from the Division of Polar Programs of the National Science Foundation. In addition Drew Bennett was supported by the M.I.T. Undergraduate Research Opportunities Program.

APPENDIX 1
THE COMPUTATION OF THE PALEODEPTH OF INDIVIDUAL SEDIMENTARY HORIZONS AT JOIDES DEEP SEA DRILLING SITES.
Prepared by Linda Meinke and Cynthia Murphy

Active spreading centers exhibit a simple relation between depth and age. It is now generally accepted that this relation is caused by the cooling and contraction of hot molten material intruded at the spreading center as it moves away from the axis of spreading. In this section, we present the theoretical relation between depth and age, show how to correct this relation for the loading effect of sediment, and finally, compute the paleodepths for JOIDES Site 14.

In the approach we outline, it is assumed that at a particular site the ocean floor subsides at the rate given by the theoretical curve. This implies a constant offset either above or below the theoretical curve which is thereby assumed to be the average of all individual points on the ocean floor. This is a reasonable assumption where the actual depth, when converted for sediment loading, is within ±400 m of the theoretical curve. As discussed in the body of the text, it is not necessarily a reasonable assumption when the offset from the theoretical curve is larger than this. In this case, it is possible that a residual depth anomaly is involved and not all of these anomalies follow the simple subsidence history of a spreading center.

THE THEORETICAL RELATION BETWEEN DEPTH AND AGE

Mid-ocean ridges, which have been created as the result of the intrusion of hot molten material at a spreading center and the consequent subsidence of the material due to cooling and contraction, exhibit a simple relation between depth and age. On the basis of a simple model compatible with the plate theory of tectonics, Parsons and Sclater (1977) have provided simple empirical formulae for this relationship. If t is the age in millions of years then:

$$\text{for } 0 < t < 70, \ d(t) = 2500 + 350(t)^{1/2} \text{ m} \tag{1}$$

$$\text{and for } t > 70, \ d(t) = 6400 - 3200e^{(-t/62.8)} \text{ m} \tag{2}$$

where d(t) is the depth of the sea floor in meters at time t. From (1) and (2), if we know the depth at a certain age, we can predict the depth at any other point in time.

THE CORRECTION FOR SEDIMENT ACCUMULATION

The above equations relating depth to age do not take into account the effect of the weight of sediment cover on the underlying crust. In using the formulae with oceanic data, the present sediment cover must first be removed and the crust allowed to rise to the depth it would lie at were the sediment removed. This depth can then be used to predict an unsedimented depth at any given time in the past. This calculation is carried out by computing the offset from the theoretical curve for the present and then calculating the paleodepths, assuming that this offset remains constant in magnitude. If the thickness of sediment at a past time is known, then we can reload the crust and calculate the water depth allowing for the accumulated sediment cover.

To solve the problem of the effect of the sediment layer, we balance the masses of columns of unit area, one with a sediment cover and the other without (Figure 7). We assume that ρ_w, ρ_1, ρ_m, and ρ_s are respectively the water, lithosphere at 0°C, mantle, and sediment densities and that dw_1, dw_2, x, l and y are respectively, the unsedimented water depth, the sedimented water depth, the sediment thickness, the thickness of the lithosphere and the amount the lithosphere sinks into the mantle under sediment load. The two columns, A and B, of equal area have equal mass and are at the same distance above the depth of compensation (Figure 1). By mass equivalence we have:

$$dw_1\,\rho_w + l\rho_1 + y_m = \rho_w dw_2 + \rho_s x + l\rho_1 \qquad (3)$$

and distance equivalence we obtain:

$$dw_1 + l + y = dw_2 + x + l \qquad (4)$$

Equating (3) and (4) we set:

$$dw_1 = dw_2 + x\,\frac{(\rho_s - \rho_m)}{(\rho_w - \rho_m)} \qquad (5)$$

Using equation (5) we can predict depth with a given sediment cover from information on the unsedimented depth or vice versa. Thus, taking information for individual sites from the JOIDES volumes about depth, age, and sediment thickness past and present, a paleodepth can be predicted for selected times in the past.

A WORKED EXAMPLE: PALEODEPTH FOR DSDP SITE 14 AT 8, 16, AND 22 MA

In order to compute the sediment correction and the paleodepths, we make some assumptions about densities.

Let $\rho_m = 3.33$ gm/cm^3
$\rho_w = 1.03$ gm/cm^3
$\rho_s = 1.9$ gm/cm^3

In the case of the DSDP sites, ρ_s can be obtained from the site description and can vary. For DSDP site 14 (Table 1), (Maxwell et al., 1970), we find that

Water depth   = 4343 m
Sediment thickness =   108 m
Depth to basement = 4451 m

From the lithologic log (Maxwell et al., 1970), we find that the sediment thickness at

8 Ma is 108 m
16 Ma is 108 m
22 Ma is 98 m.

First we find the unsedimented depth for the present. From (5)

$$dw_1 = 4343 + 108\,\frac{(1.9 - 3.33)}{(1.03 - 3.33)} = 4410 \text{ m} \qquad (6)$$

Next we need to determine the unsedimented depths for 8 Ma, 16 Ma, and 22 Ma. We start by determining the offset from the depth-age curve for $0<t<70$ for site 14. We know the age of the crust is 40 Ma and the unsedimented depth is 4410. Thus the offset can be calculated using (1).

$$d(t) = 2500 + 350(t)^{1/2} \text{ m} + \text{offset} \qquad (7)$$

or $\quad 4410 = 2500 + 350(40)^{1/2}\text{m} + \text{offset}$
offset = −305 m

Now we need to add this offset to theoretical paleocurve (1) to compute the paleodepth at times in the past.

For 8 Ma   t = 40 my − 8 my = 32
$d(t) = 2500 − 305 + 350(32)^{1/2} = 4176$ m   (8)
For 16 Ma   t = 40 my − 16 my = 24 my
$d(t) = 2500 − 305 + 350(24)^{1/2} = 3911$ m   (9)
For 22 Ma   t = 40 my − 22 my = 18 my
$d(t) = 2500 − 305(18)^{1/2} = 3681$ m   (10)

These are unsedimented depths; to find the sedimented depths, we use the sedimented thickness information for the times 8 Ma, 16 Ma, and 22 Ma and recalculate depths using the inverse of (5):

$$dw_2 = dw_1 - x\,\frac{(\rho_s - \rho_m)}{(\rho_w - \rho_m)} \qquad (11)$$

For 8 Ma dw $= 4176$ m $- 108\,\dfrac{(\rho_s - \rho_m)}{(\rho_w - \rho_m)} = 4109$ m   (12)

For 16 Ma dw $= 3911$ m $- 108\,\dfrac{(\rho_s - \rho_m)}{(\rho_w - \rho_m)} = 3844$ m   (13)

For 22 Ma dw $= 3681$ m $- 98\,\dfrac{(\rho_s - \rho_m)}{(\rho_w - \rho_m)} = 3621$ m   (14)

This procedure is easily done on a computer and can be repeated for any site which follows the simple depth-age relationship. The results have been tabulated (Table 2) and plotted (Figure 8a).

APPENDIX II. THE PRESENT WATER DEPTH AND THAT AT 8, 16, AND 22 MA FOR SELECTED DSDP SITES

Site	Location	Water depth (m)				Basement Age[1]			Comments
		Present	8 Ma	16 Ma	22 Ma	Hit	Ma	Old. Sed.	
14	South Atlantic	4346	4109	3839	3609	Yes	39	U. Eocene	-
15	South Atlantic	3938	3672	3286	2799	Yes	24	U. Oligocene	-
17A	South Atlantic	4277	4024	3714	3437	Yes	32	Oligocene	-
18	South Atlantic	4022	3751	3402	3018	Yes	26	U. Oligocene	-
22	Rio Grande Rise	2106	1998	1871	1825	No	85	-	O.K. if aseismic ridge
32	Northeastern Pacific	4758	4604	4352	4115	Yes	38	L. Oligocene	-
34	Northeastern Pacific	4322	4179	3939	3640	Yes	31	U. Oligocene	-
36	Northeastern Pacific	3273	2887	-	-	Yes	14	M. Miocene	-
55	Caroline Ridge	2850	2590	2287	1998	No	30	U. Oligocene	Age based on nearby Site 57
56	Caroline Ridge	2508	2286	1986	1696	No	30	U. Oligocene	Age based on nearby Site 57
62.1	Euripic Rise	2591	2452	2246	2011	Yes	32	M. Oligocene	-
63.1	East Caroline Basin	4472	4255	4043	3884	Yes	35	L. Oligocene	-
65	Western central Pacific	6130	6092	6056	6021	No	130	-	-
66.1	Western central Pacific	5293	5246	5185	5138	Yes	110	-	-
71	Central Pacific	4419	4247	4133	4051	No	55	-	Basement at 1000 m from seismics
73	Central Pacific	4387	4220	4025	3876	No	50	-	Basement at 750 m from seismics
74	Central Pacific	4431	4220	3981	3787	Yes	45	M. Eocene	-
77B	Central Pacific	4291	4127	3929	3732	Yes	38	U. Eocene	-
78	Central Pacific	4363	4102	3820	3600	Yes	33	L. Oligocene	-
82	East Pacific Rise	3689	3108	-	-	Yes	9	U. Miocene	-
83	Galapagos Spreading Center	3692	3186	-	-	Yes	11	M. Miocene	-
116	Rockall Bank								Continental, not attempted
149	Venezuela Basin	3972	3933	3849	3790	No?	90	-	Sed. thick, based on Site 146
155	Coiba Ridge	2752	2565	-	-	Yes	14	M. Miocene	-
157	Galapagos Ridge	1591	1310	-	-	Yes	12	M. Miocene	-
158	Cocos Ridge	1953	1592	-	-	Yes	14	M. Miocene	-
159	East Pacific Rise	4484	4216	3903	3610	Yes	30	U. Oligocene	-
171	Mid Pacific Mts.	2290	2208	2125	2055	Yes	100	Cenomanian	-
173	Off Cape Mendocino								Continental?, not attempted
178	Aleutian Trench	4218	4390	4216	4014	Yes	45	-	-
183	Pre-trench rise, Aleutians	4708	4614	4415	4244	Yes	55	L. Eocene	O.K., if effect of trench small
199	Mariana Basin	6090	6106	6089	6073	No	>160	-	Basement at 750 m from seismics
200	Mai Tai Guyot	1469	1311	1120	964	No	55	-	Age based on nearby site
205	South Fiji Basin	4320	4087	3846	3618	Yes	32	M. Oligocene	-
206	Lord Howe Rise	3196	3205	3136	3119	No	80	-	Age guessed, basement at 1000 m
207A	Lord Howe Rise	1389	1330	1208	1094	Yes?	80	-	Age guessed, hit rhyolite?
208	Lord Howe Rise	1545	1525	1492	1439	No	80	-	Age guessed
210	Coral Sea Basin	4643	4698	4512	4330	No	50	-	Age from Sclater et al. (1980)
214	Ninetyeast Ridge	1671	1578	1408	1254	Yes	60	L. Paleocene	-
216	Ninetyeast Ridge	2262	2182	2058	1869	Yes	70	U. Maestrichtian	-
217	Ninetyeast Ridge	3030	2978	2892	2797	No	80	-	Age from nearby magnetic data
219	Chagos-Laccadive Ridge	1764	1646	1446	1293	No	59	U. Paleocene	Age from nearby magnetic data
223	Owen Ridge	3633	3642	3534	3369	Yes	57	U. Paleocene	Age from nearby magnetic data
231	Gulf of Aden	2152	1901	-	-	Yes	13	M. Miocene	-
234	North Somali Basin	4721	4594	4492	4414	No	75	-	-
236	North of Seychelles	4487	4376	4197	4044	Yes	60	U. Paleocene	-
237	Mascarene Plateau	1623	1533	1352	1213	No	65	-	-
238	Central Indian Ridge	2832	2707	2516	2275	Yes	35	L. Oligocene	-
249	Mozambique Ridge	2088	2082	2067	2017	Yes	130	Neocomian	-
251	Southwest Indian Ridge	3489	3246	2786	-	Yes	18	L. Miocene	-
253	Ninetyeast Ridge	1962	1778	1575	1392	No?	50	L. Eocene	-
254	Ninetyeast Ridge	1253	1054	878	688	Yes	40	L. Oligocene	-
255	Broken Ridge	1144	1049	931	826	No	85	Santonian	Age from Santonian limestone
266	Southeast Indian Ridge	4167	3932	3510	2755	Yes	22	L. Miocene	-
267B	Southeast Indian Ridge	4495	4362	4137	3929	Yes	41	U. Eocene	-
278	McQuarie Ridge	3675	3526	3273	2994	Yes	30	M. Oligocene	O.K., if simple subsidence
279A	McQuarie Ridge	3341	2996	2554	-	Yes	20	L. Miocene	O.K., if simple subsidence
281	South Tasman Rise								Continental, not attempted
284	Challenger Plateau								Continental, not attempted
285	South Fiji Basin	4658	4522	4461	4180	Yes	35	L. Oligocene	Age from nearby site
289	Ontong Java Plateau	2206	2283	2373	2398	Yes	115	Aptian	-
292	Philippine Sea	2943	2761	2552	2366	Yes	45	M. Eocene	-
296	Sikoku Basin	2920	2782	2543	2308	No	35	L. Oligocene	-

APPENDIX II. THE PRESENT WATER DEPTH AND THAT AT 8, 16, AND 22 MA FOR SELECTED DSDP SITES (continued)

Site	Location	Water depth (m)				Basement Age[1]			Comments
		Present	8 Ma	16 Ma	22 Ma	Hit	Ma	Old. Sed.	
310	Hess Rise	3516	3474	3389	3314	No	105	U. Albian	Age based on oldest sediment drilled
317B	Manihiki Plateau	2598	2580	2551	2520	No	115	Aptian	Age based on 317A
319	East Pacific Rise	4296	3955	3492	-	Yes	20	L. Miocene	-
329	Falkland Plateau								Continental, not attempted
334	North Mid-Atlantic Ridge	2619	2185	-	-	Yes	11	U. Miocene	-
354	Amazon Cone	4052	4028	3910	3797	Yes	70	Maestrichtian	-
357	Rio Grande Rise	2086	1996	1890	1846	No	85	-	O.K., if aseismic ridge
360	Cape Rise	2949	2984	3021	2997	No	120	-	Basement at 4000 m from seismics
362	Walvis Ridge	1325	1460	1580	1580	No	120	-	Age from nearby magnetic data
366A	Sierra Leone Rise	2853	2786	2654	2514	No	72	Maestrichtian	Age from oldest sediment drilled
368	Cape Verde Plateau								Swell, no adequate model
369A	Moraccan Shelf								Continental, not attempted
372	Western Mediterranean	2699	2476	2224	1990	No	22	L. Miocene	Basement at 1500 m from seismics
391A	Western North Atlantic	4974	5044	5053	5058	No	150	L. Tithonian	Age from oldest sediment drilled
397	Moraccan Shelf								Continental, not attempted
398	Vigo Seamount (off Spain)	3910	4064	4070	4054	Yes?	125	Hauterbian	-
400A	Biscay abyssal plain	4399	4498	4513	4477	No	115	Aptian	Age from oldest sediment drilled
407	Reykjanes Ridge	2472	2313	2046	1839	Yes	35	L. Oligocene	-
408	Reykjanes Ridge	1624	1383	950	-	Yes	20	L. Miocene	-
410	North Mid-Atlantic Ridge	2975	2584	-	-	Yes	11	U. Miocene	-
436	Japan Trench, ocean	5240	5351	5323	5286	No	125	-	Basement at 625 m from seismics
438	Japan Trench, forearc								Trench wall, not attempted
439	Japan Trench, forearc								Trench wall, not attempted
445	Diato Ridge	3377	3297	3124	2994	No		M. Eocene	O.K., if simple subsidence
448	Balau Kyushu Ridge	3483	3231	2963	2725	Yes	35	L. Oligocene	O.K., if simple subsidence
468	California Borderland								Continental, not attempted
469	California Borderland, oceanic	3790	3473	2922	-	Yes	17	M. Miocene	-
470	California Borderland, oceanic	3549	3193	2250	-	Yes	16	M. Miocene	-
471	East Pacific Rise	3101	2812	-	-	Yes	15	M. Miocene	-
472	East Pacific Rise	3831	3429	-	-	Yes	15	M. Miocene	-
492	Middle America Trench								Trench wall, not attempted
493	Middle America Trench								Trench wall, not attempted
495	Middle America Tench	4140	3900	3521	2764	Yes	22	L. Miocene	Ocean side of trench
502	Columbia Basin	3052	3065	2961	2870	No	80?	-	Basement at 700 m from seismics
503A	Galapagos Spreading Center	3672	3411	-	-	No	11	-	Age from nearby site
512	Maurice Ewing Bank	1846	1788	1732	1675	No	130	-	O.K., if subsidence oceanic?
513A	South Mid-Atlantic Ridge	4373	4231	4018	3824	Yes	45	U. Eocene	-
516F	Rio Grande Rise	1313	1276	1227	1197	Yes	90	Coniacian	O.K., if aseismic ridge?
526	Walvis Ridge	1054	940	758	624	No	65	U. Maestrichtian	Age from nearby site

[1]Unless commented upon, the basement age is taken from the sediment at the basement contact or by estimation from the magnetic anomalies.

REFERENCES CITED

Barker, P. F. and Burrell, J., 1977, The opening of the Drake Passage: *Mar. Geol.* 25, p. 15–34.

Berger, W. H., Vincent, E., and Thierstein, H. R., 1981, The deep-sea record: major steps in Cenozoic ocean evolution: *Soc. Econ. Paleo and Mineral., Special Public 32,* p. 489–504.

Berggren, W. A. and Hollister, C. D., 1977, Plate tectonics and paleocirculation—commotion in the ocean: *Tectonophysics, 38,* p. 11–48.

Cande, S. C., and Mutter, J. C., 1982, A revised identification of the oldest sea-floor spreading anomalies between Australia and Antarctica: *Earth and Planetary Science Letters,* v. 58, p. 151–160.

Chase, T. E., 1975, Topography of the Oceans: IMR Technical Report Series TR57, Scripps Institution of Oceanography.

Crough, S. T., 1978, Thermal origin of mid-plate hot-spot swells: *Geophys. J. R. Astr. Soc. 55,* p. 451–470.

Crough, S. T., and Jarrard, R. D., 1981, The Marquesas-Line Swell: *J. Geophys. Res. 86,* p. 11763–11771.

Davis, E. E. and Lister, C.R.B., 1974, Fundamentals of ridge crest topography: *Earth. Planet. Sci. Lett. 21,* p. 405–413.

Detrick, R. S., Sclater, J. G. and Thiede, J., 1977, Subsidence of aseismic ridges-Geological and geophysical implications: *Earth Planet. Sci. Lett., 34, 2,* p. 185–196.

Detrick, R. S. and Crough, S. T., 1978, Island subsidence, hot spots and lithosphere thinning: *J. Geophys. Res., 83,* p. 1236–1244.

Fisher, R. L. and Sclater, J. G., 1983, Tectonic evolution of the Southwest Indian Ocean since the Mid-Cretaceous: plate motions and stability of the pole of Antarctica/Africa for at least 80 Myr: *Geophys. J. R. Astr. Soc., 73,* p. 553–576.

Handschumacher, D. W., 1976, Post-Eocene plate tectonics of the eastern Pacific: *Geophysics of the Pacific Ocean Basin and its Margin,* a volume in honour of George P. Woollard, G. H. Sutton, et al., eds., AGU (Washington) p. 177–202.

Herron, E. M., 1972, Sea-floor spreading and the Cenozoic history of the East-Central Pacific: *Geol. Soc. Amer. Bull. 83,* p. 1671–1692.

Kennett, J. P., 1977, Cenozoic-evolution of the Antarctic glaciation, the Circum-Antarctic Ocean and their impact on global paleoceanography: *J. Geophys. Res. 82,* p. 3843–3859.

Kennett, J. P., 1982, *Marine Geology:* Prentice Hall, Inc., Englewood Cliffs, N.J., p. 813.

LaBrecque, J. L., Kent, D. V., and Cande, S. C., 1977, Revised magnetic polarity time scale for late Cretaceous and Cenozoic time: *Geology, 5,* p. 330–335.

McKenzie, D. P., Watts, A., Parsons, B. and Roufosse, M., 1980, Planform convection beneath the Pacific Ocean: *Nature 288,* p. 442–446.

Maxwell, A. E., Von Herzen, R. P., Hsu, K. J., Andrews, J. E., Saito, T., Percival, S. F., Milo, E. D., and Boyce, R. E., 1970, Deep sea drilling in the South Atlantic: *Science 168,* p. 1047–1059.

Menard, H. W., 1969, Elevation and subsidence of oceanic crust: *Earth Planet. Sci. Lett. 6,* p. 275–284.

Menard, H. W. and McNutt, M., 1982, Evidence for and consequences of thermal rejuvenation: *J. Geophys. Res. 87,* B10, p. 8570–8580.

Miller, K. G., 1982, Late Paleogene (Eocene to Oligocene) paleoceanography of the northern North Atlantic [Ph.D. thesis]: Mass. Inst. Technology, p. 92.

Minster, J. B., Jordan, T. H., Molnar, P., and Haines, E., 1974, Numerical modeling of instantaneous plate tectonics: *Geophys. J. R. Astr. Soc., 36,* p. 541–576.

Norton, I. O. and Sclater, J. G. , 1979, A model for the evolution of the Indian Ocean and the breakup of Gondwanaland: *J. Geophys. Res. 84,* p. 6803–6830.

Parsons, B. and Daly, S., 1983, The relationship between surface topography, gravity anomalies and temperature structure of convection: *J. Geophys. Res. 88,* p. 1129–1144.

Parsons, B., and McKenzie, D. P., 1978, Mantle convection and the thermal structure of the plates: *J. Geophys. Res. 83,* p. 4485–4496.

Parsons, B., and Sclater, J. G., 1977, An analysis of the variation of ocean floor bathymetry and heat flow with age: J. Geophys. Res., p. 803–827.

Sclater, J. G., Anderson, R. N., and Bell, M. L., 1971, The elevation of ridges and the evolution of the central eastern Pacific: *J. Geophys. Res. 76,* p. 7888–7915.

Sclater, J. G., Lawver, L. A., and Parsons, B., 1975, Comparison of long-wavelength residual elevation and free-air gravity anomalies in the North Atlantic, and possible implications for the thickness of the lithosphere plate: *J. Geophys. Res. 80,* p. 1031–1052.

Sclater, J. G., Hellinger, S., and Tapscott, G., 1977, Paleobathymetry of the Atlantic Ocean: *J. Geol. 85,* p. 509–552.

Sclater, J. G., Jaupart, C., and Galson, D., 1980, The heat flow through oceanic and continental crust and the heat loss of the earth: *Rev. Geophys. Space Phys.* 18, p. 269–311.

Sclater, J. G., Parsons, B. and Jaupart, C., 1981, Oceans and continents: similarities and differences in the mechanisms of heat loss: *J. Geophy. Res. 86,* p. 11535–11552.

Shackleton, N. J. and Kennett, J. P., 1975, Paleotemperature history of the Cenozoic and the initiation of Antarctic Glaciation: Oxygen and carbon isotope analysis in DSDP sites 277, 279 and 281: *in* Kennett, J. P. and Hontz, R. E., et al., Initial Reports of the Deep Sea Drilling Project, Volume XXIX, Washington (U.S. Government Printing Office).

Smith, A. G., 1982, Late Cenozoic uplift of stable continents in a reference frame fixed to South America: *Nature,* p. 400–404.

Stock, J., 1981, Uncertainties in the relative positions of the Australia, Antarctica, Lord Howe and Pacific Plates during the Tertiary [M.Sc. thesis]: Mass. Inst. Technology, p. 106.

Stock, J., and Molnar, P., 1982, Uncertainties in the relative positions of the Australia, Antarctica, Lord Howe, and Pacific Plates since the late Cretaceous: *J. Geophys. Res. 87,* B6, p. 4697–4714.

van Andel, T. H., Heath, G. R., and Moore, T. C., 1975, Cenozoic history and paleoceanography of the Central Equatorial Pacific Ocean: *Geol. Soc. Amer. Memoir 143,* p. 1–34.

van Hinte, J. E., 1976a, A cretaceous time scale: *Bull. Assoc. Petrol. Geo. 60,* 4, p. 498–516.

van Hinte, J. E., 1976b, A Jurassic time scale: *Bull. Assoc. Petrol. Geo. 60,* 4, p. 489–497.

Von Herzen, R. P., Detrick, R. S., Crough, T., Epp, D., and Fehn, U., 1982, Thermal origin of the Hawaiian Swell: Heat flow evidence and thermal models: *J. Geophys. Res. 87,* p. 6711–6723.

Weissel, J. K., Hayes, D. E., and Herron, E. M., 1977, Plate tectonic synthesis: the displacements between Australia, New Zealand and Antarctica since the late Cretaceous *Mar. Geol. 25,* p. 231–277.

MANUSCRIPT ACCEPTED BY THE SOCIETY DECEMBER 17, 1984

Geological Society of America
Memoir 163
1985

A multiple microfossil biochronology for the Miocene

John A. Barron
Gerta Keller*
U.S. Geological Survey
345 Middlefield Road
Menlo Park, California 94025

Dean A. Dunn
Graduate School of Oceanography
University of Rhode Island
Narragansett, Rhode Island 02882
and
University of Southern Mississippi
Hattiesburg, Mississippi 39406

ABSTRACT

A multiple microfossil biochronology is presented for the Miocene which allows resolution of time approaching 100,000 years. Carbonate stratigraphy is integrated to greatly enhance this resolution. Graphical correlation techniques were applied to over 20 DSDP (Deep Sea Drilling Project) sections to identify 175 planktonic foraminiferal, calcareous nannofossil, radiolarian, and diatom datum levels between 24.0 and 4.3 Ma which show the most consistent (isochronous) correlations. Ages are estimated for these datum levels through 72 direct correlations to paleomagnetic stratigraphy and extrapolation between the correlation points. The resulting Miocene time scale resembles previously published time scales except for the early Miocene, where recent paleomagnetic correlations result in changes.

The three CENOP (Cenozoic Paleoceanography Project) time slices (~21, 16, and 8 Ma) are characterized biostratigraphically (planktonic foraminifers, calcareous nannofossils, radiolarians, and diatoms) and in terms of carbonate stratigraphy. The ages of the time slices are estimated as follows: the early Miocene time slice (21.2–20.1 Ma; given as 22 Ma in this volume), the late early Miocene time slice (16.4–15.2 Ma), and the late Miocene time slice (8.9–8.2 Ma).

An alternate time scale utilizing a paleomagnetic Anomaly 5-paleomagnetic Chron 11 correlation is also presented. Estimated ages for microfossil zones and datum levels in the late middle and early late Miocene (14–7 Ma) utilizing the alternate time scale are generally younger than those for the more traditional time scale. The late Miocene time slice has an estimated age of 8.0–7.0 Ma.

INTRODUCTION

During the past five years CENOP (Cenozoic Paleoceanography Project) workers have accumulated much detailed biostratigraphic and paleoecologic data on Miocene planktonic foraminifers (Srinivasan and Kennett, 1981a, b; Keller, 1980a, b, 1981a, b; Keller and Barron, 1981; Poore, 1981; Thunell, 1981; Vincent, 1981a), calcareous nannofossils (Haq et al., 1980); radiolarians (Moore and Lombari, 1981; Dunn, 1982; Romine, this volume); and diatoms (Burckle et al., 1982; Barron, 1983, in press; Barron and Keller, 1983; Burckle and Opdyke, this vol-

---
*Present address: Department of Geological and Geophysical Science, Princeton University, Princeton, New Jersey 08540.

ume). At the same time Dunn and Moore (1981), and Vincent, 1981b have developed a very detailed carbonate stratigraphy for the Miocene of the equatorial and central North Pacific which allows much greater time resolution than is possible with microfossil stratigraphies alone. These multiple microfossil and carbonate stratigraphies have been completed in as many deep-sea sequences as recovery and preservation permitted, including over 20 Miocene sequences from the Pacific Ocean (see map in volume preface). Among these latter sections, the most complete and representative biostratigraphic records are those from DSDP Sites 71 and 77 in the central equatorial Pacific and DSDP Site 289 in the western equatorial Pacific (Keller and Barron, 1983; Keller, 1981a; Saito, 1977; Ryan et al., 1974). These sites provide an excellent opportunity for correlation and integration of multiple microfossil datum events and carbonate stratigraphy and, hence, for the development of a high resolution stratigraphy for the Miocene. The expanding data base for direct ties of microfossil datum levels to paleomagnetic stratigraphy (Saito et al., 1975; Theyer et al., 1978; Burckle, 1978; Poore et al., 1983) allows correlation of this high resolution stratigraphy to radiometric time and the development of a refined Miocene chronostratigraphy (Keller et al., 1982).

The main purpose of this paper is to summarize these varied biostratigraphic data and to synthesize a high-resolution biochronology for the Miocene, based on microfossil datum events and carbonate events that are calibrated to the paleomagnetic time scale of Berggren et al. (1984). To achieve this goal we have combined microfossil datum events that are tied directly to paleomagnetic stratigraphy (Gartner, 1973; Saito et al., 1975; Burckle, 1978; Theyer et al., 1978; Poore et al., 1983) with our data set to test their isochroneity in DSDP sections analyzed in the Pacific Ocean, particularly Sites 71, 77, and 289. This approach has resulted in a reliable biochronologic model for the Miocene which yields an age resolution approaching 100,000 years. The second purpose of this paper is to define biostratigraphically the three time-slice intervals at about 21, 16, and 8 Ma which have been chosen by CENOP workers to study the paleoceanographic and paleoclimatic history of the Miocene ocean.

MIOCENE CHRONOSTRATIGRAPHY

A Miocene Time Scale

Correlation of the various low-latitude microfossil zones of the Miocene with the paleomagnetic time scale of Berggren et al. (1984) is shown in Figure 1. This figure incorporates the planktonic foraminiferal zones of Blow (1969) with subzones suggested by Srinivasan and Kennett (1981a, b) and Keller (in press), the calcareous nannofossil zones of Martini (1971) and Bukry (1973, 1975) (notation after Okada and Bukry, 1980), the radiolarian zones of Riedel and Sanfilippo (1978), and the diatom zones of Burckle (1972) and Barron (1983, in press). In addition, intervals of widespread deep-sea hiatuses (Keller and Barron, 1983) and the carbonate curves of DSDP Sites 158 (4.5–12.9 Ma) (Dunn

and Moore, 1981) and 71 (12.0–22.0 Ma) (D. Dunn, unpublished data) are included to show their relationship with the microfossil stratigraphies. The main differences between this and earlier Miocene time scales (Ryan et al., 1974; Berggren, 1981) are in the early Miocene, where new paleomagnetic correlations (Poore et al., 1983; Barron et al., in press) have resulted in changes.

Specifically, the N4/N5 planktonic foraminiferal zone boundary is correlated with mid paleomagnetic Chron 19 on the basis of comparisons with paleomagnetically tied radiolarian datum levels (Theyer et al., 1978) rather than lower Chron 20 (Berggren, 1981) or upper Chron 21 (Ryan et al., 1974; Berggren et al., 1983). Berggren et al.'s (1983) correlation of the last occurrence of the planktonic foraminifer Globorotalia kugleri (= N4/N5 boundary) relies on his interpretation of the paleomagnetic Anomaly 6C at DSDP Site 516. His interpretation of Anomaly 6C is based on the first occurrence of G. kugleri at Site 516. However, that occurrence appears to be diachronously young based on the last occurrence of the calcareous nannofossil Sphenolithus ciperoensis (Berggren et al., 1983) and the first occurrence of the diatom Rocella gelida (Gombos, 1983) at the site. Within the tropics the last occurrences of G kugleri and S. ciperoensis are nearly coincident (Berggren et al., 1984; Barron et al., in press), and the first occurrence of R. gelida is immediately below those datum levels (Barron, 1983; Barron et al., in press).

In addition, the CN1/CN2 calcareous nannofossil zone boundary is herein tied with lower Chron 17 (Barron et al., in press) rather than with lower Chron 19 as proposed by Berggren (1981). Finally, the third major early Miocene difference involves the Calocycletta costata/Stichocorys wolffii radiolarian zone boundary. This boundary is placed in lower Chron 16 rather than mid Chron 17, as proposed by Berggren (1981), following the results of DSDP Leg 85 (Barron et al., in press) and reinterpretation of the paleomagnetic assignments of Theyer et al. (1978).

Following Berggren et al. (1984), the Oligocene/Miocene boundary is placed at 23.7 Ma in lower Anomaly 6C, which corresponds with the base of the calcareous nannofossil zones NN1 and CN1 and with the first occurrence of Globorotalia kugleri (the base of Zone N4 as recognized by Keller, in press). Other workers recognize the Oligocene/Miocene boundary at the top of calcareous nannofossil Subzone CN1a (Bukry, 1975; Keller and Barron, 1983) (22.7 Ma), the first occurrence of Globigerinoides trilobus immaturus or top of planktonic foraminiferal Subzone N4a (Keller, in press) (22.1 Ma), or the first occurrence of Globoquadrina dehiscens, which defines the top of planktonic foraminiferal Subzone N4b (21.2 Ma) (Srinivasan and Kennett, 1981a, b).

The early Miocene/middle Miocene boundary is placed at 16.0 Ma in the upper part of paleomagnetic Chron 16 within the lower part of planktonic foraminiferal Zone N8 and uppermost part of calcareous nannofossil zone NN4 following Ryan et al. (1974).

The middle Miocene/late Miocene boundary falls in the upper part of paleomagnetic Chron 11 at ~11.5 Ma in the middle

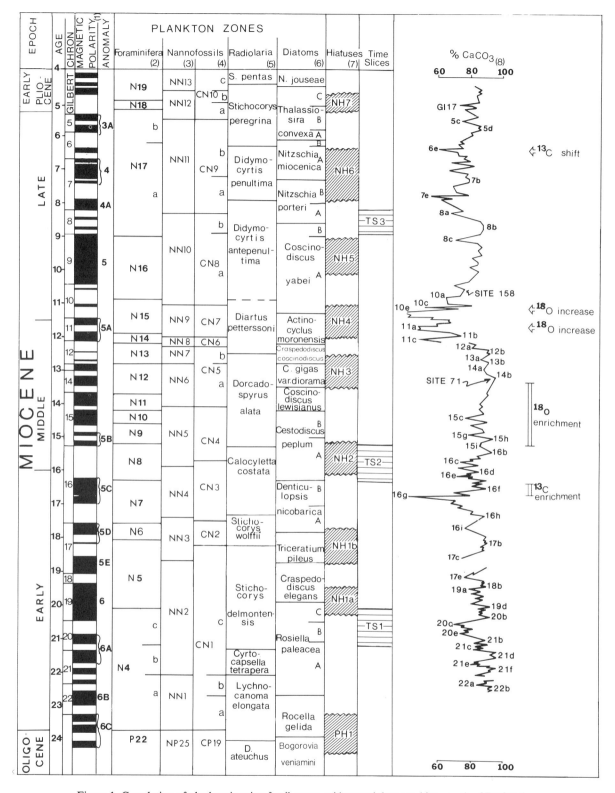

Figure 1. Correlation of planktonic microfossil zones, widespread deep sea hiatuses, the CENOP time slices, and the percent CaCO₃ curve for the eastern equatorial Pacific to the paleomagnetic time scale for the Miocene. (1) = Berggren et al., 1984; (2) = Blow, 1969; Srinivasan and Kennett, 1981a, b; Keller, in press; (3) = Martini, 1971; (4) = Bukry 1973, 1975; Okada and Bukry, 1980; (5) = Riedel and Sanfilippo, 1978; (6) = Burckle, 1972; Barron, 1983, in press; and (7) = Keller and Barron, 1983; (8) = Dunn and Moore, 1981; Dunn 1982.

of planktonic foraminiferal Zone N15, middle of calcareous nannofossil Zones NN9 and CN7, and near the first occurrence of the
radiolarian *Didymocyrtis antepenultima* according to Berggren
(1981) and Theyer et al. (1978). Ryan et al. (1974) place the
boundary slightly lower at ~11.8 Ma.

Following Cita (1975) and Burckle (1978), the Miocene/-
Pliocene boundary lies within the middle of the lower reversed
event of the Gilbert paleomagnetic Chron at ~5.2 Ma. The Mio-
cene/Pliocene boundary thus falls at the base of planktonic foraminiferal Zone N18, within calcareous nannofossil Zone NN12,
in uppermost calcareous nannofossil Subzone CN10a, and near
the Subzone A/Subzone B boundary of the *Thalassiosira convexa* Zone of diatoms.

At least five isotopic events are also useful for Miocene
stratigraphy (Fig. 1 right side). The distinctive latest Miocene
carbon shift (at 6.4 Ma on this time scale) involves an apparently
permanent −0.5 $^o/_{oo}$ depletion in $\delta^{13}C$ of the tests of benthic
foraminifers and has been shown to be isochronous across the
Pacific and Indian Oceans (Haq et al., 1980; Vincent et al.,
1980). More recently, Vincent and Killingley (in press) have
identified an isochronous +0.5 $^o/_{oo}$ enrichment in the $\delta^{13}C$ of
foraminifers which occurs in uppermost planktonic foraminiferal
Zone N7 and is correlative with paleomagnetic Anomaly 5C in
numerous cores in the Pacific and Indian Oceans.

The +1.0 $^o/_{oo}$ increase in benthic foraminiferal $\delta^{18}O$ which
occurs within planktonic foraminiferal zones N9 to lower N12 is
widely recognizable throughout the ocean basins (Woodruff et
al., 1981; Savin et al., 1981). Finally, Keigwin (1979) and
Burckle et al. (1982) have shown that two distinctive oxygen
isotope enrichment events occur in benthic foraminiferal tests
across the middle Miocene/late Miocene boundary at DSDP Site
158. The first correlates with lower paleomagnetic Chron 11
(~11.7 Ma) and involves an increase of about +0.5 $^o/_{oo}$ (11.3 Ma)
in $\delta^{18}O$. The second correlates with lower paleomagnetic Chron
10 and involves an increase of ~ +0.7 $^o/_{oo}$ in $\delta^{18}O$ (see also Keller
et al., 1982).

Datum Events

Beyond the Miocene time scale shown in Figure 1, development of a high resolution Miocene chronostratigraphy should
incorporate numerous additional microfossil datum levels within
the individual zones and subzones which have been shown to be
isochronous over widespread geographic areas. Many stratigraphically useful Miocene datum levels have been proposed for
planktonic foraminifers (Keller, 1980a,b, 1981a,b; Srinivasan
and Kennett, 1981a,b, Saito, 1977; Berggren et al., 1983), radiolarians (Theyer et al., 1978; Riedel and Sanfilippo, 1978), and
diatoms (Burckle, 1978; Barron, in press). Direct correlation of
planktonic foraminiferal (Saito et al., 1975; Poore et al., 1983),
calcareous nannofossil (Gartner, 1973; Haq et al., 1980; Poore et
al., 1983), radiolarian (Theyer et al., 1978; Johnson and Wick,
1982), and diatom (Burckle, 1978; Burckle et al., 1982) datum
levels with paleomagnetic stratigraphy and graphical correlation

plots (Shaw, 1964) provide the basis for testing the sequence and
isochroneity of various microfossil datum events in different stratigraphic sections. Numerous age estimates can be assigned to the
datum levels tied directly to paleomagnetic stratigraphy using the
paleomagnetic polarity time scale of Berggren et al. (1984), and
age versus depth plots can be constructed for individual stratigraphic sections in order to extrapolate the ages of additional
datum levels (Barron, 1980; Keller, 1981b).

Tables 1–4 list stratigraphically important planktonic foraminiferal, calcareous nannofossil, radiolarian, and diatom datum
levels for the Miocene and estimates of their ages according to the
paleomagnetic time scales of Berggren et al. (1984) and Mankinen and Dalrymple (1979). An additional column lists the ages
which result from utilizing an Anomaly 5-paleomagnetic Chron
11 correlation rather than the more traditional Anomaly 5-Chron
9 correlation. The basis for inclusion of this third column is
discussed in the addendum. All discussion prior to the addendum
utilizes the column 1 ages derived from the Berggren et al. (1984)
paleomagnetic time scale and the traditional Anomaly 5-Chron 9
correlation and constitutes the adopted CENOP time scale.

Datum levels that have been tied directly to paleomagnetic
stratigraphy are identified by an asterisk. Graphical correlation
techniques were applied to over 20 DSDP sections in the Pacific
(including DSDP Sites 71, 77, 158, 173, 289, 292, 470, and 495)
in order to construct these tables. Microfossil data used in these
compilations are predominantly from recent studies by CENOP
workers (Keller, 1980a, b, 1981a, b; Keller et al., 1982; Srinivasan and Kennett, 1981a, b; Burckle et al., 1982; Barron, 1981a,
b, 1983; Dunn, 1982). These data were supplemented by data
published in the Initial Reports of the Deep Sea Drilling Project
(Initial Reports): Volumes 7 (1971), 8 (1971), 9 (1972), 16
(1973), 17 (1973), 18 (1973), 30 (1975), 31 (1975), 32 (1975),
41 (1977), 63 (1982), and 67 (1982) as well as radiolarian data
from Westberg and Riedel (1978). Tables 1–4 represent the
composite of numerous graphical correlations which have incorporated 72 direct ties of microfossil datum levels to paleomagnetic stratigraphy between 24 and 4.3 Ma (10 planktonic
foraminiferal, 9 calcareous nannofossil, 27 radiolarian, and 26
diatom).

Published correlations of additional datum levels to paleomagnetic stratigraphy have been discarded, because they were not
consistent with graphical correlations (Keller et al., 1982). Correlation of a datum level to paleomagnetic stratigraphy in one core
is no guarantee that the correlation is consistent (isochronous)
with correlations elsewhere. For example, Theyer et al.'s (1978)
correlation of the first appearance of the radiolarian *Diartus petterssoni* with paleomagnetic Chron 11 is rejected, because graphical correlations based on diatom and planktonic foraminiferal
calibrations indicate that such a calibration date is too young.
Similarly, Johnson and Wick (1982) show that many radiolarian
datum levels vary as much as 0.6 m.y. in age in central equatorial
Pacific cores for which the paleomagnetic stratigraphy has been
measured.

Tables 1–4 are limited by the lack of total uniformity among

TABLE 1. AGES OF STRATIGRAPHICALLY-IMPORTANT MIOCENE
PLANKTONIC FORAMINIFERAL DATUM LEVELS

Datum	Paleomag. Correl.	Age (Ma) "B84"	"MD79"	"B84!"
B. Sphaeroidinella dehiscens (z)	5th(-), Gilbert	4.9*(1)	4.9*	nc
B. Globorotalia tumida (z)	5th(-), Gilbert	5.2*(1)	5.1*	nc
T. Globoquadrina dehiscens (tropics)	5th(-), Gilbert	5.25*(1)	5.2*	nc
B. Globorotalia margaritae	(-), Chron 5	5.6*(1)	5.5*	nc
B. Pulleniatina primalis (sz)		6.3	6.1	nc
B. Globorotalia conomiozea	1st(-), Chron 6	6.3*(2, 3)	6.1*	nc
B. Globigerinides conglobatus		8.0	7.7	7.1
T. G. kennetti (temp.)		8.0	7.8	7.1
B. Neogloboquadrina pachyderma (temp.)		8.9	8.5	7.8
B. G. obliquus extremus		9.0	8.6	8.0
B. Globorotalia plesiotumida (z)		9.0	8.6	8.0
B. Globigerinoides kennetti (temp.)		10.1	9.7	8.1
T. Globoquadrina dehiscens (temp.)		10.3	9.9	8.2
B. G. merotumida		10.8	10.4	8.5
B. Neogloboquadrina acostaensis (z)	2nd(-), Chron 10	10.9*(2)	10.5*	8.6*
T. Globorotalia siakensis (z)		11.9	11.7	9.9
T. Globorotalia mayeri	low, Chron 11	11.9*(2	11.7*	9.9*
B. Globigerina nepenthes (z)		12.2	12.0	10.9
B. Globorotalia menardii		12.7	12.4	11.5
T. G. fohsi robusta		12.8	12.5	11.8
B. Sphaeroidinella subdehiscens (z)		12.8	12.5	11.8
B. S. seminulina		13.1	12.8	12.4
B. G. fohsi lobata		13.5	13.2	13.0
B. G. fohsi fohsi (z)		13.7	13.5	13.5
B. G. fohsi praefoshi (z)		14.2	14.0	14.0
T. G. archaeomenardii		14.5	14.3	nc
B. G. peripheroacuta (z)		14.6	14.4	nc
B. Orbulina suturalis (z)	lowest(+), Chron 15	15.2*(4)	15.1*	nc
T. Globigerinoides diminutus		15.6	15.5	nc
B. Globorotalia archaeomenardii		15.7	15.6	nc
B. Globigerinoides mitra		16.0	15.9	nc
B. G. sicanus (z)		16.4	16.3	nc
B. G. bisphaericus		16.4	16.3	nc
B. Globorotalia peripheroronda		16.5	16.4	nc
T. Catapsydrax stainforthi		16.7	16.7	nc
T. G. zealandica incognita		17.3	17.3	nc
T. C. dissimilis (z)	uppermost, Chron 17	17.6*(5)	17.6*	nc
B. Globigerinatella insueta (z)	1st(+), Chron 17	18.1*(5)	18.1*	nc
T. Globigerina binaiensis		18.2	18.2	nc
B. Globorotalia miozea		18.8	18.9	nc
B. G. praescitula		18.8	18.8	nc
T. G. acrostoma		19.7	19.8	nc
B. G. zealandica pseudomiozea		19.9	20.0	nc
T. G. kugleri (z)		20.1	20.3	nc
B. Dentoglobigerina altispira		20.7	20.9	nc
B. Globorotalia zealandica incognita		21.1	21.3	nc
T. G. mendacis		21.2	21.4	nc
B. Globoquadrina dehiscens (sz)		21.2	21.4	nc
B. Globorotalia acrostoma		21.7	21.9	nc
B. Globigerinoides trilobus immaturus (sz)		22.1	22.3	nc
B. Globorotalia kugleri (z)	2nd(+), Anomaly 6C	23.7*(6)	24.0*	nc

(z) = zonal marker; (sz) = subzonal marker; (temp.) = temperature datum. (-) = reversed polarity
event; (+) = normal polarity event. * = direct paleomagnetic calibration. nc = no change.
Three age estimates are given based on three paleomagnetic time scales: "B84" = Berggren et al.,
1984; "MD79" = Mankinen and Dalrymple, 1979; "B84!" = Berggren et al., 1984 using an Anomaly
5-Chron 11 correlation.
References: (1) = Saito et al., 1975; (2) = Ryan et al., 1974; (3) = Loutit and Kennett, 1979;
(4) = Poore et al., 1983; (5) = Barron et al., in press; and (6) = Berggren et al., 1984.

TABLE 2. AGES OF STRATIGRAPHICALLY-IMPORTANT MIOCENE
CALCAREOUS NANNOFOSSIL DATUM LEVELS

Datum	Paleomag. Correl.	Age (Ma)		
		"B84"	"MD79"	"B84!"
B. Ceratolithus rugosus (sz) (z)	c$_2$ event, Gilbert	4.7*(1)	4.6*	nc
B. C. acutus (sz)	5th(-), Gilbert	5.0*(1)	5.0*	nc
T. Discoaster quinqueramus (z)	top of (-), Chron 5	5.5*(1)	5.4*	nc
B. Amaurolithus primus (sz)	base of (+), Chron 6	7.2#	7.0#	6.5*(2)
B. D. quinqueramus (z)	top of Chron 8†	8.3*(2)	8.0*	7.3*
B. D. neorctus (sz)		8.9	8.5	8.0
T. D. hamatus (z)	2nd (+), Chron 10	11.1*(3)	10.7*	8.8*
T. Catinaster coalitus		11.3	11.0	8.9
B. D. hamatus (z)		11.9	11.7	10.0
B. C. coalitus (z)		12.3	12.1	10.8
B. D. kugleri (sz)(z)		12.8	12.5	11.8
T. Sphenolithus heteromorphus (z)		14.0	13.9	nc
B. Calcidiscus macintyrei (z)		15.7	15.5	nc
B. D. exilis (z)	mid 1st(-), Chron 16	15.9*(4)	15.8*	nc
T. S. belemnos (z)		17.4	17.4	nc
B. S. heteromorphus (z)	lowermost Chron 16	17.5*(5)	17.5*	nc
B. S. belemnos (z)	2nd(-), Chron 17	18.2*(5)	18.2*	nc
T. Triquetrorhabdulus carinatus (z)	Anomaly 5E	18.7*(3)	18.7	nc
B. recurrent D. druggi		19.5	19.6	nc
B. D. druggii (sz)(z)		22.1	22.3	nc
T. abundant Cyclicargolithus abisectus (sz)		22.7	22.9	nc
T. Reticulofenestra bisecta	1st(+), Anomaly 6C	23.4*(4)	23.7*	nc
T. Sphenolithus ciperoensis (z)	2nd(+), Anomaly 6C	23.7*(4)	24.0*	nc

(z) = zonal marker; (sz) = subzonal marker. (+) = normal polarity event; (-) = reversed polarity event. * = direct paleomagnetic calibration. nc = no change.
Three age estimates are given based on three paleomagnetic time scales: "B84" = Berggren et al., 1984; "MD79" = Mankinen and Dalrymple, 1979; "B84!" = Berggren et al., 1984 using an Anomaly 5-Chron 11 correlation.
References: (1) = Gartner, 1973; (2) = Haq et al., 1980; (3) = Ryan et al., 1974; (4) = Poore et al., 1983; and (5) = Barron et al., in press.
= extrapolates to older age in traditional time scale (see Poore, 1981, and Keller et al., 1982).
† = correlates with lower Chron 7 (Anomaly 4) in alternate time scale.

micropaleontologists in recording the same datum levels in different cores. This is especially true for the radiolarians, where Theyer et al. (1978) and Johnson and Wick (1982) have proposed numerous datum levels that have not been recorded consistently by radiolarian workers in the Initial Reports or by Westberg and Riedel (1978). Similarly, although additional calcareous nannofossil datum levels for the Miocene have been proposed (Haq et al., 1980; Shafik, 1975), Bukry's papers in the Initial Reports typically only record his zonal and subzonal markers. In constructing Tables 1–4, we have found it preferable to use the data of as few as workers as possible for each microfossil group in order to lessen the effect of differing taxonomic concepts between workers. Planktonic foraminiferal data are mainly from the various papers of Keller and Srinivasan and Kennett, calcareous nannofossil data are derived mainly from the papers of Bukry in the Initial Reports, and diatom data are from the studies of Barron and Burckle. Radiolarian data, on the other hand, are drawn from the Initial Reports, Riedel and Sanfilippo (1978), and Dunn (1982). We acknowledge that differing taxonomic concepts by other micropaleontologists might lead to different age estimates for selected datum levels.

We have listed the datum levels which show the most consistency in the numerous cores that we have studied. We acknowledge that individual datum levels may have an imprecision of from 0.1–0.3 m.y. (Johnson and Wick, 1982), but have listed the most consistent age for each.

Sediment Accumulation Rate Curves

Sediment accumulation rate plots for the Miocene of DSDP Sites 71, 77, and 289 (Figures 2–4) show our interpretations of the record at these key sites and provide a means for assigning estimated ages to time series studies. These plots utilize the paleomagnetic time scale of Berggren et al. (1984) because it incorporates more chronological calibration points than the Mankinen and Dalrymple (1979) paleomagnetic time scale.

Site 71 generally has a good early and middle Miocene record (Fig. 2) which accumulated at relatively high (32 m/m.y.) rates. In the later part of the early Miocene (18.8–16.8 Ma) sediment accumulation rates are lower (8 m/m.y.), probably due to carbonate dissolution. However, a middle early Miocene hiatus (NH 1 b) proposed by Keller (1981b) is not apparent by the refined correlations of this paper. A hiatus occurs in the early late Miocene between 10.5 and 9.0 Ma (NH 5), and a second hiatus occurs across the Oligocene/Miocene boundary (Keller and Barron, 1983; Barron, 1983). Sediment accumulation rates in the

TABLE 3. AGES OF STRATIGRAPHICALLY-IMPORTENT MIOCENE
RADIOLARIAN DATUM LEVELS

Datum	Paleomag. Correl.	"B84"	"MD79"	"B84!"
		Age (Ma)		
T. Solenosphaera omnitubus	3rd(-), Gilbert	4.3*(1)	4.25*	nc
Spongaster berminghami→S. pentas (z)	c_1 event, Gilbert	4.45*(1)	4.4*	nc
B. Pterocanium prismatium	4th(+), Gilbert	4.6*(1)	4.5*	nc
T. Acrobotrys tritubus		5.5	5.4	nc
Stichocorys delmontensis→S. peregrina (z)	1st(-), Chron 6	6.3*(1)	6.1*	nc
B. Solenosphaera omnitubus		7.3	7.0	6.8
T. Diartus hughesi (z)		7.9	7.6	7.0
T. Didmocyrtis laticonus		8.9	8.5	7.5
B. A. tritubus		9.1	8.7	7.6
B. D. penultima		9.8†	9.5†	7.9*(1)
B. Spongaster berminghami		9.8	9.5	7.9
T. Stichocorys wolfii		10.4	10.0	8.2
T. Diartus petterssoni		10.4†	10.0†	8.2*(1)
D. petterssoni→D. hughesi (z)	2nd(-), Chron 10	10.9*(2)	10.3*	8.6*
B. D. hughesi	uppermost Chron 11	11.5*(1)	11.3*	9.0*
B. Didymocyrtis antepenultima	uppermost Chron 11	11.5*(1)	11.3*	9.0
T. Cyrtocapsella tetrapera	lower Chron 12	12.5*(2)	12.2*	11.4*
T. C. cornuta		12.5	12.2	11.4
B. Diartus petterssoni (z)	lower Chron 12	12.6*(2)	12.3*	11.5*
B. Didymocyrtis laticonus	uppermost Chron 15	13.8*(1)	13.6*	nc
B. Lithopera neotera	2nd(-), Chron 15	14.5*(1)	14.3*	nc
T. Calocyletta costata	upper (+), Anom. 5A	14.9*(1)	14.6*	nc
Dorcadospyris dentata→D. alata (z)		15.3	15.2	nc
B. D. alata	uppermost Chron 16	15.5*(1)	15.4*	nc
T. Didymocyrtis prismatica		15.8	15.7	nc
T. Lychnocanoma elongata		15.9	15.8	nc
B. Calocyletta costata (z)	lower Chron 16	17.3*(3)	17.3*	nc
B. Dorcadospyris dentata	lower Chron 16	17.3*(3)	17.3*	nc
B. Stichocorys wolfii	2nd (-), Chron 17	18.2*(3)	18.2*	nc
B. Lirospyris stauropora		18.4	18.4	nc
T. Calocyletta serrata		19.2	19.3	nc
B. Dorcadospyris forcipata		19.6	19.7	nc
T. D. ateuchus	upper Chron 19	19.7*(1)	19.8*	nc
T. Cyclamterium pegetrum	upper Chron 19	19.8*(1)	19.9*	nc
T. C. leptetrum		20.0	20.1	nc
B. Didymocyrtis violina	lower Chron 19	20.2*(1)	20.3	nc
B. D. tubaria		20.3	20.4	nc
B. Stichocorys delmontensis	(+), Chron 20	21.1*(1)	21.3*	nc
T. Theocyrtis annosa (z)	lowermost Chron 20	21.3*(1)	21.5*	nc
B. Calocyletta virginis	uppermost Chron 21	21.5*(1)	21.7*	nc
B. C. serrata	1st (+), Chron 21	21.7*(1)	21.9*	nc
B. Cyrtocapsella cornuta		22.1	22.3	nc
B. C. tetrapera (z)		22.1	22.3	nc
T. Dorcadospyris papillio		22.6	22.8	nc
T. Artophormis gracilis		22.6	22.8	nc
B. Lychnocanoma elongata (z)		24.1	24.3	nc

(z) = zonal marker. (-) = reversed polarity event; (+) = normal polarity event.
* = direct paleomagnetic calibration. † = these datums extrapolate to older relative ages under
the traditional time scale (Fig. 1). nc = no change.
Three age estimates are given based on three paleomagnetic time scales: "B84" = Berggren et al.,
1984; "MD79" = Mankinen and Dalrymple, 1979; "B84!" = Berggren et al., 1984, using an Anomaly
5-Chron 11 correlation.
References: (1) = Theyer et al., 1978; (2) = Johnson and Wick, 1982; (3) = Barron et al., in press.

proximity of the middle Miocene/late Miocene boundary (12.5.-10.5 Ma) are 16 m/m.y., whereas most of the upper Miocene (9.0–5.4 Ma) accumulated at a lower (9 m/m.y.) rate.

The record at Site 77 (Fig. 3) resembles that at Site 71 in that it contains an early late Miocene (10.5–9.0 Ma) (NH 5) hiatus. Two early Miocene hiatuses (20.3–19.8 Ma = NH 1a and 19.0–17.8 Ma = NH 1b), however, are also present at Site 77 (Keller, 1980a; Keller and Barron, 1983). The early Miocene

interval between about 23.6 and 20.3 Ma accumulated at a rate of 18 m/m.y. Similarly, most of the middle and late Miocene accumulated at a rate of 14 m/m.y., with the exception of a brief interval (13.5–12.0 Ma) which accumulated at a rate of 20 m/m.y. The near vertical drop in the Site 77 plot at about 6 Ma corresponds with Core 77B-11 where only 1.5 m of sediment was recovered in a standard 9.5 m-long core barrel. An abrupt, short-term increase in the sediment accumulation rates between 100

TABLE 4. AGES OF STRATIGRAPHICALLY-IMPORTANT MIOCENE
DIATOM DATUM LEVELS

Datum	Paleomag. Correl.	Age (Ma)		
		"B84"	"MD79"	"B84!"
B. Nitzschia jouseae (z)	3rd (-), Gilbert	4.5*(1)	4.45*	nc
T. Thalassiosira miocenica (sz)	4th (-), Gilbert	5.1*(1)	5.0*	nc
B. T. oestrupii	4th (-), Gilbert	5.1*(1)	5.0*	nc
T. Asterolampra acutiloba	top Chron 5	5.35*(1)	5.25*	nc
T. N. miocenica	mid (-), Chron 5	5.6*(1)	5.5*	nc
T. T. praeconvexa (sz)	lower Chron 5	5.8*(1)	5.7*	nc
B. T. convexa var. aspinosa (z)	upper Chron 6	6.1*(1)	6.0*	nc
B. T. miocenica	upper Chron 6	6.15*(1)	6.05*	nc
B. T. praeconvexa (sz)	1st (-), Chron 6	6.3*(1)	6.2*	nc
B. Denticulopsis kamtschatica (z) (temp.)	2nd (-), Chron 6	6.6*(2)	6.4*	nc
T. N. porteri	upper Chron 7	7.2*(1)	7.0*	6.7*
B. N. miocenica (z)	upper Chron 7	7.3*(1)	7.1*	6.8*
T. Rossiella paleacea	middle Chron 7	7.4*(1)	7.2*	6.9*
T. Thalassiosira burckliana (sz)	lower Chron 7	8.0*(1)	7.7*	7.0*
T. Coscinodiscus yabei (z)	middle Chron 8	8.6*(1)	8.3*	7.5*
B. T. antiqua (temp.)		8.7	8.4	7.6
B. T. burcklina (sz)	uppermost Chron 9	9.0*(1)	8.7*	8.0*
T. C. temperei var. delicata		9.8	9.5	8.2
T. Denticulopsis dimorpha (temp.)		10.5	10.1	8.6
T. C. vetustissimus var. javanica	upper Chron 10	10.7*(1)	10.3*	8.5
B. C. vetustissimus var. javanica	lower Chron 10	11.2*(1)	10.8*	8.8*
T. Actinocyclus moronensis(z)	lower Chron 10	11.3*(1)	11.0*	8.9*
B. D. dimorpha (temp.)		11.3	11.0	8.9
T. C. tuberculatus	lowermost Chron 12	12.0*(3)	11.7*	10.4*
T. D. punctata f. hustedtii	uppermost Chron 12	12.2*(3)	11.9*	10.7*
T. Craspedodiscus coscinodiscus (z)		12.2	11.9	10.7
B. Hemidiscus cuneiformis	middle Chron 12	12.6*(1)	12.4*	11.2*
B. Rhizosolenia barboi (temp.)		12.6	12.4	11.2
B. C. temperei var. delicata (z)	lowermost Chron 12	12.8*(3)	12.6*	11.8
T. Denticulopsis nicobarica		13.2	13.0	12.6
B. D. praedimorpha (temp.)		13.4	13.3	12.9
T. C. lewisianus (z)		13.5	13.4	12.9
B. D. hustedtii (tropics)		13.7	13.6	nc
T. Cestodiscus peplum (z)	1st (-), Chron 15	14.1*(1)	14.0*	nc
B. D. hyalina (temp.)		15.0	14.9	nc
T. Annellus californicus (sz)	3rd (-), Chron 15	15.0*(1)	14.9*	nc
B. Actinocyclus ingens (tropics)		15.5	15.4	nc
B. D. lauta (temp.)		16.1	16.0	nc
B. Cestodiscus peplum (z)	upper Anomaly 5C	16.4*(4)	16.3*	nc
T. Thalassiosira fraga		16.4	16.3	nc
T. T. bukryi (sz)	lower Chron 16	17.0*(4)	17.1*	nc
B. D. nicobarica (z)	1st (+), Chron 17	17.8*(4)	17.8*	nc
T. Actinocyclus radioovae	1st (-), Chron 17	18.0*(4)	18.9*	nc
T. Craspedodiscus elegans (z)		18.7	18.8	nc
B. Nitzschia maleinterpretaria		18.8	18.9	nc
T. Bogorovia veniamini (z)	Chron 19	19.9*(2)	20.0*	nc
B. T. fraga		19.9	20.0	nc
T. Coscinodiscus oligocenicus (sz)		20.6	20.7	nc
B. A. radionovae		21.2	21.4	nc
T. T. primalabiata (sz)		21.7	21.9	nc
B. Craspedodiscus elegans		22.2	22.4	nc
T. Coscinodiscus lewisianus var. rhomboides		22.5	22.8	nc
B. Rossiella paleacea (z)		22.7	23.0	nc
T. Rocella gelida s. ampl.		22.7	23.0	nc
B. R. gelida var. schraderi		23.6	23.9	nc
B. R. gelida s. str.		24.0	24.3	nc

(z) = zonal marker; (sz) = subzonal marker; (temp.) = temperate datum. (-) = reversed polarity
event; (+) = normal polarity event. * = direct paleomagnetic calibration. nc = no change.
Three age estimates are given based on three paleomagnetic time scales: "B84" = Berggren et al.,
1984; "MD79" = mankinen and Dalrymple, 1979; "B84!" = Berggren et al., 1984, using an Anomaly
5-Chron 11 correlation.
References: (1) = Burckle, 1978; (2) = Burckle, oral comm., 1982; (3) = Burckle et al., 1982;
and (4) = Barron et al., in press.

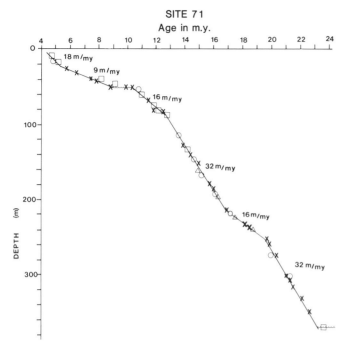

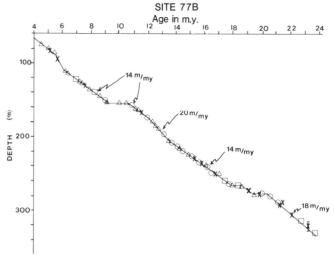

Figure 3. Sediment accumulation rate plot of the Miocene of DSDP Site 77. See Fig. 2 caption for explanation of the symbols.

Figure 2. Sediment accumulation rate plot of the Miocene of DSDP Site 71. ○ = planktonic foraminiferal datum; □ = calcareous nannofossil datum; X = radiolarian datum; Δ = diatom datum. Refer to Tables 1–4 for age of datums.

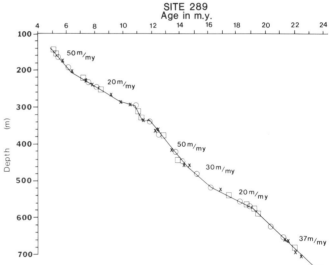

Figure 4. Sediment accumulation rate plot of the Miocene of DSDP Site 289. See Fig. 2 caption for explanation of the symbols.

and 110 m subbottom depth in Hole 77B has been plotted on Figure 3, but the results at DSDP Site 573 (Barron et al., in press), only 10 km from Site 77, do not reveal such an abrupt change in sediment accumulation rates there.

Site 289 has the most complete Miocene section of the three sections with only a short hiatus at 12.0–11.5 Ma (NH 4) (Keller, this volume) (Fig. 4). Hiatus NH 1 of Keller and Barron (1983) (~18.0–17.5 Ma) has been discounted at Site 289 by refinements in biostratigraphy. However, reduced sediment accumulation rates (20 m/m.y.) also characterize the late early Miocene (18.7–16.3 Ma) and late Miocene (10.9–6.4 Ma) at Site 289 within intervals which either contain hiatuses, or exhibit similarly reduced sedimentation rates as at Sites 71 and 77 (Figs. 2, 3). Higher accumulation rates (50 m/m.y.) characterize the upper part of the middle Miocene (14.0–12.0 Ma) and the latest part of the Miocene (6.4–5.1 Ma). Moderate rates correspond with the earliest Miocene (37 m/m.y. between 23.7 and 18.7 Ma) and the early middle Miocene (30 m/m.y. between 16.3 and 14.0 Ma). Site 289 on the Ontong-Java Plateau (2206 m water depth) apparently was above most of the erosive and corrosive effects of bottom waters which caused most of the deep-sea hiatuses of the Miocene (Keller and Barron, 1983). Comparison of the Site 289 sediment accumulation rate plot (Fig. 4) with the generalized carbonate curve for the Miocene (Fig. 1), however, shows a strong correspondence between intervals of lower accumulation rates and intervals of lower average percent carbonate values (18.7–15.3 Ma and 12.0–6.4 Ma). Presumably, this correspondence reflects greater carbonate dissolution during those intervals.

TIME SLICE INTERVALS

The early, middle, and late Miocene are each marked by major paleoclimatic and paleoceanographic events which fundamentally changed the oceanic regime as expressed by the oxygen and carbon isotope records (Woodruff et al., 1981; Savin et al., 1981) and by changing faunal and floral assemblages (Haq, 1980; Kennett, 1977; Keller and Barron, 1983). CENOP workers chose three time-slice intervals for detailed study close to the major Miocene paleoceanographic events in order to characterize the evolving Miocene ocean as well as faunal and floral responses. Each time-slice interval spans between 0.5 and 1 m.y., and uncertainties of correlation are minimized by study of multiple samples from each of the time-slice intervals in the various stratigraphic sections. At the same time, because correlations between middle and low latitudes typically have an uncertainty of at least 0.5 m.y., choice of 0.5–1 m.y.-long time slices allows a greater geographic coverage of the ocean basins. The three time slices near the beginning, the middle, and end of the Miocene provide a time lapse view of the history of the Miocene ocean.

Selecting time-slice intervals for study is difficult, primarily due to incomplete sedimentary records and imprecise biostratigraphic data and correlations. A prerequisite to choosing a time-slice interval should be that the interval be represented in as many stratigraphic sections (or DSDP holes) as possible. Certain limitations, such as the lack of a stratigraphic section in a given geographic region or the lack of recovery of a given stratigraphic interval, are unavoidable. Recognition of widespread deep sea hiatuses, which remove a specific interval from broad areas of the ocean basins, must be considered in selecting time slice intervals. Keller and Barron's (1983) study of Miocene deep sea hiatuses identified nine intervals of widespread hiatuses (Fig. 1) between the latest Oligocene and the Miocene/Pliocene boundary: 24.4–23.2 Ma (PH), 20.2–19.4 Ma (NH 1a), 18.7–17.7 Ma (NH 1b), 16.1–15.1 Ma (NH 2), 13.5–12.5 Ma (NH 3), 12.0–11.0 Ma (NH 4), 10.1–9.5 Ma (NH 5), 7.9–6.4 Ma (NH 6), and 5.2–4.7 Ma (NH 7) (Ages adjusted to the revised time scale of Fig. 1). Of these hiatuses, all except NH 2 are widespread in the Pacific and have been avoided in selecting Miocene time slices for study.

Climatically-warm intervals tend to be better represented in deep sea sequences than climatically-cool intervals which tend to correspond with widespread deep sea hiatuses (Barron and Keller, 1982). In addition, low-latitude microfossil zonations are recognizable over broader geographic regions during climatically-warm intervals than during climatically-cool intervals, and problems of correlation across latitudes are minimized. Consequently, climatically-warm intervals are better candidates for time-slice intervals than climatically-cool intervals.

The Miocene is a time of considerable paleoclimatic and paleoceanographic fluctuation (Savin et al., 1981; Haq, 1980; Barron and Keller, 1983; Keller, this volume), and ideally, time-slice intervals should be of minimal duration. The multidisciplinary time scale (Tables 1–4; Figure 1) developed here allows resolution of Miocene time approaching 100,000 years. Miocene correlations between tropical and temperate regions have been greatly improved in recent years (Keller and Barron, 1981; Keller et al., 1982; Srinivasan and Kennett, 1981b) and allow resolution of time of between 200,000 to 500,000-year range. In order to construct time slice maps that cover both high- and low-latitude areas of the oceans (the Pacific in particular), time-slice intervals of approximately 500,000 to one-million-years duration have been chosen for study. This alleviates many problems of nonrecovery of a time-slice interval and uncertainty of correlation. Undoubtedly, such lengthy intervals contain climatic fluctuations; however, an effort has been made to select generally warm and climatically-stable intervals. In addition, four or more samples per interval have been counted (measured) and averaged so as to reduce the effects of climatic fluctuation.

Early Miocene Time Slice (~21 Ma)-TS-1

The oldest time slice (~21 Ma) was chosen to characterize the early Miocene ocean prior to the major faunal turnover in planktonic foraminifers which is marked by a sharp decline and eventual extinction of Oligocene taxa and the rise of Neogene taxa (Keller, 1981a,b). The early Miocene time slice corresponds with Keller's (this volume) Subzone c of planktonic foraminiferal Zone N4, and precedes the widespread middle early Miocene deep sea hiatuses (20.2–19.4 Ma (NH 1a) and 18.7–17.7 Ma (NH 1b) (Keller, 1981b). As such, the base of the early Miocene time slice is defined by the first occurrence of *Globoquadrina dehiscens* and its top coincides with the last occurrence of *Globoquadrina kugleri* (Table 1). The early Miocene time slice also correlates with part of Martini's (1971) NN 2 Zone and with part of Subzone CN1c of Okada and Bukry (1980) (Figure 1). The base of the time slice corresponds with the last occurrence of the radiolarian *Theocyrtis annosa,* and its top is closely approximated by the first occurrence of *Didmocyrtis violina.* Consequently, the early Miocene time slice is equivalent to the lower part of the *Stichocorys delmontensis* Zone of Riedel and Sanfilippo (1978). In terms of diatoms, the time slice roughly corresponds with Subzones B and C of the *Rossiella paleacea* Zone of Barron (1983). The last occurrence of the diatom *Thalassiosira primalabiata* falls near the base of the time slice, and the last occurrence of *Bogorovia veniamini* slightly post-dates the top of the time slice. These two diatom datum levels are also recognizable in the high-latitude South Atlantic (Gombos and Ciesielski, 1983) and suggest that the early Miocene time slice (TS-1) should be recognizable in the Southern Ocean.

A distinctive carbonate dissolution interval (events 20c–20e of Dunn, 1982) characterizes the middle part of the early Miocene time slice with higher carbonate values lying at the top (events 20b–19e) and bottom (events 21c–21b) of the time slice.

Late Early Miocene Time Slice (~16 Ma)-TS-2

The major oxygen isotope event of the Miocene is the enrichment in δO^{18} by about 1.0 ‰ (cooling) which is recorded in benthic foraminifers during the early middle Miocene (planktonic foraminiferal zones N9 to N12) (Shackleton and Kennett, 1975; Woodruff et al., 1981; Savin et al., 1981). This globally recognized event has usually been interpreted as reflecting major increase in the size of the Antarctic ice sheet (Shackleton and Kennett, 1975; Woodruff et al., 1981; Savin et al., 1981) or major cooling (Matthews and Poore, 1980). The second Miocene time slice was chosen within planktonic foraminiferal Zone N8 across the early Miocene/middle Miocene boundary after the major planktonic foraminiferal faunal turnover in the middle early Miocene (Keller, 1981a) and prior to this major oxygen isotope event. It will be referred to as the late early Miocene time slice (TS-2). The base of this time slice is, therefore, defined by the first occurrence of *Globigerinoides sicanus* (16.4 Ma, Table 1) and the top by the first occurrence of *Orbulina suturalis* (15.2

Ma, Table 1). In terms of calcareous nannofossils, the late early Miocene time slice corresponds with the upper part of Zone NN4 to lower Zone NN5 of Martini (1971) and the upper part of Zone CN3 to lower part of Zone CN4 of Okada and Bukry (1980). The upper part of the *Calocyletta costata* Zone (radiolarians) of Riedel and Sanfilippo (1978) and nearly all of Subzone A of the *Cestodiscus peplum* Zone (diatoms) of Barron are also equivalent to the middle Miocene time slice.

In identifying the late early Miocene time slice it is necessary to consider the biostratigraphic data of two or more microfossil groups. Planktonic foraminiferal Zone N8 is difficult to recognize in the central and eastern equatorial Pacific, because its primary markers, *Globigerinoides sicanus* and *Orbulina suturalis,* are susceptible to dissolution (Jenkins and Orr, 1972; Keller, 1980b). Keller (1980b) suggests that the first occurrence of the resistant species *Globorotalia peripheroronda* (16.5 Ma) represents more reliable datum plane for approximating the base of Zone N8 in faunas affected by dissolution. Although Keller (1980b) also argues that the first occurrence of *G. archaeomenardii* is a secondary marker for the lower part of Zone N9, Table 1 shows that this datum (15.7 Ma) predates the *Orbulina* datum (15.2 Ma) (= base of Zone N9) by 0.5 m.y. Similarly, Bukry (1981) points out that the calcareous nannofossil *Helicosphaera ampliaperta* is sparse in deep-sea sediment, and that secondary criteria (i.e., the first occurrence of *Calcidiscus macintyrei*) had to be selected for recognition of the base of the *S. heteromorphus* Zone (CN4).

In the southwestern Pacific the time slice correlates with the uppermost part of the *Globorotalia miozea* Zone and the entire *Praeorbulina glomerosa* Zone (planktonic foraminifers) of Srinivasan and Kennett (1981b). In the middle-to-high latitude North Pacific, Subzone a of the *Denticulopsis lauta* Zone (diatoms) of Barron (1980) is equivalent to the time slice.

In the North Atlantic Ocean and Caribbean Sea, the interval of the late early Miocene time slice typically coincides with a hiatus (NH-2 of Keller and Barron, 1983). In the Pacific, however, hiatuses are relatively rare during this interval (Keller and Barron, 1983).

General carbonate dissolution (event 16e to 16c of Dunn 1982) characterizes the lower part of the early Miocene/middle Miocene time slice, whereas higher carbonate values (events 16b–15i) mark its top.

Late Miocene Time Slice (~8 Ma) -TS-3

The middle late Miocene interval correlative with paleomagnetic Chron 8 is the best represented late Miocene interval in deep-sea sections (Keller and Barron, 1983). Thus, this interval (8.9–8.2 Ma, according to the paleomagnetic time scale of Berggren et al., 1984) was chosen to characterize the late Miocene ocean after the middle Miocene oxygen isotope event and prior to the latest Miocene $\delta^{13}C$ (carbon) shift. The late Miocene time slice is equivalent to the lowermost part of planktonic foraminiferal Zone N17, the upper part of calcareous nannofossil Zone NN10 of Martini (1971), Subzone CN8b (calcareous nan-

nofossil) of Okada and Bukry (1980), the upper part of the *Didmocyrtis antepenultima* Zone (radiolarians) of Riedel and Sanfilippo (1978), and Subzone B of the *Coscinodiscus yabei* Zone to lower Subzone A of the *Nitzschia porteri* Zone (diatoms) of Burckle (1972) and Barron (in press) (Fig. 1). The first occurrences of the planktonic foraminifer *Globorotalia plesiotumida* (9.0 Ma, Table 1), the calcareous nannofossil *Discoaster neorectus* (8.9 Ma), the radiolarian *Acrobotrys tritubus* (9.1 Ma), and the diatom *Thalassiosira burckliana* (9.0 Ma) closely approximate the base of the late Miocene time slice at low latitudes. The top of the time slice is immediately below the first occurrence of *Globigerinoides conglobatus* (planktonic foraminifer) (8.0 Ma) and the last occurrences of *Diartus hughesi* (radiolarian) (7.9 Ma) and *Thalassiosira burckliana* (diatom) (8.0 Ma).

There are considerable problems with placement of the N16/N17 planktonic foraminiferal zone boundary. The base of N17, the first occurrence of *Globorotalia plesiotumida,* is not only diachronous between low and middle latitudes (Keller, 1980a, 1981b), but it appears to be diachronous across the equatorial Pacific. In the eastern equatorial Pacific, the *G. plesiotumida* datum correlates with the upper part of paleomagnetic Chron 9 (~9.0 Ma) (Keller, 1981b; Keller et al., 1982), whereas in the western equatorial Pacific that datum correlates with the paleomagnetic Chron 7 (~7–8 Ma) (Srinivasan and Kennett, 1981a, b; Keller et al., 1982). Keller (this volume) suggests that an abundance peak in *G. menardii* at about 8 Ma can be used secondarily to locate the late Miocene time slice in low-latitude sections. In the middle-latitude North Pacific the late Miocene time slice corresponds closely with the range of *Globigerinoides kennetti* (planktonic foraminifer) (Keller, 1980a) and can also be recognized by the range of *Thalassiosira burckliana* (diatom) which appears to be isochronous within low latitudes (Barron, 1980, 1981a). Relatively high carbonate values (events 8d–8b of Dunn and Moore, 1981) characterize most of the late Miocene time slice, although a marked decline in carbonate values (event 8a) begins near the top of time slice TS-3.

SUMMARY

Integration of planktonic foraminiferal, calcareous nannofossil, radiolarian, and diatom datum levels in over 20 DSDP sections has resulted in a high resolution multiple microfossil biochronology for the Miocene. Direct correlations of over 70 datum levels to paleomagnetic stratigraphy and graphical correlations allow assignment of extrapolated absolute ages to 175 datum levels between 24.0 and 4.3 Ma (Tables 1–4). The resulting resolution of time approaches 100,000 years and is enhanced considerably by integration of carbonate stratigraphy (Dunn and Moore, 1981).

A Miocene time scale correlated with the paleomagnetic time scale of Berggren et al. (1984) is presented (Fig. 1). The main differences with previously published time scales are in the early Miocene where recent paleomagnetic correlations (Poore et al., 1983; Barron et al., in press) have resulted in changes. Specifically, the N4/N5 planktonic foraminiferal zonal boundary is correlated with the middle part of paleomagnetic Chron 19 (20.1 Ma), 0.5–2.0 m.y. younger than previous correlations Ryan et al., 1974; Saito, 1977; Keller, 1980a; Berggren, 1981; Srinivasan and Kennett, 1981a; Berggren et al., 1983).

Sediment accumulation rate curves are presented for key Miocene reference sections; DSDP Sites 71 and 77B in the eastern equatorial Pacific (Figs. 2, 3) and Site 289 in the western equatorial Pacific (Fig. 4). Sites 71 and 77B contain hiatuses in the early late Miocene (Keller and Barron, 1983), but Site 289 appears to have a nearly complete Miocene section. The middle early Miocene hiatus (~18.0–17.5 Ma) of Keller and Barron (1983) at Site 289 is discounted by refined correlations. Comparison of the Site 289 sediment accumulation rate plot (Fig. 4) with the generalized carbonate curve (Fig. 1) for the Miocene shows a strong correspondence between intervals of lower accumulation rates and lower percentage carbonate values (18.7–16.3 Ma and 12.0–6.4 Ma), most likely reflecting carbonate dissolution.

The three CENOP time slices (TS-1 at 21 Ma, TS-2 at 16 Ma, and TS-3 at 8 Ma) were chosen to avoid intervals of widespread deep-sea hiatuses (Keller and Barron, 1983) and to characterize the Miocene ocean immediately prior to three major paleoceanographic/paleoecologic events: the major faunal turnover of planktonic foraminifers in the middle early Miocene (Keller, 1981a, b), the major positive $\delta^{18}O$ increase (cooling) in benthic foraminifers during planktonic foraminifer Zones N9 to lower N12 (Woodruff et al., 1981), and the carbon shift (Haq et al., 1980; Vincent et al., 1980).

The early Miocene time slice corresponds with Keller's (in press) Subzone c of planktonic foraminiferal Zone N4 (21.2–20.1 Ma; given as 22 Ma in this volume). The late early Miocene time slice is defined by planktonic foraminiferal Zone N8 (16.4–15.2 Ma). The late Miocene time slice is equivalent to paleomagnetic Chron 8 (8.9–8.2 Ma) near the N16/N17 planktonic foraminiferal zone boundary.

ADDENDUM: ALTERNATE TIME SCALE

The time scale proposed in this paper and used throughout this volume incorporates a correlation of paleomagnetic Anomaly 5 with paleomagnetic Chron 9, a correlation which has been widely accepted for more than 10 years (Berggren and Van Couvering, 1974; Theyer and Hammond, 1974; Ryan et al., 1974). Recently, however, W. A. Berggren (oral comm., 1984) has suggested that Anomaly 5 should be correlated with Chron 11 (as it has been recognized in the sediment) based on paleomagnetic-biostratigraphic correlations at DSDP Site 519 in the South Atlantic by Poore et al. (1984) and Hsu et al. (1984). Site 519 was drilled on crust slightly older than Anomaly 5, and middle Miocene calcareous nannofossils (Zones NN8 and NN9) were recorded within a relatively long normally polarized interval (10 m) immediately above the basement (Poore et al., 1984). As correlations of NN8 and NN9 (=CN6 and CN7) calcareous nannofossils with paleomagnetic Chron 11 have been well estab-

lished (Ryan et al., 1974; Berggren and Van Couvering, 1978; Keller et al., 1982), it seems apparent that the normally polarized interval containing NN8 and NN9 calcareous nannofossils at Site 519 correlates with both Anomaly 5 and Chron 11 (Hsu et al., 1984).

If one accepts an Anomaly 5-Chron 11 correlation, one must utilize an age assignment of 8.92–10.42 Ma for Chron 11 under the Berggren et al. (1984) paleomagnetic time scale. The middle Miocene/late Miocene boundary which is correlated with upper Chron 11 (Fig. 1) would then have an approximate age of 9.5 Ma. The paleomagnetic chrons above and below Chron 11 would also have to be correlated to the anomalies in a new manner (Theyer and Hammond, 1974). W. A. Berggren (oral comm., 1984) believes that the traditional correlations of chrons and anomalies (Ryan et al., 1974; LaBrecque et al., 1977; Fig. 1 of this report) above Chron 7 and below Chron 14 do not require adjustment. On Figure 5, a correlation used by Barron et al. (in press), which is basically that proposed by Theyer and Hammond (1974), is presented. The resultant polarity sequence in chrons 7 through 14 is similar to that in the more traditional time scales (Fig. 1), although Chron 7 is shortened to include only Anomaly 4 (3 normally polarized events) and Chron 14 is lengthened to include three additional normally polarized events below Anomaly 5A.

Tauxe et al. (1983) have proposed a system for labelling paleomagnetic chrons which is straightforward and avoids the nomenclatural problems with anomaly-chron correlations. Tauxe et al. (1983) label successive chrons from the top of one numbered anomaly to the top of the next oldest anomaly with the same number as the included anomaly and preface those numbers with a "C." Thus, Chron C5 includes Anomaly 5 plus the interval below Anomaly 5 and above Anomaly 5A (Fig. 5). However, most published paleomagnetic correlations of microfossil datum levels and virtually all of the papers of this volume refer exclusively to the older system of numbering chrons, so it is important to acknowledge it.

The alternate Miocene time scale utilizing an Anomaly 5-Chron 11 correlation is shown in Figure 5. The correlation of the microfossil zones and subzones with the paleomagnetic chrons on Figure 5 differs little from the correlation shown on Figure 1, but the estimated ages of zones and subzones between 6.5 and 13.8 (paleomagnetic Chrons 7–14) have been changed (compare Figs. 1 and 5). This correlation has been accomplished by reassigning ages to those microfossil datums correlated with Chrons 7 through 14 (see Tables 1–4) based on the anomaly-chron correlation model of Figure 5. Ages of additional datum levels (and zonal boundaries) have been extrapolated from replotted sediment accumulation rate plots (between 16 and 6 Ma) for Sites 71, 77, and 289 (Fig. 6) and from the sediment accumulation rate plots of Barron et al. (in press) for central Pacific DSDP Sites 572–574, which utilize the same time scale.

As a result of these changes, the estimated age of the late Miocene time slice (TS-3) becomes 8.0–7.0 Ma. Hiatuses NH 6 through NH 3 of Keller and Barron (1983) have recalculated ages

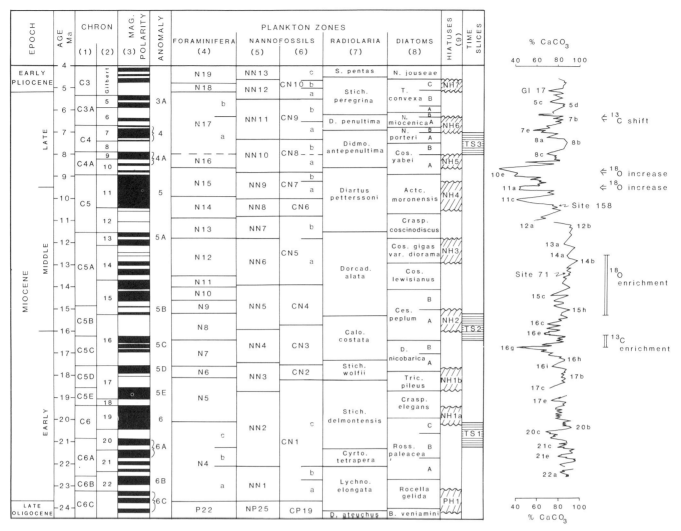

Figure 5. Alternate Miocene time scale derived from a correlation of paleomagnetic Anomaly 5 to paleomagnetic Chron 11 showing correlation of planktonic microfossil zones, widespread deep sea hiatuses, the CENOP time slices, and the percent CaCO₃ curve for the eastern equatorial Pacific to the paleomagnetic time scale. Note that differences with Fig. 1 occur between 13.8 and 6.8 Ma. (1) = Tauxe et al., 1983; (2) = LaBrecque et al., 1977 and Theyer and Hammond, 1974; (3) = Berggren et al., 1984; (4) = Blow, 1969; Srinivasan and Kennett, 1981a, b; Keller, in press; (5) = Martini, 1971; (6) = Bukry, 1973, 1975; Okada and Bukry, 1980; (7) = Riedel and Sanfilippo, 1978; (8) = Burckle, 1972; Barron, 1983, in press; and (9) = Keller and Barron, 1983.

as follows: NH 3 (12.9–11.8 Ma), NH 4 (10.5–9.2 Ma), NH 5 (8.6–8.0 Ma), and NH 6 (7.0–6.3 Ma). The major $\delta^{18}O$ enrichment of the middle Miocene has a longer estimated duration, lasting until ~12.5 Ma rather than to ~13.2 Ma in this alternate time scale, and the two $\delta^{18}O$ increases across the middle/late Miocene boundary occur at ~9.5 and 8.8 Ma (Fig. 5). Estimated ages of carbonate events 14b through 7b of Dunn and Moore (1981) also change accordingly (Fig.5) based on replotted sediment accumulation rate plots for Site 71 (Fig. 6) and Site 158.

Barron et al. (in press) notes that utilization of an Anomaly 5-Chron 11 correlation results in sediment accumulation rates that are more uniform during the late Miocene than rates estimated from the traditional time scales which use an Anomaly

5-Chron 9 correlation. Comparison of Figures 2–4 and 6 shows that such a relationship also holds true for DSDP Sites 71 and 289 and to some extent for Site 77. Both the earlier plots for Site 71 (Fig. 2) and Site 289 (Fig. 4) display middle late Miocene rates which are about one-half the rates interpreted for the middle Miocene and latest Miocene at those sites, whereas such a late Miocene kink is not apparent in the later plots (Fig. 6) that utilize the alternate time scale of Figure 5.

Thus, there is good evidence for adopting an Anomaly 5-Chron 11 correlation and accepting the time scale of Figure 5. However, this proposed time scale change is just beginning to be actively debated as this volume goes to press. The papers of this volume have utilized the more traditional time scale of Figure 1

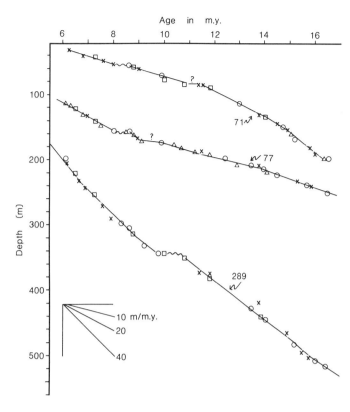

Figure 6. Sediment accumulation plots for the 16 to 6 Ma intervals of DSDP Sites 71, 77, and 289 replotted using the alternate time scale of Figure 5. See Fig. 2 caption for explanation of the symbols.

almost exclusively, but the reader should be able to utilize Tables 1–4 and Figure 5 to redetermine age estimates in this volume should they wish to accept an Anomaly 5-Chron 11 correlation.

ACKNOWLEDGMENTS

We thank David Bukry, Richard Z. Poore, and Charles A. Repenning of the U.S. Geological Survey, Fritz Theyer of the Hawaii Institute of Geophysics, and Bilal Haq of EXXON for their review of this manuscript. W. A. Berggren and D. A. Johnson were helpful in explaining the basis for the alternate time scale utilizing an Anomaly 5-Chron 11 correlation. We also are grateful to CENOP workers for their contributions, especially J. P. Kennett, E. Vincent, L. H. Burckle, K. Romine, and G. Lombari.

REFERENCES CITED

Barron, J. A., 1980, Lower Miocene to Quaternary diatom biostratigraphy of Leg 57, off northeastern Japan, Deep Sea Drilling Project, *in* Initial Reports of the Deep Sea Drilling Project: Washington, U.S. Government Office, v. 56, 57, Part II, p. 641–685.
—— 1981a, Late Cenozoic diatom biostratigraphy and paleoceanography of the middle latitude eastern North Pacific, Deep Sea Drilling Project Leg 63, *in* Initial Reports of the Deep Sea Drilling Project: Washington, U.S. Government Printing Office, v. 63, p. 507–538.
—— 1981b, Middle Miocene diatom biostratigraphy of DSDP Site 77B in the eastern equatorial Pacific: Geoscience Journal, v. 2, no. 2, p. 137–144. (Lucknow, India).
—— 1983, Latest Oligocene through early middle Miocene diatom biostratigraphy of the eastern tropical Pacific: Marine Micropaleontology, v. 7, p. 487–515.
—— in press, Miocene to Holocene planktic diatoms; *in* Saunders, J. B., and Bolli, H. M., eds., Biostratigraphy by marine plankton: Cambridge, Cambridge University Press.
Barron, J. A., and Keller, G., 1982, Widespread Miocene deep-sea hiatuses: Coincidence with periods of global cooling: Geology, v. 10, p. 577–581.
—— 1983, Paleotemperature oscillations in the middle and late Miocene of the northeastern Pacific: Micropaleontology, v. 29, no. 2, p. 150–181.
Barron, J. A., Nigrini, C. A., Pujos, A., Saito, T., Theyer, F., Thomas, E., and Weinreich, N., in press, Synthesis of central equatorial Pacific DSDP leg 85 biostratigraphy: refinement of Oligocene to Quaternary biochronology, *in* Initial Reports of the Deep Sea Drilling Project: Washington, U.S. Government Printing Office, v. 85.
Berggren, W. A., 1981, Correlation of Atlantic, Mediterranean and Indo-Pacific Neogene stratigraphies: Geochronology and chronostratigraphy, *in* Proceedings, IGCP-114 International Workshop on Pacific Neogene Biostratigraphy, Nov. 25–29, 1981, Osaka, Japan, p. 29–60.
Berggren, W. A. and Van Couvering, J. A., 1974, The Late Neogene: biostratigraphy, geochronology, and paleoclimatology of the last 15 million years in marine and continental sequences: Palaeogeography, Palaeoclimatology, and Palaeoecology, v. 16 (1-2), p. 1–216.
—— 1978, Biochronology, *in* Glassner, G. V. and Hedberg, H. D., eds., Contributions to the geologic time scale: Amer. Assoc. Petrol. Geologists, Studies in Geology, v. 6, p. 39–55.
Berggren, W. A., Aubrey, M. P., and Hamilton, N., 1983, Neogene magnetostratigraphy of Deep Sea Drilling Project Site 516 (Rio Grande Rise, South Atlantic), *in* Initial Reports of the Deep Sea Drilling Project: Washington, U.S. Government Printing Office, v. 72, p. 675–721.
Berggren, W. A., Kent, D. V., and Flynn, J. J., 1984, Paleogene geochronology and chronostratigraphy, *in* Snelling, N. J., ed., Geochronology and the geologic record: Geological Society of London, Special Paper.
Blow, W. H., 1969, Late middle Eocene to recent planktonic foraminiferal biostratigraphy, *in* First International Conference on Planktonic Microfossils, Geneva, 1967, p. 199–421.
Bukry, D., 1973, Low-latitude coccolith biostratigraphic zonation, *in* Initial Reports of the Deep Sea Drilling Project: Washington, U.S. Government Printing Office, v. 15, p. 685–703.
—— 1975, Coccolith and silicoflagellate stratigraphy, northwestern Pacific Ocean, Deep Sea Drilling Project Leg 32, *in* Initial Reports of the Deep Sea Drilling Project: Washington, U.S. Government Printing Office, v. 32, p. 677–701.
—— 1981, Cenozoic coccoliths from the Deep Sea Drilling Project: Society of Economic Paleontologists and Mineralogists Special Publication 32, p. 335–353.
Burckle, L. H., 1972, Late Cenozoic planktonic diatom zones from the eastern equatorial Pacific: Beihefte zur Nova Hedwegia, v. 39, p. 217–246.
—— 1978, Early Miocene to Pliocene diatom datum levels for the equatorial Pacific, *in* Proceedings, Second Working Group Meeting, Biostratigraphic Datum-Planes of the Pacific Neogene, IGCP Project 114: Republic of Indonesia Ministry of Mines and Energy, Directorate General of Mines, Geological Research and Development Center, Special Publication No. 1, p. 25–44.
Burckle, L. H., and Opdyke, N. D., This volume, Latest Miocene/earliest Pliocene diatom correlations in the North Pacific.
Burckle, L. H., Keigwin, L. D., and Opdyke, N. D., 1982, Middle and late Miocene stable isotope stratigraphy: Correlation to the paleomagnetic reversal record: Micropaleontology, v. 28, no. 4, p. 329–334.
Cita, M. B., 1975, The Miocene/Pliocene boundary: History and definition, *in* Saito, T. and Burckle, L. H., eds., Late Neogene epoch boundaries: New York, Micropaleontology Press, p. 1–30.

Dunn, D. A., 1982, Miocene sediments of the equatorial Pacific Ocean: Carbonate stratigraphy and dissolution history [Ph.D. Thesis]: University of Rhode Island, 302 p.

Dunn, D. A., and Moore, T. C., Jr., 1981, Late Miocene/Pliocene (Magnetic Epoch 9-Gilbert Magnetic Epoch) calcium-carbonate stratigraphy of the equatorial Pacific Ocean: Geological Society of America Bulletin, Part II, v. 92, p. 408–451.

Gartner, S., 1973, Absolute chronology of late Neogene calcareous nannofossil succession in the equatorial Pacific: Geological Society of America Bulletin, v. 84, p. 2021–2034.

Gombos, A. M., Jr., 1983, Survey of diatoms in the upper Oligocene and lower Miocene in Holes 515B and 516F, *in* Initial Reports of the Deep Sea Drilling Project: Washington, U.S. Government Printing Office, v. 72, p. 793–804.

Gombos, A. M., Jr., and Ciesielski, P. F., 1983, Late Eocene to early Miocene diatoms from the southwest Atlantic, *in* Initial Reports of the Deep Sea Drilling Project: Washington, U.S. Government Printing Office, v. 71, p. 583–634.

Haq, B. U., 1980, Biogeographic history of Miocene calcareous nannoplankton and paleoceanography of the Atlantic Ocean: Micropaleontology, v. 26, no. 4, p. 414–443.

Haq, B. U. Worsley, J. R., Burckle, L. H., Douglas, R. G., Keigwin, L. D., Opdyke, N. D., Savin, S. M., Sommer, M. A., Vincent, E., and Woodruff, F., 1980, Late Miocene marine carbon-isotope shift and synchroneity of some phytoplanktic biostratigraphic events: Geology, v. 8, p. 427–431.

Hsu, K. J., LaBrecque, J. L., and the shipboard scientific party, 1984, Site 519, *in* Initial Reports of the Deep Sea Drilling Project: Washington, U.S. Government Printing Office, v. 73, p. 27–93.

Jenkins, D. G., and Orr, W. N., 1972, Planktonic foraminiferal biostratigraphy of the eastern equatorial Pacific, Leg 9, *in* Initial Reports of the Deep Sea Drilling Project: Washington, U.S. Government Printing Office, v. 9, p. 1059–1196.

Johnson, D. A., and Wick, B. J., 1982, Precision of correlation of radiolarian datum levels in the middle Miocene equatorial Pacific: Micropaleontology, v. 28, no. 1, p. 1–30.

Keigwin, L. D., Jr., 1979, Late Cenozoic stable isotope stratigraphy and paleoceanography of DSDP site from the east equatorial and central north Pacific Ocean: Earth and Planetary Science Letters, v. 45, p. 361–382.

Keller, G., 1980a, Early to middle Miocene planktonic foraminiferal datum levels of the equatorial and subtropical Pacific: Micropaleontology, v. 26, no. 4, p. 372–391.

——1980b, Middle to late Miocene planktonic foraminiferal datum levels and paleoceanography of the north and southeastern Pacific Ocean: Marine Micropaleontology, v. 5, p. 249–281.

——1981a, Miocene biochronology and paleoceanography of the North Pacific. Marine Micropaleontology, v. 6, p. 535–551.

——1981b, Planktonic foraminiferal faunas of the equatorial Pacific suggest early Miocene origin of present oceanic circulation: Marine Micropaleontology, v. 6, p. 269–295.

——in press, The Oligocene/Miocene boundary in the equatorial Pacific, *in* Steiniger, F., and Gelati, R., eds., In search of the Palaeogene/Neogene boundary stratotype, Part 2, Rivista Italiana de Paleontologia e Stratigraphia.

Keller, G., this volume, Depth stratification of planktonic foraminifera in the Miocene Ocean.

Keller, G., and Barron, J. A., 1981, Integrated planktic foraminiferal and diatom biochronology for the northeast Pacific and the Monterey Formation, *in* Garrison, R. E., and Douglas, R. G., eds., The Monterey Formation and related siliceous rocks of California: Los Angeles, Calif., Pacific Section Society of Economic Paleontologists and Mineralogists, p. 43–54.

——1983, Paleoceanographic implications of Miocene deep-sea hiatuses: Geological Society of America Bulletin, v. 94, p. 590–613.

Keller, G., Barron, J. A., and Burckle, L. H., 1982, North Pacific late Miocene correlations using microfossils, stable, isotopes, percent Ca CO_3, and magnetostratigraphy: Marine Micropaleontology, v. 7, p. 327.

Kennett, J. P., 1977, Cenozoic evolution of Antarctic glaciation, the circum-Antarctic Ocean, and their impact on global paleoceanography: Journal of Geophysical Research, v. 82, no. 27, p. 3843–3860.

La Brecque, J. L., Kent, D. V., and Cande, S. C., 1977, Revised magnetic polarity time scale for the Cretaceous and Cenozoic. Geology, v. 5, p. 330–335.

Loutit, T. S., and Kennett, J. P., 1979, Application of carbon isotope stratigraphy to late Miocene shallow marine sediments, New Zealand: Science, v. 24, p. 1196–1199.

Mankinen, E. A., and Dalrymple, G. B., 1979, Revised geomagnetic polarity time scale for the interval 0-5 m.y. B.P.: Journal of Geophysical Research, v. 84, no. B2, p. 615–626.

Matthews, R. K., and Poore, R. Z., 1980, Tertiary ^{18}O record and glacioeustatic sea-level fluctuations: Geology, v. 8, p. 501–504.

Martini, E., 1971, Standard Tertiary and Quaternary calcareous nannoplankton zonation, *in* Farinacci, A., ed., Proceedings of the II Planktonic Conference Roma 1970: Rome, Edizioni Tecnoscienza, v. 2, p. 739–777.

Moore, T. C., Jr., and Lombari, G., 1981, Sea surface temperature changes in the North Pacific during the late Miocene: Marine Micropaleontology, v. 6, no. 5–6, p. 581–597.

Okada, H., and Bukry, D., 1980, Supplementary modification and introduction of code numbers to the low-latitude coccolith biostratigraphic zonation (Bukry 1973; 1975): Marine Micropaleontology, v. 5, p. 321–325.

Poore, R. Z., 1981, Late Miocene biogeography and paleoclimatology of the central North Atlantic: Marine Micropaleontology, v. 6, p. 599–616.

Poore, R. Z., Tauxe, L., Percival, S. F., Jr., LaBrecque, J. L., Wright, R., Petersen, N. P., Smith, C. C., Tucker, P., and Hsu, K. J., 1983, Late Cretaceous-Cenozoic magnetostratigraphic and biostratigraphic correlations of the South Atlantic Ocean: DSDP Leg 73: Palaeogeography, Palaeoclimatology, Palaeoecology, v. 42, p. 127–149.

——1984, Late Cretaceous-Cenozoic magnetostratigraphic and biostratigraphic correlations of the South Atlantic Ocean, Deep Sea Drilling Project Leg 73, *in* Initial Reports of the Deep Sea Drilling Project: Washington, U.S. Government Printing Office, v. 73, p. 645–655.

Riedel, W. R., and Sanfilippo, A., 1978, Stratigraphy and evolution of tropical Cenozoic radiolarians: Micropaleontology, v. 24, no. 1, p. 61–96.

Romine, K., this volume, Radiolarian biogeography and paleoceanography of the North Pacific at 8 MA.

Ryan, W.B.F., Cita, M. B., Rawson, M. D., Burckle, L. H., and Saito, T., 1974, A paleomagnetic assignment of Neogene stage boundaries and the development of isochronous datum planes between the Mediterranean, the Pacific and Indian Oceans in order to investigate the response of the world ocean to the Mediterranean "salinity crisis": Rivista Italiana Paleontologia, v. 80, p. 631–688.

Saito, T., 1977, Late Cenozoic planktonic foraminiferal datum levels: the present state of knowledge toward accomplishing pan-Pacific stratigraphic correlation, *in* Proceedings, First International Congress on Pacific Neogene Stratigraphy, Tokyo, 1976: Tokyo, Kaiyo Shuppan Co., p. 61–80.

Saito, T., Burckle, L. H., and Hays, J. D., 1975, Late Miocene to Pleistocene biostratigraphy of equatorial Pacific sediments, *in* Saito, T., and Burckle, L. H., eds., Late Neogene epoch boundaries: New York, Micropaleontology Press, p. 226–244.

Savin, S. M., Douglas, R. G., Keller, G., Killingley, J. S., Shaughnessy, L., Sommer, M. A., Vincent, E., and Woodruff, F., 1981, Miocene benthic foraminiferal isotope records: a synthesis: Marine Micropaleontology, v. 6, p. 423–450.

Shackleton, N. J., and Kennett, J. P., 1975, Paleotemperature history of the Cenozoic and the initiation of Antarctic glaciation: Oxygen and carbon isotope analyses in DSDP Sites 277, 279, and 281, *in* Initial Reports of the Deep Sea Drilling Project: Washington, U.S. Government Printing Office, v. 29, p. 743–755.

Shafik, S., 1975, Nannofossil biostratigraphy of the southwest Pacific, Deep Sea Drilling Project, Leg 30, *in* Initial Reports of the Deep Sea Drilling Project: Washington, U.S. Government Printing Office, v. 30, p. 549–598.

Shaw, A. B., 1964, Time in stratigraphy: New York, McGraw-Hill, 365 p.

Srinivasan, M. S., and Kennett, J. P., 1981a, A review of Neogene planktonic foraminiferal biostratigraphy: Applications in the equatorial and South Pacific, *in* Deep Sea Drilling Project: A decade of progress: Society of Economic Paleontologists and Mineralogists, Special Publication, no. 32, p. 395–432.

—— 1981b, Neogene planktonic foraminiferal biostratigraphy and evolution: equatorial to subantarctic South Pacific: Marine Micropaleontology, v. 6, p. 499–533.

Tauxe, L., Tucker, P., Petersen, N. P., and LaBrecque, J. P., 1983, The magnetostratigraphy of Leg 73 sediments: Palaeogeography, Palaeoclimatology, Palaeocology, v. 42, p. 65–90.

Theyer, F. and Hammond, S. R., 1974, Paleomagnetic polarity sequence and radiolarian zones, Brunhes to Epoch 20: Earth Planetary Science Letters, v. 22, p. 307–319.

Theyer, F., Mato, C. Y., and Hammond, S. R., 1978, Paleomagnetic and geochronologic calibration of latest Oligocene to Pliocene radiolarian events, equatorial Pacific: Marine Micropaleontology, v. 3, p. 377–395.

Thunell, R. C., 1981, Late Miocene-early Pliocene planktonic foraminiferal biostratigraphy and paleoceanography of low-latitude marine sequences: Marine Micropaleontology, v. 6, p. 71–90.

Vincent, E., 1981a, Neogene planktonic foraminifers from the central North Pacific, Deep Sea Drilling Project Leg 62, *in* Initial Reports of the Deep Sea Drilling Project: Washington, U.S. Government Printing Office, v. 62, p. 329–353.

—— 1981b, Neogene carbonate stratigraphy of Hess Rise (central North Pacific) and paleoceanographic implications, *in* Initial Reports of the Deep Sea Drilling Project: Washington, U.S. Government Printing Office, v. 62, p. 571–606.

Vincent, E., Killingley, J. S., and Berger, W. H., 1980, The magnetic epoch 6 carbon shift: A change in the ocean's $^{13}C/^{13}C$ ratio 6.2 million years ago: Marine Micropaleontology, v. 5, no. 2, p. 185–203.

Vincent, E. and Killingley, J. S., in press, Lower and middle Miocene isotope stratigraphy at DSDP Sites 573, 574, and 575, central equatorial Pacific, *in* Initial Reports of the Deep Sea Drilling Project: Washington, U.S. Government Printing Office, v. 85, p. 000–000.

Westberg, M. J., and Riedel, W. R., 1978, Accuracy of radiolarian correlations in the Pacific Miocene: Micropaleontology, v. 24, p. 1–23.

Woodruff, F., Savin, S. M., and Douglas, R. G., 1981, Miocene stable isotope record: A detailed deep Pacific Ocean study and its paleoclimatic implications: Science, v. 212, p. 665–668.

MANUSCRIPT ACCEPTED BY THE SOCIETY DECEMBER 17, 1984

Geological Society of America
Memoir 163
1985

Latest Miocene/earliest Pliocene diatom correlations in the North Pacific

Lloyd H. Burckle
Lamont-Doherty Geological Observatory
of Columbia University
Palisades, New York 10964

Neil D. Opdyke
Department of Geology
University of Florida
Gainesville, Florida 32611

ABSTRACT

Several diatom datum levels are proposed for latest Miocene/earliest Pliocene sediments in the equatorial and north Pacific. These datum levels are tied directly to the magnetostratigraphy and may be summarized as follows:

1) Last occurrence of *Denticulopsis hustedtii*—middle of Chron 10 in the equatorial Pacific.

2) First occurrence of *Thalassiosira burckliana*—lower part of Chron 9 in the equatorial Pacific.

3) First occurrence of *Coscinodiscus nodulifer* var. *cyclopus*—middle of Chron B in the equatorial Pacific.

4) Last occurrence of *Thalassiosira burckliana*—lowest part of Chron 7 in the equatorial Pacific.

5) First occurrence of *Rossiella praepaleacea*—lower part of Chron 6 in the central Pacific.

6) First occurrence of *Nitzschia miocenica* var. *elongata*—lower part of Chron 6 in the equatorial Pacific.

7) Last occurrence of *Rossiella praepaleacea*—upper part of Chron 6 in the equatorial Pacific.

8) Last occurrence of *Nitzschia miocenica* var. *elongata*—middle part of Chron 5 in the equatorial Pacific.

9) First occurrence of *Denticulopsis kamtschatica*—lower part of Chron 6 in the high latitude North Pacific.

10) Last occurrence of *Rouxia californica*—middle part of Chron 6 in the high latitude North Pacific.

Within the context of a more refined biostratigraphy and improved correlation between high and low latitudes, we can trace the latitudinal changes in distribution of selected diatom species as a function of time. *Actinocyclus ingens* is seen as a cosmopolitan species in the middle Miocene and by the Late Miocene it has retreated to the higher latitudes of the north and south Pacific. *Denticulopsis kamtschatica,* on the other hand, is shown to make its first appearance in the high latitude north Pacific at about 6.3 Ma and extend its range southward during the succeeding 1.5 Ma.

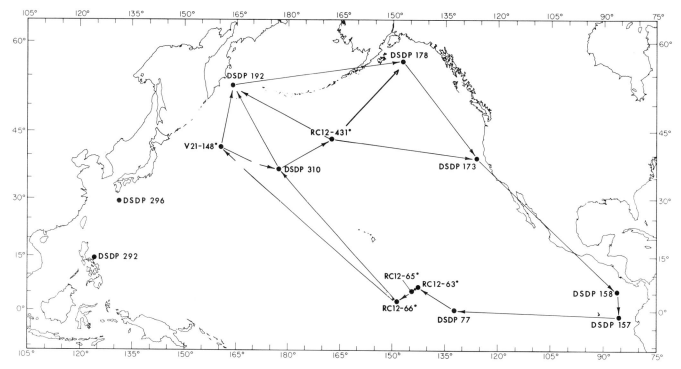

Figure 1. Distribution of piston cores and DSDP Sites used in this study. Arrows indicate correlation pathways.

INTRODUCTION

In previous studies, diatom biostratigraphy in the Pacific has been tied directly to the magnetostratigraphy (Hays et al., 1969; Burckle, 1972, 1977a, 1978; Opdyke et al., 1974; Burckle et al.; 1983). These studies were largely restricted to the tropical Pacific and, to some extent, to the northwest Pacific (Burckle and Opdyke, 1977). We have been hampered in high northern latitudes by the lack of suitable piston cores which (a) are long enough and continuous enough to provide an interpretable magnetostratigraphy and (b) are equally blessed with good silica preservation throughout the entire core. Poor silica preservation is largely the result of diatom productivity patterns and plate motion in the North Pacific. Analysis of diatoms in surface sediments demonstrates that they are either absent or very poorly preserved beneath the central water masses (Burckle, in press). Most of the cores available to us in the high-latitude North Pacific will backtrack into the central water mass and thus have poor or no late Miocene/early Pliocene diatom recovery.

In this paper, we report on several new piston cores from the North Pacific which, although not of the quality to which we are accustomed, provide us with good magneto- and diatom stratigraphy. These new data permit us to correlate between high and low latitudes in the North Pacific. Because of the nature of our material, this report is largely confined to latest Miocene/earliest Pliocene sediments and is a natural extension of a previous report (Burckle and Opdyke, 1977) which dealt with Pliocene-Pleistocene diatom correlations in the Pacific. Additional data on these datum levels and particularly their relationship to forami-

niferal and calcareous nannoplankton datums may be found in Vincent (1981) and Keller et al. (1982).

METHODS AND SOURCE OF MATERIALS

Samples were prepared using the method described by Schrader (1974a), a process which is close to that described by other diatomists. Samples were taken at 10-20 cm intervals for piston cores and 30-100 cm intervals for DSDP sites. Several scans were made along the entire length of the slide with magnifications varying between X400 and X1000 and identifications were based upon Hustedt (1930, 1958), Schrader (1973, 1974b) Barron (1975, 1980) and Burckle (1972). Both DSDP sites and wire-line supported piston cores were used in this study. All piston cores recovered from the North Pacific and presently in the Lamont-Doherty core collection were examined and suitable ones were selected for detailed study (Fig. 1).

CORRELATION STRATEGY

Figure 1 shows the location of cores used in this study as well as demonstrating the strategy used to correlate between high and low latitudes. Essentially, correlation proceeds in a clockwise direction beginning with the central Pacific, where we have three well-studied piston cores (Burckle, 1972; Gartner, 1973; Saito et al., 1975) with paleomagnetic control (Opdyke and Foster, 1970). With such good control, it is relatively easy to tie these three sites to DSDP Sites 77 and 158 to the east. To the northwest are two sites (V21-148 and DSDP Site 310), one of which has a magnetostratigrahy (V21-148). In this region both high and low latitude zonal elements are present, permitting correlation between the low latitude central Pacific and the higher latitude northwestern Pacific. Just to the north of DSDP Site 310 is RC12-431 which has a paleomagnetic record that permits us to draw a tie to site 310. Further, since RC12-431 is north of 310, it provides us with a biostratigraphic tie between Site 310 and DSDP Sites 178 and 192. These sites can then be correlated to DSDP Site 173. Since Sites 173 and 158 are both along the eastern boundary current, they have several stratigraphically useful species in common. The correlations along the eastern boundary current can further be verified by correlating Sites 158 and 77 and the central Pacific piston cores (RC12-63, 65 and 66).

EPOCH AND AGE BOUNDARIES

In previous papers (Burckle, 1972, 1978; Burckle and Opdyke, 1977) we discussed our concept of epoch and age boundaries. This paper covers the Miocene/Pliocene boundary and the Tortonian/Messinian boundary. Our recognition of the Miocene/Pliocene boundary has changed little from previous reports. We follow the generally accepted practice (Cita, 1975; Saito, et al., 1975) of placing this boundary in the lower part of the Gilbert Chron at approximately 5.3 Ma. This boundary is recognized in deep sea sediments by the first appearance of *Globorotalia tumida* (Saito, et al., 1975) and the last appearance of *Thalassiosira miocenica* (Burckle, 1978; Burckle and Opdyke, 1977). We are less sure of the precise placement of the Tortonian/Messinian boundary relative to the magnetostratigraphy. Van Couvering et al. (1976) suggested a date of less than 6.5 Ma and there seems general concensus that this boundary cannot be any older. In accepting this date, some authors have pointed to its close correspondence to the now well-known 'carbon shift' which is tied to the mid- to lower part of Chron 6 at approximately 6.1 Ma (Loutit and Kennett, 1979; Bender and Keigwin, 1979; Keigwin and Shackleton, 1980; Haq et al., 1980).

Burckle (1977b), however, has suggested that the boundary may be younger (middle part of Chron 5) on the basis of the presumed age of the diatoms in the Messinian section at Capodarso, Sicily. This point is supported by Langereis et al. (1984) but has been disputed by Gersonde (1980), among others. Bossio et al. (1976) have shown, for example, that the *Dentoglobigerina altispira* coiling change (sinstral to dextral) which Saito et al. (1975) tied to the lower normal event of Chron 5 occurs in the lower Tripoli member of the Messinian. Further, Gersonde (1980) has shown that the Tripoli member in well-exposed sections in Sicily (e.g., Falconara) contains *Thalassiosira praeconvexa*, a diatom which last occurs in the lower normal event of Chron 5 (Burckle, 1972; Saito et al., 1975). The fact that Burckle (1977b) did not have samples from near the base of the Capodarso section probably accounts for the conflict with Gersonde's results.

Since the change in coiling direction of *G. altispira* occurs in the Tripoli member of the Messinian stage and in the lower normal event of Chron 5 (Saito et al., 1975), the Tortonian/Messinian boundary must be older than approximately 5.6 m.y. On the other hand, the age of the Banqueros volcanics (6.9 m.y.) in the upper part of the Tortonian in southeast Spain (Van Couvering et al., 1976) places a lower age limit on this boundary. Extrapolating from this lower age limit, Van Couvering et al., (1976) settled on a date of approximately 6.5 m.y. for this boundary. Even if we accept an age range for this boundary of 5.7 to 6.9 m.y., the Tortonian/Messinian boundary would still fall within Chron 6, although we are not sure precisely where.

DATUM LEVELS

Figures 2 and 3 show the ranges of stratigraphically important species in the equatorial and North Pacific. The ranges of many of these species have previously been discussed by Burckle (1972, 1978) and Burckle and Opdyke (1977). In each case, these authors showed direct ties between species' first and last occurrence and the magnetostratigraphy. In the present study, most of our datum levels are tied directly to the magnetostratigraphy. In cases where they are not, we have good stratigraphic evidence which at least ties them down to one magnetic chron. Since many of the datum levels have been reported elsewhere (see above) we discuss only those that are new or presently in press.

(1.) Last occurrence of *Denticulopsis hustedtii*—Burckle et al. (1983) determined that this species last occurs in the middle of Chron 10 in the equatorial Pacific. In the mid- to high-latitude North Pacific, however, it ranges well into the late Miocene (Koizumi, 1975). This datum level is useful in subdividing the lower part of the Upper Miocene.

(2.) First occurrence datum of *Thalassiosira burckliana* Burckle (1972, 1981), Schrader (1974b) and Burckle et al. (1983) have tied this first occurrence to the lower part of Chron 8 or the uppermost part of Chron 9. The use of this datum level in the Southern Ocean is in error since this species does not occur at the levels indicated by Gombos (1979) and Ciesielski (1983).

(3.) Last occurrence datum of *Coscinodiscus plicatus*—This species name is used here in the broadest sense to include the last appearance of all plicate forms of *Coscinodiscus*. It occurs in the middle part of Chron 8 in the equatorial Pacific but there is some evidence, not entirely evaluated, to suggest that the last appearance of this form may be slightly diachronous into the higher latitude North Pacific.

(4.) First occurrence datum of *Coscinodiscus nodulifer* var. *cyclopus*—This species and the vagaries of its range have been discussed by Jouse (1974) and Burckle and Trainer (1979). In Pliocene sediment of the equatorial Pacific, this form appears and disappears at such regular intervals that its recurrences can be used as stratigraphic markers. Thus, it occurs in the lower part of the Gilbert Chron and then recurs at several intervals in the lower part of the Matuyama Chron. Its recurrence in this interval usu-

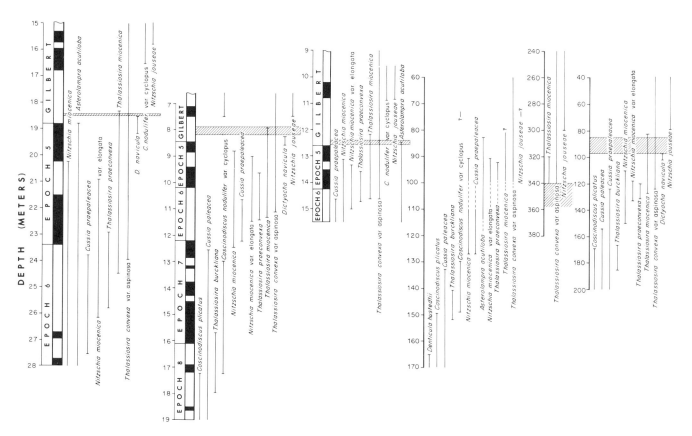

Figure 2. Proposed diatom correlation for latest Miocene/earliest Pliocene sediment in the equatorial Pacific. Note that the three piston cores (RC12-63, 65 and 66) have a magnetostratigraphy. The hatched zone in the lower part of the Gilbert Group (near the Miocene/Pliocene boundary) represents severe silica dissolution.

ally spans very little time. In the late Miocene, we note the first occurrence of this form in the middle part of Chron 8, nearly coincidental with the last occurrence of *C. plicatus.* The geographic distribution of this form appears to be quite limited, however, since we do not observe it in the easternmost equatorial Pacific nor was it seen in DSDP Site 173, beneath the eastern boundary current.

(5.) Last occurrence datum of *Thalassiosira burckliana*— The last occurrence of this species (in the lower part of Chron 7) is well-defined since it is bounded above and below by several easily recognized datum levels. Just below this last appearance is the change in dominance of *Dictyocha fibula/D. aspera,* which occurs near the base of Chron 7 (Burckle, 1981). Just above the *Th. burckliana* last occurrence datum is the first occurrence of *Nitzschia miocenica,* which takes place in the upper part of Chron 7. *Th. burckliana* is more broadly distributed throughout the Pacific region and its last occurrence seems to be isochronous. The reported occurrence of this species in the Miocene of the Southern Ocean (Gombos, 1976; Ciesielski, 1983) are in error. The Southern Ocean form is similar to, but distinct from, *Th. burckliana.*

(6.) First occurrence of *Rossiella praepaleacea*—Caution should be applied in using this datum. In the eastern equatorial Pacific, it occurs in the middle Miocene. However, in the central Pacific (piston cores RC12-65, RC12-66), its first occurrence can be tied to the lower part of Chron 6. Because this form is frequently rare, its first appearance may vary somewhat from site to site. However, as Figure 3 shows, the range in variation of first appearance within the central Pacific is slight (from the uppermost part of Chron 7 to the lowermost part of Chron 6; see Schrader, 1973 for description). This species also occurs in late Miocene sediments of southeastern Spain (Burckle, unpublished notes).

(7.) First occurrence of *Nitzschia miocenica* var. *elongata*— This species and its range is described by Burckle (submitted) and first occurs in the lower part of Chron 6 just above the *R. praepaleacea* first appearance datums. This variety is found only in the equatorial Pacific. We do not record it in higher latitudes or in the equatorial Indian Ocean in spite of the fact that the parent taxon is present in those areas.

(8.) Last occurrence of *Rossiella praepaleacea*—The last occurrence of this form appears to be isochronous across the

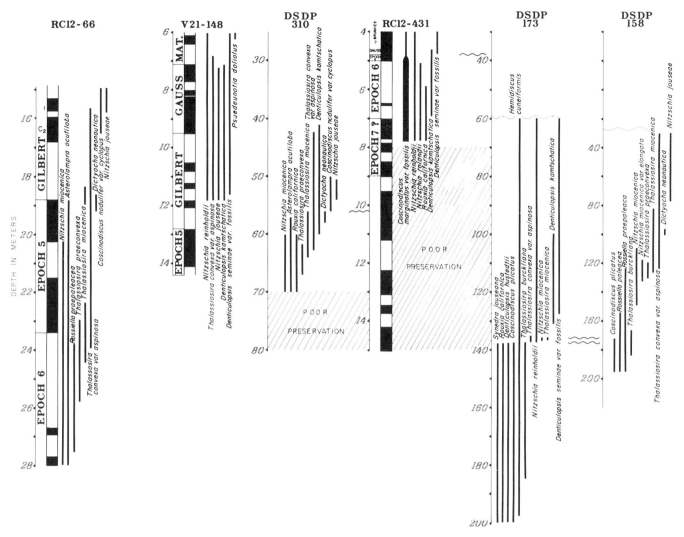

Figure 3. Proposed diatom correlation for mid-latitude North Pacific sites. Note that three piston cores (RC12-66, RC12-431 and V21-148) have a magnetostratigraphy.

eastern equatorial Pacific. As Figure 3 shows, its last occurrence is in the upper part of Chron 6, just above the first occurrence of *Thalassiosira praeconvexa* and *Th. convexa* var. *aspinosa*. We also record it in higher latitudes (DSDP site 310) as well as in southeast Spain, but there is no assurance that these occurrences are isochronous with lower latitudes. The fact that it is present in Spain but not in similar lithologies of Sicily suggests that the late Miocene Tripoli may be diachronous between these two regions.

(9.) Last occurrence of *Nitzschia miocenica* var. *elongata*—Burckle (submitted) has tied this last occurrence in the equatorial Pacific to the middle part of the middle reversed event of Chron 5 (Fig. 3). As mentioned in our previous discussion, this variety is restricted to the low latitude Pacific.

(10.) Range of *Dictyocha neonautica*—*D. neonautica* is a silicoflagellate. Its restricted range in the Pacific makes it a useful biostratigraphic marker near the Miocene/Pliocene boundary. It first appears in the upper normal event of Chron 5 and disappears in the lowermost part of the Gilbert Chron, just below the Mio-

cene/Pliocene boundary. Figure 3 shows that it also disappears just below a silica dissolution event which seems to characterize the Miocene/Pliocene boundary in equatorial regions. Fortunately, this species also occurs in the North Pacific where it can be used to identify the Miocene/Pliocene boundary interval in DSDP Site 310. We doubt that this is the true range of this species since it is rather unusual for a species to have such a restricted vertical range. We suspect that the range shown here is really a peak zone and that this species occurs elsewhere in the section, although rarely.

(11.) First occurrence of *Denticulopsis kamtschatica*—There has been some question in the literature concerning the first occurrence of this species. The fact that its first appearance is diachronous has only recently been recognized and appreciated. Harper (1977) has shown that this species appears in high northern latitudes in the late Miocene and spreads toward the mid-latitude North Pacific, arriving there by latest Miocene/earliest Pliocene time. Barron (1980) has extended this study and has shown that there are considerable variations in abundance of this species which could lead to errors concerning the age of its first occurrence. Burckle (1978), for example, placed the first occurrence of this species in the upper part of Chron 6. However, in light of the new data presented here, this is too high in the section. Fig. 3 shows that the true first appearance of *D. kamtschatica* is in the lower part of Chron 6. This agrees very well with data from DSDP Site 178 where the first appearance of this species occurs near an ash layer dated at 6.3 Ma (Scheiddeger and Kulm, 1975; Burckle, 1978).

(12.) Last occurrence of *Rouxia californica*—As Figure 3 shows, *R. californica* last occurs in the middle part of Chron 6 in RC12-431. This datum level is diachronous, however, into the northwest Pacific (Harper, 1977; Barron, 1980). Further, comparison of this datum level between DSDP Site 310 and RC12-431 also suggests that it is diachronous.

DIATOM CORRELATIONS INTO THE HIGH LATITUDE NORTH PACIFIC

Late Neogene correlations in the equatorial Pacific have previously been discussed by Burckle (1972, 1977a, 1978) and Burckle and Opdyke (1977). We have little to add to add to these studies. In the northwest Pacific, piston cores V21-148 and DSDP Site 310 contain zonal elements from both high and low latitudes. Joint occurrences of such low latitude zonal markers as *N. miocenica, N. jouseae, A. acutiloba* and *Th. praeconvexa* and high latitude and cosmopolitan forms such as *Th. convexa* var. *aspinosa* and *D. kamtschatica* allows us to tie together the equatorial Pacific and the higher latitude North Pacific. This correlation is strengthened by the fact that V21-148 has a magnetostratigraphy. In a previous publication (Burckle and Opdyke, 1977), we cited two additional Pliocene/Pleistocene cores from this region which have a magnetostratigraphy (RC12-413 and 415). To the north of DSDP Site 310 is a key piston core, RC12-431 (Fig. 3). In this core, we can identify Chrons 5, 6 and 7 as well as part of the Brunhes Chron. Its diatom stratigraphy gives us the necessary tie between DSDP Site 310 and higher latitude sites of the North Pacific (DSDP Sites 178 and 192).

SHAW DIAGRAMS

Shaw diagrams (Shaw, 1963) were constructed to compare North Pacific Late Miocene Sites. Such diagrams are useful in depicting unconformities and changes in sedimentation rates between two sites. We include only those diatom datum levels that have been directly tied to the magnetostratigraphy. Because the best paleomagnetically dated cores come from the equatorial Pacific, we begin with this region first and use the same format as with our correlation strategy.

RC12-65-DSDP Site 310

Figure 4 shows the Shaw diagram for the Late Miocene/

Early Pliocene of RC12-65 and DSDP Site 310. The Carbon-13 shift is also included (Bender and Keigwin, 1979) as are carbonate dissolution zones and silica dissolution events. Decreased rates of deposition occur in RC12-65 in three intervals. The earliest one occurs around 11 meters depth in the middle part of Chron 6. This is nearly coincidental with the Carbon-13 shift. A second, less abrupt, change in sedimentation rate occurs in the upper normal event of Chron 5, while a final decrease occurs in the earliest Pliocene near the 'C' event of the Gilbert Chron. The silica dissolution event in RC12-65 is about 8 meters depth and, more or less, coincides with an unconformity (covering some 300,000 to 400,000 years) in DSDP Site 310. The intervals of increased carbonate dissolution occur during periods of decreased sedimentation rates in Site 310. It was not possible to carry this correlation below the last appearance of *Th. burckliana* because of silica dissolution in DSDP Site 310.

DSDP Site 310-DSDP Site 158

The Shaw diagram for DSDP Sites 310 and 158 is shown in Figure 5. We note some similarities between Site 158 and RC12-65 relative to Site 310. In Site 158, a silica dissolution event occurs in the lower part of the Gilbert Chron as it does in RC12-65. In Site 158, however, the dissolved zone is much thicker, covering more than 10 meters. A carbonate dissolution event occurs between the *Th. praeconvexa* and *N. miocenica* datum levels in almost exact duplication of RC12-65. Below that, we note a split dissolution event between approximately 127-162 meters in Site 158. This is seen as a single dissolution event in RC12-65 and Site 158 suggesting unconformities in these two latter sites. There is some difference in the placement of the Carbon-13 shift in these two sites with the shift in Site 158 somewhat older stratigraphically than Site 310.

DSDP Site 173-DSDP Site 158

Since DSDP Sites 158 and 173 are both under the influence of the eastern boundary current, they should have some datum levels in common in spite of the laitudinal difference in location. As Figure 6 shows, there are major unconformities in Site 173. The younger unconformity occurs in the lowermost part of the Gilbert Chron (latest Miocene) and is coincidental with the time of silica dissolution in Site 158. The second unconformity has removed sediments equivalent to Chron 7 and is partly correlative with the two dissolution zones in Site 158 (Fig. 9). The Carbon-13 shift has not been reliably identified in Site 173.

DSDP Site 158-RC12-65

Sites 158 and 310 are important links in devising a correlation scheme between low and mid-latitudes (Fig. 7). Just as Site 310 has both low and mid-latitude stratigraphic markers, Site 158 should also contain low and mid-latitude forms. Since Site 158 is under the influence of the eastern boundary current, we found

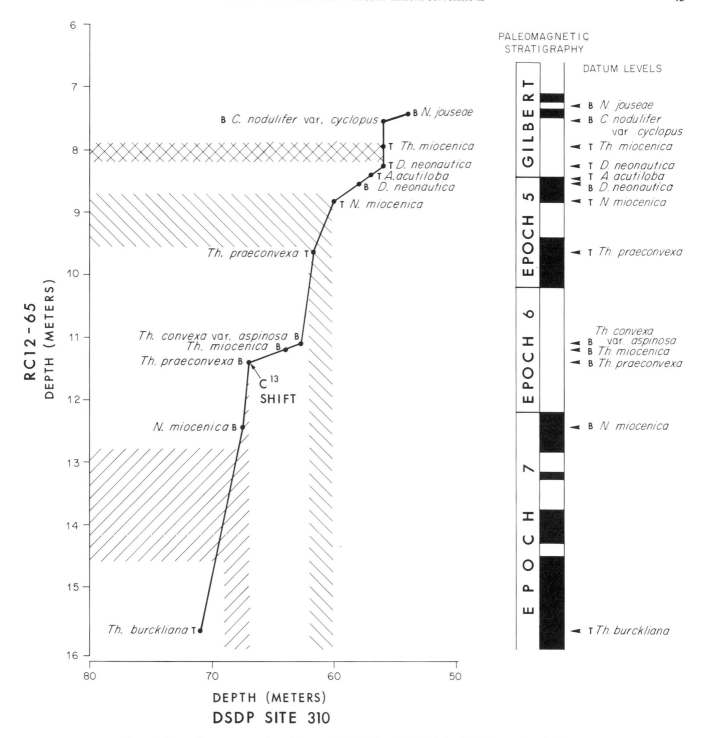

Figure 4. Shaw diagram comparing piston core RC12-65 and DSDDP site 310. The cross-hatched zone represents an interval of severe silica dissolution while the right and left hatched zones represent intervals of carbonate dissolution.

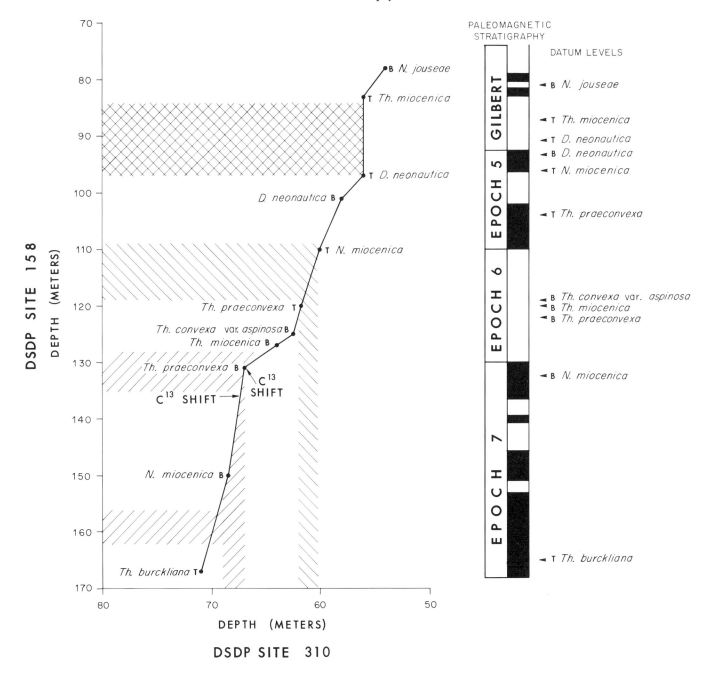

Figure 5. Shaw diagram comparing DSDP sites 158 and 310. The cross-hatched zone represents an interval of severe silica dissolution and the right and left hatched zones represent intervals of carbonate dissolution.

that it had floral elements in common with Site 173. Similarly, because of its low latitude location, it should also have floral elements in common with RC12-65. As Figure 7 shows, there is an excellent correlation between the carbonate and silica dissolution zones. Because of its position in the eastern equatorial Pacific, the sedimentation rate for 158 is much higher than for RC12-65. However, there are intervals of reduced sedimentation rates. These include the middle late Miocene (Chrons 7 to 8) and the latest late Miocene (lower Chron 5 to the middle of Chron 6).

DISCUSSION

A more refined biostratigraphy and improved correlation between high and low latitudes helps provide a framework within which paleoceanographic and paleoclimatic problems may be addressed (see, for example, CLIMAP, 1976). Here we focus on latitudinal changes in distribution of certain species as a function of time as well as changes in the temporal and spatial distribution of productivity patterns and unconformities. Figures 8 and 9

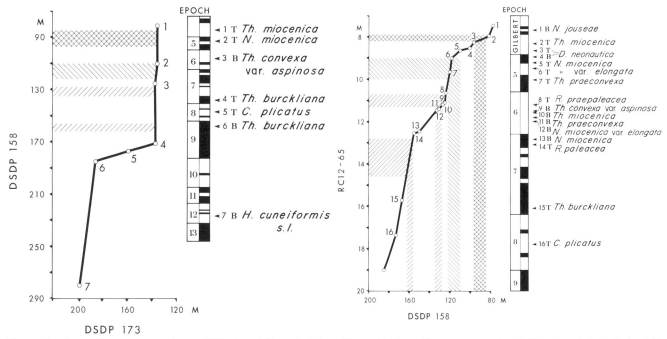

Figure 6. Shaw diagram comparing DSDP sites 158 and 173. Cross-hatched zone and right and left lined zones represent intervals of silica and carbonate dissolution, respectively, in site 158.

Figure 7. Shaw diagram comparing RC12-65 and DSDP site 158. Cross-hatched zone represents interval of silica dissolution and right and left lined zones indicate intervals of carbonate dissolution.

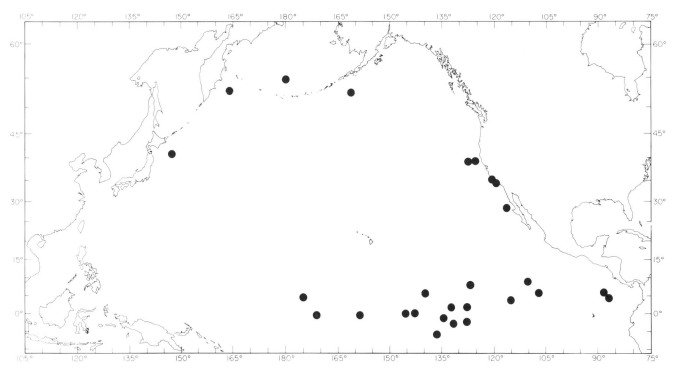

Figure 8. Distribution of *Actinocyclus ingens* in the equatorial and North Pacific during the middle middle Miocene.

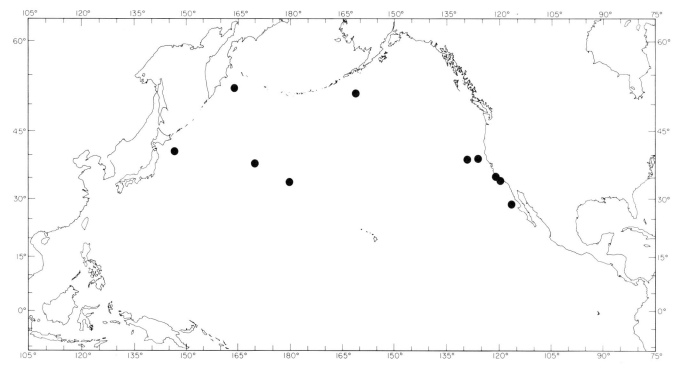

Figure 9. Distribution of *Actinocyclus ingens* in the North Pacific during the middle late Miocene.

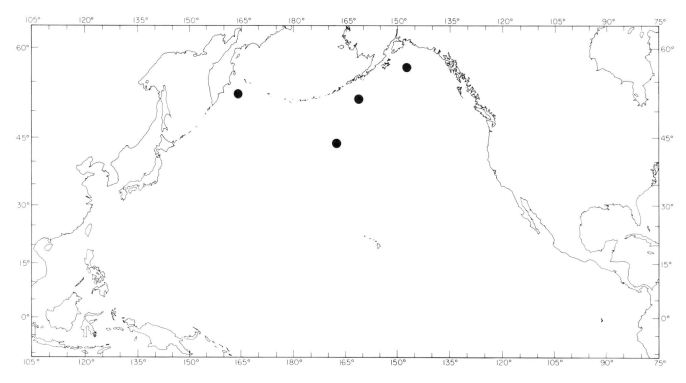

Figure 10. Distribution of *Denticulopsis kamtschatica* in the high latitude North Pacific at about 6.3 Ma.

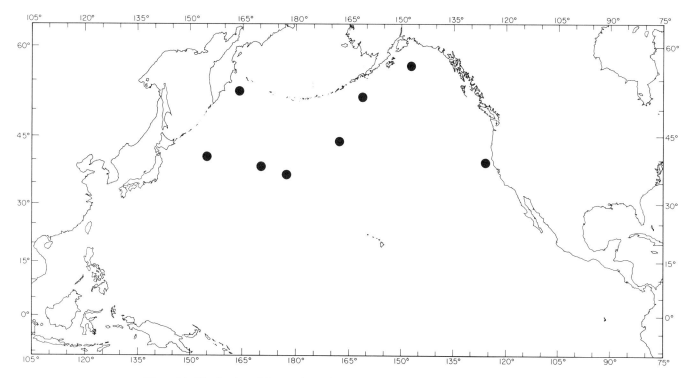

Figure 11. Distribution of *Denticulopsis kamtschatica* in the North Pacific at about 5 Ma.

show the changes in the distribution of *Actinocyclus ingens* between the middle Miocene and the middle late Miocene. This species is cosmopolitan during middle Miocene times, being found in the Southern Ocean as well as in the Bering Sea. During the middle late Miocene, it essentially disappears from the equatorial regions and is found only in high northern and southern latitudes. It ranges into the late Miocene in the North Pacific, but continues into the middle Quaternary in the Southern Ocean (Donahue, 1970).

A different scenario is played out in changes in the temporal and spatial distribution of *Denticulopsis kamtschatica*. This species first appears at about 6.3 Ma in the high latitude North Pacific. It gradually increases its range through the late Miocene until, by earliest Pliocene time, it was also present in the mid-latitude North Pacific (Figs. 10 and 11). *D. kamtschatica* appears to have been a cold-water form that originated in high latitudes and then extended its range southward. Barron (1980) notes that this species is most abundant during cool intervals in the late Miocene and early Pliocene in the North Pacific. Changes in the distribution of *A. ingens* are less easy to understand. Its disappearance from the equatorial Pacific approximately coincides with a $\delta^{18}O$ enrichment in benthic foraminifera (Woodruff et al., 1981). This isotopic shift has been attributed to the growth of the East Antarctic ice sheet (Kennett and Shackleton, 1975), although this has been disputed (Matthews and Poore, 1980). In any event, there is ample evidence to suggest that a paleoclimatic and paleoceanographic change took place during the middle Miocene. The retreat of *A. ingens* into higher latitudes during this time was apparently a response to this change.

ACKNOWLEDGMENTS

I thank D. Lazarus and C. Sancetta for reviewing this manuscript. Discussions with J. Barron and G. Keller were also extremely helpful. The Deep Sea Drilling Program, Scripps Institute of Oceanography kindly provided the DSDP samples. This research was supported by National Science Foundation Grants OCE 78-20885 and OCE 79-19092.

REFERENCES CITED

Barron, J., 1975, Late Miocene-early Pliocene marine diatoms from southern California: Paleontographica, v. 151(B), p. 97–170.

Barron, J., 1980, Lower Miocene to Quaternary diatom biostratigraphy of Leg 57, off northeast Japan, Deep Sea Drilling Project. *in*, Honza, E., ed., Initial Reports of the Deep Sea Drilling Project: Washington (U.S. Government Printing Office), v. 56, 57, Pt. 2, p. 641–685.

Bender, M. L. and Keigwin, L. D., 1979, Speculations about the upper Miocene change in abyssal Pacific dissolved bicarbonate: Earth and Planetary Science Letters, v. 45, p. 383–393.

Berggren, W. A. and Van Couvering, J., 1974, The late Neogene: Palaeogeography, Palaeoclimatology, Palaeoecology, v. 16, p. 1–216.

Bossio, A., El-Bied Rakich, K., Giannelli, L., Mazzei, R., Russo, A. and Salvatorini, G., 1976, Correlation de quelques sections stratigraphiques du Mio-Pliocene de la zone Atlantique du Maroc avec les stratotypes du Bassin Mediterraneen sur la base des foraminiferes plantoniques, nannoplancton calcaire et ostracodes: Atti della Societa Toscana Scienze Naturali Memorie Pisa, LXXXIII. p. 121–137.

Burckle, L. H., 1972, Late Cenozoic planktonic diatom zones from the eastern equatorial Pacific: Nova Hedwigia, v. 39, p. 217–248.

Burckle, L. H., 1977a, Pliocene and Pleistocene diatom datum levels from the equatorial Pacific: Quaternary Research, v. 7, p. 330–340.

Burckle, L. H., 1977b, Diatom analysis of the pre-evaporite facies in the neostratotype area of the Messinian: Messinian Seminar Abstracts, 3, Malaga, 1977.

Burckle, L. H., 1978, Early Miocene to Pliocene diatom datum levels for the equatorial Pacific. Proc. 2nd Working Group on Biostratigraphic Datums of the Pacific Neogene: Bandung, Indonesia, p. 25–44.

Burckle, L. H., 1981, Paleomagnetic date on the *Dictyocha aspera/fibula* crossover in the equatorial Pacific: Micropaleontology, v. 27, p. 332–334.

Burckle, L. H., in press, Diatoms in sediment traps and surface sediments of the equatorial Pacific: Micropaleontology.

Burckle, L. H., submitted, *Nitzschia miocenica* var. *elongata*; A new Late Miocene diatom from the equatorial Pacific: Micropaleontology.

Burckle, L. H. and Opdyke, N. D., 1977, Late Neogene diatom correlations in the circum-Pacific: Proceedings 1st International Congress Pacific Neogene Stratigraphy, Tokyo, p. 255–284.

Burckle, L. H. and Trainer, J., 1979, Middle and late Pliocene diatom datum levels from the central Pacific. Micropaleontology, v. 25, p. 281–292.

Burckle, L. H., Keigwin, L. D. and Opdyke, N. D., 1983, Middle and late Miocene stable isotope stratigraphy: Correlation to the paleomagnetic reversal record: Micropaleontology, v. 28, p. 329–334.

Ciesielski, P. F., 1983, The Neogene and Quaternary diatom biostratigraphy of subantarctic sediments: Deep Sea Drilling Project, leg 71. *in* Ludwig, W. J. and Krasheninnikov, V. A., eds., Initial Reports of the Deep Sea Drilling Project, Leg 71: Washington, U.S. Government Printing Office, v. 71, p. 635–665.

Cita, M., 1975, The Miocene/Pliocene boundary: History and definition, *in* Saito, T. and Burckle, L. H., eds., Late Neogene epoch boundaries: Micropaleontology Press, p. 1–30.

CLIMAP, 1976, The surface of the ice-age earth: Science, v. 191, p. 1131–1137.

Donahue, J. G., 1970, Pleistocene diatoms as climatic indicators in North Pacific sediments: Geological Society of America Memoir 126, p. 121–138.

Gartner, S., 1973, Absolute chronology of the late Neogene nannofossil succession in the equatorial Pacific: Geological Society of America Bulletin, v. 84, p. 2021–2034.

Gersonde, R., 1980, Palaeokogische und biostratigraphische auswertung von diatomeenassoziatonen aus dem Messinium des Caltanissettabeckens (siziliens) und einiger vergleichs profile in So. Spanien, NW Algerien und auf Freta [Ph.D. Thesis]: Universitat Kiel, F.R.G., 393 p.

Gombos, A., 1976, Paleogene and Neogene diatoms from the Falkland Plateau and Malvinas outer basin: leg 36, Deep Sea Drilling Project, *in* Barker, R. F. and Dalziel, I.W.D., eds., Initial Reports Deep Sea Drilling Project, Leg 36: Washington, U.S. Government Printing Office, v. 36, p. 575–687.

Haq, B. U., Worsley, T. R., Burckle, L. H., Douglas, R. G., Keigwin, L. D., Opdyke, N. D., Savin, S. M., Sommer, M. A., Vincent, E. and Woodruff, F., 1980, Late Miocene marine carbon-isotopic shift and synchroneity of some phytoplanktonic biostratigraphic events: Geology, v. 8, p. 427–431.

Harper, H. E., 1977, Diatom biostratigraphy of the Miocene/Pliocene boundary in marine strata of the circum-North Pacific [Ph.D. Thesis]: Harvard University, 112 p.

Hays, J. D., Saito, T., Opdyke, N. D. and Burckle, L. H., 1969, Pliocene-Pleistocene sediments of the equatorial Pacific: Their paleomagnetic, biostratigraphic and climatic record: Geological Society of America Bulletin, v. 80, p. 1481–1514.

Hustedt, F., 1930, 1958, Die Kieselalgen. In: Rabenhorst, L., ed., Kryptogramenflora von Deutschland, Oesterreich und der Schweiz. Leipzig: Akad. Verlags. (Reprint, 1962).

Jouse, A. P., 1974, Diatoms in the Oligocene-Miocene biostratigraphic zones of the tropical areas of the Pacific Ocean: Nova Hedwigia, v. 45, p. 333.

Keigwin, L. D., 1979, Late Cenozoic stable isotope stratigraphy and paleoceanography of DSDP sites from the east equatorial and central Pacific Ocean: Earth Planetary Science Letters, v. 45, p. 361–382.

Keigwin, L. D., and Shackleton, N. J., 1980, Latest uppermost Miocene carbon isotope stratigraphy of a piston core in the equatorial Pacific: Nature, v. 284, p. 613–614.

Keller, G., Barron, J. and Burckle, L. H., 1982, North Pacific late Miocene correlations using microfossils, stable isotopes, percent calcium carbonate and magnetostratigraphy: Marine Micropaleontology, v. 7, p. 327–357.

Kennett, J. P. and Shackleton, N. J., 1975, Laurentide ice sheet meltwater recorded in Gulf of Mexico deep-sea cores. Science, v. 188, p. 147–150.

Koizumi, I., 1975, Late Cenozoic diatom biostratigraphy in the circum-Pacific region: Journal Geological Society, Japan, v. 81, p. 611–627.

Langereis, C. G., Zachariasse, W. J. and Zijderveld, J.D.A., 1984, Late Miocene stratigraphy of Crete: Marine Micropaleontology, v. 8, p. 261–282.

Loutit, T. S., and Kennett, J. P., 1979, Application of carbon isotope stratigraphy to late Miocene shallow marine sediments, New Zealand: Science, v. 204, p. 1196–99.

Mathews, R. K. and Poore, R. Z., 1980, Tertiary δO record and glacioeustatic sea level fluctuations: Geology, v. 8, p. 501–504.

Opdyke, N. D. and Foster, J. H., 1970, The paleomagnetism of cores from the North Pacific, *in* Hays, J. D., ed., Geological investigations of the North Pacific: Geological Society of America Memoir, v. 126, p. 83–119.

Opdyke, N. D., Burckle, L. H. and Todd, E. A., 1974, The extension of the magnetic time scale in sediments of the central Pacific ocean: Earth Planetary Science Letters, v. 22, p. 300–306.

Saito, T., Burckle, L. H. and Hays, J. D., 1975, Late Miocene to Pleistocene biostratigraphy of equatorial Pacific sediments, *in* Saito, T. and Burckle, L. H., eds., Late Neogene epoch boundaries: Micropaleontology Press, p. 226–244.

Scheidegger, A. E. and Kulm, L. D., 1975, Late Cenozoic volcanism in the Aleutian arc: Information from ash layers in the northeastern Gulf of Alaska: Geological Society of America Bulletin, v. 86, p. 1407–1412.

Schrader, H. J., 1973, Cenozoic diatoms from the northeast Pacific Leg 18. *in* Kulm, L. D., Von Huene, R., eds., Initial Reports Deep Sea Drilling Project: Washington, U.S. Government Printing Office, v. 18, p. 673–797.

Schrader, H. J., 1974a, Proposal for a standardized method of cleaning diatombearing deep sea and land exposed marine sediments: Nova Hedwigia, v. 45, p. 403–409.

Schrader, H. J., 1974b, Cenozoic marine planktonic diatom stratigraphy of the tropical Indian Ocean, *in* Fisher, R. L., Bunce, E. T., eds., Initial Reports Deep Sea Drilling Project: Washington, U.S. Government Printing Office, v. 24, p. 887–968.

Shaw, A. B., 1963, Time in stratigraphy: New York, McGraw-Hill, 365 p.

Van Couvering, J., Berggren, W. A., Drake, R. E., Aquirre, E. and Curtis, G. H., 1976, The terminal Miocene event: Marine Micropaleontology, v. 1, p. 253–286.

Vincent, E., 1981, Neogene planktonic foraminifers from the central north Pacific, Deep Sea Drilling Project leg 62, *in* Thiede, J. and Vallier, T. L., eds., Initial Reports of the Deep Sea Drilling Project: Washington, U.S. Government Printing Office, v. 62, p. 329–353.

Woodruff, F., Savin, S. M. and Douglas, R. G., 1981, Miocene stable isotope record: A detailed deep Pacific ocean study and its paleoclimatic implications: Science, v. 22, p. 665–668.

LAMONT-DOHERTY GEOLOGICAL OBSERVATORY CONTRIBUTION
MANUSCRIPT ACCEPTED BY THE SOCIETY DECEMBER 17, 1984

Geological Society of America
Memoir 163
1985

The evolution of Miocene surface and near-surface marine temperatures: Oxygen isotopic evidence

Samuel M. Savin
Linda Abel
Enriqueta Barrera
Department of Geological Sciences
Case Western Reserve University
Cleveland, Ohio 44106

David Hodell
Graduate School of Oceanography
University of Rhode Island
Narragansett, Rhode Island 02882

James P. Kennett
Graduate School of Oceanography
University of Rhode Island
Narragansett, Rhode Island 02882

Margaret Murphy
Graduate School of Oceanography
University of Rhode Island
Narragansett, Rhode Island 02882

Gerta Keller*
U.S. Geological Survey
345 Middlefield Road
Menlo Park, California 94025

John Killingley
Scripps Institution of Oceanography
University of California, San Diego
La Jolla, California 92093

Edith Vincent
Scripps Institution of Oceanography
University of California, San Diego
La Jolla, California 92093

ABSTRACT

Oxygen isotopic analyses of planktonic foraminifera have provided a picture of many aspects of the evolution of the temperature structure of surface and near-surface oceans during the Miocene. In time slice studies oceanographic conditions have been interpreted from synoptic maps of isotopic data at between 22 and 27 locations in the Atlantic, Pacific and Indian Oceans. Three time slice intervals were examined: 22 Ma (foraminiferal zone N4B) and 16 Ma (N8) in early Miocene time; and 8 Ma (N17) in late Miocene time. In time series studies, the evolution of oceanographic conditions at single localities during an extended period of time were inferred from $\delta^{18}O$ values of planktonic foraminifera.

Surface waters warmed throughout the early Miocene at almost all localities examined. At 22 Ma, the Pacific Ocean was characterized by relatively uniform temperatures in the equatorial region but a marked east-west asymmetry in the tropical South Pacific, with higher temperatures in the west. Between 22 Ma and 16 Ma, tropical Pacific surface waters warmed, but warmed more in the east than the west. At 16 Ma, the asymmetric distribution of temperatures in the South Pacific Ocean remained, and the latitudinal temperature gradient, inferred from the isotopic data, was gentler than that of either the late Miocene or Modern ocean.

Between the late early Miocene and late Miocene, surface waters at most low-latitude Pacific sites warmed while those at high latitudes cooled or remained unchanged. However, surface waters at high northern latitudes in the Atlantic Ocean as well as in the eastern equatorial Atlantic cooled, while water temperatures remained

*Present address: Department of Geological and Geophysical Science, Princeton University, Princeton, New Jersey 08540.

relatively unchanged at most South Atlantic sites. Surface waters warmed in the south-ernmost Atlantic, off the tip of South Africa. By 8 Ma, the east-to-west asymmetry of the temperature distribution in the tropical South Pacific Ocean had lessened. Surface water temperatures had become quite similar to those of the Modern ocean except that those in the equatorial Pacific Ocean were lower than today's. This is reflected in the latitudinal gradient of surface temperatures at 8 Ma which is less steep than that of modern temperatures.

The pattern of surface temperatures and their evolution through the Miocene is consistent with the biogeographic distributions of planktonic foraminifera described by Kennett et al. (this volume). The isotopic data provide a more detailed picture of the evolution of Miocene surface temperatures than had been hitherto available, and serve as a framework against which hypotheses can be tested regarding the cause of the middle Miocene cooling of deep waters and the formation of the East Antarctic ice sheet.

INTRODUCTION

Abundant sedimentologic, paleontologic and geochemical evidence indicate that the climate of the world underwent major changes during the Miocene epoch. Among the most striking manifestations of these changes were the cooling of deep-ocean waters and the rapid growth of the East Antarctic ice sheet during the middle Miocene. Oxygen isotopic ratios of planktonic and benthic foraminifera from the Pacific Ocean have suggested that these events were accompanied by a significant decrease in merid-ional heat transport and an associated increase in the latitudinal temperature gradient. Low-latitude regions in the Pacific warmed while high-latitude regions cooled as Antarctic ice volume increased (Savin et al., 1975).

In the Cenozoic Paleoceanography Project (CENOP), the details of the evolution of Miocene oceanographic conditions and climate have been investigated using a variety of approaches. In time slice studies, an attempt has been made to construct global synoptic maps of either proxy data or inferred oceanographic conditions at selected times during the Miocene. Such studies require good stratigraphic control in order to assure that the time slice maps are indeed synoptic. In time series studies, downcore variations of stable isotopic ratios, lithology, and microfloral and microfaunal assemblages have been examined in detail at individual drilling sites. The goal of such studies is to determine how oceanographic conditions varied through time at a single locality. In this paper we present the results of three time slice studies (at 22, 16, and 8 Ma) based upon oxygen and carbon isotopic analyses of a large number of planktonic foraminifera of Miocene age from Deep Sea Drilling Project (DSDP) sites in the Atlantic, Pacific and Indian Oceans. We have related these to the results of new and previously published oxygen isotopic time series studies of Miocene planktonic foraminifera from the Atlantic and Pacific Oceans. The planktonic foraminiferal isotopic data of the time slice and time series studies are the basis of our interpretation of the evolution of the temperature structure of surface and near-surface oceans over much of the world through Miocene time. The isotopic data can also be used to evaluate and constrain interpretations based upon microfaunal and microfloral bio-geographic data.

ISOTOPIC ANALYSES AND NOTATION

Isotopic analyses that have not been previously published were carried out at one of three laboratories: Case Western Reserve University, Scripps Institution of Oceanography, or University of Rhode Island. Results are reported in the usual delta (δ) notation as deviations in per mil (parts per thousand) of the $^{18}O/^{16}O$ or $^{13}C/^{12}C$ ratio of the sample from that of the PDB standard. In each laboratory, analyses were related to PDB through repeated analyses of NBS-20 (Solenhofen Limestone), for which Craig's (1957), values, $\delta^{18}O = -4.14$ and $\delta^{13}C = -1.06$ were assumed. Analytical precision (1 standard deviation) of both oxygen and carbon isotopic ratios in each laboratory, as judged by replicate analyses of the same sample, is usually better than 0.1 per mil.

TIME SLICE STUDIES

Three Miocene time intervals were chosen for synoptic stud-ies of global oceanographic conditions from isotopic measure-ments presented in this paper as well as from sedimentologic and paleontologic data discussed in the companion papers in this volume. The time slice intervals (Figure 1) were chosen to repre-sent three oceanographic regimes which were inferred from an earlier study of benthic foraminiferal isotopic time series data (Savin et al., 1981):

1) an early early Miocene interval characterized by bottom waters warmer than those of today although cooler than those of the later early Miocene, and a relatively small volume of Antarc-tic ice (planktonic foraminiferal Zone N4B, ~22 Ma);

2) a late early Miocene interval characterized by warm bot-tom waters and minimal Antarctic ice (Zone N8, ~16 Ma);

3) a late Miocene interval characterized by cold bottom waters and a large Antarctic ice sheet (Zone N17, ~8 Ma). An attempt was made to choose time slice intervals for which the benthic oxygen isotopic record suggested relatively little variabil-ity on a time scale of about one million years. This was done to minimize the effect of small errors in stratigraphic correlation on the accuracy of the synoptic maps to be constructed. A further consideration in the choice of the time slice intervals was that the

$\delta^{18}O$

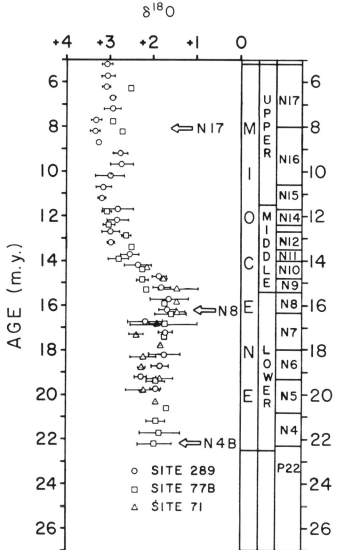

Figure 1. $\delta^{18}O$ values of benthic foraminifera from equatorial Pacific DSDP Sites 71, 77B and 289, adjusted for estimated effect of isotopic disequilibrium and averaged over 500,000-year intervals. From Savin et al. (1981). Times chosen for the CENOP time slice intervals in foraminiferal zones N4B, N8, and N17 are indicated.

TABLE 1. CENOP TIME SLICE SITES FOR
PLANKTONIC FORAMINIFERAL ISOTOPIC STUDY

| | Time Slice | | | Latitude | Longitude | Water |
	N17	N8	N4B			depth
Atlantic Ocean						
SITE 14			X	28 19.89'S	20 56.46'W	4346 m
SITE 15		X		30 53.38'S	17 58.99'W	3938 m
SITE 16	X			30 20.15'S	15 42.79'W	3526 m
SITE 18			X	27 58.72'S	08 00.70'W	4018 m
SITE 357			X	30 00.25'S	35 33.59'W	2086 m
SITE 360	X	X	X	35 50.75'S	18 05.79'E	2949 m
SITE 362	X	X	X	19 45.45'S	10 31.96'E	1325 m
SITE 366A		X	X	05 40.7'N	19 51.1'W	2853 m
SITE 369A		X		26 35.5'N	14 59'W	1752 m
SITE 391A		X		28 13.61'N	175 37.00'W	4974 m
SITE 398D		X		40 57.6'N	10 43.1'W	3910 m
SITE 407			X	63 56.32'N	30 34.56'W	2472 m
SITE 408		X		63 56.32'N	10 43.1'W	3910 m
SITE 516	X	X		30 16.59'S	35 17.11'W	1313 m
SITE 526A	X	X	X	30 7.4'S	3 8.3'E	1054 m
Pacific Ocean						
RC12-418	X			38 6'N	170 1.2'E	
SITE 55		X	X	9 18.11'N	142 32.1'E	2850 m
SITE 62.1	X			1 52.2'N	141 56.3'E	2591 m
SITE 71		X	X	4 28.28'N	140 18.91'W	4419 m
SITE 77B	X	X	X	0 28.90'N	133 13.70'W	4291 m
SITE 158	X			6 37.36'N	85 14.16'W	1953 m
SITE 173	X			39 57.71'N	125 27.12'W	2927 m
SITE 206C	X		X	32 00.75'S	165 27.15'E	3196 m
SITE 207A	X			36 57.75'S	165 26.06'E	1389 m
SITE 208	X		X	26 06.61'S	161 13.27'E	1545 m
SITE 279		X		51 20.14'S	162 38.10'E	3341 m
SITE 281	X	X		47 59.84'S	147 45.85'E	1591 m
SITE 289	X	X	X	00 29.92'S	158 30.69'E	2206 m
SITE 292	X	X	X	15 39.11'N	124 39.05'E	2943 m
SITE 296	X	X	X	29 20.41'N	133 31.52'E	2920 m
SITE 310	X			36 52.11'N	176 54.09'E	3516 m
SITE 317B	X	X		12 00.09'S	162 15.78'W	2598 m
SITE 319	X	X		13 01.04'S	101 31.46'W	4296 m
SITE 448		X	X	16 20.46'N	134 52.45'E	3483 m
SITE 470	X			28 54.56'N	117 31.11'W	3549 m
SITE 495		X	X	12 29.78'N	91 02.26'W	4140 m
Indian Ocean						
SITE 214	X	X	X	11 20.21'S	88 43.08'E	1671 m
SITE 237	X	X		7 49.99'S	58 07.48'E	1623 m
SITE 238	X	X		11 09.21'S	70 31.56'E	2832 m

species were analyzed. In most cases, isotopic analyses were made using between 0.3 and 0.8 mg of $CaCO_3$. Mean values of the isotopic compositions of each planktonic species at each site for each time slice are tabulated in Appendix I and the complete planktonic data set, except for South Atlantic sites, is presented in Appendix II (on microfiche). Data for the South Atlantic sites are included in Hodell and Kennett (this volume).

Figure 3 is a graph of $\delta^{18}O$ and $\delta^{13}C$ values of four planktonic species from the N17 time slice at Site 296 plotted against subbottom depth. The quality of the data in this figure is typical of the majority of time slice data sets. We consider it to be a data set of good quality, in that its applicability to this study is straightforward. The variability in the $\delta^{18}O$ values of any of the planktonic species, which reflects a combination of analytical errors and real oceanographic variability, is small. (In this case, the standard deviation of the $\delta^{18}O$ value of each species is less than 0.2 per mil.) Furthermore, the relative ranking of ^{18}O-enrichments of the planktonic species is consistent throughout the sequence. In most data sets, as in this one, changes in the ^{18}O-rankings of the species are rare, and involve reversals of relatively small magnitude when present.

intervals be represented in the sediments of a large and geographically well-distributed number of DSDP sites. The DSDP sites chosen for planktonic isotopic time slice study are listed in Table 1 and shown on the map in Figure 2. Each time slice study was based upon isotopic analysis of between 2 and 19 sediment samples per site. The sampled interval was chosen to represent approximately 100,000 years of sedimentation where possible, but in some instances may have been as long as several hundred thousand years (Barron et al., this volume).

For each time slice interval at each site, planktonic and benthic foraminifera were separated for isotopic analysis. Typically, between two and four species of planktonic foraminifera were separated, but occasionally as few as one or as many as eight

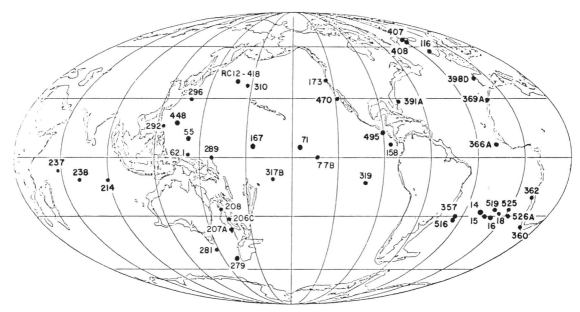

Figure 2. Map of sites from which samples were taken for use in the stable isotopic time slice study.

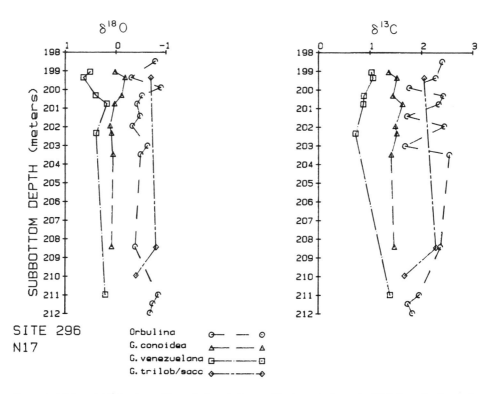

Figure 3. $\delta^{18}O$ and $\delta^{13}C$ values of four planktonic foraminiferal species from the N17 time slice interval of DSDP Site 296. The data are typical of the quality of most time slice data sets.

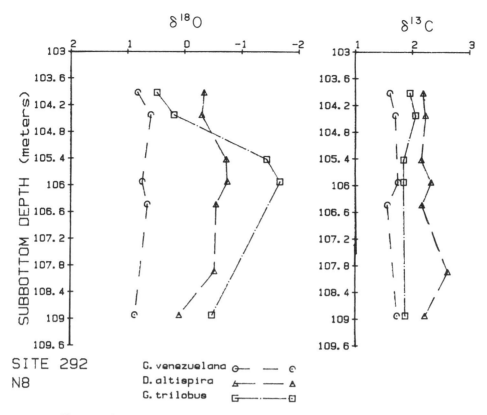

Figure 4. $\delta^{18}O$ and $\delta^{13}C$ values of three planktonic foraminiferal species from the N8 time slice interval of DSDP Site 292. The data are an example of one of the few data sets of quality less than optimum for purposes of this study.

Figure 4 illustrates one of the small number of time slice data sets of less than optimum quality, in this case the N8 data set for DSDP Site 292. We consider it to be of lesser quality because the standard deviations of some $\delta^{18}O$ values are large, the order of ^{18}O-enrichment of planktonic species changes within the sequence, and the magnitudes of reversals in the relative ^{18}O-enrichments are large. That is, the combination of analytical error plus real oceanographic variability decreases the accuracy with which oceanographic conditions at the site can be characterized by the mean $\delta^{18}O$ value of a planktonic foraminiferal species. If oceanographic conditions were, indeed, relatively stable at a locality during the time interval represented by a time slice sample set, and if the isotopic data were representative of conditions during that interval, the order of ^{18}O-enrichment of species, reflecting depth stratification and seasonal succession, should remain constant.

Each planktonic $\delta^{18}O$ value used in a time slice reconstruction is the average of all of the analyses of a species from the appropriate interval at a single site. One measure of the suitability of the mean value as an indicator of conditions is the standard deviation about the mean. Sixty-two percent of the 193 standard deviations are 0.2 per mil or less, 85 percent are 0.3 per mil or less, and 95 percent are 0.4 per mil or less. As sample sizes increase, standard deviations converge on a value of about 0.2 per mil. The observed distribution of standard deviations closely resembles the theoretical distribution of a series of measurements of identical samples with a normally distributed analytical error of 0.2 per mil and a distribution of sample sizes identical to that of the time slice data set.

A Modern Isotopic Time Slice

In this section we demonstrate, through the comparison of modern oceanographic conditions with the $\delta^{18}O$ values of shallow-dwelling Holocene planktonic foraminifera, that: a) when data from large numbers (hundreds) of sites are available, synoptic maps and latitudinal $\delta^{18}O$ gradient plots of the foraminiferal data reflect the major features of surface and near-surface oceanography; and b) a modern time slice which yields a useful, though not detailed, picture of modern marine surface temperature distribution can be generated from a relatively small number of sites. We conclude that analysis of Miocene foraminifera from a similar number of sites permits the reconstruction of Miocene oceanographic conditions with comparable detail.

Synoptic Maps. While isotopic analyses of Recent planktonic foraminifera are unavailable from the vicinity of each of the sites used in the Miocene time slice reconstructions, there are a large number of published isotopic analyses of shallow-dwelling planktonic foraminifera in core-top samples from South Pacific,

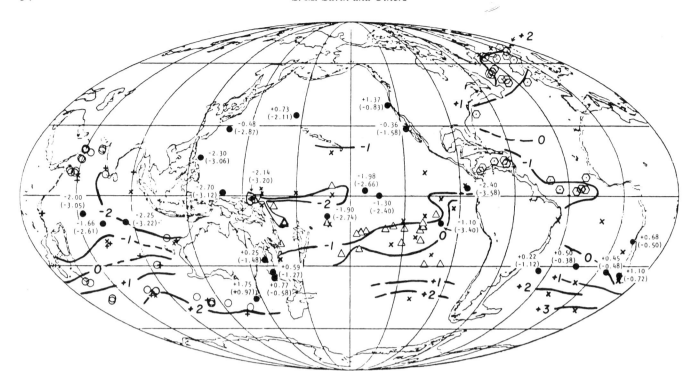

Figure 5. Contours drawn about published $\delta^{18}O$ values of core-top, and presumed Holocene, shallow-dwelling planktonic foraminifera. Data sources are: (Δ) Savin and Douglas, 1973; (X) Shackleton, 1977; (+) Williams, 1977; (O) Curry and Matthews, 1981; (·) Vincent and Shackleton, 1981; (⟨●⟩) Durazzi, 1981. Data are listed in Appendix III. Dark circles are locations of sites used in the N17 time slice reconstruction. Numbers associated with those points are calculated modern equilibrium $\delta^{18}O$ values for winter and, in parentheses, for summer as explained in text. Equilibrium $\delta^{18}O$ values were not used in drawing contours.

Indian and Atlantic Ocean sediments (Savin and Douglas, 1973; Shackleton, 1977; Williams, 1977; Curry and Matthews, 1981; Vincent and Shackleton, 1980; Durazzi, 1981; and others). Data compiled from a number of sources are contoured on the map in Figure 5 and listed in Appendix III (on microfiche). In general, where sample distribution is relatively dense, the $^{18}O/^{16}O$ ratios of shallow-dwelling Holocene planktonic foraminifera provide a reasonable picture of surface temperature distribution and circulation. For example, the expected latitudinal temperature gradients, the steepening of temperature gradients at the subtropical convergence in the South Pacific and South Atlantic, the tongue of cool water extending westward across the equatorial Pacific, and the westward increase in temperature in the equatorial Pacific are all evident from the core-top isotopic data.

Comparison of the Holocene foraminiferal data set with modern oceanographic conditions at each of the Holocene sites is now in progress. Some idea of the accuracy with which Recent oceanographic conditions are reflected in the isotopic ratios of shallow-dwelling foraminifera from core-top samples can be obtained from the comparison between modern conditions at each of the late Miocene (N17) time slice localities and the conditions inferred from the isotopic ratios of the Holocene foraminifera. Using data from the National Oceanographic Data Center (NODC), we have examined winter (minimum) and summer

(maximum) surface temperatures as well as $\delta^{18}O$ values of calcite in isotopic equilibrium with surface waters (referred to below as equilibrium $\delta^{18}O$ values). Calculations of the latter were made using relationships between salinity and surface water $\delta^{18}O$ values derived from the data of Craig and Gordon (1965) and listed in Table 2, and the relationship

$$t\ (°C) = 16.4 - 4.2 * (\delta_c - \delta_w) + 0.13 * (\delta_c - \delta_w)^2$$

TABLE 2. RELATIONSHIPS BETWEEN SALINITY AND $\delta^{18}O$ OF SURFACE WATERS USED IN CALCULATIONS OF EQUILIBRIUM $\delta^{18}O$ VALUES

Ocean	Relationship#
Equatorial Pacific	$\delta^{18}O$ = 0.222 x salinity − 7.50
Pacific E-W transect 13° S	$\delta^{18}O$ = 0.553 x salinity − 19.35
South Pacific	$\delta^{18}O$ = 0.687 x salinity − 23.74
N.E. Pacific	$\delta^{18}O$ = 0.544 x salinity − 18.63
Indian*	$\delta^{18}O$ = 0.481 x salinity − 16.53
South Atlantic	$\delta^{18}O$ = 0.106 x salinity − 3.00

\# $\delta^{18}O$ values of surface water relative to S.M.O.W.
*Relationship given by Williams (1977).

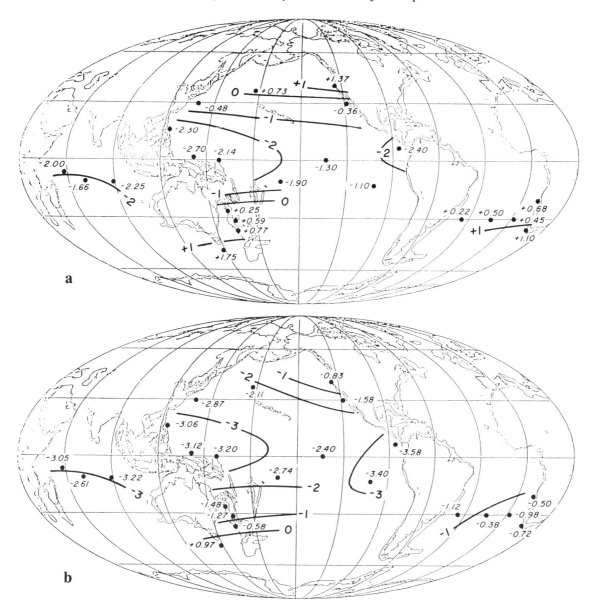

Figure 6. Equilibrium $\delta^{18}O$ values calculated for surface water at each of the sites used in the late Miocene (N17) time slice reconstruction: 6a. shows highest (coldest annual) value at each site; 6b. shows lowest (warmest annual) value.

where δ_c is the isotopic composition of CO_2 liberated from calcite at 25° C and δ_w is the isotopic composition of CO_2 in equilibrium with water at the same temperature (Epstein, unpublished manuscript).

Equilibrium $\delta^{18}O$ values for winter and summer (i.e., most positive and most negative values) at the 23 sites used in the late Miocene time slice study are plotted on maps in Figures 6a and 6b, and are superimposed on the contours drawn through the Holocene data set in Figure 5. In most cases where comparisons can be drawn, Holocene $\delta^{18}O$ values interpolated from the contours fall between the winter and summer equilibrium values as expected, indicating that the Holocene data provide useful information about the Modern oceans.

The modern oceanographic data and equilibrium $\delta^{18}O$ values also provide information about the detail with which global oceanographic conditions can be inferred from 22 to 27 data points, the number of localities examined in the Miocene time slice studies. A map of modern winter (i.e., coldest annual) surface temperature at the 23 sites used in construction of the late Miocene time slice is shown in Figure 7. Relatively gross features of the modern surface conditions can be discerned from the limited amount of data on this map or from the equilibrium $\delta^{18}O$ values plotted in Figures 6a and b. Westward-increasing surface temperatures in the western equatorial Pacific transect are evident, as is the high temperature at Site 158 in the eastern equatorial Pacific. However, the single data point in that region

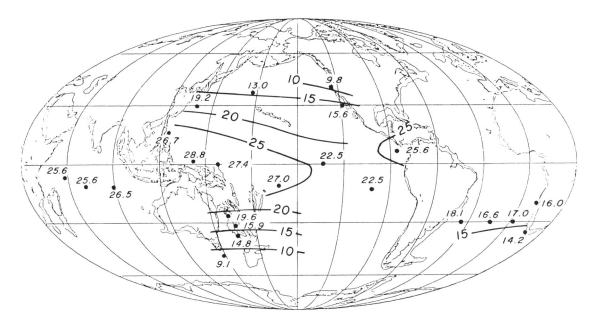

Figure 7. Modern winter (coldest annual) surface temperature at each of the 23 sites used in construction
of the late Miocene (N17) time slice.

gives no indication of the complex and convoluted shape of surface isotherms evident in the more densely sampled Holocene
planktonic data set of Figure 5. From the small number of data
points, it can be discerned, as expected, that Pacific winter
temperatures poleward of 30° latitude are markedly lower than
those at lower latitudes. Winter temperatures in the tropical Indian Ocean are similar to those in the Pacific, while those of
higher southern latitudes (30 to 45°) in the Pacific are similar to
those at comparable latitudes in the Atlantic. Most of the above
generalizations about winter surface temperatures also hold true
for summer surface temperatures (not shown).

Conclusions drawn about modern oceanographic conditions
from 23 surface temperature measurements (Figure 7) are similar
to those that can be drawn from the equilibrium $\delta^{18}O$ data
(Figures 6a and b). Because of the large contour intervals used
(1 per mil or 5°C), the effects of local variations in $\delta^{18}O$ of
sea water have been largely obscured. While relatively crude, significant details of modern oceanography can be discerned. The
resolution of Miocene oceanography from a similar number of
data points should be comparable.

Latitudinal Gradient Diagrams. In Figure 8, all of the
Holocene $\delta^{18}O$ values of shallow-dwelling planktonic foraminifera are plotted as a function of latitude. In Figures 9a and b, the
modern winter and summer equilibrium $\delta^{18}O$ values of all of the
sites used in the N17 time slice reconstruction are plotted in
similar fashion. Superimposed upon the modern equilibrium
values are envelopes about the Holocene data. The overlap between the Holocene data and the winter equilibrium $\delta^{18}O$ data is
almost complete, while many of the summer equilibrium $\delta^{18}O$
values, especially for latitudes north of 10° S, are more negative
(i.e., warmer) than the measured Holocene values. The relationship between the modern equilibrium data for the N17 sites and

the Holocene data reflects the fact that shallow-dwelling planktonic foraminifera live below the sea surface where temperatures
are lower than surface temperatures, and, especially in the case of
subtropical and higher latitudes, growth of foraminiferal species
may also be seasonally biased. Many of the summer equilibrium
$\delta^{18}O$ values at the N17 sites are markedly lower than Holocene
values at similar latitudes. Although a disproportionately large
number of the N17 sites are located near coastal currents, there is
no good correlation between salinity or proximity to coastal regions and the distance a summer equilibrium $\delta^{18}O$ value plots
above the envelope about the Holocene data. Thus it is unlikely
that the depleted values in the late Miocene are due to dilution of
surface sea water at the N17 sites with low-^{18}O fresh water.

These comparisons of modern equilibrium $\delta^{18}O$ values with
Holocene isotopic data demonstrate the usefulness of even relatively few oxygen isotopic data points in defining major regional
temperature differences and latitudinal temperature gradients.
However, detailed reconstructions of surface oceanography require large amounts of isotopic data. Furthermore, because of
factors such as seasonal growth patterns and depth stratification
of foraminiferal species, the details of the relationship between the
isotopic data and the temperature of the sea surface remain
unclear.

Presentation of Data for Miocene Time Slices

In the discussion of the Miocene synoptic reconstructions
that follows, the isotopic data of the late Miocene (N17) time
slice are compared with the contours defined by the Holocene
data set, which is taken to represent the response of shallow-
dwelling foraminifera to modern oceanographic conditions. Differences between the Holocene and the late Miocene reconstruc-

HOLOCENE PLANKTONIC δ¹⁸O

Shallow-dwelling Species

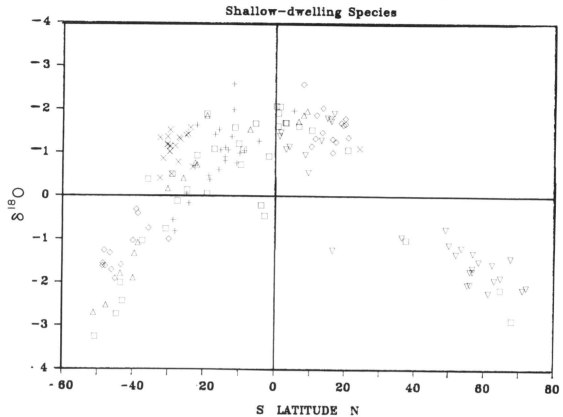

Figure 8. Latitudinal gradient of measured $\delta^{18}O$ values of Holocene shallow-dwelling planktonic foraminifera (See Figure 5). Data sources are: (□) Shackleton, 1977; (+) Savin and Douglas, 1973; (◊) Curry and Matthews, 1981; (△) Williams, 1977; (X) Vincent and Shackleton, 1980; (▽) Durazzi, 1981.

tions are then interpreted in terms of changes in oceanographic conditions between 8 Ma and the present. Similarly, the N8 synoptic reconstruction is compared with the N17 reconstruction and the N4B reconstruction is compared with the N8.

The N17 (8 Ma) Time Slice

Synoptic Map of Planktonic $\delta^{18}O$ Values. For most time slice localities, between two and five species of planktonic fora-minifera were isotopically analyzed. This was done because living planktonic foraminifera are depth-stratified within the water column, and additional data obtained by analysis of multiple species should provide information about the thermal structure of the water column, especially in tropical regions where seasonal temperature variations are small. First we have considered the $\delta^{18}O$ values of those species with the lowest $^{18}O/^{16}O$ ratios in the samples from which they were taken, i.e., those inferred to be shallowest-dwelling, providing the most accurate information about conditions in surface or near-surface waters. The $\delta^{18}O$ values of deeper-dwelling planktonic species, the interpretation of

their depth stratification and the implications for the three-dimensional temperature structure of the water columns will be discussed later. At all but one of the sites from the tropical Pacific and Indian Oceans, either *Globigerinoides sacculifer* or *Dento-globigerina altispira* was the species with the lowest (i.e., warm-est) $\delta^{18}O$ value. At Site 77B, analyses of *Gs. sacculifer* are not available, and *Globigerinoides quadrilobatus* exhibits the lowest $\delta^{18}O$ value of the species analyzed. At higher latitude Pacific sites, the species with the lowest $\delta^{18}O$ value was either *Globiger-ina nepenthes,* or *Orbulina universa.* In the South Atlantic, *Glob-igerinoides* exhibits the lowest $\delta^{18}O$ values.

Late Miocene Surface Waters. Figure 10 is a map show-ing the lowest $\delta^{18}O$ values. The resemblance is clear between the $\delta^{18}O$ values of the late Miocene (N17) shallow-dwelling forami-nifera and those of the Holocene (Figure 5), in most parts of the world where comparable data are available. Contours drawn through the Holocene Indian Ocean and South Atlantic data (Figure 7) are compatible with the late Miocene time slice data in almost every case. Woodruff et al. (1981), concluded that the average $\delta^{18}O$ value of sea water at 8 Ma did not differ from the

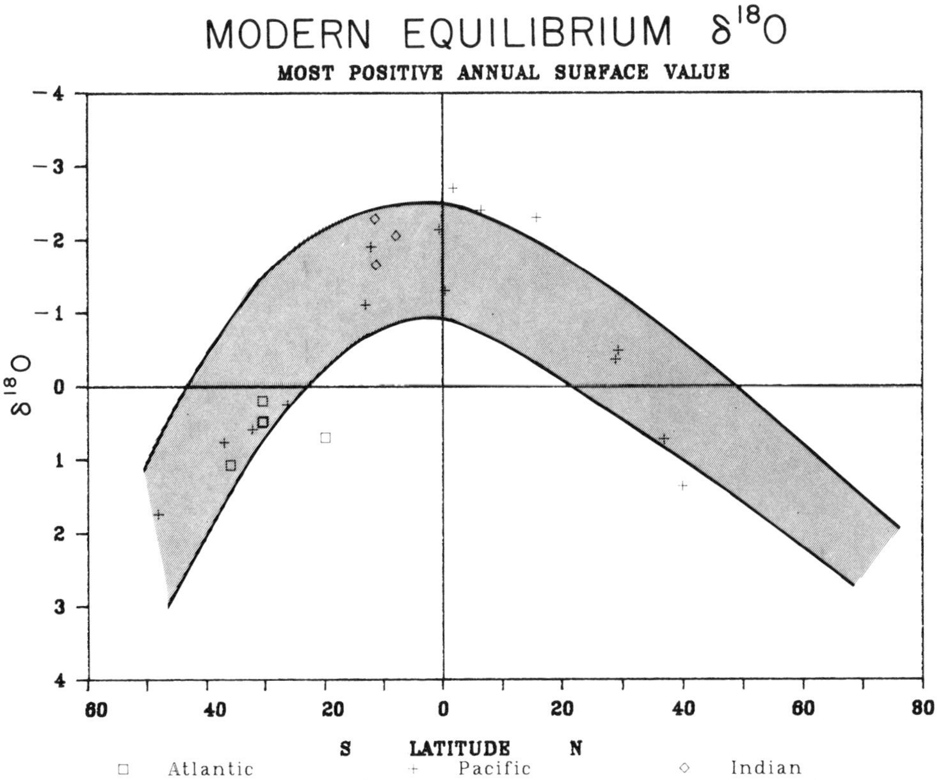

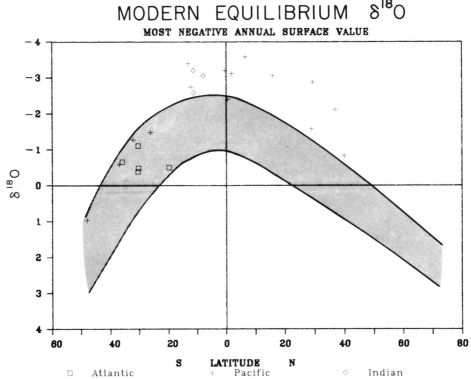

Figure 9a. Latitudinal gradient of highest (coldest annual) calculated modern equilibrium $\delta^{18}O$ values at all of the sites used in the N17 time slice reconstruction. The shaded region is an envelope about the measured Holocene data plotted in Figure 8.

Figure 9b. Latitudinal gradient of lowest (warmest annual) calculated modern equilibrium $\delta^{18}O$ values at all of the sites used in the N17 time slice reconstruction. The shaded region is an envelope about the measured Holocene data plotted in Figure 8.

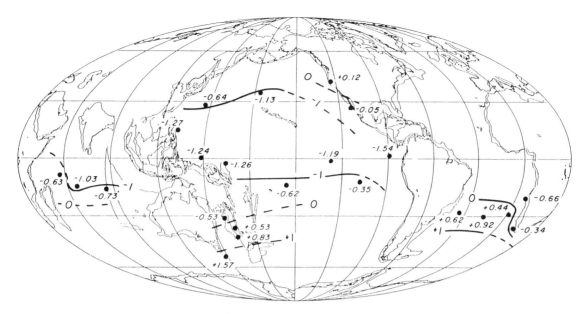

Figure 10. Map of the average $\delta^{18}O$ values of the planktonic foraminiferal species with the most negative (warmest) $\delta^{18}O$ at each of the backtracked N17 time slice sites.

modern value by more than 0.3 per mil. Barring major differences between the regional variations in $\delta^{18}O$ values of modern surface waters and those at 8 Ma, the comparable $\delta^{18}O$ values of the late Miocene (N17) and Holocene foraminifera imply comparable water temperatures. There are no northwest Pacific or southwest Pacific Holocene data sets with which to compare the N17 samples, but the steep latitudinal gradient of late Miocene $\delta^{18}O$ values (2.1 per mil) in the southwest Pacific defined by Sites 208, 206, 207, and 281 is consistent with the modern steep latitudinal temperature gradient associated with the modern subtropical convergence in that region. Only in the western equatorial Pacific is there evidence of a significant difference between late Miocene and Holocene surface temperatures. At Site 289, the N17 $\delta^{18}O$ value of *Gs. sacculifer* is −1.26 per mil while nearby Holocene values are between −1.7 and −2.1 per mil. While there are no comparisons with Holocene data at Sites 62.1 and 292, to the west of Site 289, the $\delta^{18}O$ values of shallow-dwelling N17 planktonic foraminifera are approximately 1 per mil more positive (i.e., cooler) than winter equilibrium $\delta^{18}O$ values at those sites (Figure 6a), suggesting the possibility of significant warming of the western equatorial Pacific between 8Ma and the present.

The $\delta^{18}O$ values of shallow-dwelling late Miocene (N17) foraminifera are plotted as a function of backtracked latitude (Sclater et al., this volume) in Figure 11a. Superimposed on that data is the envelope about the Holocene data from Figure 8, offset by 0.3 per mil, the value assumed for the change in the $^{18}O/^{16}O$ ratio of sea water between 8 Ma and the present. Figure 11b is a similar plot, in which modern winter and summer equilibrium $\delta^{18}O$ values, offset by 0.3 per mil, are superimposed on the N17 data for 8 Ma. We conclude that the late Miocene (N17) latitudinal temperature gradient, as inferred from the oxygen isotopic ratios of shallow-dwelling planktonic foraminifera was somewhat shallower than that of the Holocene or of today, primarily because of higher Modern equatorial temperatures.

Given the uncertainties inherent in the interpretation of foraminiferal isotopic data (including some uncertainty in the average $\delta^{18}O$ value of sea water at 8 Ma), details of differences between surface conditions at 8 Ma and those of today cannot be resolved by the relatively small N17 data set.

Depth Stratification of Late Miocene Planktonic Foraminifera. The relative rankings of ^{18}O-enrichments of species of planktonic foraminiferal species from a single sample have frequently been interpreted as reflecting the relative rankings of depth habitats of those species during test growth. Interspecific differences in ^{18}O-enrichments may also be affected by such factors as growth at different seasons, disequilibrium precipitation of calcite (Fairbanks et al., 1980), or encrustation with $CaCO_3$ associated with gametogenesis in water deeper and colder than that in which most chamber growth occurs (Duplessy et al., 1981). In addition, especially in subtropical and higher latitudes, interspecific differences may reflect the seasonal succession of foraminiferal species.

Average $\delta^{18}O$ values of each species of planktonic foraminifera at each late Miocene (N17) time slice site are shown in the histograms of Figure 12. The $\delta^{18}O$ values are a general indication of the temperature of the water inhabited by each species. (The values are, of course, also affected by regional variations in the $^{18}O/^{16}O$ ratio of sea water.) For tropical sites (30°N to 30°S) for which the data on the histogram are unshaded, the relative rankings of species primarily reflect depth stratification. However, when higher latitude samples are considered, indiscriminate comparisons of $\delta^{18}O$ values can be somewhat misleading, since

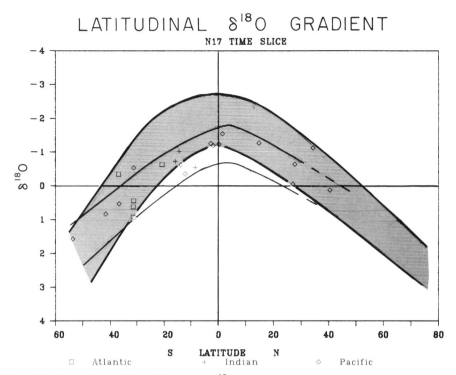

Figure 11a. Latitudinal gradient of the average $\delta^{18}O$ values of the planktonic foraminiferal species with the lowest (warmest) $\delta^{18}O$ value at each of the N17 time slice sites. Shaded region is an envelope about the measured Holocene planktonic foraminiferal $\delta^{18}O$ values of Figure 8, displaced by 0.3 per mil to adjust for the estimated effect of changing ice volume on the mean $\delta^{18}O$ isotopic ratio of the oceans between 8 Ma and the present.

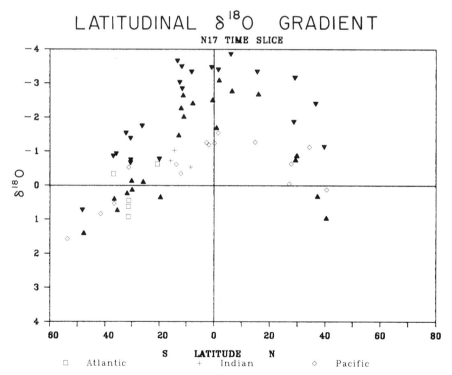

Figure 11b. Latitudinal gradient of the average $\delta^{18}O$ values of the planktonic foraminiferal species with the lowest (warmest) $\delta^{18}O$ value at each of the N17 time slice sites (squares, crosses and diamonds). Filled triangles are calculated equilibrium $\delta^{18}O$ values for modern surface waters at each of the same sites (upward pointing = winter; downward pointing = summer) displaced by 0.3 per mil.

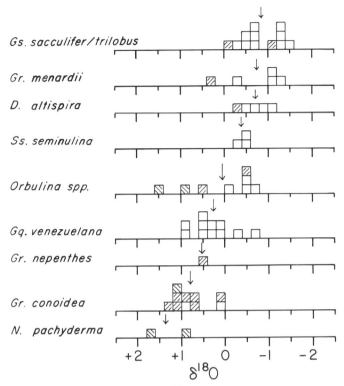

Figure 12. Histogram of average $\delta^{18}O$ values of individual species of planktonic foraminifera from each site in the N17 time slice study. Unshaded squares are data for samples with backtracked latitudes between 30°N and 30°S. (▨) indicates latitudes between 30 and 40° and (◺) indicates latitudes higher than 40°. Arrows indicate mean values of the averages for each species.

calcification may be seasonally biased and species that are deeper dwellers in the tropics may live closer to the surface in the cooler regions.

When the isotopic rankings of species are compared among sites, most species show consistent ranking from one site to another. In the tropics, *Gs. sacculifer/trilobus, D. altispira* and *Globorotalia menardii* consistently have similar low (i.e., warm) $\delta^{18}O$ values, indicative of growth near the surface. Among those, *Gs. sacculifer* usually exhibits the lowest $\delta^{18}O$ value. In tropical regions, *Globoquadrina venezuelana* invariably has the highest $\delta^{18}O$ value, indicating growth deep within the near-surface water column. At higher latitudes in the Pacific and in the South Atlantic, *Globorotalia conoidea* and *Neogloboquadrina pachyderma* consistently have the highest $\delta^{18}O$ values. Of all the species in the N17 samples analyzed, only *Orbulina universa* exhibits a wide range of $\delta^{18}O$ values, ranging from a low value, similar to that of *Gs. sacculifer* at tropical western Pacific Site 292 to high values, only slightly lower than those of *Gr. conoidea* and *N. pachyderma* at high latitude Sites 207 and 281.

Three Dimensional Temperature Structure of the Late Miocene Oceans. When isotopic data are available for several planktonic species in a sample, including both shallow-dwelling and deep-dwelling species, the range of measured $\delta^{18}O$ values of those species should be a minimum for the annual range of equilibrium $\delta^{18}O$ values in the photic zone at the time the sediment was deposited.

Isotopic rankings of planktonic foraminiferal species of the N17 time slice along west-to-east transects in the tropical Pacific and Indian Oceans are shown in Figure 13. The total range of $\delta^{18}O$ values within a sample varies from as low as 0.65 and 0.89

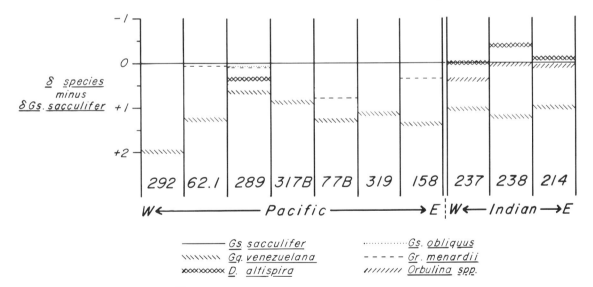

Figure 13. Ranges of $\delta^{18}O$ values of individual planktonic foraminiferal species at N17 time slice sites along west-to-east transects in the tropical Pacific and Indian Oceans. Values plotted are the differences between the $\delta^{18}O$ value of each species and that of *Gs. sacculifer* at the same site. (At Site 77B *Gs. sacculifer* was not analyzed and *Gs. quadrilobatus* was used instead.) Most values were obtained by averaging the differences in $\delta^{18}O$ values of foraminiferal species level-by-level (Appendix II) within a time slice sequence and may differ slightly from differences between mean values. Equilibrium $\delta^{18}O$ values were calculated as described in the text.

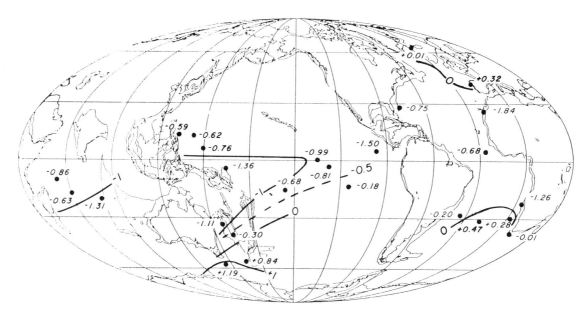

Figure 14. Map of the average $\delta^{18}O$ values of the planktonic foraminiferal species with the most negative (warmest) $\delta^{18}O$ at each of the backtracked N8 time slice sites.

per mil at Sites 289 and 317B to 2.0 per mil at Site 292. At the remaining sites, the $\delta^{18}O$ ranges from 1.13 to 1.50 per mil. Attempts to compare the temperature structure of the tropical N17 ocean with that of today by correlating the measured N17 isotopic ranges with equilibrium $\delta^{18}O$ values in the photic zone (i.e., the upper 90 or 120 m) were inconclusive. However, it is noteworthy that the largest range of N17 $\delta^{18}O$ values was found at Site 292 where the smallest range of equilibrium $\delta^{18}O$ values (1.1 per mil between 0 and 90 m and 1.65 per mil between 0 and 120 m) are found. This suggests that at this site the thermal structure of the late Miocene water column may have been different from the Modern. (The likelihood of warming of the surface waters at this site since 8 Ma has already been noted.) We speculate that the position of the Kuroshio Current shifted relative to Site 292, which is now to the east of the western boundary current, and that during late Miocene time this site lay to the west of the boundary current, where the thermal gradient in the upper portion of the water column would have been steeper.

The N8 (16Ma) Time Slice

Synoptic Map of Planktonic $\delta^{18}O$ Values. The geographic coverage of time slice locations for the late early Miocene (N8) synoptic reconstruction is more complete in the Atlantic and less complete in the Pacific than is the coverage for the late Miocene (N17) reconstruction. Figure 14 is a map showing the lowest $\delta^{18}O$ value at each of the localities examined in the study of the N8 time slice. Most features of the distribution of $\delta^{18}O$ values on this map are similar to those of the corresponding map for the late Miocene (N17) time slice reconstruction (Figure 10). However, there are two notable differences in the Pacific data. First, $\delta^{18}O$ values of the westernmost tropical Pacific sites (55,

292, and 448) indicate lower temperatures than in the central and eastern equatorial Pacific during N8. In contrast, the $\delta^{18}O$ values of the N17 planktonic foraminifera from the westernmost tropical Pacific sites are among the warmest in the entire late Miocene synoptic reconstruction. Second, $\delta^{18}O$ values (and probably surface temperatures) of N8 planktonic foraminifera from Site 208, between Australia and New Zealand, are similar to tropical values. In contrast, during N17 the surface temperature at Site 208 was intermediate between those of the tropics and higher latitudes. In addition, the data suggest the existence of an east-west temperature gradient in the South Pacific during the N8 interval which was considerably weaker during N17.

The differences between $\delta^{18}O$ values of shallow-dwelling planktonic foraminifera in the N17 reconstructions and those from the same (or in two cases, nearby) sites in the N8 synoptic reconstructions are compared on the map in Figure 15. The differences have been adjusted by 0.5 per mil to compensate for the change in the $^{18}O/^{16}O$ ratio of sea water inferred to have been caused by the growth of the Antarctic ice sheet during early middle Miocene time (Woodruff et al., 1981). On this map, sites inferred to have warmed between N8 and N17 are indicated by negative values, and sites inferred to have cooled by positive values. Changes smaller than 0.35 per mil can probably be considered insignificant.

Assuming that the ice volume adjustment of 0.5 per mil is correct, and that there have not been any major changes in the regional variation of the $^{18}O/^{16}O$ ratio of surface waters, tropical Pacific surface waters typically warmed by 2 to 5°C between 16 Ma and 8 Ma. An error in the estimate of the ice volume adjustment of 0.2 per mil (40 percent of the value of the adjustment) would cause estimated temperature changes to be in error by only about 1°C. Surface temperatures remained unchanged during the

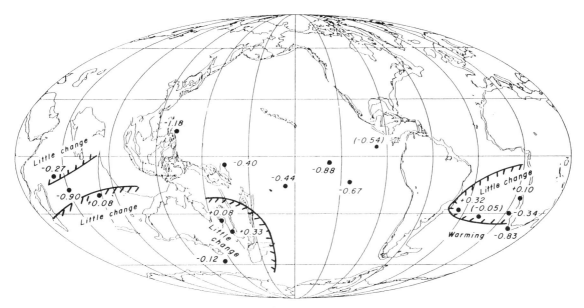

Figure 15. Map of the difference between $\delta^{18}O$ values of shallow-dwelling planktonic foraminiferal species in the N17 time slice and shallow-dwelling species in the N8 time slice. Numbers in parentheses are comparisons made between $\delta^{18}O$ values of different but nearby N17 and N8 sites. Others are comparisons made at a single site. Values have been adjusted by 0.5 per mil, the estimated change in the $\delta^{18}O$ value of sea water between 16 Ma and 8 Ma. Negative numbers imply warming surface waters and positive numbers imply cooling. Only differences greater than 0.35 per mil are considered significant.

N8 to N17 interval at the four southwestern Pacific localities. The most southerly of the South Atlantic sites (Site 360) appears to have warmed, while temperatures at the other South Atlantic sites remained essentially unchanged. There are no comparisons on Figure 15 for sites north of 30°N latitude because the N17 synoptic reconstruction included no Atlantic sites, and the N8 reconstruction included no Pacific sites in that region. Latitudinal gradients in $\delta^{18}O$ values of shallow-dwelling N8 planktonic foraminifera are shown in Figure 16, upon which is superimposed an envelope drawn about the N17 data adjusted by 0.5 per mil. The gradients drawn from the limited number of sites suggest that the late early Miocene latitudinal $\delta^{18}O$ gradient, and the temperature gradient inferred from it, were more gentle than those of the late Miocene interval. This reflects a warming of the tropics while high latitude temperatures changed little. It should be noted that while estimates of the magnitude of temperature changes are dependent upon the value assumed for the ice volume adjustment, the shapes of the latitudinal temperature gradients are not.

Depth Stratification of Late Early Miocene Planktonic Foraminifera. Average $\delta^{18}O$ values of each species of planktonic foraminifera analyzed at each of the N8 time slice localities are shown in the histograms in Figure 17. Values for samples with backtracked latitudes between 30°N and 30°S are shown as unshaded squares. The tropical species, *Gs. sacculifer/trilobus, D. altispira, Globigerinoides subquadratus, Globorotalia siakensis* and *Globorotalia peripheroronda,* have similar ranges of $\delta^{18}O$ values. This is true both when all sites are considered together and when species from individual sites are compared (Figure 18). We

conclude from the $^{18}O/^{16}O$ ratios that all of these species calcify in the upper portion of the water column and have $\delta^{18}O$ values indicative of surface or near-surface conditions.

As in the late Miocene samples, *Gq. venezuelana* is consistently the most ^{18}O-enriched of the tropical late early Miocene planktonic foraminiferal species, and indicative of a deep-water habitat. In samples from which it was analyzed, primarily at higher latitudes, *Globoquadrina dehiscens* consistently yielded high (i.e., cold) $\delta^{18}O$ values.

Synoptic Map of Planktonic $\delta^{18}O$ Values. There are reasons (discussed below) to believe that the isotopic ratios of some of the N4B samples may have been affected by diagenetic alteration of the carbonate, and therefore the N4B reconstruction is subject to greater uncertainty than either of the other two Miocene time slice reconstructions. $\delta^{18}O$ values of the N4B planktonic foraminiferal species with the lowest $^{18}O/^{16}O$ ratios at each time slice location are shown in Figure 19. The pattern of $\delta^{18}O$ values for the Pacific Ocean is generally similar to that of N8, except that in the N8 time slice equatorial $\delta^{18}O$ values were somewhat more positive (cooler) in the west than in the east, whereas in the N4B time slice they are slightly more positive in the east. In the South Atlantic, while $\delta^{18}O$ values of all of the N8 samples are similar, there are significant north-south and east-west gradients in the $\delta^{18}O$ values of the N4B samples. The South Atlantic samples are discussed in more detail by Hodell and Kennett (this volume).

Differences at individual sites between the $\delta^{18}O$ values of late early Miocene (N8) and early early Miocene (N4B) shallow-

LATITUDINAL δ¹⁸O GRADIENT

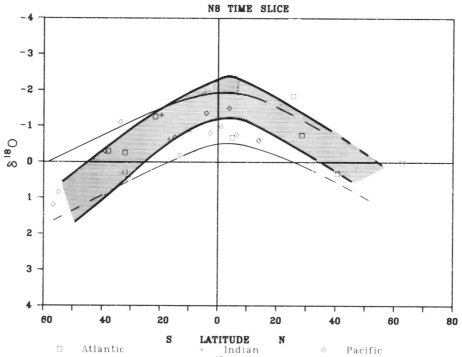

Figure 16. Latitudinal gradient of the average $\delta^{18}O$ values of the planktonic foraminiferal species with the most negative (warmest) $\delta^{18}O$ value at each of the N8 time slice sites. Shading indicates an envelope through the $\delta^{18}O$ values of the shallow-dwelling planktonic foraminifera of the N17 time slice (Figure 12) adjusted by 0.5 per mil to compensate for the estimated change in the average $\delta^{18}O$ value of sea water between 16 and 8 Ma.

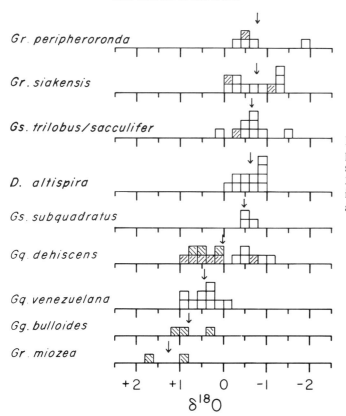

Figure 17. Histogram of average $\delta^{18}O$ values of individual species of planktonic foraminifera from each site in the N8 time slice study. Unshaded squares are data for samples with backtracked latitudes between 30°N and 30°S. (▨) indicates latitudes between 30 and 40° and (▨) indicates latitudes higher than 40°. Arrows indicate mean values of the averages for each species.

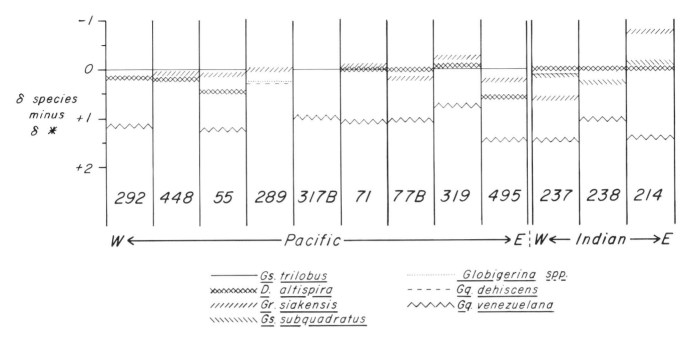

Figure 18. Ranges of $\delta^{18}O$ values of planktonic foraminiferal species at individual N8 time slice sites along west-to-east transects in the tropical Pacific and Indian Oceans. The vertical axis is the difference between the $\delta^{18}O$ value of the species in each sample with a low (warm) $^{18}O/^{16}O$ ratio, typically *Gs. trilobus, Gr. siakensis* or *D. altispira,* and the $\delta^{18}O$ value of each other species in the sample. Values plotted were obtained as described in caption of Figure 14.

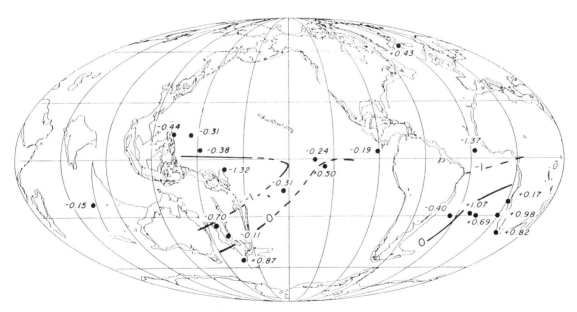

Figure 19. Map of the average $\delta^{18}O$ values of the planktonic foraminiferal species with the most negative (warmest) $\delta^{18}O$ at each of the backtracked N4B time slice sites.

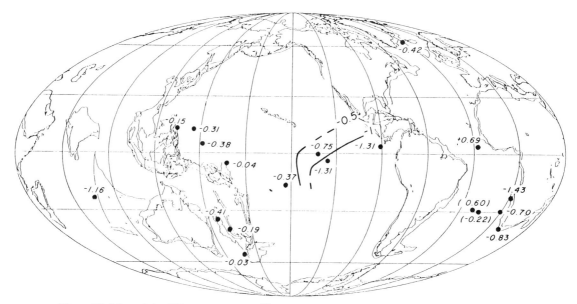

Figure 20. Map of the difference between $\delta^{18}O$ values of shallow-dwelling planktonic foraminiferal species in the N8 time slice and shallow-dwelling species in the N4B time slice. Negative numbers imply warming surface waters and positive numbers imply cooling. Only differences greater than 0.35 per mil are considered significant.

dwelling foraminifera are shown in Figure 20. On this map a negative value at a site corresponds to warming between N4B and N8 and a positive value corresponds to cooling. No adjustment was made for change in the $^{18}O/^{16}O$ ratio of sea water between N4B and N8, reflecting the conclusion of Woodruff et al. (1981) that changes in the volume of continental ice during this time period were not sufficient to cause significant changes in the isotopic composition of the oceans. However, it is conceivable that during N4B remnants of a temporary Oligocene Antarctic ice sheet were sufficiently large to affect the average $\delta^{18}O$ value of sea water, and if deglaciation did occur between 22 Ma and 16 Ma, surface waters would have warmed less or cooled more between N4B and N8 than inferred from the $\delta^{18}O$ values plotted in Figure 20.

The comparison of the two early Miocene time slices in Figure 20 shows that little temperature change (i.e., a change in $\delta^{18}O$ of less than 0.35 per mil) occurred at the sites along the western margin of the Pacific Ocean. In contrast, eastern and central Pacific surface or near-surface waters warmed significantly, as much as 4 to 6°C during that time. Waters at the one Indian Ocean time slice site and at all but two of the Atlantic sites also warmed significantly, in most cases between about 3 and 5°C from 22 Ma to 16 Ma. The N4B data do not clearly define a latitudinal $\delta^{18}O$ or temperature gradient (Figure 21).

Depth Stratification of Early Early Miocene Planktonic Foraminifera. A histogram of $\delta^{18}O$ values of N4B planktonic foraminifera is shown in Figure 22. Data for tropical sites are shown as unshaded squares. Species common to the N17 and N8 intervals apparently occupied similar depth habitats in N4B based upon the $\delta^{18}O$ depth rankings. When all sites are considered as a

group, as on the histograms, tropical species *Globorotalia kugleri, Gr. siakensis, Globigerina angustiumbilicata, and Gs. trilobus* have $\delta^{18}O$ values indicating calcification in surface or near-surface waters. When comparisons are made among different species from individual samples (Figure 23) *Gr. kugleri* invariably has a $\delta^{18}O$ value lower (i.e., warmer) than that of *Gr. siakensis* by 0.2 or 0.3 per mil and lower than that of *Gs. trilobus* by 0.3 to 0.7 per mil. Thus, while analyses of planktonic foraminifera from the N8 and N17 time slices indicated that *Gs. sacculifer/trilobus* consistently had a $\delta^{18}O$ value as low as, or lower than, any other species in a sample, this is not the case for the N4B samples.

Either *Gr. kugleri* (and perhaps also *Gr. siakensis*) secreted calcium carbonate out of isotopic equilibrium with sea water at 22 Ma or *Gs. trilobus* secreted its test at shallow depths at 22 Ma compared with 16 Ma. (The isotopic systematics of *Gr. siakensis* have been discussed in more detail by Barrera et al., this volume.) At tropical Pacific Sites 292 and 55, *Gs. trilobus* and *Gq. venezuelana* have similar $\delta^{18}O$ values, suggesting that early early Miocene *Gs. trilobus* was a deeper-dwelling species than late early Miocene or younger *Gs. trilobus*. However, there is less consistency in the N4B data than in the N8 and N17 data, making such inferences less certain. For example, at Site 71, *Gq. venezuelana, Gr. siakensis* and *Gg. angustiumbilicata* all have similar $\delta^{18}O$ values, suggesting that *Gq. venezuelana* calcified near the surface. Yet, at Site 317B, the $\delta^{18}O$ value of *Gq. venezuelana* strongly suggests a deeper-dwelling species. In the Site 77B time series data, the difference between the N4B $\delta^{18}O$ values of *Gg. venezuelana* and *Cibicidoides* spp. is extremely small, perhaps indicating diagenetic alteration of the older Miocene samples at that site. Although all samples have been examined superficially

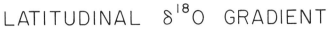

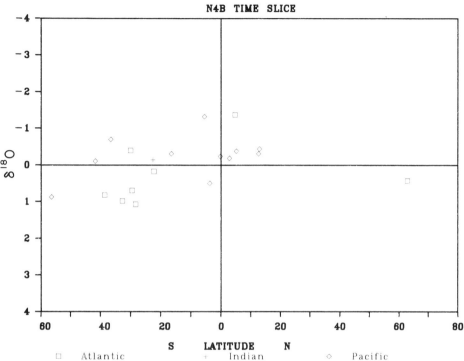

Figure 21. Latitudinal gradient of the average $\delta^{18}O$ values of the planktonic foraminiferal species with the most negative (warmest) $\delta^{18}O$ value at each of the N4B time slice sites.

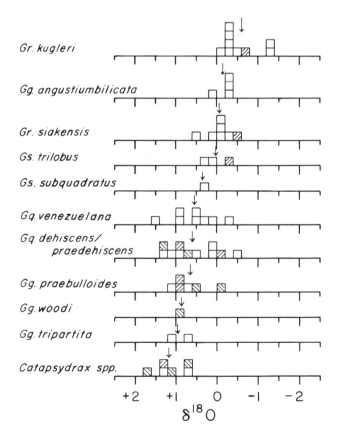

Figure 22. Histogram of average $\delta^{18}O$ values of individual species of planktonic foraminifera from each site in the N4B time slice study. Unshaded squares are data for samples with backtracked latitudes between 30°N and 30°S. (▨) indicates latitudes between 30 and 40° and (◺) indicates latitudes higher than 40°. Arrows indicate mean value of the averages for each species.

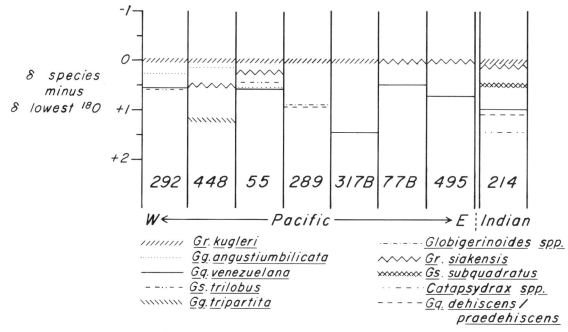

Figure 23. Ranges of $\delta^{18}O$ values of planktonic foraminiferal species at individual N4B time slice sites along a west-to-east transect in the tropical Pacific and at one site in the Indian Ocean. The vertical axis is the difference between the $\delta^{18}O$ value of the species in each sample with the lowest (warmest) $^{18}O/^{16}O$ ratio and the $\delta^{18}O$ value of each other species in the sample. Values plotted were obtained as described in caption of Figure 14.

to check for the effects of diagenesis, the extent to which diagenesis may have modified original isotopic compositions has not been examined in detail, and diagenetic alteration is more likely to have affected samples from the older, deeper portions of the sedimentary sections.

At higher latitudes, $\delta^{18}O$ values indicate that *Gg. praebulloides* was most commonly the shallowest-dwelling N4B planktonic species while either *Gq. dehiscens* or *Catapsydrax* spp. was the deepest.

TIME SERIES STUDIES

Relevant published Miocene planktonic foraminiferal isotopic time series studies from a variety of locations are listed in Table 3. Additional new time series data for planktonic foraminifera from Sites 77B and 289 are included in this paper and tabulated in Appendix IV (on microfiche).

Planktonic foraminiferal oxygen isotopic data for 19 sites are plotted in Figure 24. Where available, benthic data are also plotted for reference. Stratigraphic age assignments used in plotting the data are based on the core descriptions in the Initial Reports of DSDP, or in the case of sites which were also used in the CENOP time slice reconstructions, on the biostratigraphy of Barron et al. (this volume). While many of the time series curves span only a small portion of the Miocene epoch, the data add considerably to our understanding of the evolution of the Miocene oceans.

Where the appropriate portion of the middle Miocene sec-

TABLE 3. MIOCENE PLANKTONIC FORAMINIFERAL ISOTOPIC
TIME SERIES DATA AND SOURCES

Site	Sources
55	Douglas and Savin (1971)
77B	This paper
116	Rabussier-Lointier (1980); Blanc and Duplessy (1982)
158	Keigwin (1979)
167	Douglas and Savin (1973); Rabussier-Lointier (1980)
173	Barrera et al.
208	Loutit et al. (1983)
237	Rabussier-Lointier (1980)
238	Vincent et al. (1980)
281	Loutit (1981)
289	Shackleton (1982); This paper
310	Keigwin (1979)
357	Boersma and Shackleton (1977)
354	Boilzi (1983)
366	Rabussier-Lointier (1980)
470	Barrera et al. (this volume)
495	Barrera et al. (this volume)
519	McKenzie et al. (1984)
525	Shackleton et al. (1984)

tion was analyzed, the benthic foraminiferal time series curves show a clearly defined enrichment in $\delta^{18}O$ values between approximately 17 and 15 Ma, which Savin et al. (1975), Shackleton and Kennett (1975), Woodruff et al. (1981) and Savin et al. (1981) have interpreted as reflecting a combination of the cooling of bottom waters and the growth of the Antarctic ice sheet, and the concomitant increase in the $^{18}O/^{16}O$ ratio of sea water.

The late early Miocene interval, just prior to the early middle Miocene cooling of bottom waters, is represented in Figure 24 by data from a number of sites in tropical and high latitudes of the South Pacific. The planktonic foraminiferal $\delta^{18}O$ values indicate that, barring a significant decrease in mean oceanic $\delta^{18}O$ due to

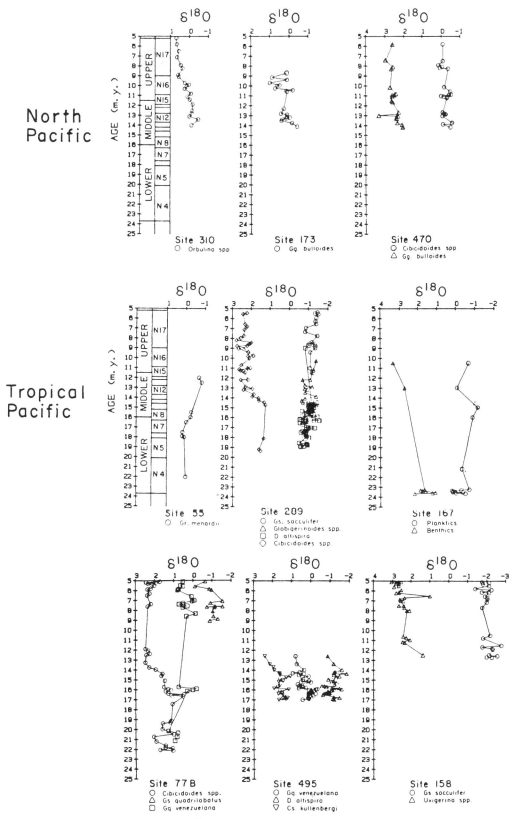

Figure 24 (this and following pages). $\delta^{18}O$ values of planktonic and benthic foraminifera from several Atlantic, Pacific and Indian Ocean sites plotted as a function of sample age. Data sources are listed in Table 3. Age assignments are discussed in the text.

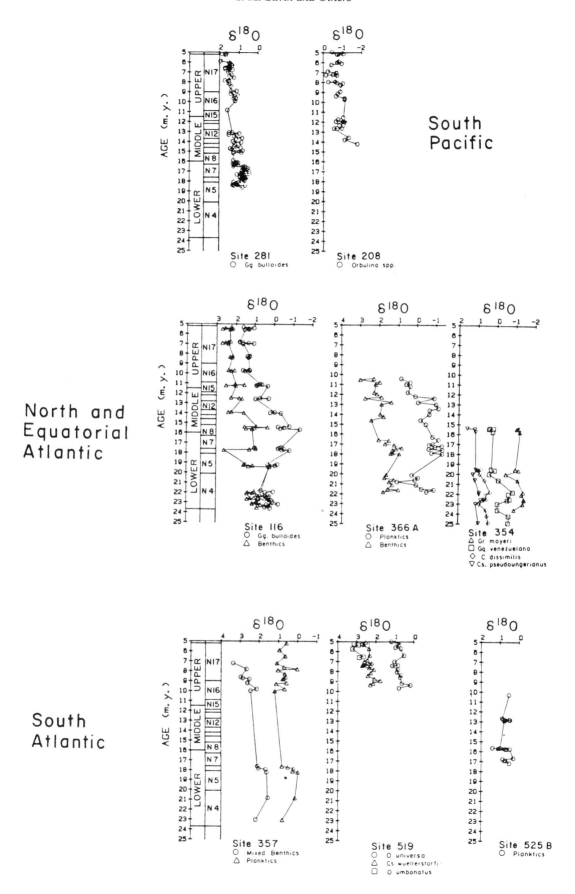

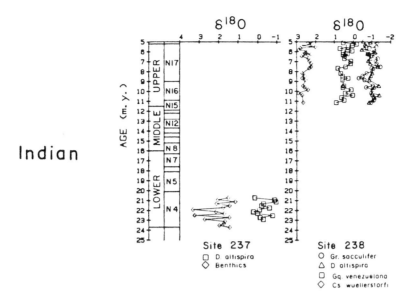

Indian

Site 237
□ D. altispira
◇ Benthics

Site 238
○ Gr. sacculifer
△ D. altispira
□ Gq. venezuelana
◇ Cs. wuellerstorfi

deglaciation, surface waters warmed over a wide range of latitudes in the Pacific (0.4 per mil between 18 and 15.5 Ma at Site 55; 0.4 per mil between 18.5 and 16.5 Ma at Site 281; 0.3 per mil between 16.5 and 14.5 Ma and 0.4 per mil between 19 and 14.5 Ma at Site 289; and 0.35 per mil between 16.8 and 14.5 Ma at Site 495).

When considered over a longer time interval, regional differences in the evolution of Pacific middle Miocene surface temperatures become apparent. The middle Miocene appears to have been a time of warming of surface waters at western tropical Pacific Sites 55, 167, and 289, and a time of cooling at high latitude Pacific Site 281. (Note that either a decrease or no change in the $\delta^{18}O$ values of a shallow-dwelling species during middle Miocene time would correspond to surface water warming because of the ice volume related change in $\delta^{18}O$ of sea water during that interval.) The latter half of the middle Miocene was a time of warming at Site 470 in the eastern Pacific (29°N) and a time of cooling at Site 310 (36°N) in the central Pacific. During the late Miocene, there was either warming or little temperature change at low latitude Pacific Sites 158, 289, and 470 and Indian Ocean Site 238, and cooling at higher latitude Sites 208, 281 and 310. Regional differences also exist in the evolution of Atlantic Ocean surface temperatures during the Miocene. Both at Site 116 in the North Atlantic and Site 366 in the eastern equatorial Atlantic, late early Miocene surface waters were warmer than those of the early early Miocene, and waters subsequently underwent considerable cooling throughout the middle Miocene. At Site 116 the cooling continued through most of the Miocene. The record is less well-defined in the South Atlantic, but the available data for Site 525B at approximately 30°S suggest little temperature change or a slight warming between 17 and 10 Ma.

DISCUSSION AND CONCLUSIONS

The time series and time slice studies described above pro-

vide a general picture of the evolution of the surface and near-surface temperature structure of the Miocene oceans. Pacific early Miocene temperature distribution patterns differed from those of today chiefly in the existence of a shallower early Miocene latitudinal temperature gradient and a marked east-to-west temperature gradient in the southwestern and south-central tropical Pacific. At most localities temperatures showed little change or increased throughout the early Miocene, with greater warming occurring in the eastern equatorial region than elsewhere in the Pacific, resulting in a lessening of the above-mentioned east-to-west gradient. Information from a single locality suggests that the tropical Indian Ocean warmed as well.

Just prior to the early middle Miocene cooling of deep waters, surface temperatures increased at all sites for which data are available. Subsequently, regional differences in the evolution of surface temperatures became pronounced. In the Pacific Ocean, surface waters at most low-latitude sites warmed while those at higher north and south latitude sites cooled or underwent little change. By 8 Ma the east-to-west temperature gradient in the southwest and south-central Pacific had largely disappeared. Pacific surface temperatures were similar to those of today except that tropical waters were cooler at 8 Ma.

At one site in the North Atlantic (Site 116) surface waters cooled significantly throughout the middle and late Miocene, as they did in the late middle and early late Miocene in the equatorial Atlantic (Site 366). At most South Atlantic sites there was little temperature change, although surface waters apparently warmed off the southern tip of South Africa.

Kennett et al. (this volume) have examined the biogeographic distribution of planktonic foraminifera in Pacific Ocean sediments during the Miocene. Their conclusions about the development of oceanographic conditions between the late early Miocene and late Miocene intervals are largely consistent with those drawn in this paper from the isotopic data. Specifically, it was concluded from the isotopic data that there was an east-to-west temperature gradient in the surface waters of the tropical Pacific during the early Miocene, and that this gradient had become markedly lessened by the late Miocene. Kennett et al. noted an east-to-west provinciality in the South Pacific early Miocene (N8 and N4B) fauna which had essentially disappeared by the late Miocene (N17).

On the basis of a limited number of isotopic analyses of planktonic and benthic foraminifera from the tropical Pacific, Savin et al. (1975) pointed out that the middle Miocene cooling of deep waters was accompanied by a warming of tropical surface waters. They concluded that this reflected a decrease in meridional heat transport. The results of the present study do not lead to a unique explanation of the causes of the marked cooling of early middle Miocene deep waters or the establishment of large Antarctic ice sheets at that time. However, the isotopic data in this paper do provide a framework against which theories of the causes of these events can be tested.

ACKNOWLEDGMENTS

We are grateful to the Deep Sea Drilling Project for providing the large number of samples used in this study and to the National Science Foundation for its financial support under the following grants: OCE 79-17017 to SMS. Michael Bender carefully reviewed an earlier version of this manuscript. We also appreciate extremely helpful reviews by Richard Fairbanks, Kenneth Miller and Nicholas Shackleton.

APPENDIX I. SUMMARIES OF ISOTOPIC DATA FOR EACH SPECIES OF PLANKTONIC FORAMINIFERA
FROM EACH SITE FOR EACH TIME SLICE

Site	Taxonomy	$\delta^{18}O$	Std. Dev.	No. of Samp.	$\delta^{13}C$	Std. Dev.	No. of Samp.	Lab.
SUMMARY N17 TIME SLICE								
RC12-418 452-55cm	Gr. conoidea	0.66	0.22	6	1.67	0.16	6	CWRU
DSDP 16 9-1 to 10-5	Gr. conoidea	1.37	0.09	5	1.90	0.09	5	URI
	Globigerinoides spp.	0.92	0.10	5	2.02	0.10	5	URI
DSDP 62.1 23-5 to 24-2	Gs. sacculifer & trilobus	-1.24	0.21	4	1.92	0.22	4	CWRU
	Gr. menardii	-1.19	0.04	3	1.44	0.08	3	CWRU
	Gq. venezuelana	0.01	0.33	6	1.48	0.21	6	CWRU
DSDP 77B 15-4 to 16-4	Gq. venezuelana	0.10	0.24	2	0.74	0.05	2	CWRU
	Gs. quadrilobatus	-1.19	0.11	2	2.18	0.20	2	CWRU
	Gr. menardii	-0.39	0.09	2	0.97	0.03	3	CWRU
DSDP 158 19-6 to 21-1	Gq. venezuelana	-0.23	0.18	7	1.16	0.18	7	URI
	Gr. menardii	-1.25	0.33	8	1.31	0.09	8	URI
	Gs. sacculifer & trilobus	-1.54	0.33	9	1.80	0.22	9	URI
DSDP 173 17CC	Gg. bulloides	0.12		1	0.84		1	CWRU
DSDP 206 21-6 to 24-3	Gg. nepenthes	0.53	0.15	6	1.42	0.22	6	URI
	Gr. conoidea	0.76	0.11	8	1.72	0.29	8	URI
DSDP 207A 6-2 to 7-3	Orbulina spp.	0.83	0.13	6	2.33	0.22	6	URI
	Gr. conoidea	1.17	0.22	7	1.61	0.24	7	URI
	N. pachyderma	0.97	0.05	2	1.38	0.00	2	URI
DSDP 208	Orbulina spp.	-0.53	0.12	2	2.00	0.36	2	URI
	Gr. conoidea	0.06	0.22	2	1.19	0.22	2	URI
	D. altispira	-0.36	0.52	3	2.24	0.38	3	URI
	Gs. sacculifer	-0.20	0.47	3	2.58	0.30	3	URI
DSDP 214 14-1 to 15-2	Gs. sacculifer	-0.57	0.17	17	2.36	0.21	17	SCRIPPS
	D. altispira	-0.68	0.10	12	2.54	0.23	12	SCRIPPS
	O. universa	-0.49	0.23	10	2.09	0.22	10	SCRIPPS
	Gq. venezuelana	0.44	0.13	17	1.51	0.08	17	SCRIPPS
	Ss. seminulina	-0.45	0.05	6	2.23	0.13	6	SCRIPPS
	Gs. conglobatus & obliquus	-0.73	0.11	5	2.45	0.04	5	SCRIPPS
DSDP 237 12-6 to 13-1	Gs. sacculifer	-0.56	0.22	6	1.84	0.39	6	SCRIPPS
	D. altispira	-0.55	0.22	6	2.07	0.09	6	SCRIPPS
	O. universa	-0.18	0.22	6	1.84	0.15	6	SCRIPPS
	Gr. limbata	-0.41	0.21	8	1.16	0.21	8	SCRIPPS
	Gq. venezuelana	0.50	0.13	5	0.95	0.17	5	SCRIPPS
	Ss. seminulina	-0.25	0.27	6	1.77	0.28	6	SCRIPPS
	Gs. conglobatus & obliquus	-0.63	0.10	6	2.09	0.24	6	SCRIPPS

Site	Taxonomy	$\delta^{18}O$	Std. Dev.	No. of Samp.	$\delta^{13}C$	Std. Dev.	No. of Samp.	Lab.
DSDP 238	Gs. sacculifer	-0.75	0.32	8	2.44	0.17	8	SCRIPPS
24-6	D. altispira	-1.03	0.18	6	2.48	0.40	6	SCRIPPS
to	O. universa	-0.70	0.56	6	2.25	0.37	5	SCRIPPS
27-5	Gr. limbata	-0.35	0.27	6	1.35	0.18	6	SCRIPPS
	Gq. venezuelana	0.47	0.15	8	1.26	0.16	8	SCRIPPS
	Ss. seminulina	-0.48	0.33	6	1.85	0.27	6	SCRIPPS
	Gq. dehiscens	0.35		1	2.18		1	SCRIPPS
DSDP 281	Orbulina spp.	1.57	0.11	3	2.22	0.18	3	URI
6-4	N. pachyderma	1.76	0.12	7	1.61	0.15	7	URI
to								
7-4								
DSDP 289	Gq. venezuelana	-0.61	0.08	5	1.38	0.05	5	CWRU
27-6	D. altispira	-0.91	0.08	2	1.89	0.02	2	CWRU
to	Gs. obliquus	-1.20	0.05	2	2.24	0.07	2	CWRU
29-2	Gr. menardii	-1.14	0.14	5	1.37	0.23	5	CWRU
	Gs. sacculifer	-1.26	0.10	6	2.68	0.13	6	CWRU
	Globigerinoides spp.	-1.23		1	2.84		1	CWRU
DSDP 292	Gq. venezuelana	0.83	0.18	4	1.32	0.18	4	CWRU
9-1	Gs. sacculifer &	-1.27	0.23	6	1.55	0.11	6	CWRU
to	trilobus							
9-5								
DSDP 296	Orbulina spp.	-0.58	0.16	13	2.11	0.32	13	CWRU
22-2	Gs. trilobus &	-0.64	0.17	3	2.00	0.25	3	CWRU
to	sacculifer							
23-4	Gr. conoidea	0.01	0.10	8	1.48	0.08	8	CWRU
	Gq. venezuelana	0.39	0.16	6	0.99	0.21	6	CWRU
DSDP 310	Orbulina spp.	0.58	0.16	4	1.51	0.10	4	CWRU
8-4	Gr. menardii	0.27		1	1.36		1	CWRU
to	Gr. menardii &	0.18		1	1.03		1	CWRU
8-6	merotumida							
	Gs. sacculifer &	-1.13		1	1.75		1	CWRU
	trilobus							
DSDP 317B	Gq. venezuelana	0.26	0.09	8	1.58	0.25	8	URI
9-4	Gs. sacculifer &	-0.62	0.28	8	2.61	0.28	8	URI
to	trilobus							
10-4								
DSDP 319	Gq. venezuelana	0.82	0.11	4	1.5	0.07	4	URI
3-1	Gs. sacculifer	-0.35	0.12	3	2.07	0.51	3	URI
to								
3CC								
DSDP 360	Gr. conoidea	1.09	0.37	9	0.98	0.36	9	URI
	Globigerinoides spp.	-0.34	0.24	8	1.36	0.30	8	URI
DSDP 362	Gr. conoidea	0.80	0.12	6	0.88	0.45	6	URI
	Globigerinoides spp.	-0.66	0.41	5	1.63	0.15	5	URI
DSDP 470	Gg. bulloides	-0.08	0.19	4	0.81	0.16	4	CWRU
9-1 to								
9CC								
DSDP 516A	Gr. conoidea	1.05	0.10	5	1.55	0.17	5	URI
13-1 to	Globigerinoides spp.	0.62	0.24	5	1.77	0.20	5	URI
14-3								

APPENDIX I (continued)

Site	Taxonomy	$\delta^{18}O$	Std. Dev.	No. of Samp.	$\delta^{13}C$	Std. Dev.	No. of Samp.	Lab.
DSDP 526A	Gr. conoidea	0.99	0.08	7	1.33	0.13	7	URI
9-1 to	Globigerinoides spp.	0.44	0.13	6	1.95	0.25	6	URI
11-3								

SUMMARY N8 TIME SLICE

Site	Taxonomy	$\delta^{18}O$	Std. Dev.	No. of Samp.	$\delta^{13}C$	Std. Dev.	No. of Samp.	Lab.
DSDP 15	Gq. dehiscens	0.93	0.02	2	2.15	0.13	2	URI
	Globigerinoides spp.	0.47	0.19	3	2.37	0.37	3	URI
DSDP 55	Gq. venezuelana	0.36	0.31	5	1.68	0.17	5	CWRU
8-5	D. altispira	-0.35	0.25	8	2.25	0.20	8	CWRU
to	Gs. trilobus	-0.76	0.31	7	2.10	0.20	7	CWRU
11-1	Gr. peripheroronda & siakensis	-0.64	0.27	5	1.68	0.19	5	CWRU
DSDP 71	Gq. venezuelana	0.22	0.16	10	2.22	0.33	10	CWRU
19-2	D. altispira	-0.91	0.16	8	3.09	0.58	8	CWRU
to	Gr. siakensis	-0.99	0.21	10	2.18	0.15	9	CWRU
22-6	Gs. trilobus	-0.94	0.22	6	2.76	0.07	6	CWRU
DSDP 77B	Gq. venezuelana	0.24	0.35	4	1.97	0.09	4	CWRU
26-2	D. altispira	-0.81	0.16	4	2.78	0.13	4	CWRU
to	Gr. siakensis	-0.62	0.02	2	1.89	0.04	2	CWRU
27-2								
DSDP 206	Gq. dehiscens	0.33	0.12	7	1.70	0.19	7	URI
31-1	Gs. sacculifer	-0.30	0.25	3	2.40	0.04	3	URI
to	Gr. siakensis	-0.14	0.15	7	1.65	0.10	7	URI
32-3	Globigerinoides spp.	-0.25	0.17	3	2.12	0.32	7	URI
DSDP 208	Gr. siakensis	-1.11	0.01	2	1.77	0.23	2	URI
21-4	Gr. peripheroronda	-0.58	0.10	2	1.67	0.13	2	URI
to	Gq. dehiscens	-0.69	0.66	3	1.62	0.39	4	URI
21-6								
DSDP 214	D. altispira	-0.57	0.21	3	2.41	0.12	3	SCRIPPS
20-4	Gq. venezuelana	0.85	0.01	2	1.56	0.10	2	SCRIPPS
to	Gs. subquadratus	-0.69	0.15	2	2.20	0.02	2	SCRIPPS
22-6	Gr. siakensis	-1.31		1	0.51		1	SCRIPPS
DSDP 237	D. altispira	-0.86	0.22	11	2.31	0.47	11	SCRIPPS
18-1	Gq. venezuelana	0.59	0.28	11	1.35	0.35	11	SCRIPPS
to	Gs. subquadratus	-0.56	0.18	4	1.97	0.32	4	SCRIPPS
19-3	Gs. trilobus & sacculifer	-0.54	0.26	3	2.02	0.15	3	SCRIPPS
	Gr. limbata	-0.61		1	0.91		1	SCRIPPS
	Gr. siakensis	-0.26		1	0.69		1	SCRIPPS
DSDP 238	D. altispira	-0.63	0.16	3	2.61	0.45	3	SCRIPPS
38-5	Gq. venezuelana	0.40	0.01	3	1.73	0.06	3	SCRIPPS
to	Gs. subquadratus	-0.51		1	2.52		1	SCRIPPS
41-2	Gq. dehiscens	-0.59		1	2.03		1	SCRIPPS
	Gr. peripheroronda	-0.27		1	1.56		1	SCRIPPS
DSDP 279A	Gq. dehiscens	0.84	0.26	6	2.32	0.13	4	URI
	Gr. miozea	0.88	0.15	6	2.52	0.14	6	URI
	Gg. bulloides	0.93	0.07	5	2.49	0.33	5	URI

APPENDIX I (continued)

Site	Taxonomy	δ¹⁸O	Std. Dev.	No. of Samp.	δ¹³C	Std. Dev.	No. of Samp.	Lab.
DSDP 281	Gr. miozea	1.61	0.08	4	2.01	0.16	4	URI
10-3	Gg. bulloides & praebulloides	1.19	0.14	3	2.05	0.07	3	URI
DSDP 289	Gr. siakensis	-1.36	0.18	6	1.57	0.20	5	URI
51-6	Gq. dehiscens	-1.06	0.28	5	1.98	0.38	5	URI
to	Globigerina spp.	-1.11	0.21	6	2.06	0.17	6	URI
55-2	Globigerinoides spp.	-1.08	0.10	3	2.26	0.23	3	CWRU
	Gs. ruber	-1.11		1	2.45		1	CWRU
DSDP 292	Gq. venezuelana	0.73	0.10	5	1.67	0.07	5	CWRU
12-2	D. altispira	-0.39	0.28	5	2.27	0.08	5	CWRU
to	Gs. trilobus	-0.59	0.85	5	1.92	0.08	5	CWRU
12-5								
DSDP 317B	Gq. venezuelana	0.13	0.24	5	2.14	0.19	5	URI
17-1	Gs. trilobus	-0.68	0.22	4	2.17	0.29	5	URI
to								
18-3								
DSDP 319	Gq. venezuelana	0.90	0.28	5	2.24	0.10	5	CWRU
11-3	D. altispira	-0.17	0.32	4	3.02	0.20	4	CWRU
to	Gs. trilobus	0.04	0.13	8	2.51	0.20	8	CWRU
12-3	Gr. siakensis	-0.18	0.22	10	1.90	0.08	10	CWRU
DSDP 360	Gq. dehiscens	0.58	0.33	9	2.32	0.16	9	URI
22-2	Globigerinoides spp.	-0.01	0.14	7	2.44	0.32	7	URI
to								
22-6								
DSDP 362	Gq. dehiscens	-0.51	0.10	5	2.15	0.13	5	URI
36CC	Globigerinoides spp.	-1.26	0.17	5	2.69	0.17	5	URI
to								
37-2								
DSDP 366A	Globigerinoides spp.	-0.56	0.16	3	2.03	0.14	3	URI
	Gr. peripheroronda	-0.62	0.12	4	1.58	0.26	4	URI
	D. altispira	-0.68	0.17	3	2.45	0.10	3	URI
DSDP 369A	Globigerinoides spp.	-1.84	0.25	4	1.98	0.31	4	URI
	Gq. dehiscens	-0.92	0.41	2	0.92	0.04	2	URI
	Gr. peripheroronda	-1.83	0.13	2	0.66	0.21	2	URI
DSDP 391	Globigerinoides spp.	-0.75	0.00	2	1.79	0.01	2	URI
10-4	Gq. dehiscens	-0.21	0.15	2	1.22	0.34	2	URI
to	Gr. peripheroronda	-0.58	0.01	2	1.07	0.05	2	URI
11-6								
DSDP 398	Globigerinoides spp.	0.32	0.39	6	2.07	0.29	6	URI
	Gq. dehiscens	0.57	0.26	9	1.53	0.22	9	URI
DSDP 408	Globigerinoides spp.	0.01		1	1.14		1	URI
	Gq. dehiscens	0.08	0.36	4	1.23	0.06	4	URI
	Gg. praebulloides	0.26	0.12	6	1.22	0.24	6	URI
DSDP 448	D. altispira	-0.49	0.12	4	2.87	0.05	4	CWRU
2CC	Gs. trilobus	-0.62	0.26	5	2.21	0.13	5	CWRU
to	Gr. siakensis	-0.35		1	2.34		1	CWRU
3-2								

APPENDIX I (continued)

Site	Taxonomy	$\delta^{18}O$	Std. Dev.	No. of Samp.	$\delta^{13}C$	Std. Dev.	No. of Samp.	Lab.
DSDP 495	Gq. venezuelana	-0.05	0.13	18	2.06	0.21	18	CWRU
26-1	D. altispira	-0.98	0.22	17	2.89	0.34	16	CWRU
to	Gr. siakensis	-1.29	0.25	17	1.93	0.23	17	CWRU
27-5	Gs. sacculifer	-1.50	0.36	12	3.02	0.23	12	CWRU
DSDP 516	Gq. dehiscens	0.06	0.11	5	1.55	0.13	5	URI
21-1	Globigerinoides spp.	-0.20	0.06	4	1.80	0.20	4	URI
to								
22-2								
DSDP 526A	Gq. dehiscens	0.74	0.19	7	1.38	0.11	7	URI
21-1	Globigerinoides spp.	0.28	0.18	4	1.99	0.08	4	URI
to								
21-4								

SUMMARY N4 TIME SLICE

Site	Taxonomy	$\delta^{18}O$	Std. Dev.	No. of Samp.	$\delta^{13}C$	Std. Dev.	No. of Samp.	Lab.
DSDP 14	Gq. dehiscens	1.38	0.10	4	1.79	0.09	4	URI
2-1	Gg. praebulloides	1.07	0.06	4	1.66	0.02	4	URI
to								
2-4								
DSDP 18	Gq. dehiscens	1.08	0.18	6	1.80	0.22	6	URI
4-2	Gg. praebulloides	0.69	0.21	4	1.69	0.24	4	URI
to								
5-5								
SITE 55	Gq. venezuelana	0.20	0.25	4	1.98	0.04	4	CWRU
12-2	Gs. trilobus	0.06	0.30	4	1.97	0.20	4	CWRU
to	Gr. kugleri	-0.38		1	1.76		1	CWRU
13-2	Gr. siakensis	-0.15		1	1.77		1	CWRU
	Gr. angustium- bilicata	0.17	0.05	4	1.79	0.17	4	CWRU
SITE 71	Gq. venezuelana	-0.21	0.46	5	1.74	0.15	5	CWRU
32-2	Gr. siakensis	-0.21	0.17	5	1.63	0.13	5	CWRU
to	Gr. angustium-	-0.24	0.14	2	1.82	0.10	2	CWRU
33-6	bilicata							
SITE 77B	Gq. venezuelana	0.99	0.19	4	1.29	0.03	4	CWRU
30-5	Gr. siakensis	0.50	0.05	3	1.32	0.06	3	CWRU
to								
31-6								
DSDP 206	Catapsydrax spp.	0.67	0.17	5	1.27	0.21	5	URI
	Gr. praebulloides	-0.11	0.18	4	0.79	0.02	3	URI
SITE 208	Gq. dehiscens	-0.14	0.21	8	1.22	0.26	8	URI
23-3	Globigerinoides	-0.70	0.20	5	1.71	0.36	5	URI
to	Gr. kugleri	-0.69	0.19	2	2.03	0.11	2	URI
24-4	Gr. siakensis	-0.50	0.14	3	1.11	0.04	3	URI
	Gs. trilobus	-0.29		1	1.23		1	URI
DSDP 214	Gr. siakensis	-0.03	0.08	2	0.65	0.35	2	SCRIPPS
23-1	Gq. venezuelana	0.83	0.10	2	1.06	0.44	2	SCRIPPS
to	Gq. subquadratus	0.34		1	1.52		1	SCRIPPS
23-6	Gr. kugleri	-0.15		1	1.34		1	SCRIPPS
	Gq. dehiscens	0.95		1	1.21		1	SCRIPPS
	Catapsydrax spp.	1.30	0.22	2	1.05	0.05	2	SCRIPPS

APPENDIX I (continued)

Site	Taxonomy	$\delta^{18}O$	Std. Dev.	No. of Samp.	$\delta^{13}C$	Std. Dev.	No. of Samp.	Lab.
SITE 279A	Gq. dehiscens	1.28	0.11	4	1.35	0.14	4	URI
10-2	Gg. woodi	0.87	0.13	4	1.74	0.24	4	URI
to	Catapsydrax spp.	1.79	0.05	4	1.11	0.05	4	URI
11-6								
SITE 289	Gq. dehiscens &	-0.52	0.19	4	1.41	0.14	4	URI
66-2	praedehiscens							
to	Gr. kugleri	-1.32	0.13	2	1.67	0.11	2	URI
69-3	Globigerinoides spp.	-0.42	0.03	3	1.94 ·	0.13	3	URI
SITE 292	Gq. venezuelana	0.16	0.17	4	1.70	0.11	4	CWRU
14-2	Gs. trilobus	0.28	0.05	5	1.72	0.05	5	CWRU
to	Gr. kugleri	-0.44	0.12	6	1.75	0.16	6	CWRU
15-4	Gg. angustium-bilicata	-0.27	0.21	6	1.59	0.16	6	CWRU
SITE 296	Gq. venezuelana	0.52	0.00	2	1.38	0.04	2	CWRU
34-3								
to								
34CC								
SITE 317B	Gq. venezuelana	1.48	0.41	3	2.21	0.19	4	URI
25-1	Gr. kugleri	-0.31	0.16	4	1.93	0.15	4	URI
to	Gg. tripartita	1.15		1	2.00		1	URI
25-6								
DSDP 357	Gq. dehiscens	0.54	0.20	11	1.28	0.34	11	URI
12-1	Gr. kugleri	-0.40	0.30	7	1.80	0.25	7	URI
to								
13.6								
DSDP 360	Gq. dehiscens	0.96	0.18	4	1.79	0.09	4	URI
26-1	Gg. praebulloides	0.82	0.14	3	1.98	0.18	3	URI
to								
26-2								
DSDP 362	Gq. dehiscens	0.17	0.13	10	1.18	0.14	10	URI
39-3	Catapsydrax spp.	0.75	0.14	8	1.13	0.14	8	URI
to								
40-6								
DSDP 366A	Globigerinoides spp.	-0.98	0.16	5	1.84	0.16	5	URI
	Gr. kugleri	-1.37	0.37	4	1.65	0.16	4	URI
to	Gq. praedehiscens	0.17	0.13	3	0.91	0.09	3	URI
DSDP 407	Catapsydrax spp.	1.16	0.15	8	0.67	0.12	8	URI
	Gg. praebulloides	0.43	0.19	8	0.38	0.13	8	URI
to	Gq. dehiscens	0.74	0.10	8	0.53	0.11	8	URI
SITE 448	Gr. siakensis	0.16	0.15	3	1.65	0.19	3	CWRU
6-1	Gg. tripartita	0.78	0.07	3	1.71	0.08	3	CWRU
to	Gr. kugleri	-0.31	0.10	5	1.64	0.05	5	CWRU
8-1	Gg. angustium-bilicata	-0.23	0.14	5	1.58	0.05	5	CWRU
DSDP 495	Gq. venezuelana	0.40	0.33	6	1.14	0.16	6	CWRU
38-1	Gr. siakensis	-0.19	0.36	6	1.01	0.14	6	CWRU
to								
39-4								

APPENDIX I (continued)

Site	Taxonomy	$\delta^{18}O$	Std. Dev.	No. of Samp.	$\delta^{13}C$	Std. Dev.	No. of Samp.	Lab.
DSDP 526A	Catapsydrax spp.	1.38	0.23	8	1.72	0.18	8	URI
27-1	Gg. praebulloides	0.98	0.08	5	1.50	0.21	5	URI
to
29-3

APPENDIX II. ISOTOPIC DATA FOR ALL PLANKTONIC FORAMINIFERAL ANALYSES FOR EACH OF THE THREE TIME SLICES
(See microfiche in pocket inside back cover.)

APPENDIX III. COMPILATION FROM PUBLISHED SOURCES OF OXYGEN ISOTOPIC COMPOSITIONS OF SHALLOW-DWELLING PLANKTONIC FORAMINIFERA OF HOLOCENE AGE
(See microfiche in pocket inside back cover.)

APPENDIX IV. ISOTOPIC TIME SERIES DATA FOR SITES 77B AND 289

SITE 289 TIME SERIES DATA (All data from CWRU except where indicated from URI)

Core/ Section	Depth (cm)	Subbottom Depth (m)	Cibicidoides species δ18O	Cibicidoides species δ13C	Globocassidulina species δ18O	Globocassidulina species δ13C	Globigerinoides acculifer δ18O	Globigerinoides acculifer δ13C	Globigerina species (URI) δ18O	Globigerina species (URI) δ13C	Globigerinoides species δ18O	Globigerinoides species δ13C	Dentoglobigerina altiapira δ18O	Dentoglobigerina altiapira δ13C	Globoquadrina Venezuelana δ18O	Globoquadrina Venezuelana δ13C	Globorotalia menardii δ18O	Globorotalia menardii δ13C	Globigerinoides obliquus δ18O	Globigerinoides obliquus δ13C	Globigerinoides ruber δ18O	Globigerinoides ruber δ13C
17-5	72-80	159.76					-1.47	2.54														
18-2	86-94	163.90					-1.34	2.53														
18-5	76-84	168.30			3.97		-1.51	2.39														
20-5	63-71	187.17				-0.37	-1.40	2.43														
22-4	69-77	204.73					-1.36	1.94														
23-1	86-94	211.44					-1.40	2.16														
24-2	81-89	220.85					-0.87	2.62														
25-2	49-57	230.07					-0.82	2.43														
26-2	22-30	239.26					-1.45	2.79														
27-5	83-91	253.87	1.99	0.76			-1.16	2.78							-0.47	1.44	-1.27	1.00	-1.15	2.17		
27-6	32-34	254.83	2.02	0.59			-1.25	2.43							-0.66	1.33	-1.31	1.22				
28-1	82-86	257.34																				
28-2	52-60	258.56					-1.05	2.68							-0.61	1.43	-0.93	1.63	-1.25	2.30		
28-3	82-86	260.34	2.13	0.79			-1.33	2.82							-0.71	1.41	-1.11	1.46				
28-4	102-104	262.03	2.21	0.81			-1.24		-1.23	2.84			-0.99	1.91								
28-5	82-90	263.36					-1.35	2.71							-0.60	1.31	-1.06	1.56				
29-2	4-8	267.56	2.30	0.81			-1.35	2.65					-0.82	1.87								
30-2	73-80	277.77					-1.09	2.71														
31-5	83-91	291.91					-1.07	2.13			-1.39	2.72										
32-4	83-91	299.87									-1.26	2.21										
33-5	52-60	310.56									-1.25	1.85										
34-5	82-90	320.36									-1.16	1.33										
35-2	76-81	325.29									-1.19	1.96										
36-4	125-132	338.29									-1.09	2.08										
37-5	75-83	348.79									-0.68	2.57										
39-5	62-70	367.66									-0.80	2.31										
40-3	82-90	374.36									-1.19	1.66										
41-2	82-90	382.36									-0.78	2.10										
42-2	86-94	391.90									-0.92	2.29										
42-5	87-93	396.90									-0.81	2.11										
44-2	75-83	410.79									-1.39	2.05									-1.03	2.46
46-2	97-99	429.79									-0.80	2.70									-0.47	3.05
47-1	97-99	437.98																				
47-2	52-60	439.06																				
48-1	97-99	447.48																				
48-2	82-90	448.86									-0.63	2.18										
49-2	82-90	458.36									-0.69	2.28									-0.74	2.47
49-5	89-97	462.93																				
50-2	82-90	467.86							-1.15	1.96	-1.40	1.93										
51-2	90-98	477.44							-1.15	2.53											-1.11	2.45
51-6	96-100	483.48							-0.94	2.28												
52-2	82-90	486.86							-1.26	1.51												
54-2	90-98	505.94																				
56-2	82-90	524.86																				
57-2	112-119	534.65									-0.94	1.74										
59-2	92-100	553.46									-0.82	1.48										
61-6	77-82	582.08	-0.08	0.40																		
66-2	82-90	619.86	0.67	0.51																		
67-6	78-83	635.29	0.53	0.35																		

APPENDIX IV (continued)

SITE 77B TIME SERIES DATA (All data from CWRU)

Core/Section	Depth (cm)	Subbottom Depth (m)	Cibicidoides species δ18O	δ13C	Mixed Benthics δ18O	δ13C	Uvigerina species δ18O	δ13C	Globorotalia plesiotumida δ18O	δ13C	Globigerinoides quadrolobatus δ18O	δ13C	Globigerina venezuelana δ18O	δ13C	Globorotalia menardii δ18O	δ13C	Dentoglogigerina altispira δ18O	δ13C	Globorotalia siakensis δ18O	δ13C
5-4	50-52	50.91							-1.63	1.34	-0.64	1.55	0.52	0.63						
9-2	46-54	84.20	2.45	-0.20																
9-3	51-53	85.72	1.94	-0.19																
9-4	55-59	87.27	2.33	-0.15																
9-5	52-56	88.74	2.01	0.01																
9-5	75-79	92.27																		
10-1	100-107	94.04					3.08	-1.13			-1.11	1.57	0.65	0.76						
10-2	106-110	95.58	2.20	-0.11									0.52	0.34						
10-3	100-107	97.40									-0.86	1.21	0.74	0.94						
10-4	104-108	98.56	2.37	-0.04																
10-6	100-108	100.04									-0.98	1.46	0.79	0.76						
11-1	78-83	101.40											0.08	0.42						
11-2	115-122	110.79																		
12-1	108-112	112.79	2.27	0.07																
12-2	107-111	115.19	2.39	0.32																
12-4	92-98	123.75																		
13-3	50-60	123.96			2.86	-0.46					-1.55	1.64	-0.06	0.63						
13-4	50-58	126.84	1.83	0.19							-0.91	1.96	0.04	0.92						
13-6	79-81	127.10	2.20	0.61									0.52	0.82						
14-1	27-29	128.28											0.73	1.48						
14-1	50-52	128.51											0.49	1.42						
14-1	100-102	129.01											0.54	1.08						
14-1	148-150	129.49											0.54	1.25						
14-2	46-48	129.97									-1.16	1.33	0.29	1.44						
14-3	28-30	131.29									-1.55	2.14	-0.35	0.93						
14-3	77-79	131.78									-0.74	2.50								
14-4	127-129	132.28									-1.13	2.57	0.53	1.28						
14-4	25-27	132.76									-1.04	1.84								
15-2	95-102	133.49																		
15-5	100-107	139.64											-0.14	0.79	-0.29	0.94				
15-6	10-14	143.22									-1.07	2.38			-0.48	1.00				
16-1	40-44	145.02											0.34	0.69						
16-2	102-110	147.26																		
16-6	94-98	148.66									-1.30	1.98	0.18	1.20						
16-6	49-56	151.23																		
16-6	51-58	154.25									-0.89	2.57								
21-6	70-74	197.22																		
26-2	95-99	237.07	1.53	1.40									0.76	1.82			-0.21	0.86		
26-4	92-96	240.04	1.31	1.12									-0.16	1.98			-1.07	2.84		
26-5	92-96	241.54	1.17	1.12									0.04	2.04			-0.81	2.56	-0.63	1.85
26-6	92-96	243.04	1.44	1.44									0.32	2.03			-0.74	2.83	-0.60	1.92
26-6	92-96	246.24	1.31	0.97													-0.63	2.87		
30-5	103-107	278.25											1.30	1.26					0.44	1.25
30-6	103-107	278.95	2.06	0.95									0.62	1.27						
31-2	103-107	282.95											0.86	1.34					0.50	-.40
31-5	103-107	287.35											0.97	1.30					-0.36	1.23
31-6	104-108	288.86	1.94	0.75															0.56	-.32

REFERENCES CITED

Barrera, E., Keller, G., and Savin, S. M., this volume, Evolution of the Miocene ocean in the eastern North Pacific as inferred from oxygen and carbon isotopic ratios of foraminifera.

Barron, J. A., Keller, G., and Dunn, D. A., this volume, A multiple microfossil biochronology for the Miocene.

Biolzi, M., 1983, Stable isotopic study of Oligocene-Miocene sediments from DSDP Site 354, Equatorial Atlantic: Marine Micropaleontology, v. 8, p. 121–139.

Blanc, P.-L., and Duplessy, J.-C., 1982, The deep-water circulation during the Neogene and the impact of the Messinian salinity crisis: Deep-Sea Research, v. 29, p. 1391–1414.

Boersma, A., and Shackleton, N. J., 1977, Tertiary oxygen and carbon isotope stratigraphy, Site 357 [mid latitude South Atlantic], in Supko, P. R., Perch-Nielsen, K. et al., eds., Initial Reports of the Deep Sea Drilling Project, v. 39, p. 911–924. U.S. Government Printing Office, Washington, D.C.

Craig, H., 1957, Isotopic standards for carbon and oxygen and correction factors for mass spectrometric analysis of carbon dioxide: Geochimica et Cosmochimica Acta, v. 12, p. 133–149.

Craig, H., and Gordon, L. I., 1965, Deuterium and Oxygen-18 variations in the ocean and the marine atmosphere, in Proceedings Spoleto Conference on Stable Isotopes in Oceanographic Studies and Paleotemperatures, v. 2, p. 1–87.

Curry, W. B., and Matthews, R. K., 1981, Paleo-oceanographic utility of oxygen isotopic measurements on planktic foraminifera: Indian Ocean core-top evidence: Paleogeography, Paleoclimatology, Paleoecology, v. 33, p. 173–191.

Douglas, R. G., and Savin, S. M., 1971, Isotopic analyses of planktonic foraminifera from the Cenozoic of the Northwest Pacific, Leg 6, in Fischer, A. G., Heezen, B. C., et al., eds., Initial Reports of the Deep Sea Drilling Project, v. 6, p. 1123–1127. U.S. Government Printing Office, Washington, D.C.

Douglas, R. G., and Savin, S. M., 1973, Oxygen and carbon isotope analyses of Cretaceous and Tertiary foraminifera from the central North Pacific, in Initial Reports of the Deep Sea Drilling Project, v. 17, p. 591–605. U.S. Government Printing Office, Washington, D.C.

Duplessy, J. C., Bé, A.W.H., and Blanc, P. L., 1981, Oxygen and carbon isotopic composition and biogeographic distribution of planktonic foraminifera in the Indian Ocean: Palaeogeography, Palaeoclimatology, Palaeoecology, v. 33, p. 9–46.

Durazzi, J. T., 1981, Stable isotope studies of planktonic foraminifera in North Atlantic core tops: Palaeogeography, Paleoclimatology, Palaeoecology, v. 33, p. 157–172.

Fairbanks, R. G., Wiebe, P. H., and Bé, A.W.H., 1980, Vertical distribution and isotopic composition of living planktonic foraminifera in the western North Atlantic: Science, v. 207, p. 61–63.

Hodell, D. A., and Kennett, J. P., this volume, Miocene planktonic foraminiferal biogeography and stable isotopes of the South Atlantic Ocean.

Keigwin, L. D., Jr., 1979, Late Cenozoic stable isotope stratigraphy and paleoceanography of DSDP sites from the east equatorial and central north Pacific Ocean: Earth and Planetary Science Letters, v. 45, p. 361–382.

Kennett, J. P., Keller, G., and Srinivasan, M. S., this volume, Miocene planktonic foraminiferal biogeography and paleoceanographic development of the Indo-Pacific region.

Loutit, T. S., 1981, Late Miocene paleoclimatology, Subantarctic water mass, southwest Pacific: Marine Micropaleontology, v. 6, p.1–27.

Loutit, T. S., Pisias, N. G., and Kennett, J. P., 1983, Pacific Miocene carbon isotope stratigraphy using benthic foraminifera: Earth and Planetary Science Letters, v. 66, p. 48–62.

McKenzie, J. A., Weissert, H., Poore, R. Z., Wright, R. C., Percival, S. F., Jr., Oberhansli, H., and Casey, M., 1984, Paleoceanographic implications of stable-isotope data from Upper Miocene-Lower Pliocene sediments from the Southeast Atlantic [Deep Sea Drilling Project Site 519], in Hsii, K. G., LaBrecque, J. L., et al., eds., Initial Reports of the Deep Sea Drilling Project, v. 73, p. 717–724, U.S. Government Printing Office, Washington, D.C.

Rabussier-Lointier, D., 1980, Variations de composition isotopique de l'oxygene et du carbone en milieu maria et coupures stratigraphiques du Cenozoique [Ph.D. thesis]: University P. and M. Curie, Paris, 182 pp.

Savin, S. M., and Douglas, R. G., 1973, Stable isotope and magnesium geochemistry of Recent planktonic foraminifera from the South Pacific: Geological Society of America Bulletin, v. 84, p. 2327–2342.

Savin, S. M., Douglas, R. G., and Stehli, F. G., 1975, Tertiary marine paleotemperatures: Geological Society of America Bulletin, v. 86, p. 1499–1510.

Savin, S. M., Douglas, R. G., Keller, G., Killingley, J. S., Shaughnessy, L., Sommer, M. A., Vincent, E., and Woodruff, F., 1981, Miocene benthic foraminiferal isotope records: A synthesis: Marine Micropaleontology, v. 6, p. 423–450.

Sclater, J. G., Meinke, L., Bennett, A., and Murphy, C., this volume, The depth of the ocean through the Neogene.

Shackleton, N. J., 1977, The oxygen isotope stratigraphic record of the Late Pleistocene: Royal Society of London Philosophical Transactions, Ser. B, v. 280, p. 169–182.

Shackleton, N. J., 1982, The deep-sea sediment record of climatic variability: Progress in Oceanography, v. 11, p. 199–218.

Shackleton, N. J., and Kennett, J. P., 1975, Paleotemperature history of the Cenozoic and the initiation of Antarctic glaciation: oxygen and carbon isotope analyses in DSDP Sites 277, 279, and 281, in Initial Reports of the Deep Sea Drilling Project, v. 29, p. 743–755. U.S. Government Printing Office, Washington, D.C.

Shackleton, N. J., Hall, M. A., and Boersma, A., 1984, Oxygen and carbon data from Leg 74 foraminifers, in Moore, T. C., Jr., Rabinowitz, P. D., et al., eds., Initial Reports of the Deep Sea Drilling Project, v. 74, p. 599–612. U.S. Government Printing Office, Washington, D.C.

Vincent, E., and Shackleton, N. J., 1980, Agulhas current temperature distribution delineated by oxygen isotope analysis of foraminifera in surface sediments: Cushman Foundation Special Publication no. 19, p. 89–95.

Vincent, E., Killingley, J. S., and Berger, W. H., 1980, The magnetic epoch-6 carbon shift: A change in the ocean's $^{13}C/^{12}C$ ratio 6.2 million years ago: Marine Micropaleontology, v. 5, p. 185–203.

Williams, D. F., 1977, Planktonic foraminiferal paleoecology in deep-sea sediments of the Indian Ocean [Ph.D. thesis]: University of Rhode Island, 283 pp.

Woodruff, F., Savin, S. M., and Douglas, R. G., 1981, Miocene stable isotope record: A detailed deep Pacific Ocean study and its paleoclimatic implications: Science, v. 212, p. 665–668.

MANUSCRIPT ACCEPTED BY THE SOCIETY DECEMBER 17, 1984
CONTRIBUTION NO. 154 OF THE DEPARTMENT OF GEOLOGICAL SCIENCES,
CASE WESTERN RESERVE UNIVERSITY

Geological Society of America
Memoir 163
1985

Evolution of the Miocene ocean in the eastern North Pacific as inferred from oxygen and carbon isotopic ratios of foraminifera

Enriqueta Barrera
Department of Geological Sciences
Case Western Reserve University
Cleveland, Ohio 44106

Gerta Keller*
U.S. Geological Survey
345 Middlefield Road
Menlo Park, California 94025

Samuel M. Savin
Department of Geological Sciences
Case Western Reserve University
Cleveland, Ohio 44106

ABSTRACT

Oxygen and carbon isotopic ratios of planktonic and benthonic foraminifera have provided information about the evolution of the oceans at low- and mid-latitude sites in the Miocene eastern North Pacific Ocean.

DSDP Site 495 (12° N; 91° W) provides a record of early and middle Miocene oceanographic conditions in the eastern equatorial Pacific. Oxygen isotopic evidence indicates that *G. sacculifer, D. altispira* and *G. siakensis* were shallow-dwelling, tropical planktonic species. *G. venezuelana* deposited its test at greater depths, probably below the thermocline. Carbon isotopic evidence conflicts with that of the oxygen isotopes in that it suggests that *G. siakensis* calcified under conditions similar to those of *G. venezuelana*.

Temperature variability at Site 495 during early and middle Miocene time was relatively small. However, while middle Miocene deep waters at this site cooled, simultaneously with a major phase of growth of the Antarctic ice sheet, surface and near-surface waters warmed.

The oxygen isotopic record at Site 470 in the eastern North Pacific (29° N; 117° W) indicates that middle and late Miocene surface temperatures at this site were relatively stable, but were probably lower than modern surface temperatures.

At Site 173 (40° N; 125° W) middle and late Miocene surface temperatures were consistently lower than those at the more southerly Site 470, and were also significantly more variable. There is no indication that surface temperatures have changed significantly at Site 173 since late Miocene time. The inferred greater variability of surface temperatures at Site 173 may reflect greater variability of the intensity of upwelling at that site than at Site 470 during Miocene time.

At Site 495 both the planktonic and benthonic foraminiferal carbon isotopic records vary sympathetically with published benthonic foraminiferal isotopic records from the Atlantic and Pacific Oceans, indicating that the carbon isotopic ratios at that site largely reflect global fluctuations in the isotopic composition of dissolved inorganic carbon.

---

*Present address: Department of Geological and Geophysical Science, Princeton University, Princeton, New Jersey 08540.

At Site 470 the planktonic carbon isotopic record fluctuates sympathetically with published benthonic records, indicating that the middle and late Miocene $^{13}C/^{12}C$ ratios of dissolved inorganic carbon in surface waters at this site reflected global fluctuations in $^{13}C/^{12}C$. The planktonic carbon isotopic record at Site 173 could not be correlated with global carbon isotopic fluctuations, indicating that, in part, local effects controlled the $^{13}C/^{12}C$ ratio of dissolved inorganic carbon in surface waters at that site.

INTRODUCTION

The evolution of Miocene surface and deep waters in the eastern equatorial Pacific and northeastern Pacific Ocean has been investigated as part of a larger study of global Miocene oceanography (Cenozoic Paleoceanography Project, CENOP). Oxygen and carbon isotopic ratios of the tests of planktonic and benthonic foraminifera from Deep Sea Drilling Project (DSDP) Site 495 in the eastern equatorial Pacific, and from DSDP Sites 173 and 470 and the onshore Miocene section at Newport Beach, California in the northeast Pacific California Current system were measured. The locations of the sites discussed in this study, as well as of other relevant sites, are shown in Figure 1. Detailed information about the sites is given in Table 1.

Numerous studies have demonstrated that the $^{18}O/^{16}O$ and $^{13}C/^{12}C$ ratios of the tests of foraminifera provide information about ocean waters at the time and place the tests were secreted (Emiliani, 1954, 1955; Savin and Douglas, 1973; Berger and Gardner, 1975; Duplessy, 1978). The $^{18}O/^{16}O$ ratio of a calcitic

foraminiferal test deposited in isotopic equilibrium with sea water reflects the temperature and the isotopic composition of the water in which the foraminifer grew. The $^{13}C/^{12}C$ ratio of a foraminiferal test reflects mainly the isotopic composition of the dissolved HCO_3^- in the water column.

Previous oxygen and carbon isotopic studies, as well as studies of the abundances of planktonic and benthonic foraminifera, have demonstrated that the Miocene epoch was a crucial episode in the Earth's climatic history. Isotopic records of benthonic and planktonic foraminifera have most often been interpreted as indicating that the early Miocene was a period characterized by limited Antarctic glaciation (Shackleton and Kennett, 1975; Woodruff et al., 1981; Savin et al., 1981), smaller equator-to-pole and tropical surface-to-bottom temperature gradients, and warmer high latitude surface waters than those of today (Savin et al., 1975). A worldwide rapid increase in $\delta^{18}O$ values of benthonic foraminifera during early middle Miocene time has gener-

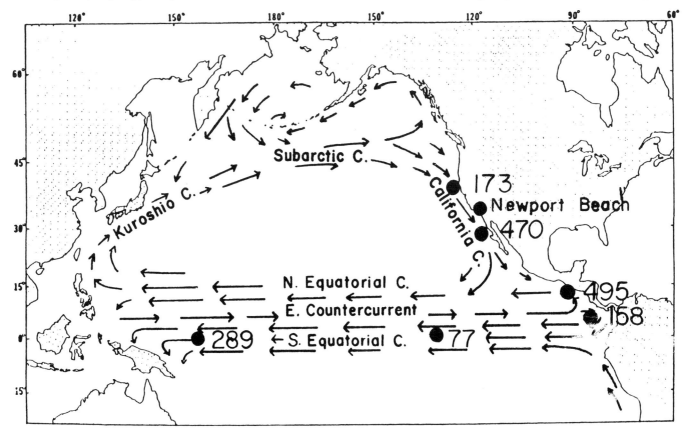

Figure 1. Locations of DSDP Sites discussed in this study.

TABLE 1. GEOGRAPHIC LOCATION OF SITES STUDIED

Site	Longitude	Latitude	Water depth
DSDP 173	125° 27.12'W	39° 57.71'N	2927 m
Newport Beach	117° 53'W	33° 38'N	onshore
DSDP 470	117° 31.11'W	28° 54.46'N	3549 m
DSDP 495	91° 2.26'W	12° 29.78'N	4150 m
DSDP 158	85° 14.16'W	6° 37.36'N	1953 m

TABLE 2. RELATIONSHIPS BETWEEN SEAWATER SALINITY AND $\delta^{18}O$*

Region	Relationship
California Current	$\delta^{18}O_{SMOW}= 0.544(salinity)-18.63$
Eastern Equatorial Pacific	$\delta^{18}O_{SMOW}= 0.222(salinity)-7.50$

*Inferred from data in Craig and Gordon (1965)

ally been interpreted as reflecting a combination of: a) an increase in the $^{18}O/^{16}O$ ratio of the oceans due to rapid growth of the Antarctic ice sheet; and b) cooling of high latitude surface waters and hence deep bottom waters (Shackleton and Kennett, 1975; Savin et al., 1975; Keigwin, 1979; Woodruff et al., 1981; Savin et al., 1981). An alternative interpretation has been proposed by Matthews and Poore (1980), who argued that the middle Miocene increase in $\delta^{18}O$ values reflects simply a drop in bottom water temperatures, and that large continental ice sheets have existed at least since early Oligocene time. Independent of these conflicting interpretations, the oxygen isotopic ratios of benthonic and planktonic foraminifera suggest a progressive increase in the equator-to-pole surface temperature gradient and the low latitude vertical thermal gradient throughout middle and late Miocene times (Savin et al., 1975, this volume).

Whereas the evolution of Miocene bottom waters has been relatively well-researched (e.g., Savin et al., 1981), there is little isotopic information documenting the details of the evolution of surface waters, particularly at mid-latitudes in the North Pacific. The isotopic data from Site 495 presented here document in detail the thermal evolution of the water column in the eastern equatorial Pacific in the interval from about 17 to 12.6 Ma. The isotopic record from this site can be correlated with those from Sites 470 and 173, which document changes in the vertical and the latitudinal thermal gradients in the California Current system during middle and late Miocene times. The ability to correlate the sites is enhanced by the availability of a high resolution biochronology (Keller and Barron, 1981).

METHODS

The techniques used in this study for processing and counting planktonic foraminifera are similar to those outlined by Keller (1980a). In this study, foraminifera larger than 150 μm were counted, and those between 150 and 250 μm were analyzed isotopically.

Isotopic analyses were performed using standard techniques (Epstein et al., 1953) and are reported in δ notation as per mil deviations from the PDB standard. Isotopic measurements were made relative to a standard CO_2 gas whose isotopic composition was related to PDB values through numerous analyses of NBS Isotopic Reference Material No. 20 (Solenhofen Limestone). NBS 20 is taken to have a $\delta^{18}O$ value of -4.14 per mil and a $\delta^{13}C$ value of -1.06 per mil (Craig, 1957). Isotopic analyses

typically have a precision of ±0.1 per mil. $\delta^{18}O$ values of calcite precipitated in equilibrium with sea water were calculated using the equation of Epstein et al. (1953) as modified by Epstein (unpublished)

$$t(°C)=16.4-4.2(\delta_c-\delta_w)+0.13(\delta_c-\delta_w)^2$$

where δ_c is the isotopic composition of the CO_2 gas evolved from the carbonate when reacted with H_3PO_4 at 25° C and δ_w is the isotopic composition of the CO_2 gas equilibrated at 25° C with the water in which the carbonate was deposited. The $\delta^{18}O$ values of present-day waters in the region were estimated using salinity and temperature data obtained from the National Oceanographic Data Center (NODC) and relationships between salinity and $\delta^{18}O$ inferred from the data of Craig and Gordon (1965) for sea water in the appropriate regions of the Pacific Ocean (Table 2).

In order to reflect fluctuations in the $^{18}O/^{16}O$ ratio of sea water caused by fluctuations in Antarctic ice volume, in some cases (where indicated) adjustments have been applied to the oxygen isotopic data. These adjustments are based on estimates by Woodruff et al. (1981), and on the suggestion of Shackleton and Kennett (1975) that pre-middle Miocene ocean water was 0.92 per mil depleted in ^{18}O relative to the present. Adjusted $\delta^{18}O$ values are intended to approximate the $\delta^{18}O$ values the foraminifera would have had if they had grown under Miocene oceanographic conditions but in sea water with today's $\delta^{18}O$ values.

ISOTOPIC SYSTEMATICS OF FORAMINIFERA

Planktonic foraminifera do not appear to deposit their tests in isotopic equilibrium with the dissolved HCO_3^- of ambient sea water (Vergnaud-Grazzini, 1976; Williams et al., 1977; Kahn and Williams, 1981). Evidence for this comes from the observation that $\delta^{13}C$ values of several planktonic species, including *Globigerinoides sacculifer,* are very close to the $\delta^{13}C$ values of total dissolved CO_2 (and HCO_3^-) in sea water at the depths at which calcification is inferred to have occurred (Williams et al., 1977). Calcite in equilibrium with sea water, however, is 1.26 to 2.38 per mil enriched in ^{13}C relative to the dissolved HCO_3^- in the temperature range from 0 to 30°C (Emrich et al., 1970).

"Vital effects" in the $^{18}O/^{16}O$ ratios of at least most planktonic foraminiferal species are apparently small (Williams et al., 1979; Curry and Matthews, 1981a). There is, however, conflicting evidence about whether *Globigerina bulloides* (which has been analyzed extensively in this study) deposits its test in isotopic equilibrium with sea water. Kahn and Williams (1981) reported that $\delta^{18}O$ values of *G. bulloides* collected in plankton tows from the northeastern Pacific Ocean are lower (i.e. warmer) than those expected for calcite in isotopic equilibrium with sea water in the upper 100 meters during the months of June and July. Other studies of this species collected in plankton tows and in surface sediments from several localities suggest $\delta^{18}O$ values close to equilibrium (Curry and Matthews, 1981a, b) and further suggest that *G. bulloides* grows predominantly during seasons of upwelling (Prell and Curry, 1981; Ganssen and Sarnthein, 1983). However, it has been pointed out by Bender (personal communication, 1984) that while the mean difference between measured $\delta^{18}O$ values of *G. bulloides* and equilibrium $\delta^{18}O$ values for surface waters in the large Indian Ocean data set of Curry and Matthews (1981a) is close to 0.00, the standard deviation of the differences is 0.24. Therefore, if a parallel can be drawn with the isotopic behavior of Miocene *G. bulloides* from the northeastern Pacific Ocean, no paleoceanographic significance can be attributed to changes smaller than 0.25 per mil in the $\delta^{18}O$ values of *G. bulloides*.

Samples for isotopic analysis from Site 173 were mixtures of *Globigerina praebulloides* and *Globigerina bulloides* because specimens of neither one of these species were present in sufficient numbers to be analyzed throughout the whole interval sampled. It was therefore necessary to investigate whether Miocene *G. praebulloides* and *G. bulloides* exhibit similar oxygen and carbon isotopic systematics. When all Site 173 samples are considered together, no relation is found between the proportions of the two species and their $\delta^{13}C$ values, suggesting no systematic difference in their carbon isotopic ratios. However, *G. bulloides* is apparently enriched in ^{18}O relative to *G. praebulloides* by 0.7 per mil in the late Miocene samples and by 0.25 per mil in the middle Miocene samples. The implications of this are not clear. It is probable that the increase in the $\delta^{18}O$ values with time and the parallel increase in the abundance of *G. bulloides* at Site 173 reflect middle and late Miocene cooling at this site rather than differences in the oxygen isotopic systematics of the two species.

The tests of most deep-water benthonic foraminiferal species are in neither oxygen nor carbon isotopic equilibrium with ambient sea water (Duplessy et al., 1970; Shackleton, 1974; Woodruff et al., 1980; Vincent et al., 1981; Graham et al., 1981). Shackleton (1974) concluded that of the commonly analyzed benthonic taxa, *Uvigerina* spp. has $\delta^{18}O$ values closest to those expected for calcite in oxygen isotopic equilibrium with sea water. Other benthonic taxa depart from equilibrium by relatively constant amounts. All species of *Cibicidoides* investigated thus far appear to have similar isotopic behavior and to be depleted in ^{18}O by approximately 0.7 per mil relative to equilibrium values (Woodruff et al., 1980; Belanger et al., 1981; Savin et al., 1981). *Cibicidoides* have $\delta^{13}C$ values similar to those of dissolved HCO_3^- in ambient sea water (Belanger et al., 1981; Graham et al., 1981).

Sample preservation and dissolution

Calcium carbonate dissolution can significantly affect both the taxonomic composition and the isotopic composition of the residual fauna. Dissolution affects biostratigraphic results by shifting the first and last appearances of species and reducing the stratigraphic ranges and abundances of dissolution-susceptible species. The isotopic signal of selectively preserved specimens can be biased in the ^{18}O-rich (cold) direction since those individuals of a species which lived higher in the water column are often more readily dissolved than deeper-dwelling individuals (Savin and Douglas, 1973).

The preservation of samples from Sites 173 and 470 has been discussed by Barron and Keller (1983). At Site 495, planktonic foraminiferal assemblages are well-preserved, except in Cores 27 and 28 (within foraminiferal Zone N7) where they have been affected moderately, and in Core 20 and upwards (Zones N12 and above) where virtually all foraminifera have been dissolved.

BIOSTRATIGRAPHY

High resolution biostratigraphic correlations have been made between the eastern equatorial Miocene Pacific sections from Sites 495 and 158 and: a) the more westerly equatorial Pacific Site 77B (Figure 2); and b) California Current Site 470 (Figure 3). Biostratigraphy and correlation between California Current Sites 470 and 173 have already been published by Keller and Barron (1981). Biostratigraphy of Site 158 was worked out by Kaneps (1973), Keigwin (1976), Keller et al. (1982) and in the present study. Biostratigraphy of Site 495 was previously examined by Thompson (1982) and was worked out in greater detail in this study.

Age assignments at Sites 495 and 158 are based on foraminiferal datum levels indirectly tied to diatom, radiolarian and coccolith datum levels calibrated with the paleomagnetic time scale as outlined by Keller (1981). Diatom datum levels of Sites 77B and 495 are from Barron (personal communication, 1983). Early and middle Miocene age assignments for Site 495 appear to be confirmed independently by comparison of the abundance curve for *Globorotalia siakensis* at that site with that for Site 77B. The faunal abundance counts for Site 77B by Keller (1980b) were used. Correlation between the eastern tropical Pacific sites and Site 470 made use of the faunal abundance data of Barron and Keller (1983) for Site 470. Faunal abundance data for Sites 495 and 158 were determined in this study and are listed in Appendix I. Only minor differences were observed between the faunal census obtained for Site 158 in this study and that published by Keigwin (1976).

The foraminiferal assemblages at Sites 495 and 77B are characterized by high abundances of warm-water species of the genera *Globorotalia* and *Globoquadrina*, in particular, *Globorotalia siakensis*, *Globorotalia peripheroronda*, *Globorotalia continuosa*, *Dentoglobigerina altispira* and *Globoquadrina venezue-*

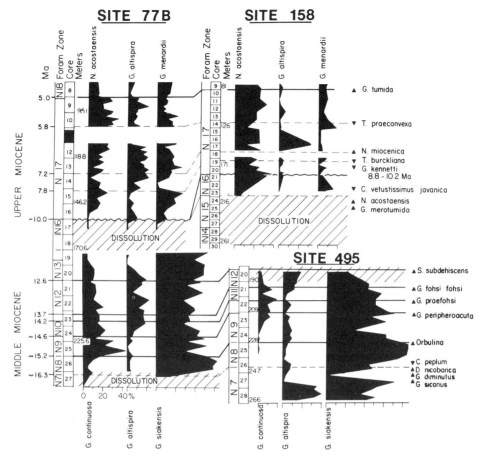

Figure 2. Correlation of the middle and upper Miocene sections, based on diatom and foraminiferal datum levels, and percent abundance of a few planktonic foraminiferal species at DSDP Sites 77B, 495 and 158. Data are from Keller (1980b), Barron (personal communication, 1983) and this study.

lana (Figure 2). The low abundance of species of the genus *Globigerinoides* reflects, in part, the effect of carbonate dissolution in the eastern equatorial Pacific during early and middle Miocene time. In the upper Miocene section at Site 77B, warm-water species such as *Dentoglobigerina altispira* and *Globorotalia menardii* are more abundant than at Site 158. The high abundances of the warm/temperate-water species *Neogloboquadrina acostaensis* and *Globigerinita glutinata* at Site 158 suggest the presence of colder near-surface water at this more easterly site during late Miocene time. The mid-latitude planktonic foraminiferal assemblages at Site 470 differ considerably from the tropical assemblages at Sites 495 and 158 (Figure 3). However, the presence of warm-water species in the middle Miocene section at Site 470 and their absence in the upper Miocene section suggest that middle Miocene surface waters were warmer than those of the late Miocene at that site (Barron and Keller, 1983).

ISOTOPIC RESULTS

$^{18}O/^{16}O$ ratios of foraminifera from DSDP Sites 495, 173, 470 and the onshore section at Newport Beach are listed in Appendix II. Results from Newport Beach are not discussed further because isotopic ratios indicate that the samples have undergone diagenetic alteration.

Benthonic Foraminiferal Isotopic Records

Composite benthonic foraminiferal isotopic records for *Cibicidoides* from Sites 173, 470 and 495 are shown in Figure 4. (Site 173 is represented by only two points.) The general appearance of the oxygen isotopic curves of the benthonic foraminifera is similar to that of other Miocene deep sea benthonic curves published by Savin et al. (1981). Early Miocene $\delta^{18}O$ values are low (indicative of high deep water temperatures and minimal ice volume), and are followed by a rapid increase in early middle Miocene time. This increase in $\delta^{18}O$ is interpreted as reflecting a combination of deep water cooling and a major growth phase of the Antarctic ice sheet (and a concomitant change in the $\delta^{18}O$ value of sea water). Middle Miocene (14 to 12 Ma) $\delta^{18}O$ values at Site 470 are more negative (i.e. warmer or indicative of less continental ice) than early late Miocene (12 to 10 Ma) $\delta^{18}O$ values.

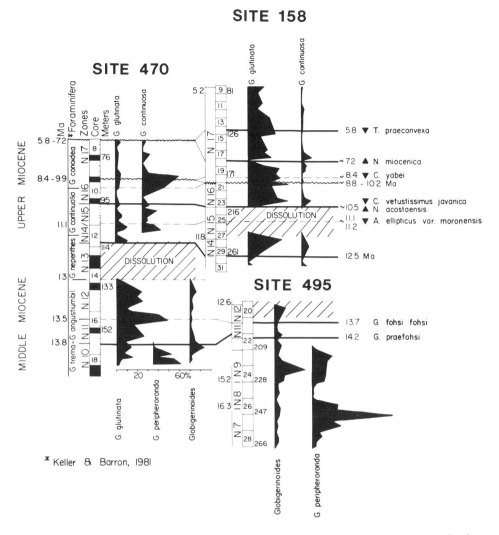

Figure 3. Correlation of the middle and upper Miocene sections and percent abundance of a few planktonic foraminiferal species at DSDP Sites 470, 495 and 158. Data are from Keller and Barron (1981), Barron and Keller (1983), Barron (personal communication, 1983) and this study.

Most of the major features of the carbon isotopic variations of the composite benthonic curve also resemble those of the benthonic deep sea curves published by Savin et al. (1981). There are some differences, however. The benthonic carbon isotopic record for the 13 to 11 Ma interval at Site 470 is especially variable. We have chosen not to correlate the eastern Pacific sites studied here with the "carbon isotopic events" defined by Loutit et al. (1983) because the eastern Pacific records did not, for the most part, have sufficiently high resolution or span sufficiently long time intervals to permit such correlations to be attempted with confidence.

Site 495

The oxygen and carbon isotopic records of foraminifera plotted in Figure 5 provide information on the depth stratification of the planktonic species. We infer, on the basis of the $\delta^{18}O$ values, that G. venezuelana was a relatively deep-dwelling planktonic species, and that the $\delta^{18}O$ values of D. altispira, G. siakensis and Globigerinoides sacculifer reflect near-surface water conditions. This is consistent with the conclusions derived from consideration of a large number of early and middle Miocene sites by Savin et al. (this volume). With the exception of G. siakensis, depth stratification of taxa inferred from $\delta^{13}C$ records is similar to that inferred from $\delta^{18}O$ records. The $\delta^{13}C$ record of G. siakensis resembles that of the deeper dwelling G. venezuelana. Assuming that the carbon isotopic systematics of G. sacculifer have not changed markedly over the past 20 m.y., it is likely that G. sacculifer and, by inference, D. altispira have $\delta^{13}C$ values close to that of the total dissolved CO_2 in the water column (Williams et al., 1977). G. siakensis is probably also a shallow-dwelling species but with quite different carbon isotopic systemat-

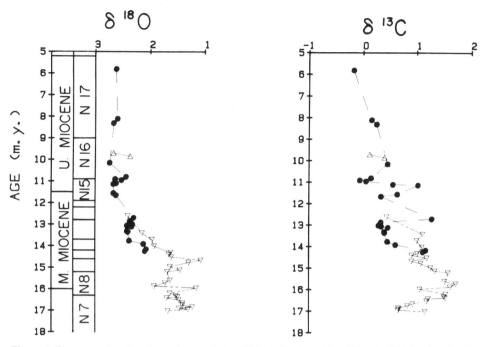

Figure 4. Oxygen and carbon isotopic records for *Cibicidoides* from Sites 173 (△), 470 (●) and 495 (∇) plotted as a function of time.

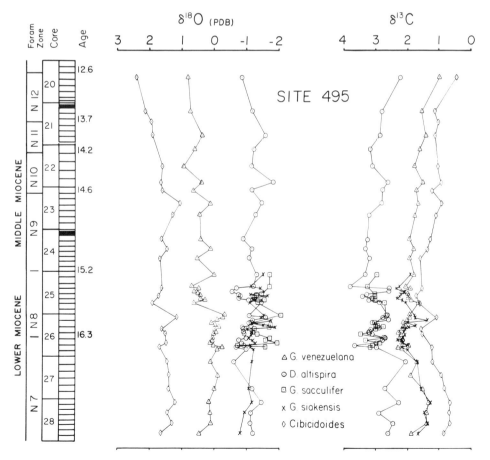

Figure 5. $\delta^{18}O$ and $\delta^{13}C$ values of selected foraminiferal species plotted as a function of depth in the sediment at Site 495.

ics than *G. sacculifer* and *D. altispira*. It is possible, but less likely, that *G. siakensis* is a deeper dweller and that it does not calcify in oxygen isotopic equilibrium with ambient sea water.

Insight into the early and middle Miocene thermal structure of the water column at Site 495 and its stability with time can be gained by comparing the total range of Miocene $\delta^{18}O$ values, adjusted to account for changes in the isotopic composition of sea water as discussed above, with $\delta^{18}O$ values of calcite precipitated in isotopic equilibrium with modern sea water (referred to in the discussions below as *equilibrium* $\delta^{18}O$ values) at several hydrographic stations near the early and middle Miocene locations of Site 495 (2 to 7°N; 99 to 108°W according to Sclater et al., this volume) (Figure 6).

Early and middle $\delta^{18}O$ values of shallow-dwelling species, adjusted for ice volume effects, vary between approximately +0.36 and –1.33 per mil. This is a small range compared to the range of equilibrium $\delta^{18}O$ values in the upper 150 meters of the modern water column (+1.1 to –2.7 per mil), and falls within the range of calculated equilibrium values within the thermocline, between 50 and 120 meters (Figure 6). The small range of $\delta^{18}O$ values of shallow-dwelling Miocene species relative to the calculated modern equilibrium $\delta^{18}O$ values suggests relatively little variation in the temperature structure of the early and early middle Miocene Oceans at Site 495. The range of measured $\delta^{18}O$ values for *G. venezuelana* is also small and falls within the range of equilibrium values calculated at or below the thermocline at depths between 100 and 300 meters.

Holocene planktonic $\delta^{18}O$ values of *G. sacculifer* in the vicinity of backtracked Site 495 are mostly around –1 per mil (Savin and Douglas, 1973; Shackleton, 1977), and hence indicate Holocene temperatures approximately 0 to 6°C higher than those indicated by the adjusted Miocene shallow-dwelling planktonic $\delta^{18}O$ values at Site 495. Evidence for warming at Site 495 during middle Miocene time comes from comparison of the benthonic and shallow-dwelling planktonic $\delta^{18}O$ values in Figures 5 and 7. $\delta^{18}O$ values of middle Miocene planktonic foraminifera show no secular change while those of benthonic foraminifera increase, reflecting both deep water cooling and increase in the $\delta^{18}O$ value of sea water as the result of growth of the Antarctic ice sheet. This implies that surface and near-surface waters warmed by an amount sufficient to balance the increase in the $^{18}O/^{16}O$ ratio of sea water. While it would be desirable to extend this argument by analyzing younger samples from Site 495, the virtually complete dissolution of the section above Core 20 prevented this. The inferred warming of surface waters at this site is consistent with the conclusion of Savin et al. (this volume), based on time-slice studies of Miocene planktonic foraminifera, that surface waters in the eastern tropical Pacific warmed between the early Miocene (N4B) and the present.

The oxygen isotopic data do reveal some variations with time in the thermal structure of the Miocene oceans at Site 495. Differences between $\delta^{18}O$ values of planktonic and benthonic species, are independent of global changes in the $\delta^{18}O$ value of the oceans and are indicative of the intensity of vertical thermal

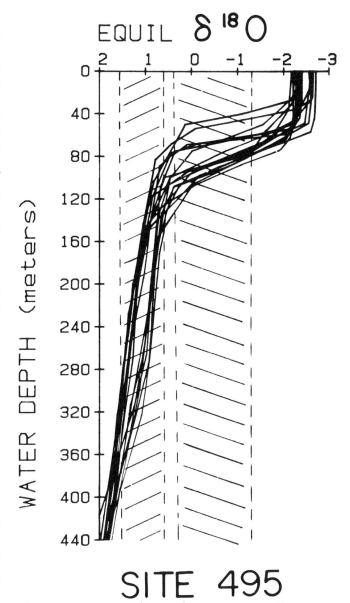

SITE 495

Figure 6. Comparison of calculated modern equilibrium $\delta^{18}O$ values for hydrographic stations near the Miocene location of Site 495 with the ranges of adjusted Miocene $\delta^{18}O$ values for *G. venezuelana* (shading sloping to the left) and for the shallow-dwelling foraminiferal species, *G. sacculifer, D. altispira* and *G. siakensis* (shading sloping to the right) from Site 495. The range of $\delta^{18}O$ values of the shallow-dwelling species falls within the range of calculated equilibrium values at depths between 50 and 120 meters, while that of the deeper-dwelling species, *G. venezuelana* falls within the range of calculated equilibrium values between 100 and 300 meters.

stratification of the water column. In Figure 7 the differences, $\delta^{18}O_{Cibicidoides}$ minus $\delta^{18}O_{D.\ altispira}$ and $\delta^{18}O_{Cibicidoides}$ minus $\delta^{18}O_{G.\ venezuelana}$ are plotted. Variations in these differences in the lower portion of the section, deposited between approximately 17 and 14.6 Ma (foraminiferal zones N7 through N9) are for the most part small and unsystematic. In the upper part of the section, however, as characterized by the uppermost four or five

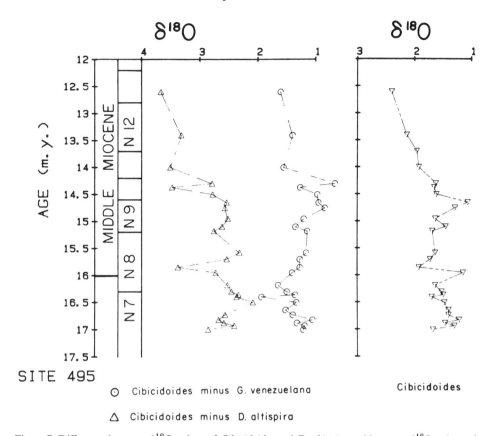

Figure 7. Difference between $\delta^{18}O$ values of *Cibicidoides* and *D. altispira* and between $\delta^{18}O$ values of *Cibicidoides* and *G. venezuelana* from Site 495 plotted as a function of age. $\delta^{18}O$ values of *Cibicidoides* are plotted on the right side of the figure.

samples representing the time period 14.6–12.5 Ma (foraminiferal zones N10 through N12), the difference between $\delta^{18}O_{Cibicidoides}$ and $\delta^{18}O_{D.\ altispira}$ is clearly greater than in the lower portion of the section (with the exception of one sample). This implies that the vertical thermal stratification of the water column became sharply intensified beginning about 14.6 Ma. The time of intensification of the vertical thermal gradient corresponds to the time of marked middle Miocene increase in benthonic foraminiferal $\delta^{18}O$ values. No significant corresponding increase is noted in $\delta^{18}O_{Cibicidoides}$ minus $\delta^{18}O_{G.\ venezuelana}$, probably because *G. venezuelana* grows below the thermocline in waters whose temperatures vary sympathetically with those of the deep waters at Site 495.

The general appearance, although not every detail, of the $\delta^{13}C$ time series curves of *D. altispira* and *G. venezuelana,* as well as *Cibicidoides* from Site 495, is similar to the pattern of $\delta^{13}C$ values of *Cibicidoides* from Site 289, as shown in Figure 8 (as well as other benthonic carbon isotopic curves shown by Savin et al., 1981). The increase in $\delta^{13}C$ values reaching a maximum between 16 and 15 Ma and the decrease in $\delta^{13}C$ values up to about 12 Ma are clear. These similarities suggest that a large part of the observed variability of both planktonic and benthonic $\delta^{13}C$ values at Site 495 reflects fluctuations in the average carbon isotopic composition of ocean waters. Furthermore, temporal var-

iations in the productivity of near-surface waters should affect $\delta^{13}C$ values of planktonic foraminifera in complex ways, perhaps causing them to increase under some circumstances and to decrease under others. The similarities in the pattern of fluctuations of the planktonic and benthonic records at Site 495 suggest that either changes in the intensity of upwelling at that site were minor or that they occurred but fortuitously produced no effect on the planktonic $\delta^{13}C$ records.

Site 173 and Site 470

Sites 173 and 470 lie within the modern California Current system (Figure 1). The Miocene planktonic and benthonic foraminiferal $\delta^{18}O$ and $\delta^{13}C$ values for those sites are plotted in Figure 9.

It is likely that at least some of the foraminifera from the lowermost four samples of Site 470 has been diagenetically altered. These samples, from Cores 17 and 18, lie within about 3 m of basalt basement. Foraminifera from Core 18 were separated from chalk. *G. bulloides* from Core 17, Section 3 has an unusually low $\delta^{18}O$ value and *G. bulloides* from Core 18, Section 1 has an unusually high $\delta^{13}C$ value. It is likely that these unusual isotopic compositions reflect diagenetic alteration of the test

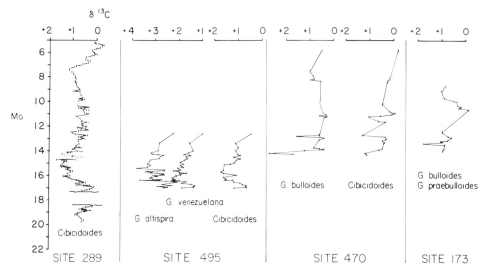

Figure 8. $\delta^{13}C$ values of *Cibicidoides* from Site 289, *Cibicidoides, G. venezuelana* and *D. altispira* from Site 495, *Cibicidoides* and *G. bulloides* from Site 470 and *G. bulloides/praebulloides* from Site 173 plotted as a function of time.

material. Therefore, little paleoenvironmental significance is attributed to the isotopic results of the lowermost four samples.

The isotopic evidence suggests that the middle/late Miocene interval sampled at Site 470 was one of stable oceanographic conditions. It is possible, however, that conditions were more variable on a short time scale than is suggested by the relatively small sample-to-sample variations in $\delta^{18}O$, since the sediments of this site were deformed, and perhaps mixed, by the drilling process (Yeats et al., 1981).

$\delta^{18}O$ values for *G. bulloides* and *G. praebulloides* from Site 173 are more positive than those of Site 470, reflecting the more northerly position of this site (Figure 9). Sample-to-sample variability of $\delta^{18}O$ values is greater in the upper Miocene section than the middle Miocene at Site 173, and the late Miocene sequence contains a number of samples that yield $\delta^{18}O$ values significantly more positive (i.e., cooler) than any middle Miocene values.

To gain insight into the oceanographic processes which could have produced the observed range of Miocene $\delta^{18}O$ values at Sites 470 and 173, adjusted $\delta^{18}O$ values of *G. bulloides* have been compared with calculated equilibrium $\delta^{18}O$ values for calcite precipitated in the upper 220 meters at several hydrographic stations (NODC) near each of the sites (Figures 10 and 11).

The range of adjusted Miocene isotopic data for *Globigerina bulloides* at Site 470, shown in Figure 10, is small (+0.52 to –0.22 per mil) in contrast to the range of equilibrium $\delta^{18}O$ values in the upper 100 meters of the water column (+1.1 to –1.6 per mil). The range of variability of the Miocene *G. bulloides* at this site is even smaller than the seasonal range of equilibrium $\delta^{18}O$ values at the surface. Thus, if the shapes (although not necessarily the absolute values) of modern bathythermograms and profiles of equilibrium $\delta^{18}O$ vs. depth approximate those of the middle and late Miocene oceans in this area, it can be concluded from the isotopic record of *G. bulloides* that there were only small variations in surface and near-surface water temperature and/or upwelling during the

middle and late Miocene interval at Site 470. (It is possible that changes in the intensity of upwelling or the water temperature were compensated by changes in the seasonality of growth. This is unlikely, however, especially in view of the conclusions of Gannsen and Sarnthein (1983) that modern *G. bulloides* records conditions during the season of upwelling.)

While water temperatures were relatively stable at Site 470 during middle and late Miocene time, they were apparently cooler than modern temperatures at the same site. All of the adjusted $\delta^{18}O$ values of *G. bulloides* from Site 470 are more positive (i.e., cooler) than any of the equilibrium $\delta^{18}O$ values for surface waters in months of significant upwelling (summer). In fact, virtually all of the differences between $\delta^{18}O$ of Miocene *G. bulloides* and modern surface water equilibrium $\delta^{18}O$ at Site 470 exceed the differences found in the modern Indian Ocean data set of Curry and Matthews (1981a). Thus, if the depth habitat of Miocene *G. bulloides* was the same as that of the modern species, it can be concluded that surface waters at Site 470 are probably warmer now than 5.8 Ma.

In a similar examination of the data from Site 173 (Figure 11), the range of adjusted Miocene *G. bulloides* and *G. praebulloides* $\delta^{18}O$ values is large (+1.15 to –0.01 per mil) relative to the range of present-day equilibrium $\delta^{18}O$ values in the upper 100 meters of the water column (+1.56 to –0.8 per mil). Thus, the observed range of Miocene oxygen isotopic values was probably produced by large fluctuations in surface or near-surface water temperatures perhaps related, at least in part, to variations in the intensity of upwelling or depth of the seasonal thermocline.

Huyer (1983) has shown that seasonal variability of upwelling in the modern ocean is greater in the region of Site 173 than in the region of Site 470. Since the seasonal variability of winds in this region is responsible for the seasonal variability in upwelling, it is likely that longer term variability of winds might also cause long-term variations in the intensity of upwelling to be great at

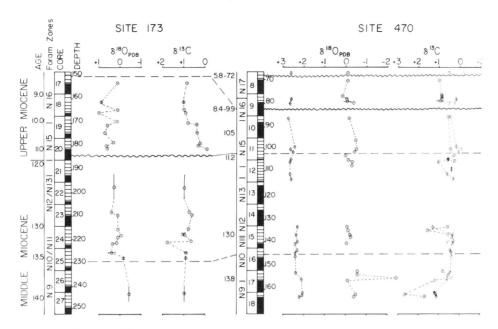

Figure 9. $\delta^{18}O$ and $\delta^{13}C$ values of *G. bulloides/praebulloides* from Site 173 and *G. bulloides* and *Cibicidoides* from Site 470 plotted versus depth in the sediment. Planktonic foraminiferal isotopic values are indicated by circles and isotopic values of *Cibicidoides* are indicated by diamonds. Biostratigraphy is from Keller and Barron (1981).

Site 173, consistent with the observed large range of measured $\delta^{18}O$ values at the site.

The adjusted $\delta^{18}O$ values of middle and late Miocene *G. bulloides/praebulloides* at Site 173 largely coincide with modern surface and mixed-layer equilibrium values at the same site and are lower by only a few per mil than modern surface equilibrium $\delta^{18}O$ values during periods of intense upwelling. There is therefore no evidence for any major change in surface temperatures at this site during the past 7.5 m.y.

Disregarding the four lowermost samples from Site 470 (because of suspected diagenetic alteration), the carbon isotopic time series curve for *G. bulloides* from Site 470 bears some resemblance to the $\delta^{13}C$ records of *Cibicidoides* from Site 289 (Figure 8) and other Pacific sites (i.e., decreasing $\delta^{13}C$ values between 14 and 11 Ma, increasing values from 11 Ma to approximately 7 Ma, and decreasing values after 7 Ma). While it is possible that this resemblance is fortuitous, it suggests that the major features of the isotopic record of *G. bulloides* at Site 470 reflect ocean-wide if not global changes in the $^{13}C/^{12}C$ ratio of dissolved inorganic carbon. It is puzzling that the benthonic $\delta^{13}C$ record of Site 289 (and most other benthonic records shown by Savin et al., 1981) agree more closely with the planktonic than the benthonic record at Site 470, and that the benthonic $\delta^{13}C$ record is especially variable in the 13 to 11 Ma interval. There is nothing in the faunal composition of the benthonic foraminifera or in their oxygen isotopic record to suggest that bottom water conditions at this site were in any way out of the ordinary.

The carbon isotopic record of *G. bulloides* and *G. praebulloides* from Site 173 is much more variable than that from Site 470, especially in the section from 11 to 9.5 Ma (Figure 8). There

is some superficial resemblance between the $\delta^{13}C$ record of *Cibicidoides* from Site 470 and that of *G. bulloides/praebulloides* from Site 173, but this is apparently fortuitous. For example, the extreme positive $\delta^{13}C$ peaks in the two records at approximately 13 Ma are diachronous according to the highly resolved biostratigraphic studies of the two sites (Keller and Barron, 1981; Barron and Keller, 1983).

The dissimilarity between the $\delta^{13}C$ record of *G. bulloides/praebulloides* from Site 173 and those of benthonic foraminifera from Site 289 and other sites probably reflects local variations in the $^{13}C/^{12}C$ ratio of surface waters at Site 173 superimposed on the average $^{13}C/^{12}C$ composition of the oceans. These local variations may reflect changes in the intensity of upwelling and productivity. However, we are not able to resolve whether decreases in ^{13}C of *G. bulloides* and *G. praebulloides* correspond to increases or decreases in upwelling intensity. This is because of uncertainty as to the extent to which the mixing of upwelled waters with surface waters lowers the $\delta^{13}C$ value of surface waters and the extent to which the enhanced biological productivity, resulting from upwelling, increases the ^{13}C content of surface waters. Prell and Curry (1981) and Gannsen and Sarnthein (1983) have found $\delta^{13}C$ values of *Globigerina bulloides* in upwelling centers to be unrelated to the intensity of upwelling, although its $^{18}O/^{16}O$ ratios record surface water temperatures during times of upwelling.

COMPARISON OF ISOTOPIC AND PALEONTOLOGIC RESULTS

Major fluctuations in the abundance of planktonic forami-

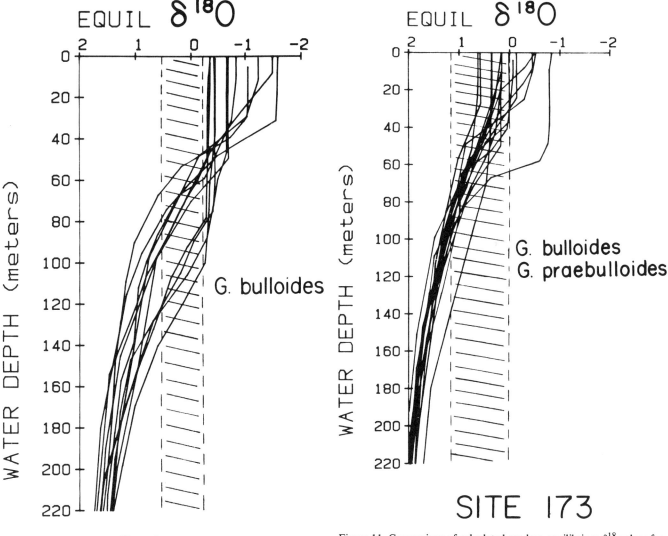

EQUIL δ¹⁸O

SITE 470

Figure 10. Comparison of calculated modern equilibrium $\delta^{18}O$ values for hydrographic stations near Site 470 with the range of adjusted middle and late Miocene $\delta^{18}O$ values for *G. bulloides* at Site 470 (shaded region). The range of $\delta^{18}O$ values of *G. bulloides* is small (+0.52 to –0.22 per mil) in contrast to the range of equilibrium $\delta^{18}O$ values in the upper 100 meters of the water column.

Figure 11. Comparison of calculated modern equilibrium δ^{18} values for hydrographic stations near Site 173 with the range of adjusted middle and late Miocene $\delta^{18}O$ values for *G. bulloides/praebulloides* at Site 173 (shaded region). The range of *G. bulloides/praebulloides* $\delta^{18}O$ values is large (+1.15 to –0.01 per mil) relative to the range of equilibrium $\delta^{18}O$ values in the upper 100 meters of the water column.

nifera inferred to be temperature-sensitive have been used by Ingle (1967, 1973a, b), Bandy and Ingle (1970), Keller (1978) and Barron and Keller (1983) to reconstruct the climatic and oceanographic history of the eastern North Pacific. Interpretations of the faunal data from these studies have been based on the assumption that in periods when high latitudes cooled or the latitudinal thermal gradient increased, the cool California Current intensified displacing cold-water faunas and floras to the south. Hence, according to this approach, paleoclimatic oscillations can be recognized by fluctuations in the relative abundances of cold to cool-water species and temperate to subtropical-water species in sediment assemblages.

The availability of foraminiferal abundance data (Barron and Keller, 1983) for exactly the same samples analyzed isotopically from Sites 173 and 470 led us to investigate whether small-scale fluctuations in the abundance of cold-water foraminiferal species (Barron and Keller, 1983) at these sites could be correlated with changes in water temperature inferred from the isotopic data. Moreover, the oxygen isotopic data were also compared with the abundance of *Globigerina bulloides* which has been identified in other studies as thriving in zones of upwelling (Zobel, 1971; Thiede, 1972; Diester-Haass et al., 1973; and Prell and Curry, 1981). Hence, if we assume that *G. bulloides* thrived during upwelling episodes in Miocene time as it does today, we might consider fluctuations in the abundance of *G. bulloides,* or perhaps species of the *bulloides* group (*Globigerina bulloides, G.*

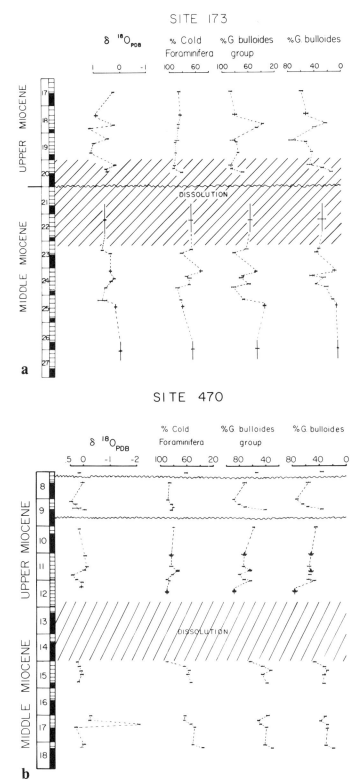

Figures 12a and 12b. The adjusted $\delta^{18}O$ time series curves of *G. bulloides/praebulloides* from Site 173 and *G. bulloides* from Site 470 are plotted versus depth in the sediments and alongside their respective foraminiferal abundance curves.

praebulloides and *G. quadrilatera*) to correspond with variations in the intensity of upwelling at the sites within the California Current system.

In Figures 12a and 12b, the oxygen isotopic time series curves for Sites 173 and 470 have been plotted alongside the foraminiferal abundance curves. In both figures, all data sets are plotted so that inferred warm conditions lie to the right and inferred cold conditions to the left. While there are some superficial resemblances between isotopic and faunal curves, at neither site can episodes of warming or cooling inferred from the isotopic data be consistently correlated with episodes of warming or cooling inferred from the faunal curves. A valid test of the relationship between the two types of records would require both faunal and isotopic analyses of larger numbers of more closely spaced samples than were examined in this study.

CONCLUSIONS

DSDP Site 495 provides a record of early and middle Miocene oceanographic conditions in the eastern equatorial Pacific. Oxygen isotopic evidence indicates that *G. sacculifer, D. altispira* and *G. siakensis* were shallow-dwelling tropical planktonic species. *G. venezuelana* deposited its test at greater depths, probably below the thermocline. With the exception of *G. siakensis,* the depth stratification of species inferred from $\delta^{18}O$ values is similar to that inferred from $\delta^{13}C$ records. $\delta^{13}C$ values of *G. siakensis* differ significantly from those of the other shallow-dwelling species and are similar to those of *G. venezuelana.*

The time series curves of the planktonic species and that of *Cibicidoides* show that temperature variability at Site 495 during early and middle Miocene time was relatively small. However $\delta^{18}O$ values of *Cibicidoides* indicate that deep waters at this site cooled while those of *D. altispira* indicate that surface and near-surface waters warmed.

Middle and late Miocene surface temperatures at Site 470 were relatively stable, but were probably lower than modern surface temperatures. At Site 173, middle and late Miocene surface temperatures were consistently lower than those at the more southerly Site 470, and were also significantly more variable. Surface temperatures at Site 173 have probably not changed significantly since late Miocene time. The inferred greater variability of surface temperatures at Site 173 may reflect greater variability of the intensity of upwelling at that site than at Site 470 during Miocene time.

The planktonic and benthonic foraminiferal carbon isotopic records at Site 495 vary sympathetically with published benthonic foraminiferal isotopic records from the Atlantic and Pacific Oceans, indicating that the carbon isotopic ratios at that site largely reflect global fluctuations in the isotopic composition of dissolved inorganic carbon.

The planktonic carbon isotopic record of Site 470 fluctuates sympathetically with published benthonic records, indicating that the middle and late Miocene $^{13}C/^{12}C$ ratio of dissolved inorganic carbon in surface waters at this site reflected carbon isotopic variations in deep waters. The planktonic carbon isotopic record at Site 173 could not be correlated with global carbon isotopic fluctuations, indicating that, in part, local effects controlled the $^{13}C/^{12}C$ ratio of dissolved inorganic carbon in surface waters at that site.

ACKNOWLEDGMENTS

We thank J. C. Ingle for his hospitality at the Department of Geology of Stanford University where part of the foraminiferal research was conducted. The laboratory assistance of Linda Abel at Case Western Reserve University is gratefully acknowledged. John Barron provided diatom data and constructive criticism of this manuscript. Kristen McDougall prepared the benthonic foraminifera analyzed isotopically in this study. Extremely helpful comments and suggestions were provided by the reviewers of this paper, Lloyd Keigwin, Michael Bender and Peter Kroopnick. Financial support was provided by the National Science Foundation under the following grants: OCE 79-17017 (CENOP) to Samuel Savin and OCE 79-18285 to Gerta Keller.

APPENDIX I. PERCENT ABUNDANCE OF PLANKTONIC FORAMINIFERA IN THE MIDDLE TO UPPER MIOCENE SECTIONS OF SITE 158
(Parentheses indicate raw counts)

DSDP SITE 158	9 2 30 32	9 6 30 32	11 2 30 32	11 6 30 32	12 6 30 32	13 6 30 32	14 6 30 32	15 6 30 32	16 6 30 32	17 6 30 32	17 6 30 32	18 6 30 32	18 6 30 32	19 6 30 32	19 6 30 32	20 6 60 64	20 3 116 120	20 3 63 67	20 4 15 19	20 6 102 106	20 6 127 131	20 6 65 69	20 6 38 42	21 2 30 32	22 6 30 32	23 6 30 32	23 5 30 32	24 1 108 110	26 6 30 32	27 2 30 32	28 6 30 32	28 1 30 32	30 6 30 32	30 5 30 32	31 1 30 32
X = <2																																				
Globigerina																																			
G. angustiumbilicata																							X								3	X			X
G. apertura	2			X	X		2	X	X	X			3	X	X	X	X	X	X	X	X	X	X	X	X	X	X		(7)	(3)	X	(3)		(1)	X 2
G. bulloides		X	X	X	X	2	2		X	X	2			X	2 4	3	X		3	X 2	X 2	X	X 2	X 2	X	X	2				2	2			3
G. ciperoensis	X	X	4 3		7	10	X	X	X	X		X X	X X	X X	X X	X X	X X	2	X X	X X	X	2	X X	X X	X X	X X	X X		(2)	(4)	X	(4)		(4)	X
G. decoraperta	X	2	X X	X X	X X	X	X X		X X				X X	X X		X X	X X	X		2		X	X X	X	X X		X X								
G. druryi												4			X	X	X	X	X	X	X	X			X						2	2			
G. falconensis										X																									
G. foliata	X	X			2	2		X		X X	X X		X	9	14	6	X	3	5	X			X	X		X					X				
G. nepenthes	X		3	3	X		X 2	X 2	X 2	X X	X		X	X	X	X	X	X	5					X X	X X	X X	X X		(7)	(6)	6	(3)		(3)	2
G. pseudociperoensis	X	X	X		X				X					X	X		X							X	X X	X X				(3)	3	X		X	
G. praebulloides	X	X X	X	X	X		2			X							X						X	X 2	X 2	X X	X X			(1)	X				
G. quadrilatera	X	X	X						X	X							X	X					X	X 5	X 5	22	X X				6	X			X
G. siphonifera	X	X	X					2			X			X	X	X							2	8	28	8	22		(12)	(6)	13	(2)			X 9
G. woodi	15	10	14	9	20	23	25	12	18	22	15	15		26	29	26	30	28	33	28	21	39	37	19	27	37	19		(10)	(16)	32				3
Globigerinita glutinata																						2				X				(10)	(1)				
G. uvula	X	X X	X X	X X	3 3	2	2	2	X		2 2	2	4 4	3 4	3 4	4	4 4	X X	3 4	3	X 2	X		4			6				X X	X			X 3
Globigerinoides bollii	X	X X	X X	X X							5 5	5	4 X	4		X	X	X		2	9	22	X						(4)		X X	X			
G. bulloideus	X	X	X	X	X		X	X	X	X	X	X																							
G. conglobatus							X															X													
G. kennetti									X	2	7	18	X	X	X	X	X	X 7	X	X	X	X 5	2	X	X	2			(1)		X				
G. obliquus obliquus	3	3	4	10	4	5	4	2	2	4	3		4	X	5	3	5 5	X	3	3	3	X 5	4	2			2		(1)	(1)	X	(6)		(1)	2
G. obliquus extremus	2	2	2	3	X	3	X	2	X	X	X	X	X	X	2	X X	X X	X	X X	X X	4	3	11	4	10	4	3		(9)	(8)	5	(18)		(18)	3
G. sacculifer	2	4	3	2	2	3	X	2	2	3	7	7	2	4	X	X 2	X X	X X	4	4	3	3	11	11	8	X	16		(1)		12				2
G. trilobus	X	3	2	X	6	2	2	X	31	2	32	2		3	X	5	2	X	9	5	X	4	X	X	X										2
Dentoglobigerina altispira			X																						X						X	(16)		(16)	X 9
Globoquadrina dehiscens	X	X	X	3	X	X	X	X	5	X	5	X	X	3	3	X	X	2	X	X	2	2	X	X	X	X	X		(3)	(1)	X				
G. venezuelana													X				X							X	X	X X	X								
Globorotalia archeomenardii	2	X	X X	3	X	X	X X	X X	2	10	3	5	4	4	4	2		X			X	X	9	3	3	9	X		(1)	(1)	4	(2)		(2)	5
G. cibaoensis	13	9	2	2	2	2	X 5	X X	2	4			X	X				X				4 4		X X	15	X X	X X			X	X	X			
G. continuosa	8	7	7	7	5	9	18	2	2	4		8	7	7	8	8 2	6 7	10 2	3 6	3 9	9	4 4		15			6			X	X	X			
G. humerosa	6	8	9	4	X	4	11	X	2	6		X 3	8	X X	X	X	X	X	X	X	X	X X	X X								4	(7)		(1)	50
G. menardii	X	X X	X	X	X	X	X	X	X	X	X		X		X	X	2	X	X	X	X	X	X	X	X	X	X							(30)	(1)
G. merotumida	X	X X	X	X	X	X	X X	X	X						X	X	X	X	X	X	X	X X	X	X	X X	X X					X 22	34			
G. plesiotumida	X	4	3	2		X	2	2	2	2		2	X	X	X	X		X	X	X	X	X	X	X	14	10	6		(2)		X	(1)			50
G. obesa	X		X	X	X								X	X X	X X			X																	
G. praemenardii							X																					(1)							
G. tumida																														(7)	4				X
G. scitula	X X	X X	X		X X	X	2	X	X	2	X	X		X	X	X	3	3	X	X	X	X	X	X	X		X			(3)	X 22	34		(1)	
G. siakensis	X	4	3	X	22	22	12	24	25	11	26	2	23	14		25	33	21	22	20	19	24	33			6					X			(30)	(1)
Globorotaloides hexagona	X	X	2	2	X	X	X	X	X	X	X	X	X	X	X	X	X	X	X	X	X	X X	X X			X					X				
G. trema	X				X																														
Hastigerina aequilateralis	X X	X	X		X	X	X	X											X			X X			X	X									
Neogloboquadrina acostaensis	21	26	25	32	22	22	12	24	25	11	26	18	23	14		25	33	21	22	20	19	24	33		14	10	6		(2)		4	(1)		(30)	50
N. pachyderma		X	X	2	X	X	X	X	X	X	X	X	X		X	X	X	X	X	X	X	X	X	X											
Orbulina universa		X			2	2	X 4	2		X	X		X X	X X		X	X X	X	X X	X X	X	2 2	2	X	2		X				X X	2		X	
O. suturalis	X	X X	X		X X	X X	X X		2	X X	X X		X								X						X							X	
Pulleniatina primalis	X X	X 4	4 2	4	X 3	2	2	3	2	X X	4	X	X X		X X	X X	X	X	X	X	X	X X	X	X X	2		X			(1)	X X	(1)			X
Sphaeroidinella seminulina	X X	X 2	2	X 2	3 4	X X	X	4	X	X X	2	X	X	X X	X X	2	X X	X X	X	X X	X	2	X X	X X			X			(1)	X	(2)			
S. subdehiscens	X	X			4	X	2		X	X X			X				X	X	X	X	X	X					X			(2)	2				
Unknown																				X	X														
No. specimens counted	397	385	408	320	342	374	430	295	331	298	338	298	418	352	376	350	403	343	398	345	327	361	293	271	268	314	276	6	59	66	336	276	302	98	302

APPENDIX I. PERCENT ABUNDANCE OF PLANKTONIC FORAMINIFERA IN THE LOWER TO MIDDLE MIOCENE SECTIONS OF SITE 495
(Parentheses indicate raw counts)

DSDP SITE 495 Core Sect. Upper cm Lower cm	17 3 84 88	20 2 70 72	20 3 70 72	21 2 67 69	21 3 67 69	21 4 67 69	22 1 70 72	22 2 70 72	22 3 35 37	23 3 75 79	23 4 75 79	23 5 75 79	24 4 75 79	24 5 75 79	24 6 75 79	25 1 98 102	25 2 98 102	25 3 75 79	25 4 75 79	25 5 98 102	25 5 140 144	25 6 98 102	25 6 140 144	26 1 75 79	26 1 98 102	26 1 140 144	26 2 92 96	26 3 140 144	26 4 98 102	26 4 140 144	26 5 75 79	26 5 95 102	26 5 136 140	26 6 52 56	26 6 103 107	27 6 75 79	27 3 75 79	27 4 75 79	27 5 75 79	27 5 75 79	28 1 75 79	28 3 75 79	28 5 64 68	28 7 44 48

x=<2

Globigerina																																												
G. angustiumbilicata	10																																											
G. bulloides		X	X		X											X					X	X		X		X	X			X			X	X	X	X		X				X		
G. ciperoensis		X	2	2																		X																						
G. sp.		X								6					X		2			2			X												X	X		2						
G. druryi		X									X					X																							X					
G. falconensis	10	X			X																				X				X													X		
G. foliata					X		X	X																														X						
G. praebulloides				X	X		X	X								X		X		X		X			X		X		X		X	X	X		X	X	X	X	X	X	X	X	X	
G. praedehiscens				X			X	X																		2			X													2		2
G. praedigitata			2					X							X																													
G. siphonifera													X																															X
G. umbilicata	X																X	X								X						X												
G. woodi	4																																											
Globigerinita glutinata	7	8		2	X		X	2		2	X	4	X	X	4	X	6	X	7	X	9	X8	X2	X5	X9	3	X8	X6	X8	6	X7	X4	X4	5	6	4	X3	X2	3	X2	2		X2
G. uvula												2	X2											X				7	8		8						X3	X3		X	X		X
Globigerinoides bollii	X										X										X			X				X			X		X					X				X	
G. bulloideus					2		X	X		X	X		X			X								X			X		X	X	X	X	X			X	X	X		X		X	
G. diminutus																										2			X		X												
G. obliquus	19																																X		X								
G. quadrilobatus	X	X		X			2	2	3	X			X		X	2		X	2		X	X	X	X		X	X	2	X	3	2	X	2	X	2	X	X	X	X	X	X		X
G. sacculifer																		X			X	X	X	X	X	X	X	X	X	X	X	X	X		X			X	X	X	X		
G. sicanus		2	2					X			X	X	X	X			X	X	X		X	X		X	X	X	X	X	X	X	X	X	X	X	X								
G. subquadratus										3	9	2	3	X	X	X		X	X	2	X	2	X	2	X	2	2	2	2	2	X	X	X	2		X		X					
G. transitoria		2						X		3	8	X	5	X		X	5	X	X	X	X	6	X	6	6	3	3	X	X	3	X	X	X			2			X	X		X	
G. trilobus	4	8		X	X		3	2	2	5	8	3	X		2		X	X	3	3	7	8	4	5	7	6	9	3	5	3	5	2	3		3	X	X	X	X	X	X	X	X
Dentoglobigerina altispira	4	18 (2)	11	X12	X4		X3	X2	5	X5	9	5	3		3	4		X	4	2	8	7	X	11	X	6	8	11	X	2	5	3	5										
Globoquadrina dehiscens		18 (2)	X					8		X	9		X		X	3		X	3				X			X	X			3		X											
G. sellii			X								X	X			X	X		X	X	X	X	8	X	X	X		X	X	X	X	X	X	X			X	X	X	X	X	X		X
G. sp.	5	11 (1)	7				4	3	X	X3	X3	X3	X2	X3	X3	X2		X3	X2	X2	8	X	X4	X5	X4	X4	X2	4	X3	X3	X5	X3	X3	X3	X3	X3	X5	X7	X7	X7	X3		
G. venezuelana							6	3	3	3	X	X	2	3	X	5	7	X	X	9	5	4	5	4	5	4	4	4	3	4	3	2	3	3	5	5	5	13		7	2		3
Globorotalia archeomenardii		3	5	14	3		6	X	X	2	X		X			X	3	2	X																									3
G. continuosa		3	X	X			2	9	9	9	2		2	9		9	12	5		2																								
G. fohsi fohsi		17			X		19	12		13										X																								
G. fohsi lobata	38									6																																	
G. menardii	X							3		X	X		2	X		2		X	X	2	X				X		X	X															X	
G. miozea								X		X				X																														
G. praemenardii			X12	X18	X13		X2	X7		X	X	X				X	X		X	X	X	X		X		X	X	X	X							X		X	X	X		X		X2
G. peripheroacuta		X	X	X	X		X	X		6	5		5	4	2	4	4	X	X	X	X	X	X	X	X		X	X	X	X	X					X	X				X		
G. peripheroronda			X	X	X		X	X		X	13		2	2			X	X		X	X	X	X	X	X		X	X																
G. praefohsi		3			X		X																				X																	
G. scitula		X	X	X	X		X	X		X	12		12	2	2	X	6		X	X		2	6	6		12	X	X	X	24	X	33	38		45	75	8	21		17	75		3
G. siakensis		23	45	43	58		51	52		70	29	62	86	87	86	79	74	67	73	86	84	79	9	49	59	47	58	61	47	48	45	44	35	45	53	71	41	48	35	51	42		
Globorotaloides hexagona	X2	X	X	X	X		X			X	X	X	X	X	X					X	X	X	X	X	X	X	X	X		X		X				X		X					
Hastigerina aequilateralis	2																																										
Neogloboq. acostaensis	5						3	13		14	14														X					X	X			X			X						
cf. acostaensis																																											
Orbulina universa	2									X		X						X				X		X					X															
O. saturalis			X								X	X		X	X	X																												
Praeorbulina sp.			8	3	6		X																				X			X		X												
Sphaeroidinella disjuncta		X (3)														X							X						X														
S. seminulina		X																																									
S. subdehiscens											X	X											X													X							
Globigerinatella insueta								X																												3							
Catapsydrax cf. stainforthi				X			X	X		X			X	X	X	X				X	X		X																X					X
Cat. variabilis															X																X													
Unknown															X	X										X					X												X	X
No. specimens counted	226 285	6 371	298	376	314	347	442	450	338	369	374	493	306	353	653	369	441	394	363	548	379	432	300	427	408	382	418	709	372	618	554	384	404	737	345	543	431	802	494	1118	298	588	763	422

APPENDIX II

OXYGEN AND CARBON ISOTOPIC COMPOSITION OF SELECTED SPECIES OF FORAMINIFERA AT SITE 173

Core/ sect	Depth (cm)	Estimated Age (m.y.)	G. bulloides G. praebulloides $\delta^{18}O$	$\delta^{13}C$	Uvigerina $\delta^{18}O$	$\delta^{13}C$	Cibicidoides kullenbergi $\delta^{18}O$	$\delta^{13}C$	Ice Effect* Adjustment
17-2 & 3	52-54	8.60			3.59	-1.00			0.14
17	CC	8.70	0.12	0.84					0.14
10-2 & 3	52-54	9.10-9.20	0.88	1.01					0.04
18-3 & 4	52-54	9.30			3.49	-1.06			0.10
18-5	52-54	9.40	0.11	0.98					0.14
18-4 &	52-54								
18	CC	9.60-9.70					2.67	0.12	0.15
18	CC	9.70	0.99	0.92					0.16
19-2	52-54	9.85					2.37	0.39	0.32
19-2 &	131-133								
19-3	52-54	9.90-10.0	0.48	0.49					0.32
19-3	52-54	10.00	0.59	0.38					0.43
19-4	122-124	10.20			3.90				0.43
19	CC	10.25	0.71	0.37					0.42
20-2	118-120	10.45	-0.19	0.22					0.41
20-3	50-52	10.50	0.14	0.31					0.41
20	CC	10.60	0.10	-0.10					0.40
21 &	CC								
22-5	122-124	12.30	0.28	0.97					0.32
23-3	52-54	12.80	0.42	0.74					0.31
23	CC	12.90	0.12	0.61					0.30
24-1 &	90-92								
24-2	15-17	13.1-13.15	0.11	0.77					0.29
24-3	15-17	13.20	-0.04	1.05					0.29
24-3	52-54	13.25	0.10	0.96					0.28
24-5	15-17	13.30	0.17	0.67					0.28
24-5	52-54	13.35	0.35	1.76					0.23
25-2	52-54	13.45	0.13	0.82					0.30
25-2	52-54	13.50	0.66	0.97					0.33
25-3 &	15-17								
25-4	52-54	13.7-13.75	-0.15	0.91					0.36
26 &	CC								
27-3	115-117	13.9-14.00	-0.41	0.97					0.40

*Described and applied as noted in text. Data in this table are not adjusted

OXYGEN AND CARBON ISOTOPIC COMPOSITION OF SELECTED SPECIES OF FORAMINIFERA AT SITE 470

Core/ sect	Depth (cm)	Estimated Age (m.y.)	G. bulloides $\delta^{18}O$	$\delta^{13}C$	Cibicidoides kullenbergi $\delta^{18}O$	$\delta^{13}C$	Uvigerina hispida $\delta^{18}O$	$\delta^{13}C$	Ice Effect* Adjustment
8-1	35-39	5.80	-0.08	0.50	2.62	-0.18	2.79	-1.04	0.24
8-3	34-39	7.40					2.97	-0.19	0.20
8	CC	7.50	-0.08	0.98					
9-1	54-59	7.90	0.11	0.87					0.32
9-1	54-59	7.90	0.20	0.87					0.32
9-2	54-59	8.10	0.01	0.78	2.60	0.16	2.51	-0.66	0.32
9-2	54-59	8.10	-0.09	0.94					0.32
9-3	54-59	8.20	0.18	0.70			3.48	-0.84	0.32
9-3	54-59	8.20	-0.32	1.22					0.32
9	CC	8.30	-0.36	0.53	2.67	0.25			0.32
10-1	54-59	10.15	-0.16	0.57	2.74	0.45			0.32
11-1	54-59	10.60	-0.47						0.42
11-3	55-60	10.80			2.44	0.15			0.40
11-4	55-60	10.90	-0.51	0.45	2.64	-0.06			0.40
11-5	55-60	10.95	-0.41		2.52	0.06			0.40
11-6	55-60	11.05	0.01	0.30					0.40
11	CC	11.10	-0.12	0.59	2.62	0.55			0.39
12-1	20-25	11.15	-0.31	0.27	2.67	1.03			0.39
12-2&3	55-60	11.25	-0.28	0.41					0.38
12-5&6	55-60	11.55			2.67	0.63			0.38
12-7	55-60	11.65			2.62	0.33			0.37
15-1	57-62	12.70	-0.07	0.59	2.30	1.29			0.31
15-2	53-58	12.80	-0.13	1.55					0.31
15-3	58-63	12.85	-0.20	0.56	2.37	0.33			0.30
15-4	60-65	12.90	-0.24	0.38					0.29
15-5	61-66	13.00			2.32	0.29			0.29
15	CC	13.05	-0.06	0.60	2.42	0.34			0.28
16-1	21-26	13.10			2.33	0.46			0.29
16-2	111-116	13.30			2.42	0.39			0.30
16-3	22-27	13.35			2.40	0.40			0.30
17-1	31-36	13.75	-0.57	0.83	2.38	0.45			0.35
17-2	31-36	13.80	-0.55	0.58					0.38
17-3	31-36	13.85	-2.43	0.40					0.40
17	CC	13.90	-0.08	1.20	2.12	0.60			0.40
18-1	107-112	14.15			2.07	1.18			0.44
18-1 &	107-112								
18-1	139-144	14.20	-0.47	2.70					0.44
18-1	139-144	14.22			2.09	1.13			0.44
18-2	52-57	14.25	-0.39	1.69	2.21	1.11			0.45

*Described and applied as noted in text. Data in this table are not adjusted.

OXYGEN AND CARBON ISOTOPIC COMPOSITION OF SELECTED SPECIES OF FORAMINIFERA AT SITE 495

Core/ sect	Depth (cm)	Estimated Age (m.y.)	G. venezuelana $\delta^{18}O$	$\delta^{13}C$	D. altispira $\delta^{18}O$	$\delta^{13}C$	G. siakensis $\delta^{18}O$	$\delta^{13}C$	G. sacculifer $\delta^{18}O$	$\delta^{13}C$	Cibicidoides kullenbergi $\delta^{18}O$	$\delta^{13}C$	Ice Effect* Adjustment
20-3	70-72	12.60	0.80	0.99	-0.87	2.24					2.40	0.44	0.31
21-2	67-69	13.40	0.74	1.54	-1.18	2.80					2.14	1.12	0.28
21-4	67-69	13.70									1.96	1.01	0.32
21-6	67-69	14.00	0.37	1.39	-1.59	2.87					1.92	1.11	0.38
22-1	70-72	14.20	0.60	1.60	-1.23	3.17							0.42
22-3	70-72	14.30	0.95	1.78	-1.17	3.09					1.62	1.02	0.46
22-5	35-37	14.37	0.38	1.51	-1.84	2.63					1.65	0.91	0.51
23-1	75-79	14.50	0.64	1.72	-1.17	2.79					1.61	1.20	0.61
23-3	75-79	14.65	0.12	1.66	-1.47	2.80					1.07	0.91	0.65
23-5	75-79	14.75	0.45	1.77	-1.28	3.20					1.29	1.09	0.70
24-2	75-79	14.95	0.42	1.95	-0.89	3.25					1.63	1.27	0.76
24-4	75-79	15.10	0.10	1.82	-1.17	3.31					1.35	1.45	0.77
24-6	75-79	15.18	0.53	1.95	-1.08	3.19					1.68	1.59	0.79
25-1	75-79	15.25	0.00	1.81	-1.32	3.33	-1.52	1.92	-1.72	2.96			0.80
25-3	75-79	15.45	0.70	2.07	-1.22	3.80	-1.15	2.26	-1.72	3.25			0.80
25-4	98-102	15.58	0.46	1.93	-0.69	2.58	-1.41	2.36			1.63	1.53	0.81
25-4	140-144	15.62	0.63	2.22	-0.56	2.58	-1.55	2.09					0.92
25-5	75-79	15.68	0.49	2.00	-1.24	3.28	-1.12	2.21	-1.46	3.02			0.92
25-5	98-102	15.70	0.44	1.98	-0.82	3.43	-1.66	2.26	-1.30	3.02	1.72	1.73	0.92
25-2	140-144	15.72	0.42	1.98	-0.77	3.13	-1.17	2.34	-1.24	3.06			0.92
25-6	98-102	15.80	0.27	1.89	-1.07	3.22	-1.26	2.17	-1.57	3.01			0.92
25-6	140-144	15.85	0.64	1.66	-1.46	2.89	-1.11	2.07	-1.35	2.73	1.92	1.63	0.92
26-1	75-79	15.93	-0.33	1.95	-1.08	2.63	-1.54	1.79	-2.06	2.69			0.92
26-1	98-102	15.95	-0.26	1.94	-1.58	2.78	-1.45	1.75			1.15	1.07	0.92
26-1	140-144	15.98	-0.09	2.03	-1.18	2.59	-1.76	1.37					0.92
26-2	98-102	16.05	0.09	2.14	-1.08	3.09	-1.18	2.02	-1.17	3.03			0.92
26-2	140-144	16.10	-0.01	2.09	-1.14	2.99	-1.52	1.74	-1.75	2.95			0.92
26-3	75-79	16.15	-0.15	2.15	-0.87	2.87	-1.89	1.97	-1.70	2.77			0.92
26-3	92-96	16.18	-0.02	2.19	-0.90	3.14	-1.02	2.08	-1.25	2.99	1.63	1.55	0.92
26-3	140-144	16.21	-0.05	2.28	-0.91	3.21	-1.17	2.13	-1.16	3.19			0.92
26-4	98-102	16.30	0.02	2.21	-0.94	3.11	-1.15	2.09	-1.33	3.46	1.52	1.52	0.92
26-4	140-144	16.33	0.04	2.25	-0.99	2.71	-1.06	2.15	-1.73	2.71			0.92
26-5	75-79	16.35	0.07	2.16	-0.76	2.79	-1.19	2.23	-1.58	3.32			0.92
26-5	98-102	16.36	0.13	2.19	-0.85	2.99	-1.27	2.14	-1.94	2.85	1.49	1.54	0.92
26-5	136-140	16.38	0.04	1.97	-0.91	3.66	-1.12	1.98	-0.73	3.07			0.92
26-6	52-56	16.40	-0.25	2.30	-0.68	2.97	-1.12	2.15	-1.58	3.16	1.69	1.23	0.92
26-6	103-107	16.43	-0.09	2.17	-0.99		-1.22	2.00					0.92
27-1	75-79	16.50	0.13	1.70	-0.62	2.06	-1.16	1.69			1.47	1.20	0.92
27-3	75-79	16.64	-0.12	1.90							1.40	0.93	0.92
27-5	75-79	16.73	0.00	1.48	-1.17	2.17	-1.06	1.57			1.40	0.83	0.92
28-1	75-79	16.82	0.16	1.32	-1.46	2.28	-1.17	1.26			1.22	0.66	0.92
28-3	75-79	16.88	0.13	1.38	-1.13	2.90	-0.94	1.45			1.46	0.67	0.92
28-5	64-68	16.93	0.10	1.37	-1.11	2.47	-1.05	1.35			1.30	0.65	0.92
28-7	44.48	17.00	0.45	1.89	-1.19	2.63	-0.08	1.67			1.67	1.17	0.92

*Described and applied as noted in text. Data in this table are not adjusted.

OXYGEN AND CARBON ISOTOPIC COMPOSITION OF SELECTED
SPECIES OF FORAMINIFERA AT NEWPORT BEACH

Sample*	G. bulloides $\delta^{18}O$	$\delta^{13}C$	Uvigerina $\delta^{18}O$	$\delta^{13}C$
N7A	0.25	0.38	1.62	-0.41
N7	-0.29	0.68	1.95	-0.59
WNPB 13	0.22	0.50		
N6	-0.03	0.05	-0.30	
N5	-0.62	-0.19		
N4A	0.27	0.24	-1.42	-0.69
N4	-0.13	0.27	1.35	-0.09
N3	-0.38	-0.23	0.92	-0.42
N2A			0.98	-0.73
NE 19	0.42	-0.21	2.33	-0.22
TM17	0.19	-0.31	-0.79	-0.49
TM14			0.37	-0.49

*For the stratigraphic position of these samples, see
Barron and Keller (1983).

REFERENCES CITED

Bandy, O. L., and Ingle, J. C., Jr., 1970, Neogene planktonic events and the radiometric scale, California, *in* Bandy, O. L., ed., Radiometric dating and paleontologic zonation: Geological Society of America Special Paper 124, p. 131–172.

Barron, J. A., and Keller, G., 1983, Paleotemperature oscillations in the middle and late Miocene of the northeastern Pacific: Micropaleontology, v. 29[2], p. 151–181.

Belanger, P. E., Curry, W. B., and Matthews, R. K., 1981, Core-top evaluation of benthic foraminiferal isotopic ratios for paleoceanographic interpretations: Palaeogeography, Palaeoclimatology, Palaeoecology, v. 33, p. 205–220.

Berger, W. H., and Gardner, J. V., 1975, On the determination of Pleistocene temperatures from planktonic foraminifera: Journal of Foraminiferal Research, v. 5, p. 102–113.

Craig, H., 1957, Isotopic standards for carbon and oxygen and correction factors for mass spectrometric analysis of carbon dioxide: Geochimica et Cosmochimica Acta, v. 12, p. 133–149.

Craig, H., and Gordon, L., 1965, Deuterium and oxygen-18 variation in the ocean and marine atmosphere: University of Rhode Island Occasional Publication 3, p. 277–374.

Curry, W. B., and Matthews, R. K., 1981a, Equilibrium ^{18}O fractionation in small size fraction planktic foraminifera: Evidence from Recent Indian Ocean sediments: Marine Micropaleontology, v. 6, p. 327–337.

Curry, W. B., and Matthews, R. K., 1981b, Paleo-oceanographic utility of oxygen isotopic measurements on planktic foraminifera: Indian Ocean core-top evidence: Paleogeography, Paleoclimatology, Paleoecology, v. 33, p. 173–191.

Diester-Haass, L., Schrader, H. J., and Thiede, J., 1973, Sedimentological and paleoclimatological investigations of two pelagic ooze cores off Cape Barbas, Northwest Africa: Meteor Forsch.-Ergebnisse, Reihe C, v. 16, p. 19–66.

Duplessy, J. C., 1978, Isotope studies, *in* Gribbin, J., ed., Climate change: Cambridge, Cambridge University Press, p. 46–67.

Duplessy, J. C., Lalou, C., and Vinot, A. C., 1970, Differential isotopic fractionation in benthic foraminifera and paleotemperatures reassessed: Science, v. 168, p. 250–251.

Emiliani, C., 1954, Depth habitats of some species of pelagic foraminifera as indicated by oxygen isotope ratios: American Journal of Science, v. 252, p. 149–158.

Emiliani, C., 1955, Pleistocene temperatures: Journal of Geology, v. 63, p. 538–578.

Emrich, K., Ehhalt, D. H., and Vogel, J. C., 1970, Carbon isotope fractionation during the precipitation of calcium carbonate: Earth and Planetary Science Letters, v. 8, p. 363–371.

Epstein, S., Buchsbaum, R., Lowenstam, H. A., and Urey, H. C., 1953, Revised carbonate-water isotopic temperature scale: Geological Society of America Bulletin, v. 64, p. 1315–1326.

Ganssen, G., and Sarnthein, M., 1983, Stable isotope composition of foraminifera: the surface and bottom water record of coastal upwelling, *in* Suess, E., and Thiede, J., eds., Coastal upwelling: Its sediment record: New York, Plenum Press, p. 99–121.

Graham, D., Corliss, B., Bender, M. L., and Keigwin, L. D., Jr., 1981, Carbon and oxygen isotopic disequilibria of recent deep-sea benthic foraminifera: Marine Micropaleontology, v. 6, p. 483–497.

Huyer, A., 1983, Coastal upwelling in the Caifornia current system: Progress in Oceanography, v. 12, p. 259–284.

Ingle, J. C., Jr., 1967, Foraminiferal biofacies variation and the Miocene-Pliocene boundary in southern California: Bulletin American Paleontologist, v. 52, p. 217–394.

Ingle, J. C., Jr., 1973a, Neogene foraminifera from the northeastern Pacific Ocean, Leg 18, Deep Sea Drilling Project, *in* Kulm, L. D., et al., Initial Reports of the Deep Sea Drilling Project, Washington, D.C., U.S. Government Printing Office, v. 18, p. 517–567.

Ingle, J. C., Jr., 1973b, Summary comments on Neogene biostratigraphy, physical stratigraphy, and paleoceanography in the marginal northeastern Pacific Ocean, *in* Kulm, L. D., et al., Initial Reports of the Deep Sea Drilling Project, Washington, D.C., U.S. Government Printing Office, v. 18, p. 949–960.

Kahn, M. I., and Williams, D. F., 1981, Oxygen and carbon isotopic composition of living planktonic foraminifera from the northeast Pacific Ocean: Paleogeography, Paleoclimatology, Paleoecology, v. 33, p. 47–69.

Kaneps, A. G., 1973, Cenozoic planktonic foraminifera from the eastern equatorial Pacific Ocean, *in* van Andel, T. H., et al., Initial Reports of the Deep Sea Drilling Project, Washington, D.C., U.S. Government Printing Office, v. 16, p. 713–745.

Keigwin, L. D., Jr., 1976, Late Cenozoic planktonic foraminiferal biostratigraphy and paleoceanography of the Panama Basin: Micropaleontology, v. 22, p. 419–442.

Keigwin, L. D., Jr., 1979, Late Cenozoic stable isotope stratigraphy and paleoceanography of DSDP sites from the east equatorial and central North Pacific Ocean: Earth and Planetary Science Letters, v. 45, p. 361–382.

Keller, G., 1978, Late Neogene planktonic foraminiferal biostratigraphy and paleoceanography of the northeastern Pacific: Evidence from DSDP Sites 173 and 310 at the North Pacific Front: Journal of Foraminiferal Research, v. 8[4], p. 332–349.

Keller, G., 1980a, Middle to late Miocene datum levels and paleoceanography of the north and southeastern Pacific Ocean: Marine Micropaleontology, v. 5, p. 249–281.

Keller, G., 1980b, Early to middle Miocene planktonic foraminiferal datum levels of the equatorial and subtropical Pacific: Micropaleontology, v. 26, p. 372–391.

Keller, G., 1981, Miocene biochronology and paleoceanography of the North Pacific: Marine Micropaleontology, v. 6, p. 535–551.

Keller, G., and Barron, J. A., 1981, Integrated planktic foraminiferal and diatom biochronology for the northeast Pacific and the Monterey Formation, *in* Garrison, R. E. et al., eds., The Monterey Formation and related siliceous rocks of California: Pacific Section, Society of Economic Paleontologists and Mineralogists, p. 43–54.

Keller, G., Barron, J. A., and Burckle, L. H., 1982, North Pacific late Miocene correlations using microfossils, stable isotopes, percent CaCO₃, and magnetostratigraphy: Marine Micropaleontology, v. 7, p. 327–357.

Loutit, T. S., Pisias, N. G., and Kennett, J. P., 1983, Pacific Miocene carbon isotope stratigraphy using benthic foraminifera: Earth and Planetary Science Letters, v. 66, p. 48–62.

Matthews, R. K., and Poore, R. Z., 1980, Tertiary δ^{18}O record and glacioeustatic sea-level fluctuations: Geology, v. 8, p. 501–504.

Prell, W. L., and Curry, W. B., 1981, Faunal and isotopic indices of monsoonal upwelling: Western Arabian Sea: Oceanologica Acta, v. 4, p. 91–98.

Savin, S. M., and Douglas, R. G., 1973, Stable isotope and magnesium geochemistry of recent planktonic foraminifera from the South Pacific: Geological Society of America Bulletin, v. 84, p. 2327–2342.

Savin, S. M., Douglas, R. G., and Stehli, F. G., 1975, Tertiary marine paleotemperatures: Geological Society of America Bulletin, v. 86, p. 1499–1510.

Savin, S. M., Douglas, R. G., Keller, G., Killingley, J. S., Shaughnessy, L., Sommer, M. A., Vincent, E., and Woodruff, F., 1981, Miocene benthic foraminiferal isotope records: A synthesis: Marine Micropaleontology, v. 6, p. 423–450.

Savin, S. M., Abel, L., Barrera, E., Bender, M., Hodell, D., Keller, G., Kennett, J. P., Killingley, J., Murphy, M., and Vincent, E., 1985, The evolution of Miocene surface and near-surface Oceanography: Oxygen Isotope Evidence: [this volume].

Sclater, J., Meinke, L., Bennett, A., and Murphy, C., 1985, The Depth of the Ocean Through the Neogene: [this volume].

Shackleton, N. J., 1974, Attainment of isotopic equilibrium between ocean water and benthonic foraminifera genus *Uvigerina*: isotopic changes in the ocean during the last glacial, *in* Les methodes quantitative d'etude des variations due climat au cours du Pleistocene: Colloques Internationaux due C.N.R.S. No. 219, p. 203–209.

Shackleton, N. J., 1977, The oxygen isotope stratigraphic record of the Late Pleistocene: Royal Society of London Philosophical Transactions, Ser. B, v. 280, p. 169–182.

Shackleton, N. J., and Kennett, J. P., 1975, Paleotemperature history of the Cenozoic and the initiation of Antarctic glaciation: oxygen and carbon isotope analyses in DSDP Sites 277, 279 and 281, *in* Kennett, J. P., et al., Initial Reports of the Deep Sea Drilling Project, Washington, D.C., U.S. Government Printing Office, v. 29, p. 743–755.

Thiede, J., 1972, Dominance and diversity of planktonic foraminiferal faunas in Atlantic Ibero-Moroccan continental slope sediment: Journal of Foraminiferal Research, v. 2[2], p. 93–102.

Thompson, P. R., 1982, Foraminifers of the Middle America Trench, *in* Aubouin, J., et al., Initial Reports of the Deep Sea Drilling Project, Washington, D.C., U.S. Government Printing Office, v. 67, p. 351–381.

Vergnaud-Grazzini, C., 1976, Non-equilibrium isotopic compositions of shells of planktonic foraminifera in the Mediterranean Sea: Palaeogeography, Palaeoclimatology, Palaeoecology, v. 20, p. 263–276.

Vincent, E., Killingley, J. S., and Berger, W. H., 1981, Stable isotope composition of benthic foraminifera from the equatorial Pacific: Nature, v. 289, p. 639–643.

Williams, D. F., Sommer II, M. A., and Bender, M. L., 1977, Carbon isotopic composition of recent planktonic foraminifera of the Indian Ocean: Earth and Planetary Science Letters, v. 36, p. 391–403.

Williams, D. F., Bé, A.W.H., and Fairbanks, R. G., 1979, Seasonal oxygen isotopic variations in living plantonic foraminifera off Bermuda: Science, v. 206, p. 447–449.

Woodruff, F., Savin, S. M., and Douglas, R. G., 1980, Biological fractionation of oxygen and carbon isotopes in Recent benthic foraminifera: Marine Micropaleontology, v. 5, p. 3–11.

Woodruff, F., Savin, S. M., and Douglas, R. G., 1981, Miocene stable isotope record: A detailed deep Pacific Ocean study and its paleoclimatic implications: Science, v. 212, p. 665–668.

Yeats, R. S., et al., 1981, Site 470: Off Guadalupe Island, *in* Initial Reports of the Deep Sea Drilling Project, Washington, D.C., U.S. Government Printing Office, v. 63, p. 227–268.

Zobel, B., 1971, Foraminifera from plankton tows, Arabian Sea: Areal distribution as influenced by ocean water masses: Proceedings of II Plankton Conference, 2, p. 1323–1335.

MANUSCRIPT ACCEPTED BY THE SOCIETY DECEMBER 17, 1984
CONTRIBUTION NO. 153 OF THE DEPARTMENT OF GEOLOGICAL SCIENCES, CASE WESTERN RESERVE UNIVERSITY

Geological Society of America
Memoir 163
1985

Miocene oxygen and carbon isotope stratigraphy of the tropical Indian Ocean

Edith Vincent
John S. Killingley
Wolfgang H. Berger
Scripps Institution of Oceanography
University of California, San Diego
La Jolla, California 92093

ABSTRACT

Oxygen and carbon isotope stratigraphies are presented for planktonic and benthic foraminifera from Miocene sediments of the tropical Indian Ocean (DSDP Sites 214, 216, 237, 238) and placed into a biostratigraphic framework. Approximately 1000 foraminiferal analyses were performed to determine both oxygen and carbon isotopic ratios. The species analyzed were *Globigerinoides sacculifer* and *Dentoglobigerina altispira* (shallow-dwelling planktonic), *Globoquadrina venezuelana* (deep-dwelling planktonic), and *Oridorsalis umbonatus* (benthic). We identified a number of isotopic signals which appear to be synchronous with previously recognized signals in the Pacific Ocean, and thus provide useful tools for chronostratigraphic correlations.

The oxygen isotope record is dominated by a permanent increase in benthic foraminiferal $\delta^{18}O$ values of $1^0/_{00}$ or more in the middle Miocene, between approximately 16.5 and 12.6 Ma. The 0–18 enrichment in planktonic foraminifera is less distinct and temporary. Thus these data do not unequivocally support the notion of a large ice buildup at this time. A large post-shift separation of benthic and planktonic $\delta^{18}O$ values indicates a more stably stratified ocean, presumably with a stronger thermocline, and hence increased regional upwelling.

The carbon isotopic record is dominated by a broad early-to-middle Miocene positive excursion seen in both planktonic and benthic records. The excursion begins with a shift toward greater $\delta^{18}C$ values by about $1^0/_{00}$ (the Chron-16 Carbon Shift) in the latest early Miocene between approximately 17.5 and 16.5 Ma and ends with a rather gradual decline toward initial (early Miocene) values. The excursion terminates at approximately 13.5 Ma. The next large disturbance is the Chron-6 event, with a $\delta^{13}C$ shift to low values, by about $1^0/_{00}$ in all three foraminiferal records.

We propose that the Chron-16 shift, and subsequent "heavy carbon" state of the ocean, was caused by the extraction of isotopically light carbon into the organic-rich Monterey Formation and its equivalent sediment bodies rimming the Pacific. A decrease in the rate of carbon extraction (due to exhaustion of nutrient phosphorus) would have allowed relaxation toward initial values. We further propose that the Chron-6 carbon shift, toward low δ-values, was due to regression and contribution of organic matter to the ocean from erosion. This would be the reverse process from the Monterey buildup.

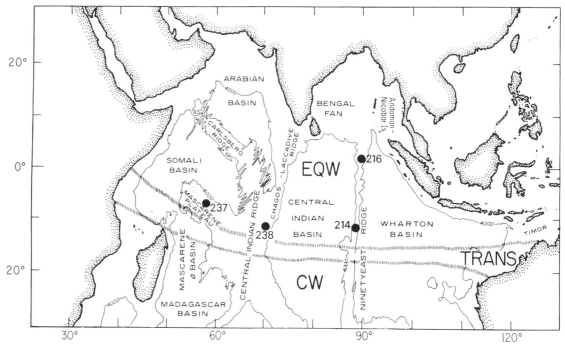

Figure 1. DSDP site locations. Major water masses: EQW, Equatorial Water; CW, Central Water; TRANS, Tropical-Subtropical Transition Zone (after Bé and Hutson, 1977; modified from Sverdrup et al., 1942).

INTRODUCTION

Many questions regarding the paleoceanographic and climatic history of the Cenozoic can be resolved only by multiple stratigraphic approaches; that is, the simultaneous consideration of several signals. The ocean-atmosphere system consists of a great number of variables, each of different importance at different times. In Pleistocene deep-sea stratigraphy the classic example for multiple integrated stratigraphies was set by Imbrie and co-workers. By combining oxygen isotope stratigraphy with faunally-derived paleotemperature estimates, they showed that, in principle, the ice volume effects could be separated from the effect of temperature on $\delta^{18}O$ values (Imbrie et al., 1973).

Tertiary stratigraphy has been at a disadvantage with respect to the Pleistocene methodology, in that time-control remains unsatisfactory (despite much progress in recent years) and paleotemperature calculations, by the transfer-function method, are not possible. Our approach to this difficulty has been to use multiple isotope signals. We have made increasing use of shifts in stable isotope ratios within the record, as correlatable time markers. This technique was introduced by Emiliani (1955), and has since been applied successfully to both Pleistocene and pre-Pleistocene sediments (e.g. Thierstein et al., 1977; Haq et al., 1980, among others). We have identified a number of isotopic markers in the Miocene record from four DSDP sites in the tropical Indian Ocean, which correlate with those previously recognized in the Pacific Ocean.

We have routinely determined the stable isotopic composition of several species in each sample in order to glean informa-

tion simultaneously on oceanic conditions at the surface, in the thermocline, and at the seafloor, as a function of geologic time. We hoped that this would provide information about the development of the thermocline, the overall temperature gradient, the oxygen minimum layer, and the general level of productivity. We present multispecies isotope stratigraphies for the entire Miocene at four DSDP sites. These records greatly increase the number of Miocene stable isotope stratigraphies which have both planktonic and benthic data. Only a few such records are available (e.g. Keigwin, 1979; Kennett et al., 1979; Vincent et al., 1980).

DATA

Results of isotopic analyses from DSDP Sites 214, 216, 237 and 238 (see site locations in Table 1 and Figure 1) are given in Tables 2 to 5 and are plotted against depth below sea floor at each site, with their biostratigraphy, in Figures 2 to 9. Approximately 1000 foraminiferal analyses were performed to determine both $^{18}O/^{16}O$, and $^{13}C/^{12}C$ ratios between 1978 and 1984 (135 of these analyses have been published previously; Vincent et al., 1980).

Our main purpose here is to make these data generally available, so they can be brought to bear on major paleoceanographic questions, such as Antarctic ice-buildup in the mid-Miocene (Savin et al., 1975; Shackleton and Kennett, 1975; Matthews and Poore, 1980) and the problem of step-like climate transitions (Berger, et al., 1981; Vincent and Berger, 1981; Berger, 1982). At this juncture, our analysis is of a preliminary nature: the four records are not yet well integrated with each other, or with

TABLE 1. LOCATION OF DSDP SITES

Hole	Latitude	Longitude	Water Depth (m)	Physiographic Feature
214	11°20.21'S	88°43.08'E	1665	Ninetyeast Ridge
216, 216A	1°27.73'N	90°12.48'E	2247	Ninetyeast Ridge
237	7°04.99'S	58°07.48'E	1640	Mascarene Plateau
238	11°09.20'S	70°31.56'E	2844	Central Indian Ridge

information from other ocean basins published in recent studies (Savin, 1977; Keigwin, 1979; Woodruff et al., 1981; Savin et al., 1981; Barron and Keller, 1982).

GENERAL SETTING OF DSDP SITES STUDIED

All four sites lie under equatorial water masses. Site 216, however, is located in the central area of the Equatorial Water whereas the three other sites are located near the transition between Equatorial Water and Central Water (Sverdrup et al., 1942). Rochford (1962) described a "Tropical-Subtropical Transition Zone" between these two water masses (see Fig. 1). Two distinct circulation systems are delineated: 1) the seasonally changing monsoon gyre in the equatorial region and 2) the subtropical anticyclonic gyre of the Central Water (Wyrtki 1971, 1973). Wyrtki described a hydrochemical front separating high-nutrient, low-oxygen waters of the monsoon gyre in the northern Indian Ocean from low-nutrient, high-oxygen waters in the subtropical gyre. The boundary between the areas of high and low productivity is reflected by the present pattern of sediment distribution in the deep basins on each side of the Ninetyeast Ridge. In these basins, siliceous sediments are associated with the equatorial high-productivity area north of about 15°S, whereas sediments south of this latitude are pelagic clays (see Davies and Kidd, 1977).

Bé and Hutson (1977) have shown that faunal provinces of planktonic foraminifera in the Indian Ocean generally follow the patterns of nutrient concentrations and planktonic productivity, which in turn are closely linked to water masses and circulation patterns. In terms of faunal provinces recognized by these authors in surface sediment assemblages, Site 216 is in the central area of the "Tropical Assemblage" whereas the three other sites are in the vicinity of the boundary between the "Tropical Assemblage" and the "Tropical-Subtropical Assemblage."

In Miocene time the sites were further south. Owing to plate motions, all sites moved toward the northeast through the Cenozoic. The extent of the motion, however, has been much less for Site 237 located on the African plate than for the three other sites located on the Indian plate (see paleoposition reconstruction maps, Sclater et al., this volume). At 22 Ma, in the earliest Miocene, Site 216 was at a latitude similar to that of Site 214 today and was apparently the only one of the four sites in the equatorial area of high productivity at that time. Evidence for this is the presence of a continuous Miocene siliceous record at Site 216 whereas at the three other sites the siliceous record starts in the Middle or Upper Miocene portion of the section.

All four sites were located well above the Calcite Compensation Depth throughout the Miocene. In the tropical Indian Ocean the CCD was near 4000 meters in the Oligocene, deepening progressively through the Miocene (Sclater et al., 1977). Today it lies near 5000 meters (Kolla et al., 1976). From paleodepth reconstructions, the depth of the sea floor in the Early Miocene at 20 Ma was about 2000 meters at Site 238, the deepest of the four sites. The sites appear to have crossed the foraminiferal lysocline during Miocene time as evidenced by poor foraminiferal preservation in Lower Miocene sediments. The foraminiferal lysocline in the tropical Indian Ocean lies near 3600 m today (Coulbourn and Parker, in prep.). Thus it was significantly shallower in the early Miocene.

At each of the four sites continuous coring provided good recovery and obtained an apparently continuous Miocene record, conformably overlying Oligocene sediments. At Site 216, however, the uppermost 20 meters of Miocene sediments were not recovered owing to discontinuous coring in that interval. Each Miocene sequence is in excess of 100 m thick. The sediments consist of calcareous ooze with a calcium carbonate content typically greater than 85%. Calcareous plankton is abundant throughout each sequence. Calcareous nannofossils are well- to moderately-well preserved. Foraminiferal faunas are well-preserved in the Upper Miocene sediments and moderately- to poorly-preserved in the Middle and Lower Miocene.

Sediment accumulation rates are low at Sites 214, 216, and 237 and higher at the deepest site 238, where they are near three times as high as at the other sites. Average values for Lower and Middle Miocene sediments at Sites 214, 216 and 237 are close to 5 m/m.y., increasing to near 20 m/m.y. for the Upper Miocene. At Site 238 average values are about 12 m/m.y. for the Lower-Middle Miocene, increasing to 50 m/m.y. for the Upper Miocene. The increased depositional rate at Site 238 possibly reflects penecontemporaneous redeposition at this site, a result of its location close to the ridge crest and to an active transform fault (Vincent, 1974).

SAMPLING STRATEGY AND SAMPLE PREPARATION

Sampling intervals vary within the different stratigraphic

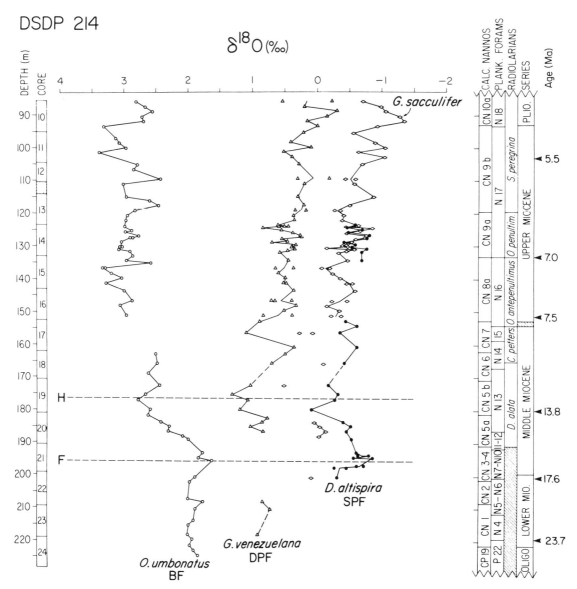

Figure 2. Miocene oxygen isotope stratigraphies of the benthic foraminifera (BF) *Oridorsalis umbonatus*
(○), the shallow-dwelling planktonic foraminifera (SPF) *Dentoglobigerina altispira* (●) and *Globigerinoides sacculifer* (◇) and the deep-dwelling planktonic species (DPF) *Globoquadrina venezuelana* (△)
at DSDP Site 214. Data from Table 2. Where there are duplicate analyses, lines are drawn through the
average. Biostratigraphy after Vincent (1977). Biostratigraphic zonations according to the zonal schemes
of Okada and Bukry (1980) for calcareous nannofossils, Blow (1969) for planktonic foraminifera, and
Riedel and Sanfilippo (1971) for radiolarians. Hachured areas in the column at left represent un-
recovered intervals and in columns at right unzoned intervals. Numerical ages at right are derived from
the position in cores of biostratigraphic datums calibrated elsewhere to the paleomagnetic time scale (see
text). Ages follow those of Barron et al. (this volume) based on Berggren et al. (in press) paleomagnetic
time scale. Horizons A to H are defined in Figure 11.

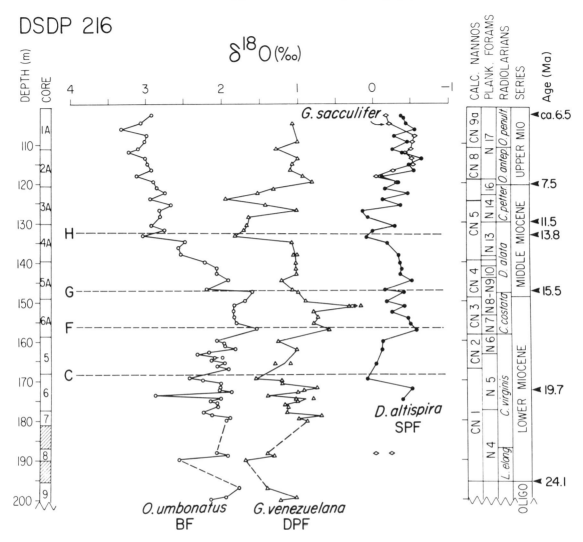

Figure 3. Miocene oxygen isotope stratigraphies at DSDP Site 216. Data from Table 3. See rest of caption in Figure 2.

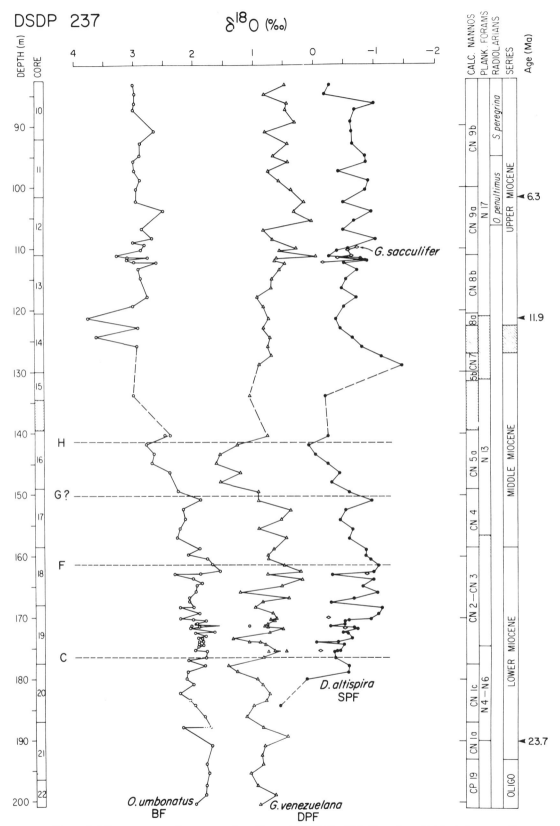

Figure 4. Miocene oxygen isotope stratigraphies at DSDP Site 237. Data from Table 4. See rest of caption in Figure 2.

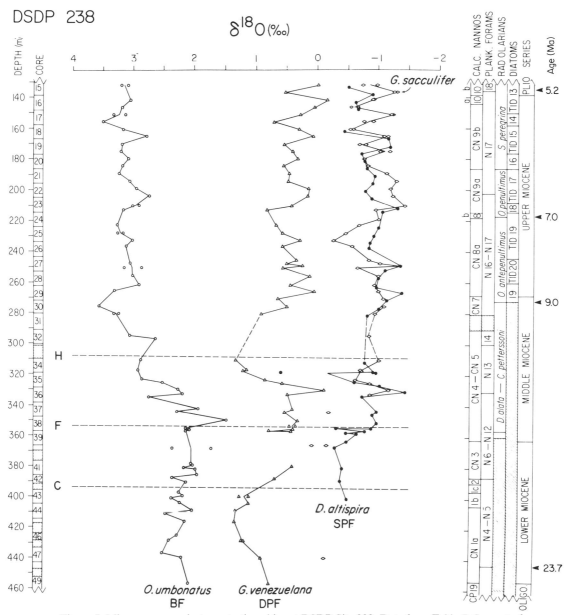

Figure 5. Miocene oxygen isotope stratigraphies at DSDP Site 238. Data from Table 5. See rest of caption in Figure 2. Diatom zonation after Schrader (1974).

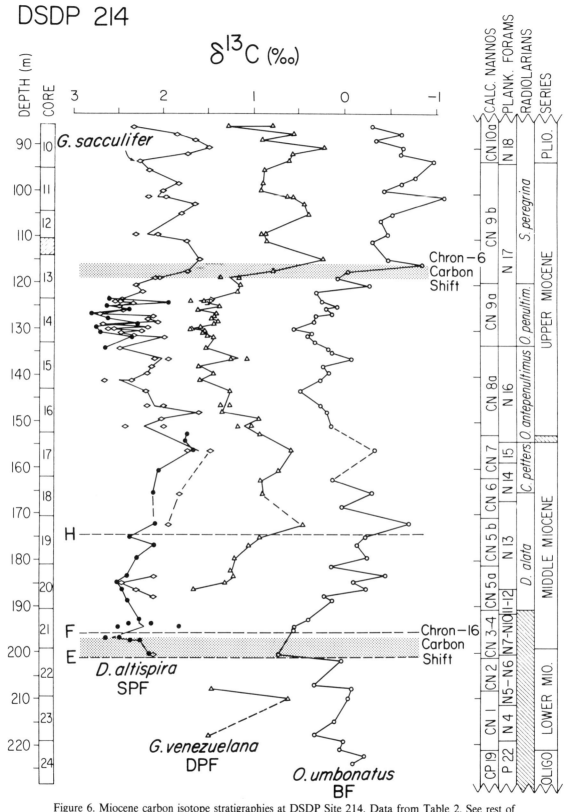

Figure 6. Miocene carbon isotope stratigraphies at DSDP Site 214. Data from Table 2. See rest of caption in Figure 2.

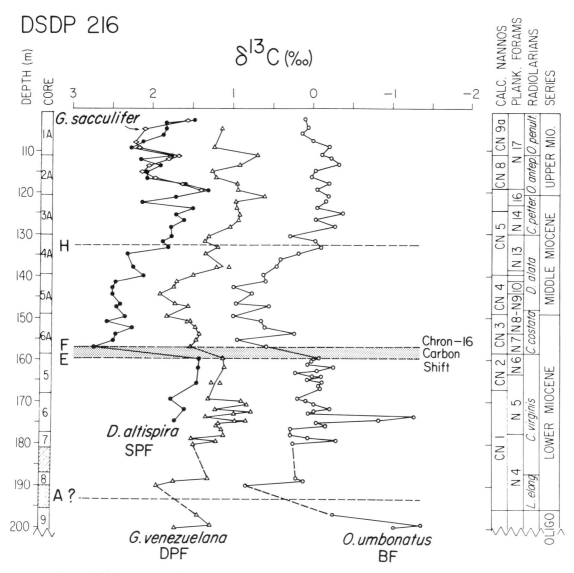

Figure 7. Miocene carbon isotope stratigraphies at DSDP Site 216. Data from Table 3. See rest of caption in Figure 2.

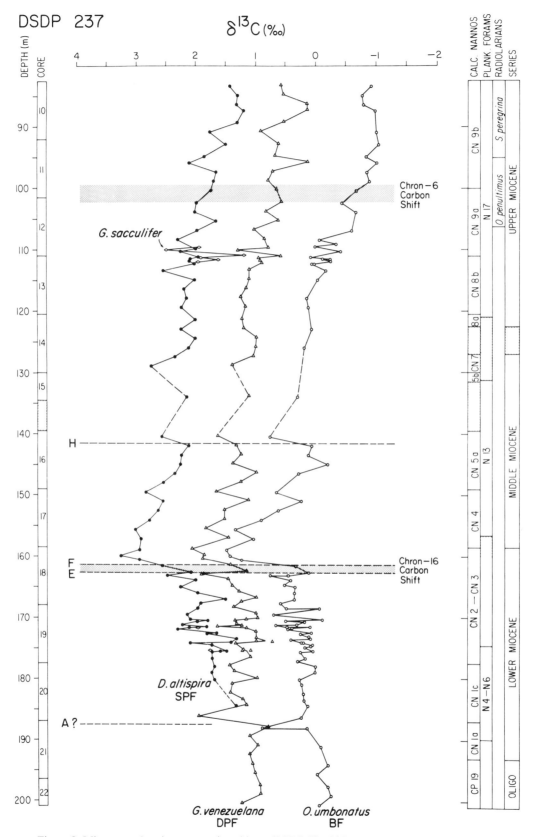

Figure 8. Miocene carbon isotope stratigraphies at DSDP Site 237. Data from Table 4. See rest of caption in Figure 2.

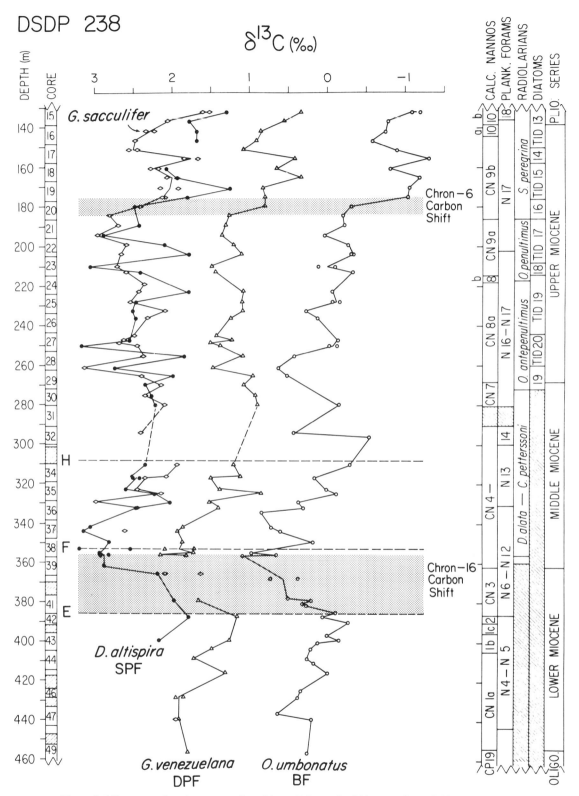

Figure 9. Miocene carbon isotope stratigraphies at DSDP Site 238. Data from Table 5. See rest of caption in Figure 2 and 5.

TABLE 2. OXYGEN AND CARBON ISOTOPE COMPOSITION (⁰/oo, PDB) OF GLOBIGERINOIDES SACCULIFER (350-420 μm), DENTOGLOBIGERINA ALTISPIRA (350-420 μm), GLOBOQUADRINA VENEZUELANA (500-600 μm), AND ORIDORSALIS UMBONATUS (>250 μm) AT DSDP SITE 214

Hole	Core	Section	Interval (cm)		Depth below sea floor middle interval (cm)	G. sacculifer		D. altispira		G. venezuela		O. umbonatus	
						$\delta^{18}O$	$\delta^{13}C$	$\delta^{18}O$	$\delta^{13}C$	$\delta^{18}O$	$\delta^{13}C$	$\delta^{18}O$	$\delta^{13}C$
214	10	1	83	85	86.34	-0.72	2.34			-0.22	0.82	2.83	-0.30
214	10	1	83	85	86.34					0.54	1.31		
214	10	2	97	99	87.98	-1.00	1.85			0.20	0.58	2.69	-0.62
214	10	3	80	82	89.31	-1.07	1.65			-0.30	0.93	2.58	-0.34
214	10	4	82	84	90.93	-1.29	1.50			-0.14	0.25	2.75	-0.64
214	10	5	81	83	92.32	-1.36	1.74			0.16	0.60	2.72	-0.61
214	10	6	84	86	93.85	-0.94	2.26			0.00	0.64	3.34	-0.97
214	11	1	82	84	95.83	-0.57	2.16			0.22	0.91		
214	11	2	89	91	97.40							3.15	-0.77
214	11	3	78	80	98.79	-0.91	1.84			0.42	0.93	3.10	-0.62
214	11	4	82	84	100.33	-1.07	2.01			0.10	0.95	2.99	-0.43
214	11	5	66	68	101.67	-0.65	2.17			0.53	0.65	3.41	-1.09
214	11	5	66	68	101.67	-0.66	1.98			0.52	0.59		
214	11	6	73	75	103.24	-1.06	1.65			0.40	0.47		
214	12	1	88	90	105.39	-0.71	1.80			0.29	0.42	2.81	-0.52
214	12	4	80	82	109.81	-0.45	2.07			-0.18	0.94	2.45	-0.47
214	12	4	80	82	109.81	-0.61	2.31			0.31	0.90		
214	12	5	81	83	111.32	-0.59	1.74			0.20	0.88	3.03	-0.30
214	13	1	118	120	115.19	-0.89	1.60			0.31	0.26	2.98	-0.47
214	13	2	80	82	116.31							2.62	-0.84
214	13	3	80	82	117.81	-0.52	1.73			021	0.81	2.48	-0.03
214	13	4	80	82	119.31	-0.28	2.05			0.34	1.39	2.85	0.08
214	13	4	80	82	119.31	-0.37	2.09			0.17	1.19		
214	13	6	82	84	122.33	-0.40	2.23			0.36	1.20	2.99	0.32
214	14	1	40	44	123.92	-0.66	2.46	-0.61	2.61	0.61	1.49		
214	14	1	80	82	124.31	-0.49	2.48			0.45	1.49	3.01	0.25
214	14	1	80	82	124.31	-0.46	2.53			0.51	1.70		
214	14	1	80	82	124.31					0.59	1.55		
214	14	1	120	124	124.72	-0.87	2.33	-0.48	1.96	0.84	1.52		
214	14	2	43	47	125.45	-0.69	2.49	-0.72	2.64	0.56	1.38	2.90	0.08
214	14	2	82	84	125.83							3.00	0.21
214	14	2	120	124	126.22	-0.47	2.44	-0.50	2.39	0.38	1.62		
214	14	3	56	60	127.08	-0.77	2.70	-0.83	2.80	0.26	1.42	2.79	0.14
214	14	3	104	106	127.55	-0.76	2.11			0.24	1.44	2.88	
214	14	3	104	106	127.55							2.92	0.32
214	14	3	130	134	127.82								
214	14	4	10	14	128.12	-0.61	2.17	-0.79	2.62	0.56	1.48		
214	14	4	80	82	128.81	-0.55	2.06			0.46	1.40	3.06	0.34
214	14	4	120	124	129.22	-0.51	2.67	-0.43	2.30	0.70	1.44		
214	14	5	44	48	129.96	-0.40	2.17	-0.61	2.75	0.33	1.58		
214	14	5	83	85	130.34	-0.55	2.61				1.70	3.08	
214	14	5	83	85	130.34	-0.49	2.54			0.39	1.67	3.05	0.56
214	14	5	123	127	130.75	-0.16	2.24	-0.56	2.71	0.41	1.54		
214	14	6	18	22	131.20	-0.59	2.47	-0.79	2.71	0.35	1.56	3.10	0.36
214	14	6	80	82	131.81	-0.47	2.33			0.46	1.52	2.93	0.41
214	14	6	119	123	132.21	-0.34	1.99	-0.71	2.36	0.59	1.45		
214	15	1	21	23	133.22							2.88	0.33
214	15	2	10	14	134.62	-0.49	2.48	-0.71	2.66	0.45	1.54	2.98	0.18
214	15	2	87	89	135.38							2.60	0.14
214					136.42								
214	15	3	78	80	136.79	-0.08	1.95			0.37	1.26	3.31	
214	15	3	78	80	136.79	-0.21	2.10			0.65	1.08	3.35	-0.07
214	15	4	10	14	137.62								
214	15	4	84	86	138.35	-0.25	2.13			0.61	1.61	3.21	0.24
214	15	5	10	14	139.12								
214	15	5	80	82	139.81	-0.37	2.18			0.50	1.44	3.05	0.18
214	15	6	81	83	141.32	-0.47	2.35			0.52	1.60	3.30	0.27
214	15	6	81	83	141.32	-0.56	2.65			0.48	1.58		
214	16	1	64	66	143.65	-0.61	2.19			0.37	1.27	3.01	
214	16	3	77	79	146.78	-0.48	2.00			0.71		2.89	0.27
214	16	3	77	79	146.78	-0.24	2.18			0.65	1.38		

TABLE 2. OXYGEN AND CARBON ISOTOPE COMPOSITION (O/oo, PDB) OF GLOBIGERINOIDES SACCULIFER (350-420 μm), DENTOGLOBIGERINA ALTISPIRA (350-420 μm), GLOBOQUADRINA VENEZUELANA (500-600 μm), AND ORIDORSALIS UMBONATUS (>250 μm) AT DSDP SITE 214 (continued)

Hole	Core	Section	Interval (cm)		Depth below sea floor middle interval (cm)	G. sacculifer		D. altispira		G. venezuela		O. umbonatus	
						δ18O	δ13C	δ18O	δ13C	δ18O	δ13C	δ18O	δ13C
214	16	3	77	79	146.78					0.40	1.27		
214	16	4	75	77	148.26	-0.16	1.60			0.33	1.36	3.08	0.20
214	16	5	69	71	149.70	-0.36	2.02			0.52	0.95		
214	16	6	79	81	151.30	-0.39	2.00			0.39	1.03	2.98	0.15
214	16	6	79	81	151.30	-0.23	2.42			0.83	1.18		
214	17	1	58	60	153.09			-0.46	1.75	0.89	0.94		
214	17	2	58	60	154.59			-0.64	1.77				
214	17	3	104	106	156.55	0.06	1.73	-0.37	1.68	1.10	0.59	2.22	-0.32
214	17	3	104	106	156.55	0.26	1.48						
214	17	6	95	97	160.96			-0.64	2.07	0.36	0.74		
214	18	1	95	97	162.96					0.50	0.93	2.53	0.14
214	18	3	89	91	165.90	-0.11	1.82	-0.44	2.13	0.70	0.91	2.50	-0.29
214	18	5	86	88	168.87							2.64	0.04
214	19	1	100	102	172.51	0.50	1.95	-0.19	2.11	1.03	0.46	2.46	-0.69
214	19	3	79	81	175.30			-0.34	2.39	1.32	0.94	2.68	-0.21
214	19	4	100	102	177.01			-0.29	2.12	1.07	1.06	2.80	-0.12
214	19	6	82	84	179.83			0.07	2.31	1.20	1.22	2.61	-0.23
214	20	1	60	62	181.61							2.64	0.16
214	20	1	139	141	182.40					0.77	1.26		
214	20	2	113	115	183.64	0.03	2.11	-0.42	2.42	0.85	1.22	2.44	-0.43
214	20	3	99	101	185.00	-0.06	2.46	-0.54	2.53	1.03	1.32	2.31	-0.08
214	20	4	100	102	186.51	-0.15	2.30	-0.47	2.47	0.84	1.66	2.33	-0.22
214	20	5	99	101	188.00	-0.04	2.11					2.10	0.24
214	20	6	41	43	188.92			-0.55	2.41			2.01	0.15
214	21	2	101	103	193.02			-0.63	2.28			1.78	0.41
214	21	3	39	41	193.90			-0.65	2.15				
214	21	3	39	41	193.90			-0.82	2.40				
214	21	3	110	112	194.61			-0.58	1.84			1.85	0.56
214	21	3	110	112	194.61			-0.88	2.52				
214	21	4	40	42	195.41							1.64	0.57
214	21	5	46	48	196.97			-0.63	2.50				
214	21	5	46	48	196.97			-0.74	1.65				
214	21	5	113	105	197.54			-0.28	2.38				
214	21	5	103	105	197.54			-0.47	2.27				
214	22	1	47	49	200.48	0.08	2.11	-0.32	2.17			1.90	0.74
214	22	2	48	50	201.99							1.99	0.05
214	22	3	100	102	204.01								
214	22	5	110	112	207.11							2.02	0.35
214	22	6	47	49	207.98					0.85	1.46	1.78	-0.06
214	23	1	69	71	210.20					0.73	0.62	1.90	-0.02
214	23	2	47	49	211.48								
214	23	2	110	112	212.11								
214	23	3	120	122	213.71							1.93	
214	23	4	110	112	215.11							2.01	0.13
214	23	5	50	52	216.01								
214	23	6	94	96	217.95					0.93	1.50	2.03	0.35
214	24	1	40	42	219.41							1.95	0.03
214	24	1	98	100	219.99								
214	24	2	71	73	221.22							1.99	0.07
214	24	3	70	72	222.71							1.93	-0.20
214	24	4	80	84	224.33							1.86	-0.07

TABLE 3. OXYGEN AND CARBON ISOTOPE COMPOSITION (O/oo, PDB) OF GLOBIGERINOIDES SACCULIFER (350-420 μm), DENTOGLOBIGERINA ALTISPIRA (350-420 μm), GLOBOQUADRINA VENEZUELANA (500-600 μm), AND ORIDORSALIS UMBONATUS (>250 μm) AT DSDP SITE 216 (continued)

Hole	Core	Section	Interval (cm)		Depth below sea floor middle interval (cm)	G. sacculifer		D. altispira		G. venezuela		O. umbonatus	
						$\delta^{18}O$	$\delta^{13}C$	$\delta^{18}O$	$\delta^{13}C$	$\delta^{18}O$	$\delta^{13}C$	$\delta^{18}O$	$\delta^{13}C$
216A	1	6	29	35	109.32	-0.53	2.17	-0.45	2.26	0.99	1.23	3.02	-0.22
216A	2	1	18	24	111.21	-0.51	1.68	-0.25	1.75	1.27	0.69	3.11	-0.13
216A	2	1	104	110	112.07	-0.43	1.80	-0.38	2.14			3.23	-0.24
216A	2	2	103	109	113.56	-0.52	2.05	-0.64	1.89	0.99	0.91	3.01	-0.34
216A	2	3	104	110	115.07	-0.54	2.13	-0.47	2.06	1.07	1.25	2.98	-0.05
216A	2	4	104	110	116.57	-0.27	1.97	-0.54	2.06	1.10	1.20	2.93	0.00
216A	2	5	104	110	118.07	-0.05	1.63	-0.11	1.58	0.92	0.94	3.13	-0.22
216A	2	6	109	115	119.52	-0.33	1.40	-0.33	1.30	0.80	0.93	2.90	-0.07
216A	3	1	58	64				-0.15	1.70	1.31	0.60	2.86	-0.21
216A	3	2	30	36	122.33			-0.46	2.12	1.52	0.96	2.75	-0.18
216A	3	3	30	36	123.83			-0.12	1.49	1.93	0.95	2.95	-0.07
216A	3	4	30	36	125.33			-0.36	1.70	1.41	0.91	2.67	-0.39
216A	3	5	19	25	126.72			0.15	1.60	1.00	0.93	2.83	-0.05
216A	3	6	33	39	128.36			0.08	1.76	1.63	1.03	2.82	-0.29
216A	4	1	60	66	130.63			-0.28	1.75	1.66	1.29	2.94	0.27
216A	4	2	30	36	131.83			0.02	1.86	1.70	1.34	2.76	-0.05
216A	4	3	30	36	133.33			0.10	1.79	1.82	1.18	3.05	-0.12
216A	4	4	30	36	134.83			-0.18	2.30	1.07	1.33	2.49	0.16
216A	4	5	19	25	136.22							2.58	0.39
216A	4	6	30	36	137.83			-0.33	2.23	0.99	1.19	2.55	0.44
216A	4	6	30	36	137.83					1.05	1.04		
216A	5	1	31	37	139.84			-0.35	2.10	1.02	1.48	2.23	0.60
216A	5	2	28	34	141.31			-0.37	2.45	1.02	1.68	2.08	0.57
216A	5	3	18	24	142.71			-0.35	2.49	1.01	1.72	2.08	0.97
216A	5	4	30	36	144.33			-0.51	2.49	1.21	1.89	1.93	0.74
216A	5	5	115	121	146.68			-0.15	2.39	1.06	1.70	2.21	0.94
216A	5	6	30	36	147.33			0.40	2.44	0.99	1.54	1.61	0.53
216A	6	1	63	69	149.66			-0.17	2.33	0.89	1.80	1.71	0.97
216A	6	2	30	36	150.83	0.25	2.04	-0.41	2.56	0.31	1.51	1.85	0.63
216A	6	2	30	36	150.83	0.27	1.74			0.15	1.54		
216A	6	3	30	36	152.33			-0.24	2.24	0.78	1.45	1.86	0.58
216A	6	4	29	35	153.82			-0.46	2.45	0.72	1.40	1.85	0.21
216A	6	5	28	34	155.31			-0.49	2.48	0.78	1.44	1.82	0.92
216A	6	6	30	36	156.83			-0.57	2.72	0.57	1.51	1.55	0.56
216	5	1	122	125	159.74			-0.12	1.41	1.25	1.11	2.08	-0.09
216	5	2	39	43	160.41							1.98	0.00
216	5	2	115	121	161.18							1.97	0.05
216	5	3	34	40	161.87			-0.11	1.42	1.00	1.09	1.83	-0.27
216	5	3	114	120	162.67							2.18	-0.07
216	5	4	19	25	163.22							2.33	0.20
216	5	4	100	106	164.03							2.00	-0.01
216	5	4	100	106	164.03							2.11	-0.12
216	5	5	25	31	164.78							2.15	0.06
216	5	5	94	100	165.47			-0.03	1.44	1.09	1.25	1.97	-0.13
216	5	5	94	100	165.47					1.29	1.14		
216	5	6	25	31	166.28							2.07	-0.09
216	5	6	94	100	166.97							1.92	-0.11
216	6	1	121	127	169.24			0.09	1.76	1.54	1.29	2.43	0.18
216	6	2	23	29	169.76					1.20	0.88	2.26	0.08
216	6	2	103	109	170.56					1.20	0.81	2.02	-0.03
216	6	3	64	70	171.67			-0.51	1.59	0.73	1.20	2.03	-0.23
216	6	3	123	130	172.26					0.91	0.76	2.04	-0.03
216	6	4	24	30	172.77					0.99	0.97	1.88	0.04
216	6	4	100	106	173.53					1.38	1.32	2.88	-1.29
216	6	5	36	42	174.39			-0.38	1.71	0.78	0.96	2.02	-0.84
216	6	5	36	42	174.39					1.02	0.82		
216	6	5	102	108	175.05					0.99	1.16	2.16	-0.06
216	6	6	23	29	175.76					1.16	1.18	2.08	-0.18

TABLE 3. OXYGEN AND CARBON ISOTOPE COMPOSITION (O/oo, PDB) OF GLOBIGERINOIDES SACCULIFER (350-420 μm), DENTOGLOBIGERINA ALTISPIRA (350-420 μm), GLOBOQUADRINA VENEZUELANA (500-600 μm), AND ORIDORSALIS UMBONATUS (>250 μm) AT DSDP SITE 216 (continued)

Hole	Core	Section	Interval (cm)		Depth below sea floor middle interval (cm)	G. sacculifer		D. altispira		G. venezuela		O. umbonatus	
						$\delta^{18}O$	$\delta^{13}C$	$\delta^{18}O$	$\delta^{13}C$	$\delta^{18}O$	$\delta^{13}C$	$\delta^{18}O$	$\delta^{13}C$
216	6	6	94	100	176.47					1.12	1.12	2.06	0.26
216	7	1	34	40	177.87					1.14	1.10	2.25	0.26
216	7	1	110	116	178.63					0.67	1.49	2.14	0.04
216	7	2	25	31	179.28					0.98	1.19	1.90	-0.31
216	7	2	93	99	179.96					0.86	1.47	1.95	0.23
216	8	1	107	113	188.10	-0.24	2.02			1.39	1.29	2.08	0.19
216	8	1	107	113	188.10	-0.03	1.78						
216	8	2	23	29	188.76					1.30	1.71	1.93	0.10
216	8	2	128	134	189.81					1.67	1.92	2.57	0.81
216	9	1	27	33	196.80					1.39	1.41	1.79	-0.26
216	9	2	115	121	199.18					1.01	1.25	1.96	-1.37
216	9	3	25	31	199.78					1.22	1.69	2.15	-1.02

E. Vincent and Others

TABLE 4. OXYGEN AND CARBON ISOTOPE COMPOSITION (O/oo, PDB) OF GLOBIGERINOIDES SACCULIFER (350-420 μm), DENTOGLOBIGERINA ALTISPIRA (350-420 μm), GLOBOQUADRINA VENEZUELANA (500-600 μm), AND ORIDORSALIS UMBONATUS (>250 μm) AT DSDP SITE 237

Hole	Core	Section	Interval (cm)		Depth below sea floor middle interval (cm)	G. sacculifer		D. altispira		G. venezuela		O. umbonatus	
						$\delta^{18}O$	$\delta^{13}C$	$\delta^{18}O$	$\delta^{13}C$	$\delta^{18}O$	$\delta^{13}C$	$\delta^{18}O$	$\delta^{13}C$
237	10	1	80	84	83.32			-0.25	1.44	0.47	0.56	3.01	-0.92
237	10	2	80	84	84.82			-0.17	1.31	0.81	0.52	2.99	-0.77
237	10	3	80	84	86.32			-0.98	1.33	0.43	0.13	2.99	-0.79
237	10	3	80	84	86.32								
237	10	4	80	84	87.32			-0.67	1.21	0.45	0.13	3.01	-0.98
237	10	5	75	84	89.27			-0.60	1.32	0.29	0.51		
237	10	6	80	84	90.82			-0.62	1.78	0.79	0.90	2.66	-1.00
237	11	1	80	84	92.82								
237	11	1	80	84	92,82			-0.63	1.51	0.42	0.60	2.90	-1.04
237	11	2	80	84	94.82			-0.84	1.87	0.66	0.66	2.90	-0.83
237	11	3	80	84	95.82			-0.85	2.11	0.41	0.12	3.02	-1.01
237	11	3	80	84	95.82								
237	11	4	80	84	97.32			-0.40	1.67	0.75	0.70	3.00	-0.84
237	11	5	80	84	98.82			-0.89	1.71	0.56	0.78	2.89	-0.89
237	11	6	80	84	100.32			-0.84	1.76	0.36	0.63	2.96	-0.67
237	12	1	80	84	102.32			-0.48	1.99	0.15	0.55	2.97	-0.44
237	12	2	80	84	103.82			-0.94	2.02	0.31	0.81	2.51	-0.67
237	12	3	80	84	105.32			-0.66	1.67	0.02	0.60		
237	12	4	80	84	106.82			-0.48	1.98	0.81	1.01	2.86	-0.60
237	12	5	80	84	108.32								
237	12	5	80	84	108.32			-1.01	2.30	0.66	0.84	2.70	-0.07
237	12	6	5	9	109.01							3.02	-0.35
237	12	6	36	40	109.38								
237	12	6	36	40	109.38							2.82	0.01
237	12	6	80	84	109.82	-0.72	1.94	-0.56	2.00	0.27	0.77		
237	12	6	120	124	110.22	-0.57	2.48	-0.38	2.25	0.56	1.27	2.88	-0.42
237	13	1	10	14	111.12	-0.63	1.19	-0.24	1.96	-0.04	0.56	3.28	0.08
237	13	1	41	45	111.43							3.11	-0.25
237	13	1	41	45	111.43	-0.40	1.89	-0.77	2.09	0.60	0.93	2.76	-0.12
237	13	1	80	84	111.82	-0.85	1.61	-0.87	2.11	0.63	0.91	3.12	-0.26
237	13	1	80	84	111.82								
237	13	1	120	124	112.22							3.00	0.02
237	13	1	120	124	112.22	-0.16	1.95	-0.49	2.02	0.46	0.88	2.62	0.06
237	13	2	80	84	113.32			-0.71	2.55	0.56	1.09	2.93	-0.18
237	13	2	80	84	113.32								
237	13	3	80	84	114.82			-0.53	2.02	0.69	1.09	2.89	-0.04
237	13	4	80	84	116.32			-0.45	2.19	0.70	1.14		
237	13	5	80	84	117.82			-0.70	2.15	0.93	1.23	2.77	0.14
237	13	6	80	84	119.32			-0.49	2.24	0.82	1.15	3.01	0.11
237	14	1	80	84	121.32			-0.36	2.01	0.73	1.21	3.76	
237	14	2	80	84	122.82			-0.43	2.24	0.83	1.18	2.93	0.06
237	14	3	80	84	124.32			-0.63	2.01	0.71	0.97	3.63	
237	14	4	80	84	125.82			-0.79	2.12	0.75	0.98	2.95	0.18
237	14	5	80	84	127.32			-1.10	2.35	0.69	1.02		
237	14	6	80	84	128.82			-1.44	2.76	0.90	1.37		
237	15	3	80	84	133.82			-0.18	2.15	1.06	1.09	3.01	0.29
237	16	1	80	84	140.32							2.40	0.74
237	16	1	80	84	140.32			-0.23	2.58	0.76	1.61	2.49	
237	16	2	80	84	141.82			0.09	2.12	1.26	1.30	2.80	0.05
237	16	3	80	84	143.32			-0.02	2.24	1.55	1.22	2.68	0.11
237	16	4	80	84	144.82			-0.23	2.26	1.62	1.36	2.70	-0.22
237	16	5	80	84	146.32			-0.42	2.35	1.21	0.97	2.41	0.27
237	16	6	80	84	147.82			-0.29	2.55	1.55	1.23		
237	17	1	38	42	149.40			-0.58	2.85	0.91	1.64	2.26	0.63
237	17	2	33	37	150.85			-0.95	2.55	0.91	1.10	1.89	0.22
237	17	3	38	42	152.40			-0.52	2.64	0.37	1.50	2.18	0.60
237	17	4	38	42	153.90			-0.43	2.79	0.53	1.50	2.14	0.89
237	17	5	38	42	155.40			-0.64	3.03	0.91	1.82	2.23	1.32

TABLE 4. OXYGEN AND CARBON ISOTOPE COMPOSITION (°/oo, PDB) OF GLOBIGERINOIDES SACCULIFER (350-420 μm), DENTOGLOBIGERINA ALTISPIRA (350-420 μm), GLOBOQUADRINA VENEZUELANA (500-600 μm), AND ORIDORSALIS UMBONATUS (>250 μm) AT DSDP SITE 237 (continued)

Hole	Core	Section	Interval (cm)		Depth below sea floor middle interval (cm)	G. sacculifer		D. altispira		G. venezuela		O. umbonatus	
						$\delta^{18}O$	$\delta^{13}C$	$\delta^{18}O$	$\delta^{13}C$	$\delta^{18}O$	$\delta^{13}C$	$\delta^{18}O$	$\delta^{13}C$
237	17	5	38	42	155.40								
237	17	6	38	42	156.90			-0.58	2.93	0.44	1.43	2.28	1.02
237	18	1	18	26	158.72			-0.86	2.96	0.65	2.04	1.88	1.15
237	18	1	117	123	159.70			-0.85	3.27	0.75	1.84	2.07	1.41
237	18	2	32	36	160.34			-0.93	2.96	0.74	1.87	1.75	1.22
237	18	2	131	138	161.34			-1.06	2.57	0.47	1.40	1.67	0.31
237	18	3	88	96	162.42			-0.98	2.09	0.20	1.13	1.55	0.10
237	18	3	138	140	162.89							1.87	0.44
237	18	3	138	140	162.89	-0.87	1.87	-0.30	2.48	0.76	1.89	2.30	0.74
237	18	4	62	68	163.65			-0.98	2.01	0.18	1.44	1.99	0.40
237	18	4	138	142	164.40							1.84	0.50
237	18	5	21	29	164.75			-0.80	2.26	0.52	1.37	1.93	0.33
237	18	5	121	129	165.75			-1.04	1.98	1.21	1.26	1.95	0.34
237	18	6	71	79	166.75			-0.66	1.51	0.40	0.97	2.05	0.34
237	18	6	138	142	167.40			-0.18	1.93	0.84	1.22	2.05	0.57
237	19	1	20	28	168.24							1.98	0.47
237	19	1	20	28	168.24			-1.12	1.98	0.97	1.35	2.21	-0.08
237	19	1	121	129	169.25			-1.06	2.15	0.67	0.98	1.88	0.68
237	19	2	53	57	170.05	-0.24	2.22	-0.92	2.10	0.60	0.95	2.20	-0.13
237	19	2	72	80	170.26			-0.57	1.81	0.71	1.14	1.99	0.48
237	19	2	92	96	170.44			-0.51	1.99	0.64	1.33	1.78	0.17
237	19	2	138	142	170.90							1.93	0.30
237	19	2	142	146	170.94			-0.50	2.23	0.76	1.30	1.95	0.65
237	19	3	22	28	171.26							2.02	0.46
237	19	3	22	28	171.26			-0.27	1.83	0.81	1.21	1.06	0.08
237	19	3	40	44	171.42			-0.66	2.12	0.75	1.62	1.89	0.38
237	19	3	65	69	171.67	-0.51	1.97	-0.72	2.31	0.49	1.13	2.04	0.45
237	19	3	122	128	172.26			-0.47	1.66	0.72	0.98	1.64	0.06
237	19	3	140	144	172.42			-0.56	1.83			1.95	0.24
237	19	4	40	44	172.92							1.78	0.10
237	19	4	71	78	173.25			-0.63	1.33	1.33	0.98	1.91	0.07
237	19	4	84	88	173.36							1.84	0.40
237	19	4	131	135	173.83			-0.40	1.42	1.06	0.98	1.82	0.22
237	19	4	138	142	173.90			-0.03	2.10	0.88	0.70	1.89	0.19
237	19	5	21	28	174.25			-0.49	1.74	0.79	1.32	1.90	0.04
237	19	5	40	44	174.42							1.82	0.07
237	19	5	64	68	174.66							1.90	0.17
237	19	5	123	130	175.27			-0.43	1.49			1.96	0.03
237	19	5	123	130	175.27			-0.38	1.60	0.63	1.18		
237	19	5	140	144	175.42	-0.10	1.76	-0.33	1.74	0.44	1.07	1.76	0.24
237	19	5	140	144	175.42					0.74	1.21		
237	19	6	72	79	176.45			-0.35	1.74	0.83	1.07	1.78	0.16
237	19	6	138	142	176.90							2.07	0.29
237	20	1	21	28	177.75			-0.58	1.70	1.41	1.42	1.79	-0.01
237	20	1	120	127	178.74			-0.56	1.75	1.26	1.34	2.08	0.00
237	20	2	88	94	179.91			0.12	1.70	0.93	0.96	2.10	0.26
237	20	2	88	94	179.91								
237	20	3	21	28	180.74					0.84	1.38	1.98	0.22
237	20	4	21	28	182.24					0.72	1.41	2.21	0.19
237	20	4	131	138	183.35					0.78	1.20	2.04	0.18
237	20	5	71	78	184.24			0.56	1.34	0.99	1.14	1.96	0.12
237	20	6	107	114	186.11					1.11	1.94	1.80	0.23
237	21	1	80	84	187.82					0.83	0.77	1.70	0.88
237	21	1	80	84	187.82							2.16	0.13
237	21	2	80	84	189.32					0.42	1.09		
237	21	3	80	84	190.82					0.80	0.95	1.68	-0.10
237	21	4	80	84	192.32					0.85	1.09		
237	21	5	80	84	193.82					0.83	1.04	1.77	-0.22
237	21	6	80	84	195.32					1.04	1.00	1.73	-0.04
237	22	1	80	84	197.32					0.92	0.91	1.77	-0.22
237	22	2	80	84	198.82					0.62	0.90	1.78	-0.27
237	22	3	80	84	200.32					0.88	1.22	1.95	-0.07

TABLE 5. OXYGEN AND CARBON ISOTOPE COMPOSITION (O/oo, PDB) OF GLOBIGERINOIDES SACCULIFER (350-420 μm), DENTOGLOBIGERINA ALTISPIRA (350-420 μm), GLOBOQUADRINA VENEZUELANA (500-600 μm), AND ORIDORSALIS UMBONATUS (>250 μm) AT DSDP SITE 238

Hole	Core	Section	Interval (cm)		Depth below sea floor middle interval (cm)	G. sacculifer		D. altispira		G. venezuela		O. umbonatus	
						$\delta^{18}O$	$\delta^{13}C$	$\delta^{18}O$	$\delta^{13}C$	$\delta^{18}O$	$\delta^{13}C$	$\delta^{18}O$	$\delta^{13}C$
238	15	3	12	20	132.66	-0.76	1.51	-0.51	1.30	-0.01	0.33	3.09	-1.17
238	15	3	12	20	132.66	-0.99	1.60					3.20	-1.07
238	15	6	16	24	137.2	-1.27	2.05	-0.91	1.78	0.53	0.55		-0.76
238	15	6	16	24	137.2	-1.31	2.06						
238	16	3	23	31	142.27	-0.93	2.33	-0.63	1.68	-0.16	0.85	3.05	-0.73
238	16	3	23	31	142.27	-0.92	2.22						
238	16	6	42	50	146.96	-0.67	2.47	-0.67	1.68	0.05	0.91	3.19	-0.56
238	16	6	42	50	146.96	-0.56							
238	17	3	16	24	151.7	-1.23	2.44			0.28	1.08	3.33	
238	17	3	16	24	151.7	-1.25	2.55					3.13	-0.88
238	17	6	16	24	156.2	-0.93	1.66			0.71	0.41	3.50	-1.29
238	17	6	16	24	156.2	-0.78	1.84						
238	18	3	16	24	161.2	-0.55	2.27	-0.44	2.07	0.30	0.65	3.17	-0.79
238	18	3	16	24	161.2	-0.60	2.17						
238	18	6	16	24	165.70	-1.17	1.93	-1.17	1.93	0.06	0.33	2.78	-1.17
238	18	6	16	24	165.70	-1.11	2.06			0.08			
238	19	3	50	58	171.04	-0.70	1.91	-1.20	1.25	0.54	0.83	3.17	-1.04
238	19	3	50	58	171.04	-0.81	2.15						
238	19	6	46	54	175.50	-1.20	2.09	-0.73	1.80	0.41	0.80	3.18	-1.02
238	19	6	46	54	175.50	-1.04	2.12						
238	20	3	16	24	180.2	-0.78	2.41	-0.78	2.48	0.32	0.80	3.05	-0.30
238	20	6	16	24	184.7	-0.84	2.80	-0.82		0.55	1.26	3.15	-0.19
238	21	3	16	24	189.7	-1.15	2.68	-0.95	2.42	0.46	1.31	3.21	-0.21
238	21	6	46	54	194.7	-1.31	2.95	-0.91	2.88	0.48	1.36	3.03	0.05
238	22	3	46	54	199.5	-1.21	2.58	-0.79	2.09	0.14	1.21	2.92	-0.26
238	22	3	46	54	199.5								
238	22	6	70	78	201.24	-1.25	2.65	-0.90	1.78	0.16	1.10		-0.30
238	22	6	70	78	204.24							2.71	-0.33
238	23	4	46	54	210.50	-1.45	2.70	-1.32	3.05	0.42	1.49	2.88	-0.09
238	23	4	46	54	210.50							2.99	0.12
238	23	6	16	24	213.20	-0.97	2.58	-1.07	2.40	0.82	1.44	3.14	-0.32
238	24	3	120	128	219.24	-1.02	2.35						
238	24	6	51	59	223.05	-0.70	2.42	-1.01	1.78	0.67	1.08	3.24	-0.05
238	24	6	51	59	223.05								
238	25	3	70	78	228.24	-0.47	2.53	-0.93	2.46	0.57	1.10	3.24	-0.15
238	25	3	70	78	228.24							3.14	-0.06
238	25	6	76	84	232.80	-0.27	2.08	-0.87	2.50	0.28	1.09	2.99	0.28
238	26	2	90	98	236.44	-0.58	2.31	-0.85	2.46	0.57	1.25	3.10	0.13
238	27	2	33	39	245.36	-0.86	2.48			0.34	1.43		
238	27	2	105	111	246.08								
238	27	3	21	27	246.74								
238	27	3	102	110	247.56	-1.04	2.62	-1.37	2.54	0.50	1.234	3.03	-0.13
238	27	4	20	26	248.23								
238	27	4	93	99	248.96	-1.37	2.67			0.24	1.51		
238	27	5	86	94	250.40	-0.67	2.44	-1.12	3.16	0.56	1.38	2.84	-0.02
238	27	5	86	94	250.40							3.13	-0.12
238	28	2	120	128	255.74	-1.02	2.36	-1.01	1.84	0.12	1.09	2.99	0.43
238	28	6	112	120	261.66	-0.95	3.12	-0.98	2.73	0.44	1.48	2.88	0.64
238	28	6	112	120	261.66								
238	29	3	16	24	265.70	-1.02	2.38	-1.40	1.98	0.04	0.96	3.29	0.52
238	29	6	16	24	270.20	-1.10	2.13	-1.15	2.34	0.64	1.08		
238	30	3	36	44	275.40	-1.11	2.34	-1.02	2.26	0.49	0.93	3.55	
238	30	6	52	60	280.06	-0.96	2.09	-0.83	2.21	0.91	0.90	3.30	-0.14

TABLE 5. OXYGEN AND CARBON ISOTOPE COMPOSITION (O/oo, PDB) OF GLOBIGERINOIDES SACCULIFER (350-420 μm), DENTOGLOBIGERINA ALTISPIRA (350-420 μm), GLOBOQUADRINA VENEZUELANA (500-600 μm), AND ORIDORSALIS UMBONATUS (>250 μm) AT DSDP SITE 238 (continued)

Hole	Core	Section	Interval (cm)		Depth below sea floor middle interval (cm)	G. sacculifer		D. altispira		G. venezuela		O. umbonatus	
						$\delta^{18}O$	$\delta^{13}C$	$\delta^{18}O$	$\delta^{13}C$	$\delta^{18}O$	$\delta^{13}C$	$\delta^{18}O$	$\delta^{13}C$
238	30	6	52	60	280.06							3.22	-0.15
238	32	3	29	36	294.33	-0.86	2.40					3.04	0.44
238	32	4	92	98	296.45							2.61	-0.53
238	34	1	30	36	310.33	-1.03	1.93	-0.79	2.34	1.32	1.21	2.86	-0.28
238	34	5	28	34	316.31	-0.72	2.07	-0.93	2.51	1.14	1.12		
238	34	5	103	109	317.03			0.58	2.50				
238	34	5	103	109	317.03	-0.74	2.34	-0.98	2.41	1.20	1.50	2.91	0.17
238	35	3	28	34	322.81	-0.64	2.44	-0.62	2.59	0.84	1.38	2.84	0.02
238	35	3	28	34	322.81								
238	35	4	103	109	325.06	-0.88	2.13	-1.04	2.22	0.56	0.85	2.50	-0.11
238	35	4	103	109	325.06								
238	36	1	27	33	329.30	-1.19	2.97	-1.46	2.02	-0.12	1.52	2.24	0.38
238	36	3	27	33	332.30	-0.88	2.44	-0.75	2.46	0.48	1.40	2.17	0.31
238	36	4	91	97	334.44							2.73	0.86
238	37	3	41	47	341.94			-0.99	3.05	0.40	1.86	1.91	0.73
238	37	4	91	97	343.94	-0.20	2.60	-0.92	3.14	0.53	1.93	2.26	0.61
238	38	1	131	137	349.34			-0.99	2.80	0.31	1.87	1.45	0.19
238	38	3	27	33	351.30								
238	38	4	30	36	352.83			-0.90	3.19	0.35	1.71		
238	38	4	30	36	352.83			-0.32	2.53	0.45	2.09		
238	38	5	80	84	354.82			-0.80	2.93	0.39	1.72	2.12	0.99
238	38	6	30	36	355.83			-0.48	2.81	0.78	2.15	2.05	0.66
238	38	6	67	71	356.19			-0.66	2.92	0.42	1.81	2.12	1.10
238	39	3	90	96	361.43			-0.49	2.87				
238	39	3	90	96	361.43								
238	39	6	32	38	365.34	0.07	2.08	-0.30	2.18				
238	39	6	32	38	365.34	-0.17	1.62						
238	40	1	80	86	367.83							1.70	0.38
238	40	1	88	92	367.90							2.34	0.74
238	41	1	118	124	377.71							2.03	0.51
238	41	2	83	87	378.85			-0.42	1.97	0.40	1.66	2.00	0.21
238	41	3	91	97	380.44							2.15	0.33
238	41	4	30	36	381.33							1.96	0.28
238	41	6	97	103	385.00							1.93	-0.10
238	42	1	102	108	387.05			-0.39	1.78	0.69	1.16	2.34	0.07
238	42	3	93	99	389.96							2.11	-0.26
238	43	1	93	99	396.46							2.23	0.01
238	43	3	33	39	398.86			-0.50	2.16	1.11		2.16	-0.14
238	43	3	33	39	398.86					1.26	1.26		
238	43	4	37	33	400.30							2.35	0.13
238	43	6	28	34	403.31					1.11	1.49	2.21	0.22
238	44	3	23	29	408.26					1.33	1.72	2.01	0.27
238	44	4	89	95	410.42							2.45	0.18
238	45	1	89	95	415.42					1.34	1.31	2.13	0.00
238	46	1	31	37	424.34							2.26	0.35
238	46	3	86	92	427.89					1.23	1.85	2.39	0.39
238	46	3	86	92	427.89					1.19	1.95		
238	47	2	108	114	436.11							2.50	0.65
238	47	4	100	106	439.03	-0.13	1.94			0.91	1.91	2.19	0.21
238	49	3	30	36	455.83					0.78	1.80	2.07	0.27

intervals, according to a changing focus of interest through the course of this study. This study was undertaken within the CENOP Project and followed its goals: first to obtain time-series studies for the early and late Miocene, and then to study in greater detail three time slice intervals. The densest sampling was conducted in these time slice intervals. Middle Miocene sections were usually more sparsely sampled than the Lower and Upper Miocene, this stratigraphic interval having been studied at a later stage of the investigation. Portions of the Lower Miocene showed very poor foraminiferal preservation, and are not well represented.

Sample size (thickness) and sample spacing are given in Tables 2-5. Sample thickness varies from 2 to 8 cm and sampling intervals vary from 40 cm to 10 m.

The variable sampling density presents special problems for the statistical analysis of stratigraphic trends and variability. High resolution studies in Miocene sediments (e.g. Woodruff et. al., 1981; Shackleton, 1982) show that isotopic values can fluctuate rather rapidly over a considerable range (order of $1^0/_{00}$). This can produce spurious trends when sampling intervals are large. Also, it can produce, superficially, the appearance of alternations between climatically quiet and noisy zones, for variable sampling density. One (partial) solution to these problems is careful stacking of stratigraphic signals, from different cores. We are working on the biostratigraphic refinements which will allow such stacking. At this point, we are satisfied that the sampling strategy used, albeit less than optimal, will allow the identification of sections which are worth the effort of high resolution study.

Samples were disaggregated by soaking in buffered Calgon solution at room temperature, washed over a 63-μm sieve and cleaned ultrasonically to remove fine particles from inside the foraminiferal tests. Foraminiferal species were hand-picked from well-defined size fractions and analyzed for oxygen and carbon isotopic composition. Isotopic analysis followed our standard procedure (Berger and Killingley, 1977). Approximately 0.5 mg of material was used for isotopic analysis which usually consisted of 10 to 20 specimens. The standard error for the laboratory standard was about $0.1^0/_{00}$ for both oxygen and carbon.

SPECIES ANALYZED AND MEANING OF SIGNALS

We chose *Globigerinoides sacculifer,* an extant species known to inhabit the shallow-water environment, or mixed layer. Its $^{18}O/^{16}O$ ratio is among the lowest of modern planktonic foraminifera and has been used extensively as a tool for sea-surface temperature reconstruction in the Pleistocene. *G. sacculifer* retains its low $\delta^{18}O$ values back into the Miocene (Vincent et al., 1980; Savin et al., this volume). This species is not abundant enough throughout the sequences and we therefore chose *Dentoglobigerina altispira* as another shallow-dwelling indicator to obtain continuous shallow-water records. *D. altispira* became extinct at the end of the Pliocene and has no modern representative. In sections for which we have isotopic data on both species, both oxygen and carbon isotopic signatures are quite similar throughout the Upper and Middle Miocene (see Figures 2–9). In

the Lower Miocene, however, *G. sacculifer* is on the whole slightly enriched in ^{18}O and slightly depleted in ^{13}C. The incompleteness of the Lower Miocene record, owing to the rarity of both taxa, precludes an interpretation of this separation between the isotopic signals.

We chose *Globoquadrina venezuelana* as a deep-dwelling planktonic foraminifera. This is an extinct species. Its direct descendant *G. conglomerata,* has adult populations living predominantly in deep waters (deeper than 100 m: Bé, 1977) and $^{18}O/^{16}O$ ratios which are among the highest of Recent planktonic foraminifera (Berger et al., 1978). *G. venezuelana* has been shown to be the most reliable deep-water indicator among planktonic foraminifera for Neogene paleoceanographic reconstructions retaining an enriched ^{18}O composition throughout the Miocene (Vincent et al., 1980; Savin et al., this volume).

We chose *Oridorsalis umbonatus,* a rather common species of the benthic faunas throughout the sequences, to represent the sea-floor environment. *O. umbonatus* is an extant species. Its $\delta^{18}O$ values in Miocene deep-sea sediments are quite close to coexisting *Uvigerina* spp. (Vincent et al., 1980; Savin et al., 1981), although there does seem to be a difference in recent sediments (Woodruff et al., 1980; Graham et al., 1981). If *U. peregrina* is taken as reflecting 'equilibrium' values (i.e., values conforming to accepted regression equations; see Shackleton, 1974) then *O. umbonatus* is close to equilibrium, in our data. Unlike previous studies (e.g. Belanger et al., 1981; Savin et al., 1981, among others), we found *O. umbonatus* to be a reliable carrier of information, by comparison with the two other benthic species *Cibicidoides wuellerstorfi* and *Globocassidulina subglobosa* which we also analyzed in these sequences. This may be because we took care to interpret the taxonomy of *O. umbonatus* narrowly and to exclude morphotypes which did not clearly belong to this taxon. Records of mixed species, of course, can show spurious variations owing to changes in the mixture rather than in isotopic composition of components.

The various factors which must be considered when interpreting stable isotopic compositions of foraminifera have been summarized in several recent reviews (Berger, 1981; Vincent and Berger, 1981; Douglas and Woodruff, 1981). Here we keep to the first-order factors. Thus, following Emiliani (1954) we interpret changes in $\delta^{18}O$ as changes in temperature, and changes in $\delta^{18}O$ differences as changes in the temperature profile. Effect of ice growth and decay seem to be of secondary importance only, as we shall see. Changes in $\delta^{13}C$ values are seen as: 1) global changes in the input/output ratios of carbonate and organic carbon, and 2) changes in the degree of biological pumping, which enriches subsurface waters in ^{12}C at the expense of the mixed layer. These mechanisms, in principle, were recognized by Tappan (1968) and by Deuser and Hunt (1969). They were formulated in useful conceptual models by Broecker (1970, 1973, 1982), Craig (1970), Garrels and Perry (1974), Fischer and Arthur (1977), Shackleton (1977), Bender and Keigwin (1979), Scholle and Arthur (1980), Vincent et al. (1980), and Garrels and Lerman (1981) among others (see Arthur, 1982).

"Vital effects" may be defined as anything which causes isotopic values to deviate from "equilibrium"; that is, the conventional best-fit between composition and external environmental factors. Such effects are ubiquitous in planktonic as well as in benthic foraminifera. We have previously spent considerable effort on elucidating their importance (Berger et al., 1978; Vincent et al., 1981), as have others (Williams et al., 1977; Belanger et al., 1981; Graham et al., 1981, among others; see papers in Berger et al., 1981). As long as vital effects result in a constant deviation of isotopic values from the true signal, they interfere little or not at all with paleoceanographic and paleoecologic interpretation. However, if the magnitude of vital effects changes through geologic time, or responds in a non-linear manner to changes in the environment, misinterpretations become likely. The detection of changes in vital effect requires multiple species isotope stratigraphy. As we found in the earlier study (Vincent et al., 1980) the three benthic species, *O. umbonatus, C. wuellerstorfi* and *G. subglobosa,* on the whole show isotopic values which remain parallel over long periods of geologic time, suggesting that vital effects are indeed more or less constant. However, as we have argued previously (Vincent et al., 1981), vital effects do appear to respond to changes in ocean productivity, in a manner which varies from species to species.

To minimize the effects of changing depth preferences during planktonic life cycles on the isotopic composition of the test (Berger et al., 1978), we have analyzed rather narrowly defined size fractions of planktonic foraminifera. We chose for each species the optimum size class for the maximum abundance of adult specimens: 350-420 μm for *G. sacculifer* and *D. altispira* and 500-600 μm for *G. venezuelana.* Exceptions were made where these species are very rare and more widely defined size fractions were picked (as indicated in the data listings in Tables 2–5). Isotopic composition of benthic foraminifera does not appear to vary with test size (Vincent et al., 1981) and we have analyzed the fraction greater than 250 μm. Species other than those mentioned above were also analyzed in several intervals, in particular for the CENOP time slices. These data appear in Savin et al., elsewhere in this volume.

BIOSTRATIGRAPHY

Biostratigraphic zonations established by various authors on the basis of calcareous and siliceous planktonic microfossils in the four sedimentary sequences were summarized by Vincent (1977; see references therein) in her synthesis of Indian Ocean Neogene biostratigraphy. They are shown on the right-hand side of Figures 2–9. The wide sampling intervals often used for biostratigraphic studies (which are mainly those published in the DSDP Initial Reports) precluded in many cases the exact placement of zonal boundaries. In Figures 2–9 zonal boundaries are drawn as a line in the middle of interval of uncertainty for clarity of the figures on this scale. However, the low biostratigraphic resolution, which may be improved with further detailed study, should be kept in mind.

Blow's (1969) zonation for tropical planktonic foraminifera was applied. Zonal assignments, however, were commonly complicated by: 1) the lack or paucity of zonal nominate taxa, and/or 2) the co-occurrence of certain species which, according to Blow's zonal scheme, should not co-occur. All these problems, as well as the position of foraminiferal zonal boundaries relative to the zonations of other microfossils at the various sites, are discussed in some detail by Vincent (1977). In many instances, when zonal boundaries could not be located, zones were combined into wider intervals. This is often the case in particular in the Lower and Middle Miocene interval encompassing zones N5 to N12 where poor preservation eliminated many of the solution-susceptible species characterizing zones of this interval. Considering the short time span of several of these zones and the condensed nature of the sedimentary sequences, these sections are not well-suited for a detailed foraminiferal zonation in the upper Lower Miocene and the lower Middle Miocene.

Planktonic foraminifera zonal assignment near the Middle/ Late Miocene boundary was not possible at Sites 237 and 238 or at other western Indian Ocean sites. Several zonal boundaries in the interval encompassing Zones N14 to N16 could not be identified owing to the co-occurrence of species characterizing these zones. Vincent (1974, 1977) interpreted the telescoping of strata and the presence of mixed foraminiferal assemblages in this interval throughout the northwestern Indian Ocean as reworking of older faunas into younger sediments, following an unconformity spanning the Middle/Late Miocene boundary. She suggested that it was possibly related to increased tectonic activity and circulation readjustments in the northwest Indian Ocean after the opening of the Gulf of Aden and the closure of the Indo-Tethyan seaway. Because an almost complete calcareous nannoplankton sequence is apparently present at all sites within the disturbed interval, Vincent suggested that this fossil group shows much less reworking than did the foraminifera and that sediments may have been winnowed during transport. There is evidence, however, of condensing of nannofossil zones (CN6 and CN7) in this stratigraphic interval at several sites, including the eastern sites 214 and 216. At the latter two sites, radiolarian data also suggest the occurrence of a hiatus in the interval between 7.5–9 Ma. The low overall accumulation rate, however, precludes here the recognition of a hiatus with certainty. The occurrence of an interval containing stratigraphic gaps, spanning the Middle/Late Miocene boundary at various localities of the Indian Ocean, may result from an oceanographic event of greater extent than a regional effect. It is in this stratigraphic interval that Barron and Keller (1982) identified their hiatuses NH4 and NH5, widespread in deep-sea sequences of the Pacific Ocean.

Because of the problems encountered in zoning the foraminiferal sequences and the inconsistency from site to site of foraminiferal zonal boundaries relative to other fossil zonations, Vincent (1977) used, for the sake of consistency, the nannofossil zonation to draw the threefold division of the Miocene throughout the Indian Ocean. Her divisions are followed in this paper.

We have critically evaluated the positions in cores at our

four Indian Ocean sites of various biostratigraphic datums correlated to paleomagnetostratigraphy elsewhere (see references in Vincent, 1981; Berggren et al., in press; Barron et al., this volume). A number of those, which appear as reliable biostratigraphic markers here are given in Figure 10 and are identified with triangular tick marks in Figures 2–9. Their numerical ages expressed in Ma (million of years) are those given by Barron et al. (this volume), based on the paleomagnetic timescale of Berggren et al. (in press). The ages given in our study for the interval 6.5–13.8 Ma (encompassing paleomagnetic Chrons 7–14) are those revised by Barron et al. (this volume) in their addendum, utilizing an Anomaly 5 - Chron 11 correlation, as recently proposed by Berggren et al. (in press) (following the original proposition of Foster and Opdyke, 1970). This replaces the Anomaly 5 - Chron 9 correlation traditionally used for the past ten years. The change results in a shift of approximately 1.5 to 2 million years toward younger age estimates for biostratigraphic datum levels in the interval between approximately 6.5 and 13.8 Ma and implies considerable changes in sedimentation rate profiles previously published.

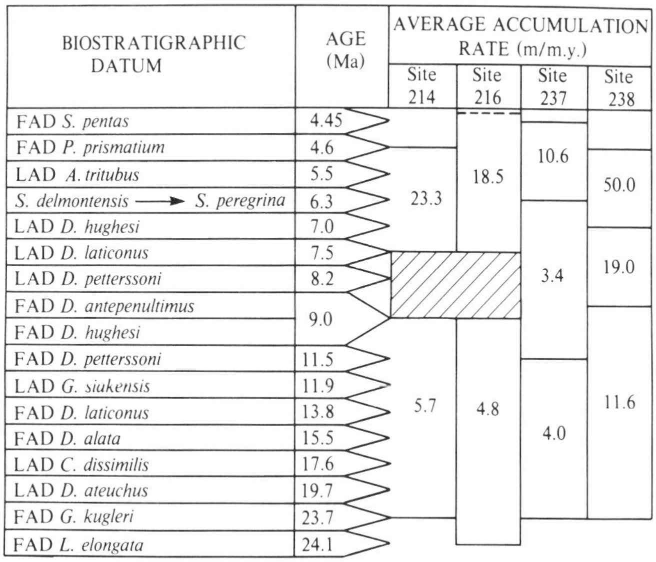

ACCUMULATION RATES

Average accumulation rates between selected biostratigraphic datums are given in Figure 10. We have adopted here a conservative approach—integrating time and sediment thickness over long periods and assuming little short-term change in rate.

From an examination of smoothed accumulation-rate curves (not shown) constructed by drawing a visual best-fit line through reliable biostratigraphic datums (see also data in Figure 10), an overall pattern of sediment accumulation throughout the Miocene results, which is similar at the four Indian Ocean sites. Accumulation rates for sediments older than about 9.0 Ma (Lower Miocene to lowermost Upper Miocene) are low, with average values of about 4 to 6 m/m.y. at Sites 237, 214 and 216 and about 12 m/m.y. at Site 238. A sedimentary gap probably occurs at Sites 214 and 216 corresponding to the interval 9–7.5 Ma. Upper Miocene sediments younger than about 7.5 Ma accumulated at much higher rates, with average values three to four times as high as those of the Lower-Middle Miocene (about 11 m/m.y. at Site 237, 19 and 23 m/m.y. at Sites 214 and 216 respectively, and 50 m/m.y. at Site 238). The 9–7.5 Ma hiatus, inferred at the eastern sites on the basis of radiolarian data, cannot be identified from the siliceous record at Sites 237 and 238. It is noteworthy, however, that the inflection point on smoothed accumulation-rate curves at these two sites (between slow sedimentation below and fast sedimentation above) occurs near the 7.5 Ma level.

ISOTOPIC CHRONOSTRATIGRAPHIC MARKERS

We have identified a number of features of stratigraphic value in the isotopic and calcium carbonate records of continuous expanded Lower and Middle Miocene sequences in the central

BIOSTRATIGRAPHIC DATUM	AGE (Ma)	AVERAGE ACCUMULATION RATE (m/m.y.)			
		Site 214	Site 216	Site 237	Site 238
FAD *S. pentas*	4.45				
FAD *P. prismatium*	4.6			10.6	
LAD *A. tritubus*	5.5		18.5		50.0
S. delmontensis → *S. peregrina*	6.3	23.3			
LAD *D. hughesi*	7.0				
LAD *D. laticonus*	7.5				19.0
LAD *D. petterssoni*	8.2			3.4	
FAD *D. antepenultimus*	9.0				
FAD *D. hughesi*					
FAD *D. petterssoni*	11.5				
LAD *G. siakensis*	11.9	5.7	4.8		11.6
FAD *D. laticonus*	13.8				
FAD *D. alata*	15.5			4.0	
LAD *C. dissimilis*	17.6				
LAD *D. ateuchus*	19.7				
FAD *G. kugleri*	23.7				
FAD *L. elongata*	24.1				

Figure 10. Average accumulation rates of Miocene sediments at four DSDP sites in the tropical Indian Ocean. Dashed area: hiatus. Ages from Barron et al. (this volume). The 18.5 m/m.y. value at Site 216 is obtained from data in the interval 7.0–7.5 Ma and in the middle-upper Pliocene.

equatorial Pacific at DSDP Sites 574 and 575 (Vincent and Killingley, in press). These expanded sequences, recovered with minimum disturbance (Hydraulic Piston Coring was performed at Site 575), yielded excellent siliceous biostratigraphic control and some paleomagnetic data, allowing us to make a confident age estimate for our markers (labelled in a stratigraphic order from A to H; see Figure 11). We correlated these signals with those identified in DSDP Site 289 on the Ontong-Java Plateau in the western equatorial Pacific (Woodruff et al., 1981). Using biostratigraphic correlations, we were able to recognize a number of these synchronous Pacific Early and Middle Miocene signals in our Indian Ocean records as indicated in Figures 2–9. Some of these identifications remain tentative due to the condensed nature of the Indian Ocean sequences. The depth below sea floor of these signals plotted for each site against time (using ages given in Figure 11) falls approximately on the smoothed sedimentation-rate curves.

Two distinct oxygen isotopic enrichments identified by Keigwin (1979) in the expanded Middle to Upper Miocene section of DSDP Site 158 in the Panama Basin have been used as stratigraphic markers by Burckle et al., (1982) and Barron et al. (this volume). These two isotopic signals are close to the Middle/ Late Miocene boundary, within the stratigraphic interval which is here highly condensed and disturbed, and are not recognizable in our records.

The distinct shift toward lower $\delta^{13}C$ values recorded in the uppermost Miocene of various benthic records from the Indo-Pacific region (Keigwin, 1979; Vincent et al., 1980; Haq et al., 1980) is well-identified at Sites 214, 237 and 238 in the benthic records, as well as in both shallow- and deep-dwelling planktonic foraminifera records (see Figs. 6, 8, 9). The top of the isotopic

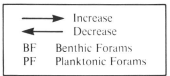

| Increase |
| Decrease |
| BF | Benthic Forams |
| PF | Planktonic Forams |

LEVEL	BIOSTRATIGRAPHIC AND MAGNETOSTRATIGRAPHIC POSITION	APPROX AGE (Ma)	CHEMICAL SIGNAL		
			$\delta^{18}O$	$\delta^{13}C$	$CaCO_3$
H	Within *C. gigas* var. *diorama* zone Within upper *D. alata* zone Within CN5a Within N12	12.5	PF		
G	Within subzone B of *C. peplum* zone Within lower *D. alata* zone Within CN4 Approximates the N9/N10 boundary	14.6	BF		
F	Approximates the *D. nicobarica/C. peplum* boundary Within the middle of the *C. costata* zone Within CN3 Within undifferentiated interval N6-N8 Within magnetic anomaly 5C	16.5	BF Minimum		
E	Within subzone A of *D. nicobarica* zone Approximates the *S. wolfii/C. costata* boundary Approximates the CN2/CN3 boundary Within undifferentiated interval N6-N8 Within lower reversed interval of magnetic anomaly 5c	17.5		BF + PF	
D	Approximates the *C. elegans/T. pileus* boundary Within uppermost *S. delmontensis* zone Within uppermost CN1c Within N5	18.7			
C	Within lowermost *C. elegans* zone Within *S. delmontensis* zone Within CN1c Within N5	19.5	BF + PF		
B	Slightly below the B/C boundary in *R. paleacea* zone Within lower *S. delmontensis* zone Within interval CN1b-CN1c Within upper N4	20.7		PF	
A	Approximates the *R. gelida/R. palacea* boundary Within *L. elongata* zone Within lower CN1b Within N4	22.7		PF	

Figure 11. Biostratigraphic and magnetostratigraphic position of levels marking changes in the signature of the isotopic and carbonate records at DSDP Sites 573, 574 and 575 in the central equatorial Pacific (modified from Vincent and Killingley, in press).

record at Site 216 is stratigraphically below this carbon isotopic marker. We have previously described this carbon shift at Site 238 and discussed in detail its biostratigraphic position (Vincent et al., 1980). This shift, which has been shown to be isochronous throughout the Indo-Pacific and has been tied directly to paleomagnetic Chron 6 in Piston Core RC12-66 (Keigwin and Shackleton, 1980), has been used extensively as a chronostratigraphic marker in various areas. Its age is given as 6.1–6.2 Ma by Berggren et al. (in press) and 6.4 by Barron et al. (this volume). Its age derived from the sedimentation-rate curves at our three Indian Ocean sites is approximately 6.2–6.3 Ma.

OXYGEN ISOTOPE STRATIGRAPHY

In the following we shall refer to shallow-dwelling planktonic foraminifera as SPF, to deep-dwelling ones as DPF, and to benthic species as BF. The oxygen isotopic record is dominated by a change in the BF from near $2^0/_{00}$ in the Lower Miocene to values near $3^0/_{00}$ in the Upper Miocene. The transition is quite rapid, occurring between levels F and H (between 16.5 and 12.5 Ma).

This major change in the benthic isotopic signal was first recognized by Savin et al. (1975) and by Shackleton and Kennett (1975). These authors proposed a substantial buildup of Antarctic ice at that time to explain the feature. Their interpretation was challenged by Matthews and Poore (1980) who suggested that bottom-water cooling alone could produce the effect. More recently, DSDP Site 289 provided a detailed record of the mid-Miocene oxygen step (Woodruff, et al., 1981).

The $\delta^{18}O$ records of SPF and DPF do not show the mid-Miocene step as clearly as BF. If ice buildup caused the change, it must have been at least partially compensated by warming of tropical surface waters. This is quite possible, if increased thermal isolation of Antarctica (through development of the Circumpolar Current) is the basic reason for the event (Savin et al., 1975; Savin, 1977). However, in the absence of such compensation, no more than about one third of the $1^0/_{00}$ step can be assigned to an ice effect. The ice effect, of course, changes the average composition of the entire water column, and must affect BF, DPF, and SPF in like fashion.

A major feature of the $\delta^{18}O$ record, which derives directly from the BF step, is the increasing separation of the $\delta^{18}O$ values of SPF and BF. It reflects an increase in the planetary temperature gradient, from early to late Miocene. We think it is tied to increased albedo in high latitudes, due to sea-ice formation and snow-cover on land. The evidence for this suggestion is that $\delta^{18}O$ values between 3 and $3.5^0/_{00}$ would indicate rather frigid temperatures, say, around 5°C. At depths greater than those of our sites the water temperature was presumably even lower than this, indicating icy conditions near the bottom water sources. The temperature gradient drives zonal winds, and winds drive surface currents. Hence, the intensity of wind-driven currents must have increased throughout the Miocene, and the same is true for up-

welling. These trends have been discussed previously (Berger et al., 1981; Vincent and Berger, 1981).

The records of SPF and DPF are, on the whole, parallel. Only in Site 216 is there an indication of increased separation between SPF and DPF, from the middle Miocene on. Thus in this site, but not in the others, the DPF record appears as intermediate between SPF and BF. The significance of this signal is not obvious to us. We suggest that the near-equatorial position of Site 216 may be responsible in some fashion. Equatorial upwelling appears reflected in the good siliceous fossil record at the site. Alternatively, the increased separation of DPF and SPF in Site 216 only may indicate a shift of the upper boundary of cold waters to shallower levels, that is, a strengthening of the thermocline.

The following observations seem noteworthy, but have no obvious explanation: 1) The lightest $\delta^{18}O$ values of BF, of the entire Neogene, occur at the onset of the mid-Miocene cooling step. Apparently a warming trend in high latitudes preceded the rapid cooling initiating the transition. 2) In one of our records (Site 216, BF) there is an abrupt step *within* the mid-Miocene oxygen enrichment, at level G. The same abrupt shift appears in equatorial Pacific benthic records at DSDP Sites 289 (Savin et al. 1981), 574 (Pisias and Shackleton, in press) and 575 (our unpublished data). Slight discrepancies in the biostratigraphic position of this level relative to the foraminiferal zonation at the various sites have been discussed by Vincent and Killingley (in press). They appear to result from differing biostratigraphic interpretations of foraminiferal data by various workers.

We hesitate to extract any short-term signals from these records because of variable sampling density, the low biostratigraphic resolution, and the differences in preservation, which need further attention. A rather simple statistical treatment will be offered below, for Site 216, which has the best biostratigraphic control and apparently the least disturbed record.

CARBON-ISOTOPE STRATIGRAPHY

The dominant feature of the Miocene carbon isotope stratigraphy is the broad Lower-to-Middle-Miocene maximum, which we call the "Monterey Carbon Excursion" (Vincent and Berger, in press). The excursion begins with a shift toward high $\delta^{13}C$ values (the Chron-16 Carbon Shift) in the upper portion of the Lower Miocene, and ends with a rather gradual decline toward background values within the mid-Miocene. The Monterey event is readily identified in previously published records (Shackleton and Kennett, 1975; Boersma and Shackleton, 1977; Vergnaud-Grazzini and Rabussier-Lointier, 1980; Savin et al., 1981; Woodruff et al., 1981). It seems that this excursion toward heavy values is correlated with the deposition of organic-rich sediments around the margin of the North Pacific, in the Monterey Formation of California and its equivalents (Vincent et al. 1983).

The initiation of the carbon excursion precedes the major mid-Miocene oxygen shift. Thus, there is a possibility that extraction of organic carbon preconditions the ocean-atmosphere system for subsequent cooling. It has been proposed by various authors that the buildup and erosion of readily accessible reservoirs of organic carbon, and of nutrients, provides opportunities for large-scale climatic changes (Berger, 1982). This would be true also in ice-free situations (the "Greenhouse State" of Fischer, 1984), when the climate-albedo-feedback is weak. E. Barron (personal communication, Chapman Conf., 1983) argues, from climate modeling, that a high atmospheric CO_2 content is necessary to keep polar regions warm and the planetary temperature gradient weak during ice-free periods of Earth's history (see also Barron and Washington, 1984). If true, it follows that the pCO_2 must first be decreased to allow sufficient cooling for polar whitening. There are three obvious ways to accomplish this: 1) By a decrease of global input of volcanogenic acids, 2) by accelerated weathering of continental crust and 3) by additional sequestration of organic carbon. Judging from available indices of volcanic activity (Kennett, 1982, p. 386), the acid input in the Miocene was higher than average. Also, from sea-level variation (Vail and Hardenbol, 1979) there was no unusual mountain-building activity (or exposure of continental crust) during the mid-Miocene. Neither decreased volcanism nor increased weathering, then, can be called upon as a cause for strong pCO_2 drawdown. We prefer, therefore, the concept of organic carbon buildup.

The amount of organic carbon extracted into additional reservoirs can be estimated as follows: 1) The $\delta^{13}C$ excursion has a typical magnitude of $0.5^0/_{00}$ (max $1^0/_{00}$). 2) It lasts about 4 million years (17.5 My–13.5 My). 3) Extraction of 1% of the ocean's carbon as organic carbon ($\delta^{13}C = -20^0/_{00}$) leaves the remaining carbon with a $\delta^{13}C$ of $+0.2^0/_{00}$. 4) For a sustained deviation by $+0.5^0/_{00}$ from steady state, 2.5% of the ocean's reservoir has to be extracted, in excess over steady state extraction. This has to happen once per flushing time. 5) The ocean has a carbon mass of about 60 present-day atmospheric carbon masses (ACM). 6) The ocean's dissolved carbon has a residence time on the order of 10^5 years (Garrels et al., 1973). 7) 2.5% of 60 ACM is 1.5 ACM, and 4 million years contains 40 flushing times. The estimate is: Excess Carbon Extracted = 1.5 ACM $\times$ 40 = 60 ACM.

Thus, the answer is that the system sequesters roughly *one oceanic carbon mass* in the course of the Monterey Event. This may not be pure coincidence, but may point to a clue for the origin of the Monterey Formation and equivalent deposits around the Pacific rim. Presumably, these formations arose because of the onset of coastal upwelling on a large scale, depleting a previously sated ocean of its excess carbon and phosphorus. We have elaborated elsewhere on this hypothesis (Vincent and Berger, in press).

We have focused on the input/output portion of the carbon system (rather than on internal distribution) because $\delta^{13}C$ values of BF, DPF, and SPF change in parallel fashion, even in detail. All four sites show this very well. However, there is also a productivity effect. If the ocean became indeed more productive during the mid-Miocene, as suggested by the oxygen isotope data, then the difference between $\delta^{13}C$ of SPF and BF should increase

upsection. The reason is that biological transfer of organic carbon ($\delta^{13}C = -20^0/_{00}$) to deep waters, and concomitant oxidation of carbon at depth, causes deep water $\delta^{13}C$ values to be relatively low.

There is indeed some evidence for increased separation in the $\delta^{13}C$ signals, between BF and SPF, especially at the initiation of the Monterey Excursion, and continuing, on the whole, upward in the section.

In an earlier study of the meaning of rapid shifts in carbon signals, we focused on the Chron-6 Carbon Shift in the latest Miocene, a shift toward negative values (Vincent et al., 1980). In that report, we suggested the following rank-order of factors: 1) excess organic carbon input to the system; 2) productivity change; and 3) changes in exchange patterns between Pacific and Atlantic (basin-basin fractionation). We propose the same rank order here; except that in (1) "input" must be changed to "output" to reverse the sign of the shift. In our earlier study, we considered only the problem of the shift, treating it as an instantaneous phenomenon. In analogy to the above outline of calculations, it is evident that a permanent change in input/output ratio of organic carbon to inorganic carbon is required to maintain the new state at lowered $\delta^{13}C$ values, after the shift.

One note of caution: the possible effect of diagenetic recrystallization cannot be ruled out as contributing to the convergence of the various isotopic signals in the lower part of the sections (Killingley, 1983).

CORRELATION MATRIX, SITE 216

We have extracted, from Table 4, $\delta^{18}O$ and $\delta^{13}C$ values for those levels for which measurements are available for all three signal carriers. The correlation matrix for these data is given as Table 6. The full utility of such tables will emerge when comparative matrices become available for other sites and time-spans. A few features stand out. Without elaboration (most of the trends have been discussed already) they may be summarized as follows: 1) Within the oxygen isotope signals, the low correlation between benthic and shallow planktonic signal suggests that ice-buildup and decay was of secondary importance in producing oxygen isotope variations. Temperature fluctuations apparently dominated. 2) Within the carbon signals, the high positive correlations (.73; .76; .85) indicate the dominance of whole water-column phenomena. That is, global changes in dissolved carbon composition are reflected, rather than changes in water-mass fractionation. 3) Correlations between oxygen and carbon isotope records are surprisingly high and there are two sets of such correlations. 4) The first set of correlations is between the shallow-water carbon signal and each oxygen isotope signal (−.49; −.45; −.49). It indicates that temperature fluctuations at all latitudes are coupled to the carbon budget of surface waters, and hence to the CO_2-exchange between ocean and atmosphere. 5) The second set of correlations consists of those between the benthic oxygen signal and each carbon signal (−.49; −.66; −.62), signifying a strong link between temperatures at high latitudes and bottom-water forma-

TABLE 6. CORRELATION MATRIX FOR ISOTOPIC DATA AT DSDP SITE 216

	OXSPF	CSPF	OXDPF	CDPF	OXBF	CBF	OXD-S	OXB-S	CD-S	CB-S
OXSPF	1.00	-0.49	0.51	-0.20	0.18	-0.20	-0.09	-0.22	0.48	0.37
CSPF	-0.49	1.00	-0.45	0.73	-0.49	0.76	-0.18	-0.29	-0.62	-0.25
OXDPF	0.51	-0.45	1.00	-0.30	0.54	-0.37	0.81	0.33	0.31	0.07
CDPF	-0.20	0.73	-0.30	1.00	-0.66	0.85	-0.21	-0.57	0.09	0.26
OXBF	0.18	-0.49	0.54	-0.66	1.00	-0.62	0.50	0.92	-0.04	-0.24
CBF	-0.20	0.76	-0.37	0.85	-0.62	1.00	-0.29	-0.53	-0.13	0.44
OXD-S	-0.09	-0.18	0.81	-0.21	0.50	-0.29	1.00	0.53	0.03	-0.18
OXB-S	-0.22	-0.29	0.33	-0.57	0.92	-0.53	0.53	1.00	-0.23	-0.39
CD-S	0.48	-0.62	0.31	0.09	-0.04	-0.13	0.03	-0.23	1.00	0.66
CB-S	0.37	-0.25	0.07	0.26	-0.24	0.44	-0.18	-0.39	0.66	1.00

Key: OX = $\delta^{18}O$; C = $\delta^{13}C$; SPF = shallow planktonic foraminifera;
DPF = deep-living planktonic foraminifera; BF = benthic foraminifera;
OXD-S = difference in $\delta^{18}O$ between deep and shallow planktonic f.;
OXB-S = difference in $\delta^{18}O$ between benthic and shallow planktonic f.;
CD-S and CB-S = analogous to previous, for $\delta^{13}C$. Numbers are correlation
coefficients, for values at those levels in Site 216 where data are
available for all three signal carriers (Table 4).

tion on the one hand, and the global carbon cycle on the other. 6) The highest correlations between differences and straight isotope signals are between oxygen in benthics with the difference of benthics to shallow planktonics (.92), which derives from the lack of response of the shallow water tropical temperature to high latitude cooling. 7) Oxygen in deep planktonic foraminifera is highly correlated with the difference between deep and shallow planktonics, presumably for a similar reason as in 6. 8) The carbon difference benthic-shallow correlates well with carbon difference deep-shallow (.66), which is derived from a high correlation between benthic and deep carbon signals (.85). 9) The carbon difference deep-shallow correlates with shallow $\delta^{13}C$ (−.62), signifying less variability in the deep signal than in the shallow one (standard deviation of .30 vs. .38). 10) Finally, oxygen difference benthic-shallow correlates with carbon of deep planktonics (−.57), which suggests a link between variation in temperature gradient and strength of oxygen minimum.

This simple preliminary analysis is not meant to replace detailed bio- and chemo-stratigraphic analysis, but to illustrate some of the possibilities in multispecies isotope stratigraphy. Much more needs to be done, and much remains to be extracted from these data. It would be interesting to analyze the correlation matrices as a function of time period considered. While oxygen and carbon signals may be positively correlated on a long time scale, they may be negatively correlated on a short one. This has implications for the reaction time of climatic and geochemical mechanism which can be invoked to explain the observations.

CONCLUSIONS

Regarding Miocene paleoceanography, perhaps the most important result of our study is additional evidence that major changes in the carbon system evolved parallel with, and pre-

ceded, the mid-Miocene increase in the planetary temperature gradient. Awareness that the carbon cycle is intimately intertwined with climate evolution has much increased recently, owing in large part to the finding, from ice core analyses, that atmospheric carbon dioxide varied greatly during the Pleistocene (Berner et al., 1980; see Broecker 1982, and articles in Hansen and Takahashi, 1984). The integration of oxygen and carbon isotope signals with biostratigraphic information holds the key to understanding the climate-carbon cycle interaction.

ACKNOWLEDGMENTS

We gratefully acknowledge the careful reading of our manuscript by David A. Hodell, Lloyd D. Keigwin, and Samuel M. Savin. Their criticisms and suggestions greatly improved the paper. The remaining flaws, of course, are our own responsibility. We also owe thanks to the editor, J. Kennett, for unusual patience and endurance.

REFERENCES CITED

Arthur, M. A., 1982, The carbon cycle: Controls on atmospheric CO_2 and climate in the geologic past, in Berger, W. H. and Crowell, J. C., eds., Climate in earth history: National Research Council Studies in Geophysics, National Academy Press, p. 55–67.

Barron, E. J., and Washington, W. M., 1984, The role of geographic variables in explaining paleoclimates: results from Cretaceous climate sensitivity studies. Journal Geophysical Research, v. 89, p. 1267–1279.

Barron, J. A., and Keller, G., 1982, Widespread Miocene deep-sea hiatuses: coincidence with periods of global cooling: Geology, v. 10, p. 577–581.

Barron, J. A., Keller, G., and Dunn, D. A., this volume, A multiple microfossil biochronology for the Miocene.

Bé, A.W.H., 1977, An ecological, zoogeographic and taxonomic review of recent planktonic foraminifera, in Ramsay, A.T.S., ed., Oceanic Micropaleontology: Academic Press, London, v. 1, p. 1–100.

Bé, A.W.H., and Hutson, W.H., 1977, Ecology of planktonic foraminifera and biogeographic patterns of life and fossil assemblages in the Indian Ocean: Micropaleontology, v. 23, p. 369–414.

Belanger, P. E., Curry, W. B. and Matthews, R. K., 1981, Core-top evaluation of benthic foraminiferal isotopic ratios for paleo-oceanographic interpretations: Palaeogeography, Palaeoclimatology, Palaeoecology, v. 33, p. 205–220.

Bender, M. L. and Keigwin, L. D., Jr., 1979, Speculations about the Upper Miocene change in abyssal Pacific dissolved bicarbonate $\delta^{13}C$: Earth Planetary Science Letters, v. 45, p. 383–393.

Berger, W. H., 1981, Paleoceanography: the deep-sea record, *in* Emiliani, C., ed., The oceanic lithosphere: The sea, Vol. 7: Wiley Interscience, New York, p. 1437–1519.

Berger, W. H., 1982, Deep-sea stratigraphy: Cenozoic climate steps and the search for chemo-climatic feedback *in* Einsele, G., and Seilacher, A., eds., Cyclic and event stratification: Springer-Verlag, p. 121–157.

Berger, W. H., and Killingley, J. S., 1977, Glacial-Holocene transition in deep-sea carbonates: selective dissolution and the stable isotope signal: Science, v. 197, p. 563–566.

Berger, W. H., Bé, A.W.H. and Vincent, E., eds., 1981, Oxygen and carbon isotopes in foraminifera: Palaeogeography, Palaeoclimatology, Palaeoecology, v. 33, p. 1–278.

Berger, W. H., Killingley, J. S., and Vincent, E., 1978, Stable isotopes in deep-sea carbonates: Box Core ERDC-92, west equatorial Pacific: Oceanologica Acta, v. 1, p. 203–216.

Berger, W. H., Vincent, E., and Thierstein, H. R., 1981, The deep-sea record: major steps in Cenozoic ocean evolution: Society of Economic Paleontologists and Mineralogists Special Publication, v. 32, p. 489–504.

Berggren, W. A., Kent, D. V., and Van Couvering, J. A., in press, Neogene geochronology and chronostratigraphy *in* Snelling, N. J., ed., Geochronology and the geological record: Geological Society of London Special Paper.

Berner, W., Oeschger, H., and Stauffer, B., 1980, Information on the CO_2 cycle from ice core studies: Radiocarbon, v. 22, p. 227–235.

Blow, W. H., 1969, Late middle Eocene to Recent planktonic foraminiferal biostratigraphy, *in* First International Conference on Planktonic Microfossils, Geneva, 1967, p. 199–421.

Boersma, A. and Shackleton, N., 1977, Tertiary oxygen and carbon isotope stratigraphy, Site 357 (Mid–latitude South Atlantic), *in* Initial Reports Deep Sea Drilling Project, v. 39, p. 911–924.

Broecker, W. S., 1970, A boundary condition on the evolution of atmospheric oxygen: Journal Geophysical Research, v. 75, 3553–3557.

Broecker, W. S., 1973, Factors controlling CO_2 content in the oceans and atmosphere *in* Woodwell, G. M., and Pecan, E. V., eds., Carbon and the biosphere: AFC Symposium, 30, p. 32–50.

Broecker, W. S., 1982, Glacial to interglacial changes in ocean chemistry: Progress in Oceanography, v. 11, p. 151–197.

Burckle, L. H., Keigwin, L. D., and Opdyke, N. D., 1982, Middle and late Miocene stable isotope stratigraphy: Correlation to the paleomagnetic reversal record: Micropaleontology, v. 28, p. 329–334.

Craig, H., 1970, Abyssal Carbon-13 in the South Pacific: Journal Geophysical Research, v. 75, 691–695.

Davies, T. A., and Kidd, R. B., 1977, Sedimentation in the Indian Ocean through time, *in* Heirtzler et al., eds., Indian Ocean geology and biostratigraphy: American Geophysical Union, p. 61–85.

Deuser, W. G. and Hunt, T. M., 1969, Stable isotope ratios of dissolved inorganic carbon in the Atlantic: Deep Sea Research, v. 16, 221–225.

Douglas, R., and Woodruff, F., 1981, Deep-sea benthic foraminifera *in* Emiliani, C., ed., The oceanic lithosphere: The sea, Vol. 7: Wiley Interscience, New York, p. 1233–1327.

Emiliani, C., 1954, Depth habitats of some species of pelagic foraminifera as indicated by oxygen isotope ratios: American Journal of Science, v. 252, p. 149–159.

Emiliani, C., 1955, Peistocene temperatures: Journal of Geology, v. 63, p. 538–578.

Fischer, A. G., 1984, The two Phanerozoic supercycles. *in:* Berggren, W. A. and VanCouvering, J. A., eds., Catastrophes and earth history: Princeton University Press, New Jersey, p. 129–150.

Fischer, A. G., and Arthur, M. A., 1977, Secular variations in the pelagic realm: Society Economic Paleontologists and Mineralogists, Special Publication, n. 24, p. 19–50.

Foster, J. H., and Opdyke, N. D., 1970, Upper Miocene to Recent magnetic stratigraphy in deep-sea sediments: Journal of Geophysical Research, v. 75, p. 4465–4473.

Garrels, R. M., and Lerman, A., 1981, Phaneozoic cycles of sedimentary carbon and sulfur. Proceedings National Academy of Science, v. 78, p. 4652–4656.

Garrels, R. M. and Perry, E. A., 1974, Cycling of carbon, sulfur, and oxygen through geologic time, *in* Goldberg, E. D., ed., The sea, Vol. 5: Wiley Interscience, New York, p. 303–336.

Garrels, R. M., Perry, E. A., Jr., and Mackenzie, F. T., 1973, Genesis of Precambrian iron-formations and the development of atmospheric oxygen, *in* Precambrian iron formations of the world: Economic Geology, v. 68, no. 7, p. 1173–1179.

Graham, D., Corliss, B., Bender, M. L. and Keigwin, L. D., Jr., 1981, Carbon and oxygen isotopic disequilibria of recent deep-sea benthic foraminifera: Marine Micropaleontology, v. 6, p. 483–498.

Hansen, J. E., and Takahashi, T., eds., 1984, Climate processes and climate sensitivity, Geophysical Monograph, v. 29, American Geophysical Union, Washington, D.C., 368 pp.

Haq, B. R., Worsley, T. R., Burckle, L. H., Douglas, R. G., Keigwin, L. D., Jr., Opdyke, N. D., Savin, S. M., Sommer, M. A. II, Vincent, E., and Woodruff, F., 1980, Late Miocene marine carbon-isotope shift and synchroneity of some phytoplanktonic biostratigraphic events: Geology, v. 8, p. 427–431.

Imbrie, J., Van Donk, J., Kipp, N. G., 1973, Paleoclimatic investigation of a late Pleistocene Caribbean deep-sea core: comparison of isotopic and faunal methods: Quaternary Research, v. 3, p. 10–38.

Keigwin, L. D., Jr., 1979, Late Cenozoic stable isotope stratigraphy and paleoceanography of DSDP Sites from the east equatorial and north central Pacific Ocean: Earth and Planetary Science Letters, v. 45, p. 361–382.

Keigwin, L. D., Jr., and Shackleton, N. J., 1980, Uppermost Miocene carbon isotope stratigraphy of a piston core in the equatorial Pacific: Nature, v. 284, p. 613–614.

Kennett, J. P., 1982, Marine Geology: Prentice-Hall, Englewood Cliffs, N.J., 813 pp.

Kennett, J. P., Shackleton, N. J., Margolis, S. V., Goodney, D. E., Dudley, W. C. and Kroopnick, P. M., 1979, Late Cenozoic oxygen and carbon isotopic history and volcanic ash stratigraphy: DSDP Site 284, South Pacific: American Journal of Science, v. 279, p. 52–69.

Killingley, J. S., 1983, Effects of diagenetic recrystallization on the $^{18}O/^{16}O$ values of deep-sea sediments: Nature, v. 301, p. 594–597.

Kolla, V., Bé, A., and Biscay, P. E., 1976, Calcium carbonate distribution in the surface sediments of the Indian Ocean: Journal of Geophysical Research, v. 81, p. 2605–2616.

Matthews, R. K., and Poore, R. Z., 1980, Tertiary $\delta^{18}O$ record and glacio-eustatic sea-level fluctuations: Geology, v. 8, p. 501–504.

Okada, H., and Bukry, D., 1980, Supplementary modification and introduction of code numbers to the low-latitude coccolith biostratigraphic zonation (Bukry, 1973; 1975): Marine Micropaleontology, v. 5, p. 321–325.

Pisias, N. G. and Shackleton, N. J., in press, Stable isotope and calcium carbonate records from HPC Site 574A: High resolution records from the Middle Miocene, *in* Initial Reports Deep Sea Drilling Project, v. 85.

Riedel, W. R., and A. Sanfilippo, 1971, Cenozoic radiolaria from the western tropical Pacific, Leg 7, *in* Initial Reports of the Deep Sea Drilling Project, v. 7, p. 1529–1672.

Rochford, D. J., 1962, Hydrology of the Indian Ocean. II. The surface waters of the south-east Indian Ocean and Arafura Sea in the spring and summer: Australian Journal of Marine Freshwater Research, v. 13, p. 226–251.

Savin, S. M., 1977, The history of the Earth's surface temperature during the past 100 million years: Annual Review Earth Planetary Science, v. 5,

p. 319–355.

Savin, S. M., Douglas, R. G., and Stehli, F. G., 1975, Tertiary marine paleo-temperatures: Geological Society of America Bulletin, v. 86, p. 1499–1510.

Savin, S. M., Douglas, R. G., Keller, G., Killingley, J. S., Shaughnessy, L., Sommer, M. A., Vincent, E., and Woodruff, F., 1981, Miocene benthic foraminiferal isotope records: a synthesis: Marine Micropalentology, v. 6, p. 423–450.

Savin, S. M., Abel, L., Barrera, E., Hodell, D. A., Keller, G., Kennett, J. P., Killingley, J., Murphy, M., and Vincent, E., 1985, The evolution of Miocene surface and near-surface marine temperatures: Oxygen isotopic evidence, this volume.

Scholle, P. A., and Arthur, M. A., 1980, Carbon isotope fluctuations in Cretaceous pelagic limestones: potential stratigraphic and petroleum exploration tool: American Association Petroleum Geologists Bulletin, v. 64, p. 67–87.

Schrader, H. J., 1974, Cenozoic marine planktonic diatom stratigraphy in the tropical Indian Ocean, in Initial Reports of the Deep Sea Drilling Project, v. 24, p. 887–967.

Sclater, J. G., Abbott, D., and Thiede, J., 1977, Paleobathymetry and sediments of the Indian Ocean, in Heirtzler et al., eds., Indian Ocean geology and biostratigraphy: American Geophysical Union, p. 25–59.

Sclater, J. G., Meinke, L., Bennett, A., and Murphy, C., 1985, The depth of the ocean through the Neogene, this volume.

Shackleton, N. J., 1974, Attainment of isotopic equilibrium between ocean water and the benthonic foraminifera genus Uvigerina: isotopic changes in the ocean during the last glacial: C.N.R.S. Colloquium, 219, p. 203–209.

Shackleton, N. J., 1977, Carbon-13 in Uvigerina; tropical rainforest history and the equatorial Pacific carbonate dissolution cycles, in Andersen, N. R., and Malahoff, A., eds., The fate of fossil fuel CO_2 in the oceans: Plenum Press, New York, p. 401–427.

Shackleton, N. J., 1982, The deep-sea sediment record of climate variability: Progress in Oceanography, v. 11, p. 199–218.

Shackleton, N. J., and Kennett, J. P., 1975, Paleotemperature history of the Cenozoic and the initiation of Antarctic glaciation: Oxygen and carbon isotope analyses in DSDP Sites 277, 279 and 281: Initial Reports of the Deep Sea Drilling Project, v. 29, p. 743–756.

Sverdrup, H. W., Johnson, M. W., Fleming, R. H., 1942, The oceans, their physics, chemistry and general biology: Prentice Hall Inc., New York, 1987 pp.

Tappan, H., 1968, Primary production, isotopes, extinctions and the atmosphere: Palaeogeography, Palaeoclimatology, Palaeoecology, v. 4, p. 187–210.

Thierstein, H. R., Geitzenauer, K., Molfino, B., Shackleton, N. J., 1977, Global synchroneity of late Quaternary coccolith datums: Validation by oxygen isotopes: Geology, v. 5, p. 400–404.

Vail, P. R. and Hardenbol, J., 1979, Sea-level changes during the Tertiary: Oceanus, v. 22, p. 71–79.

Vergnaud-Grazzini, C. and Rabussier-Lointier, D., 1980, Essai de correlation stratigraphique par le moyen des isotopes de l'oxygène et du carbone: Bul-

letin Societé Géologie France, (7) 22, p. 719–730.

Vincent, E., 1974, Cenozoic planktonic biostratigraphy and paleooceanography of the tropical western Indian Ocean: in Initial Reports of the Deep Sea Drilling Project, v. 24, p 1111–1154.

Vincent, E., 1977, Indian Ocean Neogene planktonic foraminiferal biostratigraphy and its paleoceanographic implications in Heirtzler et al., eds., Indian Ocean geology and biostratigraphy: American Geophysical Union, p. 469–584.

Vincent, E., 1981, Neogene carbonate stratigraphy of Hess Rise (central North Pacific) and paleoceanographic implications, in Initial Reports of the Deep sea Drilling Project, v. 62, p. 571–606.

Vincent, E., and Berger, W. H., 1981, Planktonic foraminifera and their use in paleoceanography, in Emiliani, C., ed., The Oceanic Lithosphere: The sea, Vol. 7: Wiley Interscience, New York, p. 1025–1119.

Vincent, E., and Berger, W. H., in press, Carbon dioxide and polar cooling in the Miocene: The Monterey hypothesis in Sundquist, E. T., and Broecker, W. S., eds., Natural Variations in Carbon Dioxide and the Carbon Cycle: American Geophysical Union Geophysical Monograph, v. 32.

Vincent, E., and Killingley, J. S., in press, Oxygen and carbon isotope record for the early and middle Miocene in the Central Equatorial Pacific (DSDP Leg 85) and paleoceanographic implications, in Initial Reports of the Deep Sea Drilling Project, v. 85.

Vincent, E., Killingley, J. S., and Berger, W. H., 1980. The Magnetic Epoch-6 carbon shift: a change in the ocean's $^{13}C/^{12}C$ ratio 6.2 million years ago: Marine Micropaleontology, v. 5, p. 185–203.

Vincent, E., Killingley, J. S., and Berger, W. H., 1981, Stable isotope composition of benthic foraminifera from the equatorial Pacific: Nature, v. 289, p. 639–643.

Vincent, E., Killingley, J. S., and Berger, W. H., 1983, The Chron-16 carbon event in the Miocene Indian Ocean: Geological Society of America, Abstracts with programs, v. 15, p. 712.

Williams, D. F., Sommer, M. A. and Bender, M. L., 1977, Carbon isotopic compositions of recent planktonic foraminifera of the Indian Ocean: Earth Planetary Sciences Letters, v. 36, p. 391–403.

Woodruff, F., Savin, S. M., Douglas, R. G., 1980, Biological fractionation of oxygen and carbon isotopes by recent benthic foraminifera: Marine Micropaleontology, v. 5, p. 3–11.

Woodruff, F., Savin, S. M., and Douglas, R. G., 1981, Miocene stable isotope record: a detailed deep Pacific Ocean study and its paleoclimatic implications: Science, v. 212, p. 665–668.

Wyrtki, K., 1971, Oceanographic atlas of the International Indian Ocean Expedition: National Science Foundation, Washington, D.C., 531 pp.

Wyrtki, K., 1973, Physical oceanography of the Indian Ocean: in Zeitzschel, B., ed., Springer–Verlag, The biology of the Indian Ocean, N.Y., p. 18–36.

MANUSCRIPT ACCEPTED BY THE SOCIETY DECEMBER 17, 1984

Geological Society of America
Memoir 163
1985

Changes in Miocene deep-sea benthic foraminiferal distribution in the Pacific Ocean: Relationship to paleoceanography

Fay Woodruff
Department of Geological Sciences
University of Southern California
Los Angeles, California 90089-0741

ABSTRACT

This paper reports the results of a detailed analysis of 25 Miocene Pacific Deep Sea Drilling Project (DSDP), sites at which 110 benthic foraminiferal species in 255 DSDP samples were quantified. The sites range in depth from 1.5 to 4.5 km between 45 degrees north and south latitudes. The faunal distributions were studied for many Pacific Ocean sites at different water depths through time; diagrams depicting both areal and water depth distributions are presented.

Miocene deep-sea benthic foraminifera form clear distribution patterns. Some species appear to have reacted to changes in water mass quality; others appear to reflect changes in food availability and sample preservation resulting from changes in surface productivity related to paleoceanographic change.

Many changes in faunal distribution, as well as evolutionary originations and extinctions, occurred between 13 and 16 Ma, concomitant with paleoclimatic and oceanographic changes presumably related to Antarctic glacier expansion and cooling of the deep Pacific Ocean. Late Miocene faunal changes between 8 and 10 Ma were primarily changes in species abundances and depth distribution which appear to be related to the effects of further Antarctic glaciation.

The oxygen isotopic record for most of the faunal samples is shown for different water depths through time. The data show clear isotopic changes 14.5 Ma and 8–10 Ma, concurrent with many changes in the foraminiferal distribution patterns.

A new species, *Cibicidoides cenop* n. sp. is named (Appendix I). It evolved in the early Miocene and is apparently an indicator of cold-water conditions in the Miocene Pacific Ocean.

The benthic foraminiferal data are compatible with the following paleoceanographic interpretations: (1) a warm early Miocene deep ocean with a less well-developed surface-to-bottom thermal-gradient that in mid-late Miocene time; 2) an increase in ocean margin surface productivity after 16 Ma related to intensification of atmospheric-oceanic circulation and upwelling; 3) a change in the deep ocean water mass after 15 Ma, including a cooling and a shallowing of deep-ocean conditions related to permanent establishment of the Antarctic ice sheet; 4) an increase in surface productivity along the equator after 15 Ma, related to intensification of the latitudinal thermal gradient, more vigorous deep-ocean circulation and perhaps establishment of the Equatorial Counter-current and Undercurrent systems; 5) an increase in sediment organic carbon and intensification of the low oxygen zone after 8–10 Ma as well as an increase in deep-ocean dissolution and Antarctic bottom water influence related to further Antarctic ice sheet expansion in latest Miocene time; and 6) a deep water corridor across the Indo-Pacific until at least 21 Ma.

INTRODUCTION

This is the first large-scale study on the response of deep sea benthic foraminifera to paleoceanographic changes in the Miocene Pacific Ocean, 22.5 to 5.2 million years ago (Ma). The study evaluates the relationship of benthic foraminiferal species, genera and assemblages to large-scale Pacific paleoceanographic changes during Miocene time. The study is divided into four parts: a section on the effects of sample preservation on species distributional patterns; a section on the distribution of individual species through space and time; a study of the changes in distribution through time of the key Miocene assemblages identified by R-mode principal component analysis; and a summary discussion of paleoceanographic implications.

Benthic foraminifera are good indicators of deep-sea environmental conditions. They constitute the largest biomass in the modern lower bathyal (>1000 m) and abyssal environment (Paul and Menzies, 1973; Hessler and Jumars, 1974) and are often the only carbonate shells robust enough to be preserved in deep-ocean sediments. Benthic and planktonic foraminifera of the same age accumulate in the same samples so that bottom water and surface conditions may be interpreted and correlated together. Although deep-sea benthic foraminifera have lower rates of faunal turnover than planktonic foraminifera, many benthic species have discrete ranges and their evolutionary and migratory patterns give valuable biostratigraphic and paleoecologic information about the deep-sea environment (Douglas and Woodruff, 1981).

Many modern benthic foraminifera have been correlated to specific water masses (Streeter, 1973, Schnitker, 1974; Lohmann, 1978; Corliss, 1979; Gofas 1978; Walch, 1978; Burke, 1981; Resig, 1981), hence many species should react to changes in water mass location when paleoceanographic changes occur. Recent species-environment correlations are poorly known, so that the reasons behind the changes in benthic distributions are not always apparent. However, Miocene as well as Recent foraminifera form clear distributional patterns.

The Miocene was a period of major oceanographic change. During this time the Antarctic ice sheet became permanently established (Savin, Douglas and Stehli, 1975; Shackleton and Kennett, 1975b). A 1.5 per mil increase in Pacific Ocean $\delta^{18}O$ 15.5–14.0 Ma (Woodruff et al., 1981), is attributed to cooling of deep Pacific water and oceanic enrichment in oxygen-18 due to Antarctic ice uptake of oxygen-16. While benthic foraminiferal isotopic temperatures cooled in middle Miocene time, planktonic foraminiferal isotopic temperatures warmed in the equatorial regions (Savin et al., 1975; Savin, 1977; Shackleton and Kennett, 1975b; Shackleton, 1982), signifying an increase in the surface-to-ocean floor, as well as the equatorial-to-polar, thermal gradient after 16 Ma. An increase in the number of deep ocean hiatuses occurred 15–9 Ma, due in part to Antarctic corrosive bottom waters (van Andel et al., 1975; Keller and Barron, 1983). Neogene diatomites appeared around the Pacific rim in the middle Miocene, as a result of increasingly vigorous atmospheric-oceanic

circulation 15–16 Ma associated with climatic deterioration and Antarctic glacial buildup (Ingle, 1981). Calcium carbonate "supply rates" increased steadily in the central equatorial Pacific, reaching a Tertiary maximum in the late Miocene (van Andel et al., 1975). There may have been an increase in the amount of North Atlantic Deep Water (NADW) introduced into the Pacific via the circumpolar current between 13.2 and 6.2 Ma (Blanc and Duplessy, 1982; Blanc et al., 1980; Schnitker, 1980). Circulation through the Indo-Pacific and the Panama region may have terminated during middle Miocene time (Edwards, 1975; Weyl, 1980).

Other tectonic changes including changes in spreading rates and marginal basin formation around the Pacific Ocean also occurred at this time. After 15 Ma there was a major increase in denudation of the continents associated with climatic deterioration (Donnelly, 1982). There was also a significant increase in eolian deposition to the deep sea 13 Ma in response to steepening thermal gradients and intensifying atmospheric circulation associated with global aridity and expansion of the Antarctic ice sheet (Janecek and Rea, 1983). In the late Miocene there was a major increase in sedimentation in highly productive regions (Barron, 1980; Keller, 1980a, Vincent et al., 1980) and a 1.0 per mil permanent negative shift in $\delta^{13}C$ of deep water HCO_3 (Bender and Keigwin, 1979; Keigwin, 1979; Haq et al., 1980; Vincent et al., 1980). The sedimentation increase, the $\delta^{13}C$ depletion 7–6 Ma, the Messinian event 7–5 Ma (Van Couvering et al., 1976), the coastal offlap (Vail and Hardenbol, 1979) and major hiatuses 9.8 and 6.6 Ma (Vail and Hardenbol, 1979; Keller and Barron, 1981) may all be directly or indirectly related to major Antarctic glacial expansion after 10 Ma (Shackleton and Kennett, 1975a and b; Hamilton, 1972; Ciesielski and Weaver, in press; Savage and Ciesielski, 1983).

PROCEDURE

The benthic foraminifera in 255 Miocene samples (Appendix 2) were studied from 25 Miocene Pacific DSDP sites (Figs. 1a, b, Table I) between 1.5 and 4.5 km water depth, 45° north and south latitudes and 22–5 Ma. Large (30 cc) samples were used and total samples or splits containing a minimum of 300 specimens were counted unless there were not 300 benthics in the total sample. The benthic species in the >150 micron fraction were quantified (Appendix 3). This size fraction was studied so that the results would be comparable with other laboratories and so that small species, such as *Epistominella umbonifera* and *E. exigua* are represented. Many monospecific *Cibicidoides* samples were sent to Dr. S. M. Savin of Case Western Reserve University for oxygen and carbon isotopic analysis. Samples from the Experimental Mohole Expedition (EM) are used interchangeably with its recently redrilled counterpart, DSDP Site 470.

The paleodepths of the DSDP sites were estimated (Thierstein, 1979; Berger, 1973; and Sclater et al., 1971) and the areal maps are backtracked location maps generated by John Sclater (this volume). The planktonic biostratigraphy used was generated

TABLE 1. SITE DATA AND REFERENCES

DSDP Site	Present Latitude Longitude	Location	Paleo-depth (m)	Benthic Faunal	Planktonic Biostratigraphy
55	9°18'N 142°32'E	Flank of Caroline Ridge	2850	This paper	DSDP V. 6
62.1	1°52'N 141°56'E	Eauripik Ridge	2591	This paper	DSDP V. 7 Keller, unpubl.
71	4°28'N 140°19'W	Central Equatorial Pacific Floor	4100	This paper	Keller, 1981b
77B	00°28'N 133°13'W	Central Equatorial Pacific Floor	3900 - 4100	This paper	Keller, 1981b and Keller in Savin et al., 1981
158	6°37'N 85°14'W	Crest of Cocos Ridge	1900	This paper	Keller et al., 1982
159	12°19'N 122°17'W	West flank East Pacific Rise	3000 - 3500	Quinn, 1982	DSDP V. 16
167	7°4'N 176°49'W	Crest of Megellan Rise	3400	This paper; Douglas, 1973	Douglas, 1973
171	19°07'N 169°27'W	Saddle at Crest of Horizon Guyot	2200	This paper	Douglas, 1973
173	35°57'N 125°27'W	Continental Slope off Cape Mendocino	3000	This paper	Keller et al., 1982
178	56°57'N 147°07'W	Western Alaskan Abyssal Plain	4200	This paper	DSDP V. 43
206	32°0.1'S 165°27'E	Bottom of Caledonia Basin	3200	This paper	Srinivasan and Kennett, 1981
208	26°06'S 161°13'E	Near Crest of Lord Howe Rise	1500	This paper	Srinivasan and Kennett, 1981
281	47°59'S 147°45'E	Near Crest of So. Tasmin Rise	1600	This paper	Srinivasan and Kennett, 1981
288	5°58'S 161°49'E	Flank of Ontong-Java Plateau	3100	This paper	DSDP V. 30
289	0°29'S 158°30'E	Near crese Ontong-Java Plateau	2300 - 2500	Woodruff, et al., 1981; and this paper	Srinivasan and Kennett, 1981 Keller in Savin et al., 1981
292	15°49'N 124°39'E	Benham Rise W. Philippine Basin	2400 - 2800	This paper	Keller, 1981b
296	29°20'N 133°31'E	Terrance on Palau-Kyushu Ridge	2800 2900	This paper	Keller, 1980a,b
310	36°52'N 176°54'E	Near crest of Hess Rise	3500	This paper	Keller, et al., 1982
317B	11°0'S 162°15'W	Crest of Manihiki Plateau	2700	This paper	DSDP V. 33
319	12°1'S 101°31'W	Bauer Basin Deep	4296	This paper	Keller, 1980a,b
320	9°0.4'W 83°31'W	Peru Basin	3500 - 3600	This paper	Resig, 1976
448	16°20'N 134°52'E	Saddle Near Crest Palau-Kyushu Ridge W. Philippine Basin	3100 - 3300	This paper	Keller, unpubl.
470	28°46'N 117°31'W	Abyssal seafloor off Baja	3600	K. McDougal Unpubl. and	Keller and Barron, 1981
Experi-mental Mohole	28°59'N 17°30'W	Abyssal seafloor off Baja	3600	This paper; Parker, 1964	Bandy, 1971
495	12°30'N 9°2'W	Rise on seaward slope of Middle America Trench	3600 -	This paper	Keller, unpubl.

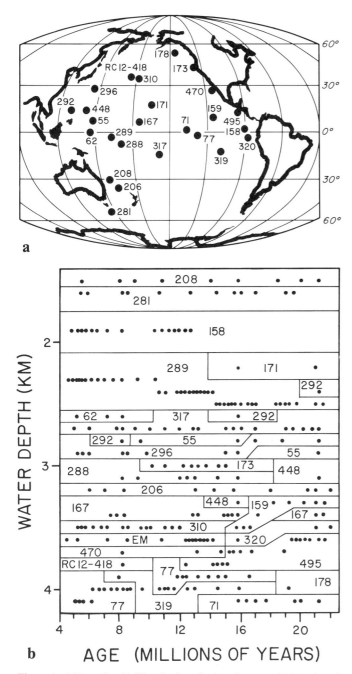

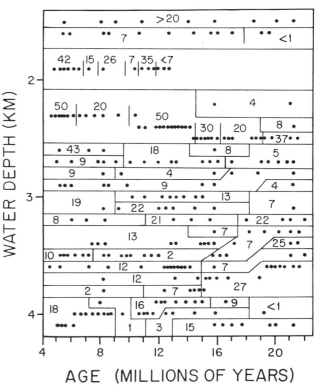

SEDIMENT ACCUMULATION RATES (M/My)

Figure 2. Sediment accumulation rates in m/Ma. Data compiled from Initial Reports of the Deep Sea Drilling Project and Vincent (1981), Barron et al. (this volume).

Figure 1. a) Deep Sea Drilling Project site location map; b) location of samples at paleo depths, 22-5 Ma. Paleodepth data is from Initial Reports of the Deep Sea Drilling Project, Thierstein (1979), Berger (1973), Sclater et al. (1971). Many sites did not significantly subside or become uplifted during the Miocene.

primarily by the (CENOP) team. It is listed in Table 1 and is more detailed than that published in the *Initial Reports of the Deep Sea Drilling Project*. The sediment accumulation rates were compiled (Fig. 2) primarily from information given in the DSDP volumes, Vincent (1981), Barron et al. (in press).

The species identified are listed in Appendix 4. The most helpful taxonomic references include: Parker, 1964; Cushman

and Stainforth, 1945; Douglas, 1973; Beckmann, 1953; Burke, 1981; Tjalsma and Lohmann, 1983; Resig, 1976, 1981; Lohmann, 1978; Boltovskoy 1978; Bermudez 1949; Barker, 1960 (Brady plates); Phleger, Parker and Pierson, 1953; Belford, 1966; Berggren, 1972; Corliss, 1979; Cushman and Parker, 1947; Schnitker, 1979; Finlay, 1939 a, b, c, 1940; Miller et al. (in press); Galloway and Heminway, 1941; Nuttall, 1932; Galloway and Morrey, 1929.

An R-mode principal component analysis (PCA) was performed on the data set in order to identify the major assemblage groupings. The species and station scores are listed in Appendices 5 and 6.

EFFECTS OF SAMPLE PRESERVATION ON SPECIES DISTRIBUTION

Aspects of sample preservation considered here include dissolution, diagenesis and sample contamination. The methods used to evaluate sample preservation included diversity-abundance relationships, planktonic to benthic (P/B) ratios, evaluation of the fragile planktonic specimens, fragmentation, specimen luster, and miliolid preservation. The study of sample and specimen preservation is important because it affects the species distributional patterns. Fortunately, in this study of predominantly carbonate oozes and chalk-oozes, preservation was generally very good so that the basic faunal patterns have been maintained.

Diagenesis and dissolution, the two main factors affecting specimen preservation, are inversely correlated in the deep-sea, and both are related to oceanographic conditions. Dissolution, however corrosive the water mass or deep the station, is more intense when the specimen remains exposed for long periods at the sediment/ocean interface, which occurs when sediment accumulation rates are low. Diagenesis is more intense when the specimen is deeply buried, which occurs more quickly when the sediment accumulation rates are high.

Diagenesis has affected the preservation of the early Miocene samples. In general, the more deeply a sample is buried below the sea floor, the lower the porosity, the more indurated and diagenetically altered the sample, and the more mechanically compacted and chemically altered the specimens (Schlanger and Douglas, 1974). Species which withstand diagenetic destruction become concentrated with increased burial depth and become a larger percent of the more deeply buried samples. Fragile species are weakened, crushed, become rare or are eliminated from the fossil record. Depth of burial is dependent on the rate of sediment accumulation, which varies between sites from 1 to 50 m/my (Fig. 2-3). Most of the dewatering and porosity reduction takes place before the transition from the "Shallow-burial Realm" (above 200 m) to the "Deep-burial Realm," below which cementation becomes the dominant process (Schlanger and Douglas, 1974). None of the species or assemblage patterns in this paper show patterns similar to the 200 m depth contour shown in Figure 3. Indeed, one of the best preserved sites, Site 289, is one of the most deeply buried. Degree of diagenesis and depth of burial are apparently less important to specimen preservation than dissolution at the sediment/ocean interface.

Dissolution of the specimens as they lie exposed to corrosion by the overlying water masses and acidic water at the sediment/ocean interface is apparently a major factor. Most specimen dissolution occurs at the sediment/ocean interface and, therefore, specimen preservation is enhanced by high rates of sedimentation which quickly bury the specimens. Specimen preservation along the equator improved after 15 Ma apparently due to accelerated rates of burial. This is compatible with progressively increasing Miocene "extrapolated surface carbonate supply rates" with a peak in "carbonate accumulation rates" 14–15 Ma (van Andel et al., 1975). It is also compatible with the major increase in biogenic silica accumulation rates (which are an indicator of surface productivity without being severely affected by dissolution) after 16 Ma in the eastern equatorial Pacific (Leinen, 1979). The major mid to late Miocene hiatuses occurred later, between 13.5 and 9 Ma, associated with intensified deep ocean current activity (Keller and Barron, 1983), although P/B ratios at Sites 77 and 289 suggest severe dissolution started around 14.3 Ma. A good supply of planktonic carbonate to the sea floor would improve the chances for benthic preservation because planktonic carbonate is more dissolution-prone.

The dissolution of calcium carbonate associated with the

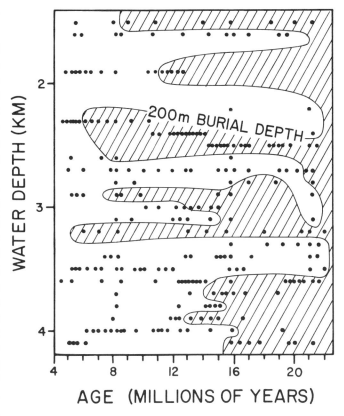

Figure 3. Distribution of samples buried below 200 m beneath the sea floor (shaded area). Burial depth to some degree controls sample diagenesis and hence preservation.

oxidation of large amounts of organic matter in the sediments (Berger and Winterer, 1974) would be important primarily in the coastal areas near sources of terrigenous debris and under areas of high coastal productivity rather than the open ocean where little organic matter reaches the sea floor. The low P/B values at Site 158 (~1:1), below the eastern equatorial upwelling zone, are partially attributed to the relatively acidic water associated with the oxidation of organic matter on the sea floor.

The diversity of species in a sample is lowered by the compaction and diagenetic alteration of the more fragile species with increasing depth of burial. The diversity used in Figure 4 is the number of species based only on the 110 species monitored in this study, not the total diversity of the sample. In most cases, the more specimens counted, the higher the diversity (Fig. 4a). In very poorly-preserved samples there may be many residual specimens of one species, all else having been dissolved, or a few specimens with only sand grains or the like remaining. Each site forms a distinctive cluster of points (Fig. 4b). The realms depicted are related primarily to the degree of sample preservation, which generally increases from the lower left to the upper right quadrant. Early Miocene sites (triangles) generally fall to the lower left in the realm of that particular site, which indicates poorer preservation, not lower actual early Miocene diversity. Sites 173 and 178 contained the most poorly preserved faunas, with P/B ratios

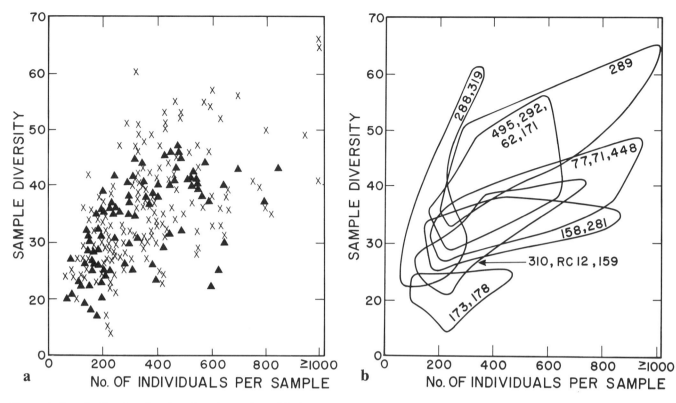

Figure 4. Sample diversity plotted against the number of individuals counted per sample. a) The early Miocene (triangles) samples have lower diversities than the mid-late Miocene samples (x), due to poorer preservation. The more specimens counted, the more species found. b) The sites fall into distinct realms with the samples in the lower left quadrant less well preserved than those toward the upper right quadrant. The samples toward the upper left have higher diversities than expected due to downslope transport.

generally less than 30:1 while the preservation at Site 289 was excellent, with P/B ratios generally greater than 500:1. It should be noted, however, that P/B ratios are somewhat controlled by surface productivity and planktonic sedimentation rates and are not always a good indicator of relative sample preservation. Samples from subantarctic Site 281 were somewhat better preserved than Figure 4b indicates because they contained a number of unmonitored species which were not common to the area emphasized in this study, the deep equatorial Pacific.

Sample diversity is affected by contamination from transported microfossils (Fig. 4b). Sites 288 and 319, and to a lesser degree, Site 495 fall off the general trend. This is apparently due to an increase in diversity due to downslope contamination in otherwise poorly preserved samples. Site 288 is fairly far down the flank of the Ontong-Java Plateau and Site 319 is near the bottom of the Bauer Basin, so both are likely to collect downslope contaminants; Site 495, off Central America, contains turbidites of a similar age but from shallower environs (Thompson, 1982).

Grain size of the deep-ocean samples is related to diagenesis and dissolution as well as to downslope contamination. The deep-sea benthic assemblages were not noticeably related to sample grain size (compiled from the Initial Reports of the DSDP) probably because grain size was originally fairly uniform, being primarily the result of pelagic organisms raining to the sea floor creating nannofossil and foraminiferal oozes. As these sediments

were more deeply buried, they were diagenetically altered and crystallized and they often progressed from an ooze with predominantly clay-sized grains to a chalk with larger silt-sized grains. The sand-size fraction, generally less than 5% in the Miocene, was sometimes increased by influxes of volcanic glass, glauconite, zeolites, radiolarians, diatoms, or more important, downslope contamination from slumpage or turbidites. Sites which have unusually high percentages of the sand-size fractions at erratic intervals (such as 288 and 495) also tend to be those with turbidites and downslope contaminated faunas.

The many distinctly different distributional patterns (see next section) for species with apparently similar dissolution indices suggest that many distributional patterns are not related to preservation. Corliss and Honjo (1981) list five levels of Recent benthic species or species groups based on their relative susceptibility to dissolution: 1) *Amphistigina* (most easily dissolved, prefers depths too shallow for this study); 2) *Pyrgo murrhina* (a miliolid); 3) *Planulina wuellerstorfi (Cibicidoides wuellerstorfi)*; 4) *Cibicidoides kullenbergi, Epistominella umbonifera, Hoeglundina elegans, Oridorsalis tener*; 5) *Gyroidinoides orbicularis* and *G. soldanii* (least easily dissolved). The flatter, thinner-walled species (group 2) dissolve before the round, robust species (group 4-5), as might be expected. This Miocene study will show that *Epistominella umbonifera, C. kullenbergi* and *Oridorsalis* (all of group 4) display obviously different distributional patterns. *Ori-*

dorsalis is ubiquitous, averaging from 7-10% of the fauna. *Gyroidinoides soldanii* (group 5) is ubiquitous but rarely reaches concentrations greater than 4%, not what one would expect of a species being concentrated as a residue in poorly preserved samples. Miliolid species (group 2) equally overlap the realms of *E. umbonifera* (group 4) and the somewhat dissolution-prone *Uvigerina* species. Delicate *Stilostomella* species *S. annulifera*, *S. spinata*, *S. curvatura*, *S. modesta* (not the more resistant species *S. abyssorum*, *S. subspinosa*) overlap the realms of *C kullenbergi* and *G. subglobosa*, shown to be more concentrated in more lithified samples (Woodruff and Douglas, 1981). The six common species of the *Cibicidoides* genus found in this study probably range from group 3 to 4 in the Corliss and Honjo study and yet they show distinctly different distributional patterns. The implication is that although preservation is undoubtedly affecting faunal distribution, the basic distributional patterns are being maintained.

CHANGES IN SPECIES DISTRIBUTION THROUGH TIME

Extinctions

Benthic evolution accelerated in the middle Miocene between 13 and 16 Ma, a period of major increase in $\delta^{18}O$ values (Savin et al., 1981). A similar faunal change was also noted 15 Ma in the Atlantic Ocean (Berggren, 1972). Documenting evolutionary changes (that is, evolutionary first and last occurrences) is difficult because of the widespread downslope displacement, reworking, downcore contamination, generally poorer early Miocene preservation and scarcity of some species. For these reasons, only those species which were fairly abundant and were present at several sites are reported (Table 2, 3).

Eleven species became extinct (Table 2). Four *Bulimina/Buliminella* species which preferred bathyal (0.5–2.5 km) to upper abyssal (2.5–3.2 km) water depths (Fig. 5) became extinct 14 Ma: *Bulimina miolaevis; B. jarvisi; B glomarchallengeri* and *Buliminella grata.* Boltovskoy (1980) also noted these approximate biostratigraphic ranges (*Bulimina* sp. cf. *B markisi* of Boltovoskoy = *B. glomarchallengeri*). Their preference for shallow water is reflected in their western Pacific distribution where most of the relatively shallow sites are located (Fig. 6). *Bulimina jarvisi* and *B. grata* have shown an upward trend in depth preference since the Eocene (Douglas and Woodruff, 1981) and perhaps now reached their shallowest limit.

Three *Cibicidoides* species which preferred abyssal (2.5–5 km) water depths became extinct approximately 14 Ma: *C. grimsdalei, C. dohmi,* and *C. havanensis* (Fig. 5c). These species prefer deeper water and therefore were found in the eastern Pacific where most of the deeper sites are located (Fig. 6).

Other species which apparently became extinct include: *Anomalinoides* sp. 1 of Douglas (1973); *Cassidulina* sp. cf. C. *angulosa; Hanzawaia cushmani; Melonis* sp. cf. *Anomalina regina. Vulvulina miocenica* may have become extinct 8 Ma. It is difficult to rule out the possibility that some of the more robust species which are presumed to have become extinct are reworked older specimens despite the abundance of apparently well-preserved juveniles. Some of the specimens appear too fragile to be reworked. If the specimens are reworked, it is significant that there was a major decrease in reworked specimens in the deep ocean after 14 Ma (Fig. 5). A decrease in reworked radiolaria was noted after approximately 14 Ma in the southwest Pacific (Moore et al., 1978.

Originations

Three *Cibicidoides* species first appeared after 16 Ma: *C. wuellerstorfi, C.* sp. C. and *C. cenop* (Figs. 4, 5, plates 1-2). *Cibicidoides* sp. C may have been the parent stock as it resembles both *C. wuellerstorfi* and *C. cenop,* n. sp., the latter of which may have evolved slightly later. The newly evolved *Cibicidoides* are flatter, more porous, more abundant and larger in diameter, if not in volume, than the extinct *Cibicidoides.* Foraminiferal shell shape is related to environmental parameters (Bandy, 1960). It is possible that the larger but flatter shape was an adaptation for hugging the sea floor during increased deep-ocean current activity, or for increasing the surface area to facilitate the absorption of some dissolved nutrient, as in the case of *Bolivina* species which exhibit increased surface area in lower oxygen environments (Douglas, 1981).

There are many species which may have originated during the middle to late Miocene but which are too rare to evaluate at this time. These include *Ehrenbergina trigona, E.* sp cf E. *hystrix, E. glabra, E. bosoensis, Rotorbinella lobatula, Bolivina* sp cf. *B. punctata, Bulimina rostrata, Favocassidulina favus* and several other *Cibicidoides* species.

Changes in Species Distributional Patterns

Figures 7-10 show clearly different distributional patterns in space and time for individual species and genera through the Miocene. The percentage data are based on the 110 species monitored in each sample. The diversities average over 35 so that one species rarely dominates a sample or even comprises as much as 10 percent of the sample.

Four common distribution patterns are exemplified by *Cibicidoides cenop* n. sp., *C. kullenbergi, Uvigerina* species and *Epistominella umbonifera* (Fig. 7). Abundant early Miocene species (*C. kullenbergi, G. subglobosa, Stilostomella* spp.) were replaced, or became less abundant, below 2 km. By the late Miocene *Uvigerina* species became very abundant in the bathyal realm (0.5–2.5 km). *Epistominella umbonifera, Melonis barleeanum* and the miliolids became increasingly important at abyssal depths and portray a shallowing trend after 16 Ma.

Epistominella umbonifera clearly shallowed starting 16 Ma, culminating after 8.5 Ma (Fig. 7). Its distribution may be related to increasing amounts of Antarctic Bottom Water (AABW) during maximum glaciation in the latest Miocene (Shackleton

TABLE 2. SPECIES EXTINCTIONS

	Extinction age (Ma)	DSDP Site occurrence	References
Bulimina jarvisi	14	171, 208, 292, 289, 317, 55, 296, 288, 206, 448, 159, 320.	Douglas, 1973 (Eocene-early Miocene); Boltovskoy, 1980 (Oligocene-Middle Miocene).
Bulimina glomarchallengeri	15	171, 292, 208, 317, 296.	Boltovskoy, 1980 (Oligocene-early Miocene; = B. cf. markisi).
Bulimina miolaevis	14	289, 292, 55, 317, 238, 448, 171, 208, 231, 71, 167.	Boltovskoy, 1980 (Oligocene-middle Miocene).
Buliminella grata	14	171, 281, 289, 292, 317, 296, 55, 288, 448, 206, 159, 167, 320, 495, 319.	Beckman, 1953 (Eocene-Oligocene); Boltovskoy, 1980 (Oligocene-early Miocene); Douglas, 1973 (Olig-early Miocene; = B. ovata); Resig, 1976
Cibicidoides dohmi	14	77, 495, 167, 289, 320, 178, 71	Bermudez, 1949 (Oligocene-early Miocene; = Cibicides dohmi).
Cibicidoides grimsdalei	14	77, 71, 495, 159, 167, 320, 288, 289.	Beckman, 1953 (Eocene-Oligocene); Douglas, 1973 (Eocene-early Miocene); Resig, 1976 (Eocene-early Miocene).
Cibicidoides havanensis	13	77, 71, 495, 159, 167, 288, 317, 289, 178, 319, 320.	Beckman, 1953 (Eocene-Oligocene); Resig, 1976 (Eoc-early Miocene); Douglas, 1973 (Oligocene).
Cassidulina sp cf. C. angulosa	? 14	77, 71, 159, 448, 238, 55, 296, 317, 292, 289, 171, 281, 208.	Douglas, 1973 (Oligocene-early Miocene; = C. angulosa).
Anomalinoides sp 1	14	208, 495, 71, 77, 319, 178, 206, 448, 320, 159, 288, 173, 55, 296, 292, 317, 289, 167.	Douglas, 1973 (Oligocene); Resig, 1976 (Oligocene-early Miocene).
Melonis sp cf. Anomalina regina	? 13	239, 208, 206.	
Lenticulina sp cf. L. clerrici	? 15	289, 281, 292, 296.	
Hanzawaia cushmani	? 14	289, 292, 288.	
Vulvulina miocenica	? 8	289, 208, 158, 62, 55, 288, 448, 292, 171, 159, 206, 167, 310, RC12-418, 470, 495, 77, 71, 319, 320.	Resig, 1976 ((Vulvulina jarvisi, Eocene-early Miocene).
Cibicidoides wuellerstorfi	16	289, 281, 62, 206, 173, 470, 310, RC12-418, 208, 158, 292, 296, 288, Mohole, 470, 495, 77.	Boltovskoy, 1980 (middle Miocene-Recent); Douglas, 1973 (middle Moicene -Recent), Resig, 1976 (not in Eocene-early Miocene).
Cibicidoides sp C.	16	208, 289, 77, 495, 310, Mohole, 470, 167, 296, 317, 158.	
Cibicidoides cenop	16	289, 158, 310, RC12-418, 167, 77, 317, 208, 62, 296, 288, 206, 319, Mohole, 470, 495,	
Bulimina rostrata	? 14	208, 158, 289, 55, 62, 281, 296, 173, 238, 206, 310.	
Ehrenbergina sp cf. E. hystrix	? 16	319, 77, 495, 289, 167, 288, Mohole.	Douglas, 1973, (middle Miocene; = E. hytrix).
Ehrenbergina bosoensis	? 16	319, 77, 495, 289, 167.	

TABLE 3. SPECIES ORIGINATIONS

	Origination age, (Ma)	DSDP Site occurrence	References
Cibicidoides wuellerstorfi	16	289, 281, 62, 206, 173, 470, 310, RC12-418, 208, 158, 292, 296, 288, Mohole, 470, 495, 77.	Boltovskoy, 1980 (middle Miocene-Recent); Douglas, 1973 (middle Moicene -Recent), Resig, 1976 (not in Eocene-early Miocene).
Cibicidoides sp C.	16	208, 289, 77, 495 310, Mohole, 470, 167, 296, 317, 158.	
Cibicidoides cenop	16	289, 158, 310, RC12-418, 167, 77, 317, 208, 62, 296, 288, 206, 319, Mohole, 470, 495,	
Bulimina rostrata	? 14	208, 158, 289, 55 62, 281, 296, 173, 288, 206, 310.	
Ehrenbergina sp cf. E. hystrix	? 16	319, 77, 495, 289, 167, 288, Mohole.	Douglas, 1973, (middle Miocene; = E. hytrix).
Ehrenbergina bosoensis	? 16	319, 77, 495, 289, 167.	
Ehrenbergina trigona	? 13	158, 281, 208, 289, 288, 319, 77.	
Ehrenbergina amina	? 13	289, 317, 288	Bermudez, 1949, (middle Miocene).
Uvigerina hispido-costata	? 15	281, 289, 158, 62.1 317, 296, 288, 173, 206, 310, RC12-418, 470.	Bermudez, 1949, (mid-late Miocene).
Melonis pompilioides	? 14 (rare)	289, 158, 62, 317, 296, 208, 288, Mohole, 470, 319, RC12-418.	Resig, 1976 (not in Eocene-early Miocene).
Uvigerina vadescens	? 16	158, 289, 62, 317, (rare at 281, 171, 288, 292, 448, 167, 206, 310, 319.	Bermudez 1949, (late Miocene).
Rotorbinella lobatula	? 14	281, 208, 158, 289, 621, 317, 288.	
Favocassidulina favus	? 10	289, 62.1, 317, 55, 288, 206, 167	

and Kennett, 1975b). *Epistominella umbonifera* is today associated with AABW (Corliss 1979; Gofas, 1978; Burke, 1981), and with less than 2°C water (Streeter, 1973). It was found commonly below 4 km in the early Miocene, but it appeared at shallower depths intermittantly between 15–8 Ma. High percentages of *E. umbonifera* sometimes alternated with high percentages of *Epistominella exigua,* (associated today with Arctic Bottom Water (BW), (Schnitker, 1974; Indian Bottom Water (IBW), Corliss 1979; Pacific Bottom Water (PBW), Burke 1981; North Atlantic Deep Water (NADW), 2–3°, Streeter 1973; oxygenated water, Streeter and Shackleton 1979, Schnitker, 1980) a species which shallowed 14–8 Ma (Fig. 7). This species is often found in poorly-preserved samples and, although perhaps more dissolution-prone than *E. umbonifera,* it is most common in Re-

cent PBW in and above the *E. umbonifera* realm (Burke, 1981). Its intermittent abundant appearances (>10%) at 2.7 km after 14 Ma (DSDP Site 317) alternated with the appearance of abundant *E. umbonifera,* perhaps because of the intermittent effects of AABW on depths as shallow as 2.7 km in the Pacific bottom water corridor south of the equator. *Epistominella exigua,* associated in the Atlantic with interglacial periods (Schnitker, 1979; Streeter and Shackleton, 1979), NADW (Streeter and Shackleton, 1979; Hodell and Kennett, 1982) and ABW where AABW loses its character (north of 30°N latitude; Schnitker, 1974), may reflect a period (13–8 Ma) when PBW was not dominated by AABW (similar to today) or it may reflect the influence of NADW.

Species associated with low oxygen water today, such as

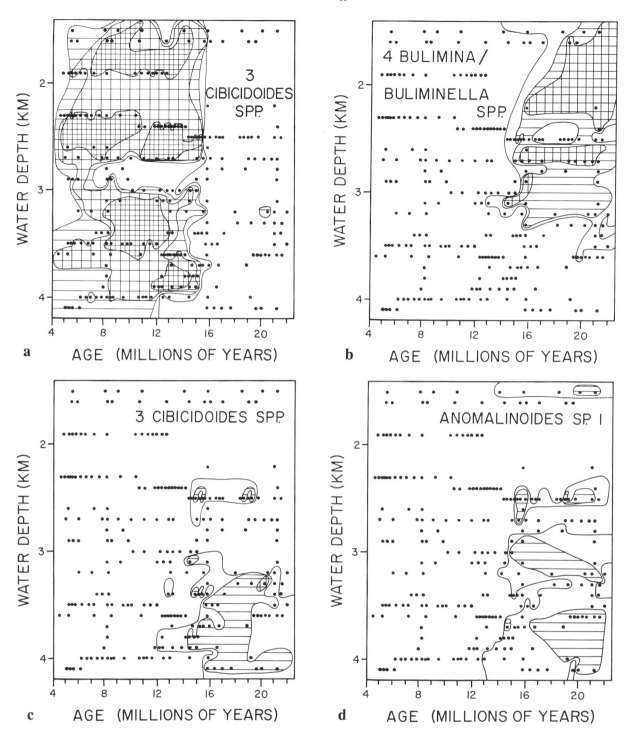

Figure 5. a) Distribution of mid-late Miocene *Cibicidoides* species (*C. wuellerstorfi, C. sp. C., C. cenop* n. sp.) evolving 16 Ma. Contour intervals are 0%, 2%, 6%, 10% of the 110 species quantified. b-d) Distribution of early-mid Miocene species which became extinct 14 Ma (*Bulimina jarvisi, B. glomar-challengeri, B. miolaevis, Buliminella grata, Anomalinoides* sp *1, Cibicidoides dohmi, C. grimsdalei, C. havanensis*). Contour intervals are 0, 2%, 6%.

22–14 Ma

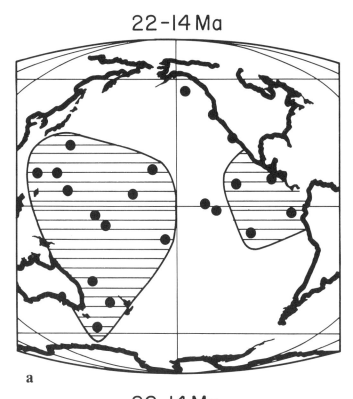

a

22–14 Ma

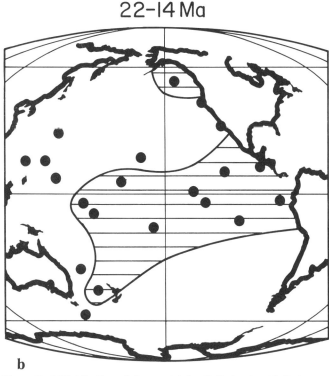

b

Figure 6. a) Distribution of sites containing *Bulimina jarvisi, B glomar-challengeri, B. miolaevis, Buliminella grata.* b) Distribution of sites containing the species *Cibicidoides dohmi, C. grimsdalei, C. havanensis.* These species, which preferred deeper water than the *Bulimina* species, are located more in the eastern Pacific where most of the deepest sites are located.

Uvigerina vadescens, U. proboscidea, U. hispido-costata, Ehrenbergina trigona, Bulimina rostrata, Rotorbinella lobatula, became increasingly common at shallow depths after 12 Ma (Figs. 8-9). Most of these species are associated with organic carbon-rich sediments or with the intermediate water mass (Table 4). The *Bulimina* species as a group were more common in the early Miocene ocean but were replaced in abundance by *Uvigerina* species in the middle to late Miocene (Figs. 8-9). Miller and Lohmann (1982) find that a modern *Bulimina/Globobulimina* assemblage is associated with low oxygen content of the water column whereas *U. peregrina* is associated with increased organic carbon in the sediments at slightly deeper water depths. The Miocene distribution of these species and genera is evidence for shoaling of the low oxygen zone after the early Miocene and an increase in sediment organic carbon in the late Miocene (Fig. 5b, 7a, 8a, b, 9).

The *Uvigerina* species became more abundant after 10 Ma but were only common around the Pacific rim (Fig. 10), especially in areas of high sedimentation. The percent of organic carbon is generally highest in areas of high sediment accumulation and high surface productivity (Heath, Moore and Dauphin, 1977). *Uvigerina* prefer organic carbon-rich sediments (Douglas, 1981; Miller and Lohmann, 1982; Lutze and Coulbourn, 1984) and their increase after 10 Ma may be related to an increase in organic carbon in the sediments associated with intensification of the oxygen minimum zone to depths of 2.5 km, 10–6 Ma in the Pacific (Summerhayes, 1981). *Uvigerina peregrina* (related to *U. hispido-costata*) are found associated with less than 1 ml/L oxygen and between 2 an d 4.5% organic carbon in Recent sediments of the California Borderland (Douglas, 1981). *Uvigerina peregrina* also has been associated with glacial periods (Schnitker 1979; Corliss 1982; Lutze, 1978) and the relationship between particular *Uvigerina* species and glacial periods may be due to an increase in organic food availability during periods of accelerated atmospheric-oceanic circulation and regressive periods when continental erosion is strong. Because the *Uvigerina* species show overlapping patterns (Fig. 9) as well as a progressive change in species dominance through time, their distribution may reflect changing environmental conditions. *Uvigerina multicostata,* a costate form, was progressively overshadowed through time by the hispid species *U. proboscidea* and *U. vadescens* and eventually *U. hispido-costata.* The increase in abundance of *U. hispido-costata* in the latest Miocene may reflect an increase in the availability of organic carbon in the sediments related to latest Miocene glaciation.

Cibicidoides species exhibit somewhat mutually exclusive distributions which suggest different environmental preferences (Figs. 7, 11). Between 2.6 and 3.4 km water depth, *C. sp cf C. pseudoungerianus* (plate 1) prefers the deep central Pacific bottom water corridor (DSDP Sites 167, 317) while *C. sp N* prefers the ocean margin sites (288, 296, 206). *Cibicidoides sp C,* most common 12–16 Ma, evolved after 16 Ma, as did *C. wuellerstorfi,* and both are rare in the deep, open ocean. *Cibicidoides wuellerstorfi* became more common after 11 Ma and is more abundant

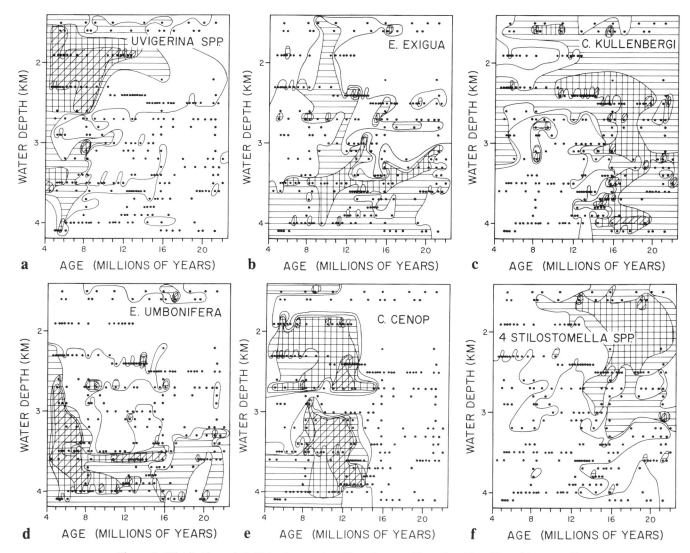

Figure 7. Distribution of a) *Uvigerina* species (*U. vadescens, U. proboscidea, U. multicostata, U. hispido-costata, U. hispida*); b) *Epistominella exigua;* c) *Cibicidoides kullenbergi;* d) *E. umbonifera;* e) *C. cenop; and* f) *4 Stilostomella* species (*S. annulifera, S. spinata, S. curvatura, S. modesta*). Contour intervals are 0%, 2%, 6%, 10% of total abundance.

at ocean margin sites (Figure 8). *Cibicidoides kullenbergi* dominated the deep-ocean during the early Miocene (Fig. 7, Plate 3), especially during the period of warmest planktonic and benthic $\delta^{18}O$ paleotemperatures, 17–15 Ma (Savin et al., 1981) but was replaced in the deep ocean 14 Ma when *C. cenop* n. sp. became dominant (Fig. 12, plates 1-2). After 14 Ma *C. kullenbergi* is common only at the shallower sites of the western Pacific. The change in species distribution of *C. kullenbergi* and *C. cenop* appears to characterize the change which occurred in the deep sea environment 13–14 Ma.

Bolivina species (Fig. 8), a group strongly associated with the oxygen minimum zone (Streeter, 1972; Douglas, 1981; Resig, 1981; Ingle, 1981), are rare in the deep ocean. They tend to be found today at depths shallower (100–1000 m) than this study's limits. Their low numbers suggest that no well-developed oxygen minimum zone extended into the deep ocean during the Miocene.

R-MODE PRINCIPAL COMPONENT ANALYSIS

Methods

An R-mode principal component analysis (PCA; Legendre and Legendre, 1979) was performed using the 110 most abundant species in the 255 samples. This analysis was used to identify the major components of variability for the foraminiferal assemblages. A study of the distribution of key assemblages through time helps to differentiate faunal boundaries. Assuming Miocene assemblages reflect environmental conditions, changes in the distribution of key assemblages would indicate environmental changes through time. The distribution of the assemblages is displayed by water depth and on areal distribution maps by contouring the scores attained for the assemblages at each station. Each factor is composed of two assemblages because in a diverse data-

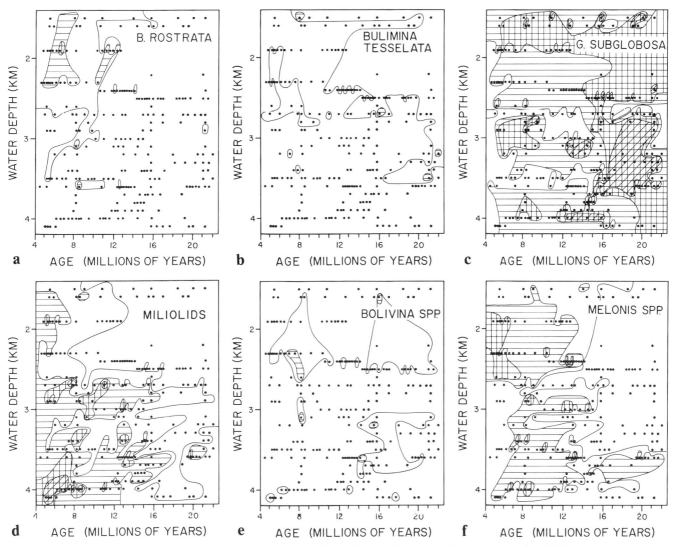

Figure 8. Distribution of a) *Bulimina rostrata;* b) *tesselata;* c) *Globocassidulina subglobosa;* d) miliolid species, *(Pyrgo murrhina, Pyrgo depressa, Sigmoilina tenuis, Quinqueloculina* and *Triloculina spp.);* e) *Bolivina* species *(B. optima, B. cf punctata)* and f) *Melonis* species *(M. pompilioides, M. soldani, M. barleeanum, M. affinis).* Contour intervals are 0, 2%, 6%, 10% of total abundance.

set the opposing counterpart assemblages can generally be found. The factor scores are listed in Appendices 5 and 6; the sample number key is found in Appendix 2; the species number key is found in Appendix 4; the resultant assemblages are listed in Tables 5 and 6.

The data were first transformed with a natural log transformation because biological populations are log normally distributed (Blasco et al. 1980). Then a linear correlation matrix, the standardized covariance matrix, was computed to define the components. Principal component scores were calculated for each sample so that the distribution of the components throughout the sampling space could be examined.

The data were analyzed in 3 different ways: as percent data, as raw counts and as raw counts multiplied to account for the sample splits (samples of equal volume were used). Species which

were highly scored in all three analyses are starred in Table 5. The assemblages in all three analyses showed similar assemblage distribution patterns and therefore suggested similar conclusions. However, only the unmultiplied raw count data were easily contoured for all factors and for both the areal and vertical distributions. This is attributed to the extra weight given to the better preserved samples (from which more specimens were counted), which should improve the results because better preserved samples better reflect the live assemblage. At one extreme, the percent data clearly emphasized the poorly preserved samples. This was apparently because resistant species composed a relatively high percentage of the more poorly preserved samples, thus emphasizing those species/samples which poorly reflect the live assemblage. For example, Site 319, one of the most poorly preserved sites, was highly significant in almost every factor of the percent

F. Woodruff

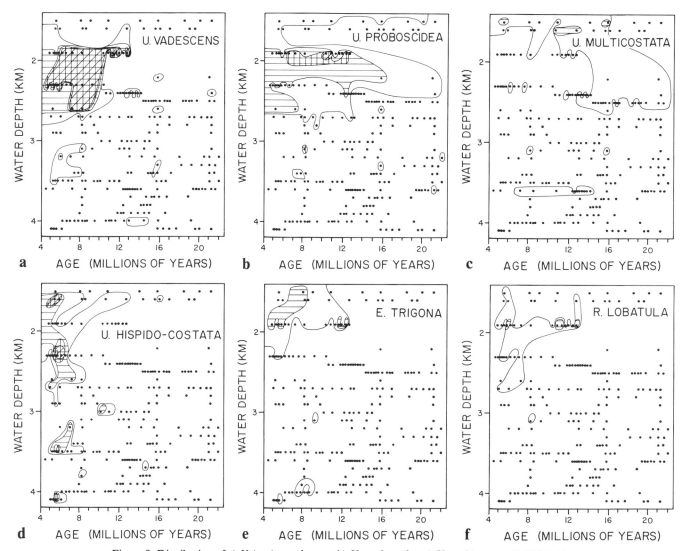

Figure 9. Distribution of a) *Uvigerina vadescens,* b) *U. proboscidea,* c) *U. multicostata,* d) *U. hispido-costata,* e) *Ehrenbergina trigona* and f) *Rotorbinella lobatula.* Contour intervals are 0%, 2%, 6%, 10% of total abundance.

data, whereas Site 289, the best preserved site, was never significantly scored. At the other extreme, when the raw counts were multiplied by as much as 4 to account for the sample splits (resulting in some samples having 20 times more total specimens, especially those at Site 289), the assemblage and sample scores of the split-multiplied data totally dominated every factor and virtually excluded not only the other sites but also the counterpart or opposing assemblages. The raw count data falls between the two extremes, with the counts generally ranging from 150 to 450 per sample. Therefore, the raw count data is presented because it gives greater weight to the better preserved portion of the data set while keeping the whole data set in the analysis. Support for this decision is given by the fact that the resultant contoured patterns reflect the individual species distributions figured. Appendices 2 and 3 list the raw data and the split corrections.

In addition to the analysis on the whole data set, a second analysis was run in which the data were subdivided into two sets.

This subdivision was based on the results of the first PCA which showed a strong demarcation in the first component at about 15 Ma. Therefore, for the second analysis the data were separated into the set for 4–15 Ma and the set for 15–23 Ma. The results were substantially the same as those of the whole data set because the period of transition actually spans 13–16 Ma, overlapping both sets. Only factor three from the divided data set, is reported here (Table 6).

Results of Principal Component Analysis

The patterns displayed by the PCA were in most cases anticipated by the individual species distributions. Although the groupings were more complex than had been anticipated, species of the same genera, extant or not, tended to fall within the same assemblage, supporting evidence that today's genera can be a key to the past. For example, all of the early Miocene *Bulimina* species fall

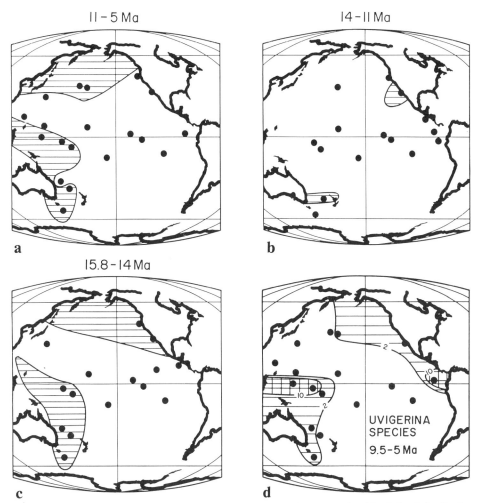

Figure 10. a-c) Areal distribution of *Cibicidoides wuellerstorfi*. Shaded area is >2%. d) Areal distribution of *Uvigerina* spp. The distribution below the equatorial high productivity zone suggests a preference for high organic carbon in the sediments.

into the assemblage found at relatively shallow depths, as they do today. Species which appeared to be contaminants were rarely important in the factor analysis apparently because they did not consistently occur in conjunction with any other particular species.

The assemblage patterns appear to show the following: Factor 1 shows the faunal change between 16 and 13 Ma previously seen in Figure 5. Factor 2 delineates sample diversity which is in part related to sample preservation. Factor 3 shows the distribution of early Miocene intermediate and bottom water assemblages. Factor 4 shows the ocean margin assemblage becoming important 16–12 Ma. Factor 5 shows the expansion of a bottom water corridor assemblage 15 Ma. Factors 1, 3, 4 and 5 show assemblage changes between 13 and 16 Ma. Factors 1 and 5 suggest a second faunal change occurred 8–10 Ma. The effects of changes in surface productivity can be seen in Factors 1 and 4. The effects of a major deep-ocean change are seen in Factors 1, 3 and 5.

Factors 1 through 5 account for 9%, 9%, 5%, 4%, and 3% of the variance, respectively, or 30% of the total variance. This rela-

tively low variance is to be expected considering the large size of the data set, the numerous species, stations, millions of years, and the many different environments, latitudes, and ocean depths involved.

Factor 1

Factor 1, the biostratigraphic factor, depicts an assemblage change dividing the stations into early and mid-late Miocene clusters (Figs. 13, 14). The faunal change was apparently related to an oceanwide paleoceanographic change since it occurred at all sites between 13 and 16 Ma. It was also related to an increase in sample preservation and an increase in food supply to the ocean floor under the equatorial high productivity zone. Subdivision of the data set into two separate principal component factor analyses, 4–15 and 23–15 Ma (not shown) reveals a subdivision of the data at 13–14 Ma and at 15.5–16 Ma showing that the division 13–16 Ma does not simply divide the data set in half, but represents a clear faunal change.

Early Miocene stations (negatively scored) are characterized

TABLE 4. MODERN LOW OXYGEN SPECIES

	Factor Scores					Reference
	1	2	3	4	5	
Rotorbinella lobatula	+0.36	+0.17	+0.22	-0.08	-0.28	Deep oxygen minimum, Burke 1981, Pacific Ocean
Uvigerina vadescens	+0.56	+0.50	+0.26	+0.03	-0.14	Deep oxygen minimum (= Siphouvigerina interrupta, Burke 1981), Pacific Ocean
Uvigerina proboscidea	+0.52	+0.49	+0.34	+0.01	-0.22	Deep oxygen minimum (= Uvigerina asperula, Burke 1981), Pacific Ocean
Uvigerina hispido-costata	+0.45	+0.14	+0.11	-0.05	-0.17	Related to U. peregina (see below)
Bulimina rostrata	+0.53	+0.40	+0.24	-0.10	-0.22	Deep oxygen minimum, Burke 1981 (one station), Pacific Ocean
Ehrenbergina trigona	+0.40	+0.12	+0.23	-0.14	-0.43	Decreased oxygen, Lohmann 1978 (= Ehrenbergina sp), Atlantic Ocean
Bolivina cf. punctata	+0.36	+0.31	0.0	+0.16	-0.03	Bolivina species are abundant in shallow Pacific oxygen minimum environments (Douglas, 1981; Resig, 1981; Ingle, 1981). However they are rare in the deep ocean and probably often transported.
Bolivina optima	-0.03	+0.36	+0.03	+0.12	-0.06	
Favocassidulina australia	+0.03	+0.54	+0.07	-0.16	+0.33	(? = Cassidulina sp C., Burke 1981, deep oxygen minimum, Pacific)
Bulimina jarvisi	-0.41	+0.14	+0.33	+0.27	-0.03	Lower Miocene extinct species, see Bulimina assemblage below
Bulimina miolaevis	-0.34	+0.11	+0.28	+0.19	-0.03	
Bulimina glomarchallengeri	-0.29	+0.03	+0.34	+0.36	-0.04	" " "
Buliminella grata	-0.50	+0.05	+0.35	+0.28	-0.02	" " "
Bulimina tesselata	-0.14	+0.27	+0.34	+0.22	+0.03	See Bulimina/Globobulimina assemblage below
Globocassidulina subglobosa	-0.15	+0.52	-0.24	+0.17	-0.26	Decreased oxygen, Atlantic; Lohmann, 1978
Uvigerina peregrina	--	--	--	--	--	High organic carbon, low oxygen in the sediments (Miller & Lohmann 1982). High sediment organic carbon, low oxygen, Pacific (Douglas, 1981)).
Bulimina/Golobobulimina assemblage	--	--	--	--	--	Related to low oxygen in the water column, Atlantic Ocean (Miller and Lohmann, 1982).

by an assemblage of species, many of which became extinct approximately 14 Ma (*Anomalinoides sp. 1, Buliminella grata, Bulimina jarvisi, B. miolaevis, Cibicidoides havanensis, C. grimsdalei*, see Table 2, 5) while others lived on but were not common in the deep ocean after 14 Ma (*Cibicidoides kullenbergi*, Fig. 7). This assemblage is dissolution-resistant and contains five of the six agglutinated species, all of which are robust. Diagenesis and to a lesser extent dissolution is responsible for the greater abundance of some species in the early Miocene assemblage (see section on sample preservation).

The mid-late Miocene assemblage (positive scores) is composed primarily of *Cibicidoides* species (*C. cenop, C. wuellerstorfi*) which evolved in the middle Miocene and of *Uvigerina*,

miliolid, *Melonis* species, *Epistominella umbonifera*, all of which were rare or absent in the early Miocene. As a group *Epistominella umbonifera, Melonis*, and miliolid species exhibit a shallowing trend as previously discussed (Figs. 7–8). They were rare, found only at abyssal depths in the early Miocene, then became more common at abyssal depths in the middle Miocene, and were common in lower and middle bathyal depths by the late Miocene. Miliolid species are dissolution-prone but *E. umbonifera* and *Melonis* species probably are not, so that the shallowing of the latter two species realms (which causes them to be relevant to this assemblage) appears real, rather than preservation-related. This suggests that some deep-water conditions may have expanded into shallower realms after 14 Ma. *Uvigerina* became

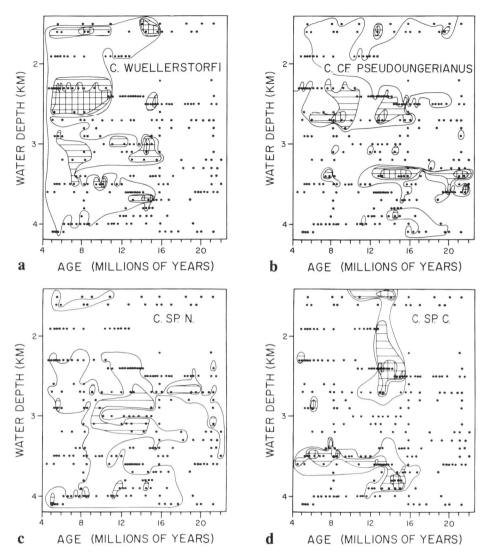

Figure 11. a-d) Distribution of four *Cibicidoides* species. The species have somewhat mutually exclusive distributions. Contour intervals are 0, 2%, 6% of total abundance.

more abundant after 10 Ma, but were common only around the Pacific rim (Fig. 10), perhaps because of an increase in organic carbon in the sediments (see section on species distribution).

Study of the six areal distribution maps (Fig. 14) shows that the 21 and 16 Ma maps are very similar; all stations are negatively scored, with no particular pattern emerging; however, 15-12 Ma the positive factor begins to dominate the western boundary and equatorial area. The positively scored assemblage pattern continues to become more pronounced along the equator, and progresses toward more westerly sites, culminating in the latest Miocene, when the whole equatorial area and Site 281 become highly positively scored. The stations remaining negatively scored the longest, and hence perhaps remaining most similar to the early Miocene ocean, are in the backarc or marginal basins of the Philippine Sea, where sediment accumulation was low. The distribution of the areal patterns suggests that after 16 Ma the effects of an increase in the productivity of the surface

waters produced an increase in food availability for the benthos and/or improved sample preservation beneath the highly productive zone. The rate of sediment accumulation in high-productivity regions dramatically increased in the latest Miocene (Barron, 1980; Keller, 1980b; Vincent et al., 1980) and the "carbonate supply" rate increased steadily throughout the Miocene (van Andel et al., 1975). Calcium carbonate dissolution occurs in deeper water below areas of increased surface productivity and biogenic carbonate sedimentation so that sample preservation would be improved at many sites particularly along the equator. An increase in the surface productivity and sedimentation rates in the Southern Ocean during the latest Miocene (Kennett, 1977) is compatible with the high positive scores at Site 281 (located near the Subtropical Convergence) in the latest Miocene.

The apparent increase in surface productivity in the equatorial region may have been the result of steepening of the latitudinal thermal gradient (Savin, 1977), intensification of atmospheric

TABLE 5. ASSEMBLAGES AND SPECIES FACTOR SCORES

The numbers to the left are the species numbers keyed to Appendix 4. The numbers to the right are the species factor scores of Appendix 5. The stars indicate species highly scored in all three runs of the principal component analysis discussed in Part 3.

------------Positive Scores-------------   -----------------Negative Scores-----------------

Factor I

		(Mid-Late Miocene)			(Early Miocene)	
*	28	Cibicidoides cenop n. sp.	+0.66	*	1 Anomalinoides sp 1.	-0.54
*	65	Melonis soldanii	+0.66	*	13 Buliminella grata	-0.50
*	66	Melonis affinis	+0.63	*	109 Vulvulina miocenica	-0.48
*	101	Pyrgo Spp.	+0.58	*	32 Cibicidoides havanensis	-0.48
*	96	Uvigerina vadescens	+0.56	*	24 Cibicidoides kullenbergi	-0.48
*	102	Quinqueloculina/Triloculina spp	+0.54	*	10 Bulimina jarvisi	-0.41
	9	Bulimina rostrata	+0.53	*	89 Stilostomella curvatura	-0.41
	97	Uvigerina proboscidea	+0.52		20 Cassidulina cf. C. angulosa	-0.39
	67	Melonis barleeanum	+0.49		68 Melonis sp cf. A. regina	-0.39
	26	Cibicidoides sp. B.	+0.49		76 Pleurostomella naranjoensis	-0.37
	25	Cibicidoides wuellerstorfi	+0.47		33 Cibicidoides grimsdalei	-0.38
	62	Laticarinina pauperata	+0.46		36 Cibicidoides sp cf. C. trincherasensis	-0.34
	64	Melonis pompilioides	+0.45		14 Buliminella cf. grata	-0.34
	99	Uvigerina hispido-costata	+0.45		12 Bulimina miolaevis	-0.34
	103	Sigmoilina tenuis	+0.43		37 Anomalinoides trinitatensis	-0.31
					5 Anomalinoides cf. alazanensis	-0.30
	[44	Epistominella umbonifera	+0.27]			

Factor II

		(High Diversity)			(Low Diversity)	
*	22	Chrysalogonium tenuicostata and longicostata	+0.64	*	31 Cibicidoides sp R	-0.35
	24	Cibicidoides kullenbergi	+0.59	*	110 Uvigerina hispida	-0.24
*	86	Sphaerodina bulloides	+0.58	*	82 Pullenia quinqueloba	-0.15
	57	Gyroidina torulus	+0.58	*	44 Epistominella umbonifera	-0.14
*	100	Vaginulina pseudoclavata	+0.58		101 Pyrgo sp	-0.14
	69	Nonion havanensis	+0.53	*	94 Stetsonia sp	-0.13
*	47	Favocassidulina australis	+0.54	*	14 Buliminella sp cf. B. grata	-0.12
	51	Globocassidulina subglobosa	+0.52		52 Gyroidina io	-0.11
*	96	Uvigerina vadescens	+0.50		103 Sigmoilina tenuis	-0.11
	109	Vulvulina miocenica	+0.50			
	83	Pullenia quadriloba	+0.50			
*	97	Uvigerina proboscidea	+0.49			
	35	Cibicidoides bradyi	+0.49			
	[70	Oridorsalis umbonatus	+0.38]			

Factor III

		(Early Miocene Intermediate Water)			(Early Miocene Bottom Water)	
*	91	Stilostomella spinata	+0.44		44 Epistominella umbonifera	-0.60
	72	Osangularia culter	+0.39	*	33 Cibicidoides grimsdalei	-0.59
	88	Stilostomella annulifera	+0.36	*	32 Cibicidoides havanensis	-0.49
	92	Stilostomella modesta	+0.36		43 Epistominella exigua	-0.49
*	13	Buliminella grata	+0.35		34 Cibicidoides dohmi	-0.49
*	11	Bulimina tesselata	+0.34		103 Sigmoilina tenuis	-0.46
*	15	Bulimina glomarchallergeri	+0.34		102 Quinqueloculina/Triloculina spp	-0.45
	97	Uvigerina proboscidea	+0.34	*	49 Favocassidulina elegantissima	-0.45
*	10	Bulimina jarvisi	+0.33	*	56 Gyroidina planulata	-0.40
	71	Orthomorphina spp	+0.31		48 Favocassidulina sp cf. F. favus	-0.32
					32 Pullenia quinqueloba	-0.31
	[96	Uvigerina vadescens	+0.26]		[67 Melonis barleeanum	-0.22]
	[95	Rotorbinella lobatula	+0.22]		[65 Melonis soldanii	-0.15]

circulation (Janecek and Rea, 1983), and perhaps to development of the equatorial undercurrent (Leinen, 1979) and counter-current following closure of the Indo-Pacific (Tethys) corridor in the middle Miocene (Edwards, 1975). Part of the pattern could be attributed to stronger deep western equatorial dissolution until 9 Ma resulting from a corrosive watermass near the Pacific bottom water corridor (western Pacific). The balance between the corrosiveness of bottom waters on the carbonate flux and the sedimentation rate affects sample preservation (Berger, 1970) but is difficult to evaluate. Van Andel et al. (1975) suggest that increasing Miocene deep ocean carbonate dissolution was balanced by increasing equatorial carbonate supply rates throughout the Miocene.

Factor 2

Factor 2, the diversity factor (Figs. 15, 16), differentiates high diversity samples (40-60 species per sample) from low diver-

TABLE 5. ASSEMBLAGES AND SPECIES FACTOR SCORES (continued)

```
------------Positive Scores-------------    ------------------Negative Scores-----------------
```

Factor IV

(Central Gyre/Ocean Basin/Low Sedimentation)		(Ocean Margin/High Sedimentation)	
* 14 Buliminella sp cf. B. grata	+0.50	* 27 Cibicidoides sp C.	-0.37
* 91 Stilostomella spinata	+0.46	86 Sphaerodina bulloides	-0.34
53 Gyroidina sp cf. G. io	+0.41	68 Melonis sp cf. A. regina	-0.32
15 Bulimina glomarchallengeri	+0.36	* 31 Cibicidoides sp R	-0.31
20 Cassidulina angulosa	+0.35	4 Anomalinoides cicatricosa	-0.31
29 Cibicidoides sp N	+0.35	* 98 Uvigerina multicostata	-0.28
45 Eponides sp E	+0.35	35 Cibicidoides bradyi	-0.27
* 94 Stetsonia sp	+0.30	2 Anomalinoides semicribrata	-0.25
77 Pleurostomella acuminata	+0.29	85 Siphogenerina pohana	-0.25
* 92 Stilostomella modesta	+0.29	59 Hanzawaia cushmani	-0.22
13 Buliminella grata	+0.28	60 Heterolepa spp	-0.22
10 Bulimina jarvisi	+0.27	* 3 Anomalinoides globulosa	-0.21
81 Pullenia angusta	+0.25	25 Cibicidoides wuellerstorfi	-0.19
* 75 Pleurostomella alternans	+0.24	6 Astrononion guadalupae	-0.19
* 19 Cassidulina tricamerata	+0.23	47 Favocassidulina australis	-0.16
		* 37 Anomalinoides trinitatensis	-0.15
		72 Osangularia culter	-0.15

Factor V

(Deep-Ocean, Deep Pacific Bottom Water Corridor)		(Shallow and Early Miocene)	
* 3 Anomalinoides globulosa	+0.39	* 39 Ehrenbergina trigona	-0.43
* 28 Cibicidoides cenop n. sp.	+0.37	104 Bolivinopsis cubensis	-0.33
90 Stilostomella subspinosa	+0.37	* 26 Cibicidoides sp B	-0.32
* 23 Chrysalogonium lanceolum	+0.37	* 6 Astrononion guadalupae	-0.29
* 30 Cibicidoides sp cf C. pseudo-ungereanus	+0.35	95 Rotorbinella lobatula	-0.28
		51 Globocassidulina subglobosa	-0.26
* 78 Pleurostomella recens	+0.34	33 Cibicidoides grimsdalei	-0.25
* 47 Favocassidulina australis	+0.33	36 Cibicidoides trincherasensis	-0.25
100 Vaginulina pseudoclavata	+0.31	1 Anomalinoides sp 1	-0.24
45 Eponides sp E	+0.30	53 Gyroidina sp cf G. io	-0.24
* 27 Chibicidoides sp C.	+0.30	* 55 Gyroidina neosoldanii	-0.24
[43 Epistominella exigua	+0.24]		

TABLE 6. ASSEMBLAGE AND SPECIES SCORES FOR FACTORS IIIa AND IIIb
BASED ON A SUBDIVISION OF THE DATA AT 15 Ma

Factor IIIa (23-15 Ma)

(Early Miocene Intermediate Water)		(Early Miocene Bottom Water)	
Buliminella grata	+0.59	Epistominella umbonifera	0.70
Stilostomella spinata	+0.59	Cibicidoides grimsdalei	0.70
Bulimina jarvisi	+0.47	Laticarinina pauperata	0.68
Bulimina glomarchallengeri	+0.45	Melonis barleeanum	0.63
Bulimina tessellata	+0.42	Cibicidoides bradyi	0.61
Stilostomella annulifera	+0.42	Bolivinopsis cubensis	0.61
Stilostomella modesta	+0.39	Cibicidoides havanensis	0.59
Uvigerina proboscidea	+0.38		
Uvigerina vadescens	+0.37		

Factor IIIb (15-4 Ma)

(Mid-Late Miocene Intermediate Water)		(Mid-Late Miocene Bottom Water)	
Osangularia culter	0.45	Epistominella umbonifera	0.54
Ehrenbergina trigona	0.43	Quinqueloculina/Triloculina spp	0.52
Uvigerina proboscidea	0.37	Epistominella exigua	0.50
Rotorbinella lobatula	0.35	Sigmoilina tenuis	0.50
Bulimina rostrata	0.35	Pleurostomella naranjoensis	0.49
Uvigerina multicostata	0.32	Cibicidoides sp N	0.45
Stilostomella annulifera	0.31	Eponides sp E	0.45
Stilostomella curvatura	0.31		
Cibicidoides wuellerstorfi	0.31		
Uvigerina vadescens	0.28		

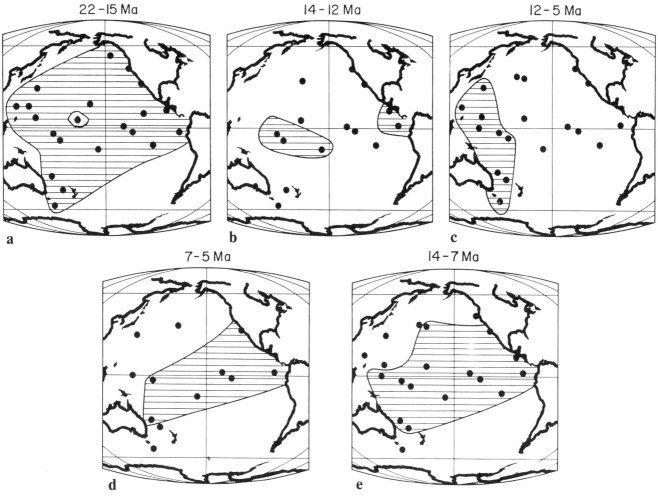

Figure 12. Areal distribution of *C. kullenbergi* (greater than 2%), a) 22–15 Ma, b) 14–12 Ma, c) 12–5 Ma. The species is prevalent in the deep ocean until 15 Ma. It moves to the shallower ocean margin after 13 Ma. d-e) Distribution of *Cibicidoides cenop* n. sp. (greater than 2%), 14–7 Ma, 7–5 Ma. The species is first seen after 15 Ma when it becomes prevalent in the deep ocean.

sity samples (20-39 species per sample). Factor 2 reflects both the high real diversity of the shallow and highly productive equatorial regions as well as the somewhat misleading lower diversity caused by the increased dissolution at increased depth in the water column. The species characterizing the high diversity samples are abundant and/or cosmopolitan species such as *Globocassidulina subglobosa* and *Oridorsalis umbonatus* or species common to shallow depths. The species characterizing the low diversity samples tend to be deep-dwelling, rare species. The anomalous positively-scored area at Site 288 (3 km, 8 Ma) is attributed to the addition of species from the top of the Ontong-Java Plateau as downslope displacement. The positively scored area at DSDP Site 495 in the lower Miocene deep ocean is partially attributable to downslope displacement by turbidites (Thompson, 1982) and occasional specimens of shallower specimens were seen at sites 71 and 77, perhaps displaced from the Christmas Island Ridge.

The areal distribution (Fig. 17) of these assemblages remains similar throughout the Miocene with sample diversity being generally highest (positive scores) at the shallowest (281, 158, 171, 289) and equatorial (289, 158) sites. Northward migration of the Pacific Plate beneath the equator places different sites beneath the equatorial high productivity zone at different times. This is nicely reflected in migration of the diversity maximum through time (figure 16 a-c), which is seen in the change in high factor scores through time as Site 55 migrates from the equatorial zone and Sites 288 and 289 pass beneath it.

Factor 3

Factor 3, the early Miocene factor (Figs. 17, 18), depicts an "intermediate water assemblage" above 2.6 km and a "bottom water assemblage" below 3.2 km, water depths which are similar to modern water mass depths in the western equatorial Pacific (Table 7; Wyrtki, 1962). The early Miocene "intermediate water assemblage" was apparently located at slightly deeper ocean

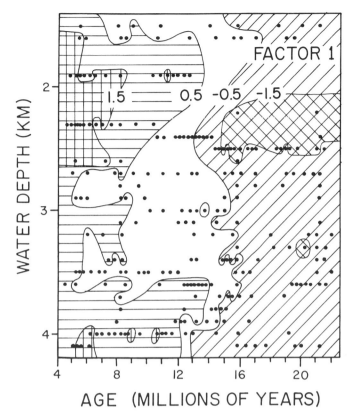

Figure 13. Vertical distribution of the two assemblages based on PCA, Factor 1. The species in the two assemblages are listed in Table 5. These assemblages appear to be controlled by a change in water quality, species evolution as well as sample preservation and nutrient availability. Contour intervals for the factor scores are ±0.5, ±1.5. Negative scores are diagonally hatched; positive scores are horizontally and vertically hatched.

depths than its late Miocene counterpart, but the evidence is based primarily on Site 317. Both the assemblages of Factor 3 are dominated by species which became extinct approximately 14 Ma: *Buliminella grata, Bulimina glomarchallengeri, B. jarvisi, Cibicidoides grimsdalei, C. havanensis, C. dohmi* (Table 2). The areal distribution (Fig. 18) of the assemblages is primarily governed by site depth and shows no notable change through time.

The shallower stations are associated by an assemblage of *Bulimina, Uvigerina, Osangularia* and shallow *Stilostomella* species which as a group are best correlated to intermediate water today. This assemblage contains many species associated with low oxygen according to the literature (Table 4) but it is difficult to separate species which prefer the intermediate water mass from those which prefer shallow depths and from those which prefer poorly-oxygenated water, since all three are found at relatively shallow depths. Pacific shallow waters are less well-oxygenated because new waters enter the Pacific from the south, sinking into the deep and bottom water below approximately 2.5 km and return above 2.5 km as older, less well-oxygenated, intermediate water. Today the oxygen minimum zone extends deepest in the north Pacific, away from the equator and the eastern coastal

margin (Wyrtki, 1966). Therefore it is not surprising that the highest scores for this assemblage are found at Site 171 in the central North Pacific. The positive scores in the Bauer Basin which appear strangely placed at 4 km (Site 319; Figs. 1b and 17) are probably due to oxygen-poor water in the isolated Bauer Deep, but may also result from downslope contamination by shallow species onto the basin floor.

The deepest stations are grouped by an assemblage of *E. umbonifera* (highest negative score, Table 5), a species correlated with bottom waters, especially AABW, in all ocean basins (Schnitker, 1980; Lohmann, 1978; Corliss, 1979; Gofas, 1978) and by *E. exigua,* miliolids, and three abyssal *Cibicidoides* species which had become extinct by 14 Ma. The *Cibicidoides* species becoming extinct dominate the early Miocene bottom water assemblage but *E. umbonifera* is more abundant later, suggesting more AABW influence on the deep water assemblage after 16 Ma but especially after 8 Ma (Figs. 7).

Subdivision of the data set into two separate Principal Component Factor Analyses at 23–15 Ma and 15–4 Ma reveals the faunal changes (Table 6 and Fig. 19). Both early and mid-late Miocene bottom water assemblages are characterized by *Epistominella umbonifera,* with *E. exigua* and the miliolids becoming more prevalent in the mid-late Miocene assemblage. Both early and mid-late Miocene intermediate water assemblages are dominated by *Uvigerina, Bulimina* and *Stilostomella* species with *Rotorbinella lobatula, Ehrenbergina trigona, Osangularia culter, Bulimina rostrata* and *Uvigerina proboscidea* becoming more dominant in mid-late Miocene time. *Bulimina* species were more prevalent in early Miocene time while *Uvigerina* species became more prevalent later.

There is a close similarity between modern intermediate water and Miocene intermediate water species, as seen by comparing Burke's (1981) factor analysis of piston and box core tops from the Recent Ontong-Java plateau (Western Equatorial Pacific). She identifies an intermediate water/oxygen minimum assemblage which includes: *Siphouvigerina interrupta,* (= *Uvigerina vadescens*), *U. asperula* (= *U. proboscidea*), *Rotorbinella lobatula,* and a *Bulimina* species not seen in the deep Miocene ocean. *Stilostomella* species are less common in modern Pacific samples but the relationship between the Miocene intermediate water assemblage, especially the late Miocene assemblage, and the Recent are clear.

The top of the "bottom water assemblage" at 3.4 km may be coincident with the lysocline (the abrupt decrease in calcium carbonate preservation) which today is found at 3.5 km (Valencia, 1973, western equatorial Pacific) at approximately the top of the Pacific bottom water mass (Parker and Berger, 1971). No marked shallowing of the bottom water assemblage is seen after 6 Ma as had been expected by the distribution of *E. umbonifera* (Fig. 7), although values near zero show an undepth trend (Fig. 17). This is probably because no other species follow this trend (factor analysis deals with assemblages) and also because the early Miocene *Cibicidoides* species which became extinct 14 Ma dominate the distribution of this early Miocene assemblage.

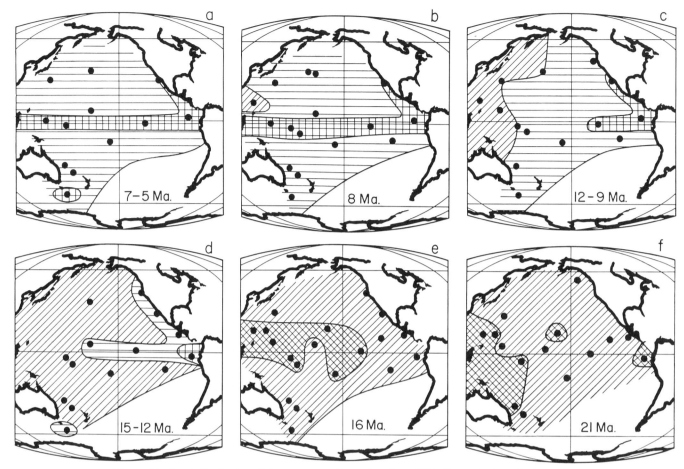

Figure 14. Areal distribution of the two assemblages based on PCA Factor 1 through time. The equatorial pattern which develops may be related to the effects of increased surface productivity following an increase in the latitudinal thermal gradient. Negative scores are diagonally hatched; positive scores are horizontally and vertically hatched. Contour intervals are zero and ±1.0.

TABLE 7. WESTERN SOUTH PACIFIC MODERN WATER MASSES
(from Wyrtki, 1962)

	Temperature (°C)	Salinity (°/oo)	Oxygen (ml/l)
Deep oxygen minimum, intermediate water 1.2-2.4 km	1.9-3.9	34.58-34.65	2.76-3.94
Deep water 2.5-3 km	1.4-1.9	34.74 (salinity maximum)	4.0-4.4
Bottom water >3 km	1.5-1.75	34.7	3.4-4.6 (low values are in deep sea basins)

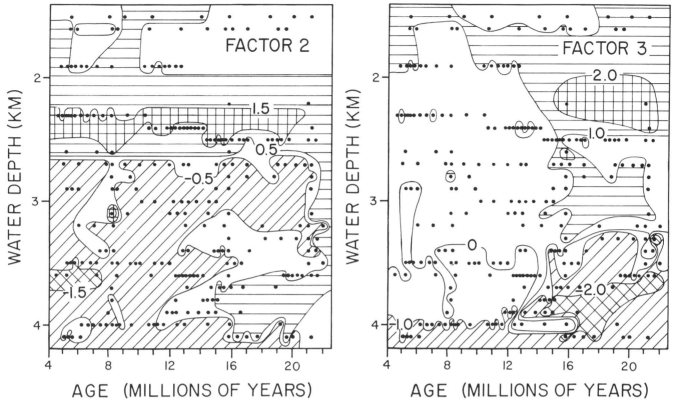

Figure 15. Vertical distribution of the assemblages of the diversity factor, Factor 2. Samples with the highest diversities tend to be located in shallow equatorial regions. Contour intervals are ±0.5 and ±1.5. Hatched patterns are as in Figure 13.

Figure 17. Vertical distribution of the assemblages based on a PCA, Factor 3 (see Table 5). These assemblages characterize intermediate and bottom water, primarily during early Miocene time. Hatched patterns are as in Figure 13. Contour intervals are 0, ±2.

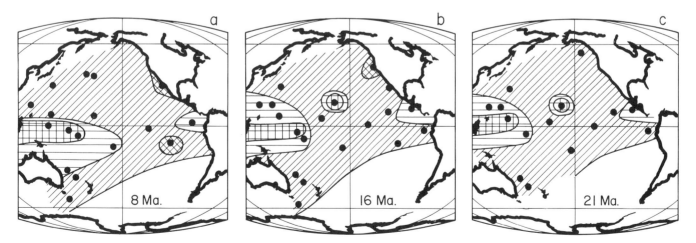

Figure 16. Areal distribution of the assemblages of Factor 2. The sites containing the highest diversities change as they migrate beneath the equator. Hatched patterns are as in Figure 10. Contour intervals are as in Figure 14. Assemblages are in Table 5.

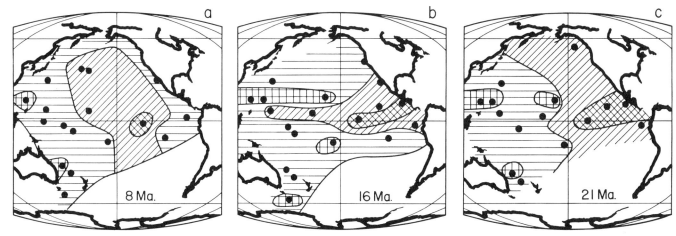

Figure 18. The areal distribution of the assemblages based on a PCA, Factor 3 (see Table 5). These assemblages are controlled primarily by site depth. Hatched patterns are as in Figure 13. Contour intervals are as in Figure 14. Assemblages are in Table 5.

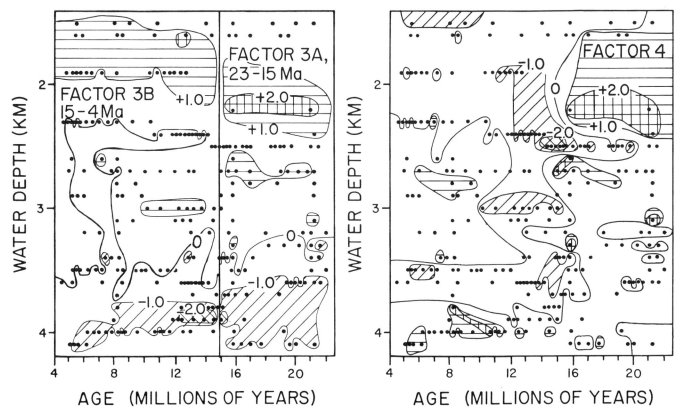

Figure 19. Distribution of the two assemblages of Factor 3a (22-15 Ma) and the 2 assemblages of Factor 3b (15-5 Ma). These were derived by using the same data set, subdivided at 15 Ma, and running 2 separate principal component factor analyses so that the assemblages characterizing the early versus the mid to late Miocene intermediate and bottom water species could be compared. The species of the assemblages are listed in Table 6. Hatched patterns are as in Figure 13.

Figure 20. Vertical distribution of the two assemblages based on a PCA, Factor 4. An ocean margin assemblage becomes more prevalent in middle Miocene time. The open ocean/central gyre basinal assemblage is associated with low sedimentation rates. Contour interval: 0, ±1, ±2. Hatched patterns are as in Figure 13.

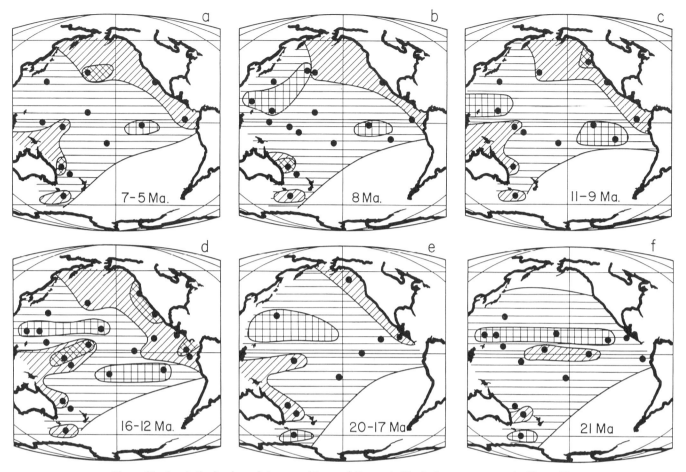

Figure 21. Areal distribution of the assemblages of Factor 4. Hatched patterns are as in Figure 13. Contour intervals are as in Figure 14. Assemblages are in Table 5.

Factor 4

Factor 4, the ocean basin/ocean margin factor (Figs. 20, 21) divides species from areas of low surface productivity and low sediment accumulation rates (1–9 m/my), such as in central gyres and restricted basins, from areas of high sediment accumulation rates (13–40 m/my; Fig. 2) and high surface productivity in the ocean margins. The increase in dominance of the ocean-margin assemblage (negative scores) at specific sites and ages (particularly 16–11 Ma) is attributed to the effects of increased surface productivity and associated carbonate and siliceous sedimentation. It appears that atmospheric-oceanic circulation became more vigorous, bringing food to the surface along the ocean margins and areas of upwelling after 16 Ma. If the ocean margin assemblage is recording an increase in food availability due in part to coastal upwelling and erosion, the data suggest that the upwelling started 1 my before the period of most rapid $\delta^{18}O$ "cooling" 14.5 Ma.

The ocean margin assemblage is composed of dissolution resistant species (Table 5) which may prefer readily available organic nutrients beneath areas of upwelling and high surface productivity. These include *C. bradyi* and *Uvigerina multicostata*.

Burke (1981) finds *Cibicidoides bradyi* associated with upwelling (ridging) in the Recent western equatorial Pacific. Some *Uvigerina* species have been shown to prefer high organic nutrient availability (Douglas, 1981; Miller and Lohmann, 1982) as would be expected under highly productive zones.

The ocean margin assemblage was particularly prevalent 11–16 Ma, with the period of highest scores 14.3–15.8 Ma at Site 289 (Figs. 20, 22). At Site 289 this period was marked by the highest planktonic to benthic ratios (2400:1), highest benthic diversities (60 to 70 species), and high sedimentation rates (30 to 70 m/my) (Andrews et al., 1975; Barron et al., this volume). Some dissolution is apparent at Site 289 despite the high P/B ratios and high sediment accumulation rates because the increase in parallel laminae and percentage of sand-size grains after 16 Ma (Fig. 23) suggest that slope dissolution has led to slumpage and size-sorting (Berger and Johnson, 1976). This period coincides with a moderately severe hiatus 15–16 Ma which was mainly seen in the areas of low sedimentation apparently because the zone of high productivity (which promotes preservation of the underlying sediments) was particularly extensive and had expanded to include the northeast Pacific boundary (Keller and Barron, 1983).

High silica and/or carbonate accumulation rates are related

DSDP SITE 289

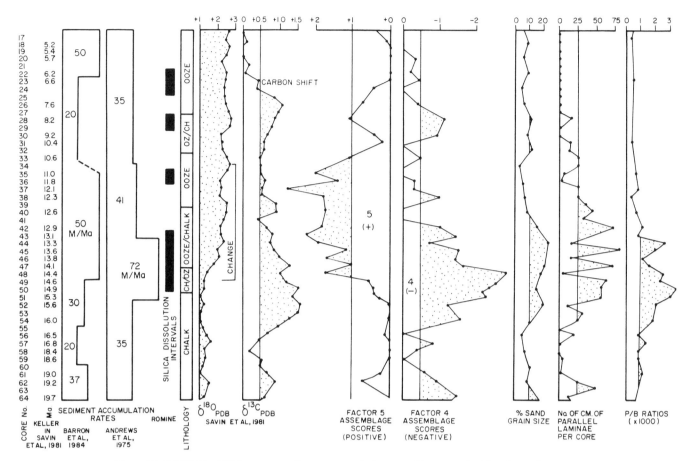

Figure 22. DSDP Site 289 downcore sediment accumulation rates (Barron et al., this volume; Andrews et al., 1975), and grain size, laminae, lithology (Andrews, *et al.,* 1975), silica dissolution intervals (Romine, this volume), and planktonic to benthic foraminiferal ratios (Woodruff and Douglas, 1981) and stable isotopes compared to the factor scores of assemblages in Factors 4 and 5. The isotopic values in each core are averaged (from Savin et al., 1981).

to high ocean margin assemblage scores. Increased silica accumulation rates generally indicate high carbonate and siliceous rain from productive surface areas (Leinen, 1979). The highly scored sites (158, 470, 173) are found in the equatorial and northeast Pacific margin and have high percentages of siliceous microfossils. Likewise, the highest silica accumulation rates at Site 289 occur 13–17 Ma (Romine, this volume) at the time of the highest ocean margin assemblage scores at this site. At Site 310, the highest scores, 7–5.5 Ma co-occur with increased biogenic opal accumulation (6.5–5.5 Ma) (Leinen, personal communication), and the first appearance of siliceous diatoms 7.6 Ma. This change 7.6 Ma is attributed to an increase in surface productivity as Site 310 migrates from the central gyre to beneath the Kuroshio Current (Burckle, personal communication) when it also exhibits a several-fold increase in carbonate accumulation (Fig. 2; Vincent, 1981). The highest P/B ratios and highest sediment accumulation rates occur with the highest ocean margin assemblage scores at both Site 289 and Site 317.

Factor 5 and the Oxygen Isotopic Record

An assemblage (positive scores) characterizing the deep central Pacific bottom water corridor 22–5 Ma (Figs. 23 to 26) expanded its realm to include the whole Pacific Basin 14.5 Ma, the time of most rapid isotopic paleotemperature cooling (Woodruff et al., 1981). The expansion clearly coincided with the change in $\delta^{18}O$ at Site 289 (Fig. 22). The assemblage was most prevalent (scores greater than +1.2) at sites in the Pacific bottom water corridor where today's temperatures are colder than 1.1°C (Fig. 25). The implication is that this assemblage expanded its range during the influx of cold dense bottom waters into the Pacific Ocean during middle Miocene glacial expansion. The assemblage was most prevalent between 14.5 and 11 Ma, which coincides with a period of hiatuses 13.5–11 Ma attributed to corrosion by bottom waters associated with buildup of Antarctic ice (van Andel et al., 1975; Keller and Barron, 1983).

The oxygen isotopic values shown in Figure 26 are from

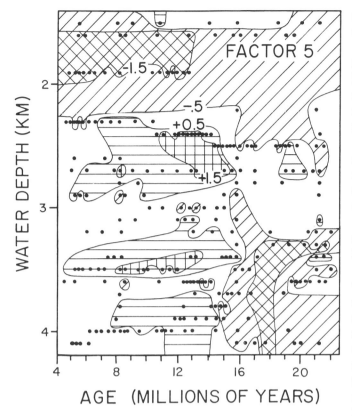

Figure 23. Vertical distribution of the two assemblages based on PCA, Factor 5. They are a deep ocean Pacific bottom water corridor assemblage and a shallow and early Miocene assemblage. The deep ocean assemblage expands its range during the period of deep ocean cooling, 15-9 Ma. Contour intervals are ±5, ±1.5. Hatched patterns are as in Figure 13. Assemblages are in Table 5.

monospecific benthic foraminifera from the same faunal samples used in this study) or occasionally the average values of adjacent samples). As the benthic foraminifera grows, it incorporates proportions of the isotopes oxygen-18 and oxygen-16 ($\delta^{18}O$) into its calcium carbonate test at values systematically near equilibrium with the surrounding water. The $\delta^{18}O$ in the ocean primarily reflects the amount of ^{16}O tied up in glaciers and the temperature of the ocean water. All data have been correlated to *Cibicidoides* species by data given in Graham et al. (1981). Most but not all of the isotopic data have been published (Keigwin 1979; Loutit 1981; Savin et al., 1981; Savin et al., unpublished). Data originally correlated to the B-1 standard have been recorrelated to PDB by subtracting .06 per mil (personal communication, L. Keigwin).

The data show a major increase in $\delta^{18}O$, 14.5 Ma at all sites below 2 km (attributed to both ocean cooling and ice expansion) with the highest values occurring after 10 Ma, a major glacial period. The "warmest" isotopic values occur above 2.5 km approximately 15–17 Ma. The unexpectedly "warm" values at Site 158 (1.9 km) are attributed to the severe diagenesis found at this site causing overgrowths of recrystallized planktonic calcite

(grown in warm surface waters) on the benthic foraminifera (Keigwin, 1979; personal communication).

The correlation of the positive assemblage and the "cooling" isotopic values 14.5 Ma is clear. It is also evident that the assemblage was not dominant during the period of the "coldest" isotopic values (9 Ma) when *Epistominella umbonifera* (the species associated with Antarctic Bottom Water, Figure 7) and the species of late Miocene "glacial" assemblage of Factor 1 (Figure 13) increased in abundance. When found in numbers, the *Cibicidoides* species of the bottom water corridor assemblage of Factor 5, *C. cenop* n. sp, and *C.* cf sp *C. pseudoungerianus* (Figures 7 and 11 and plates 1-3) are apparently good indicators of a cold deep-ocean, but not necessarily major glacial conditions.

The data clearly suggest a deep-water Tethys corridor north of New Guinea 21 Ma. It is striking that the sites which show bottom water corridor assemblage affinities 21 Ma (292, 55, 448, 167; positive assemblage scores; Figs. 23-26) are all located in the western Pacific north of New Guinea and all have "cold" oxygen isotopic values (average 2.1 per mil), whereas those with negative scores 21 Ma (208, 171, 289, 317, 296, 206, 159, 320, 495, 77) are located throughout the rest of the Pacific and have "warm" isotopic values (average 1.7 per mil). Furthermore, Figure 21f suggests that equatorial surface productivity was more prevalent than ocean margin surface productivity 21 Ma, which supports the idea of an open Tethys rather than strong gyral circulation. Nevertheless, due to the gaps in the early Miocene record the data may be registering a cooling in the earliest Miocene.

There is some evidence that species of the bottom water corridor assemblage preferred more oxygenated water, but it is not conclusive (Table 8). *Epistominella exigua* (+0.27) has been associated with young oxygenated water in the Atlantic (Schnitker, 1979; Streeter and Shackleton, 1979), however deep-ocean species-environment correlations are poorly known for highly scored species such as *Anomalinoides globulosa* and *Cibicidoides cenop* n. sp. Species of the counterpart assemblage include *Ehrenbergina trigona,* (the highest negative score, –0.43), and *Globocassidulina subglobosa* (–0.26) both of which have been associated with less well-oxygenated water in the Atlantic (Lohmann, 1978), as has *Rotorbinella lobatula* (–0.28) in the Pacific (Burke, 1981). Furthermore, sites with the most negative scores tend to be farthest from the bottom water corridor source of dense oxygenated water (159, 495, 77) or shallow (281) or both (158).

RELATIONSHIPS TO PALEOCEANOGRAPHY

Benthic foraminifera form clear distributional patterns in the deep ocean. Some species appear to reflect changes in water mass quality, while others react to changes in food availability and sample preservation resulting from changes in surface productivity, in turn related to paleoceanographic change. Many changes in faunal distribution occurred in the deep Pacific Ocean 16–13 Ma at the same time as paleoclimatic and oceanographic changes that presumably occurred as a result of Antarctic glacial expansion

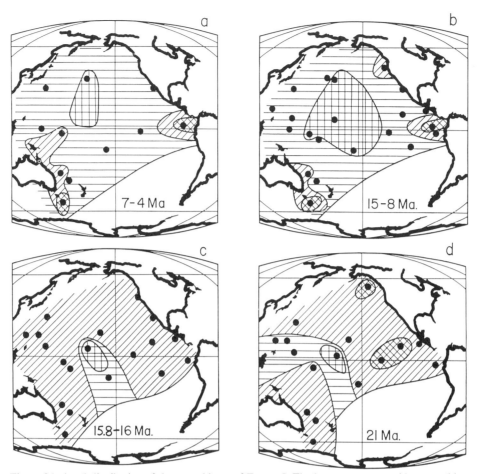

Figure 24. Areal distribution of the assemblages of Factor 5. The bottom water corridor assemblage expands its range after 15 Ma. An Indo-Pacific corridor is evident 21 Ma. Contour intervals are as in Figure 14. Hatched patterns are as in Figure 13.

TABLE 8. MODERN HIGH OXYGEN SPECIES

Species	Factor Scores					Reference
	1	2	3	4	5	
Pyrgo Spp	+0.58	-0.14	-0.21	+0.07	+0.01	miliolids; Lohmann, 1978
Triloculina/						
Quinqueloculina Spp	+0.54	-0.03	-0.45	+0.23	0.0	" " "
Sigmoilina tenuis	+0.43	-0.11	-0.46	+0.10	-0.04	" " "
Epistominella exigua	+0.08	-0.01	-0.49	+0.17	+0.24	Streeter and Shackleton, 1979; Schnitker, 1979b
Epistominella umbonifera	+0.27	-0.14	-0.60	+0.16	+0.10	Burke, 1981
Cibicidoides wuellerstorfi	+0.47	+0.27	+0.23	-0.19	-0.11	Well oxygenated water, Atlantic, Lohmann, 1978
Oridorsalis umbonatus	+0.03	+0.38	-0.28	+0.12	-0.03	Schnitker, 1979b

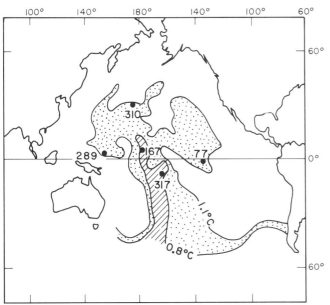

Figure 25. Distribution of the DSDP sites with significantly positive scores in factor 5 (greater than ±1.2), all of which fall within the corridor of today's coldest bottom waters (temperature data from van Andel et al., 1975).

and cooling of the deep Pacific Ocean. Late Miocene faunal changes appear related to expanded Antarctic glaciation. The general paleoceanographic history of benthic foraminifera is not obscured by the overprint caused by dissolution and diagenesis.

Early Miocene

The early Miocene benthic fauna 22–16 Ma exhibit relatively uniform distribution patterns with depth, and few changes through time, as expected of species living in a relatively warm, less well-stratified early Miocene ocean. The early Miocene fauna is dominated (15–30%) by *Cibicidoides kullenbergi, Globocassidulina subglobosa* and *Oridorsalis,* species which have similar percentages at all depths between 1.5 and 4 km. After 15 Ma the faunal boundaries became more distinct and numerous (Figs. 7–11). This increase in species depth-stratification was presumably related to an increase in thermal stratification after 15 Ma, as would occur with the introduction of colder deep waters as suggested by the oxygen isotopic data (Figure 26). Increased *C. kullenbergi* abundances are associated with relatively warm isotopic periods in the Atlantic Ocean (Lutze, 1979) and a similar association may be seen in the Miocene by comparing figures 7 and 26 where *C. kullenbergi* was most abundant during the period of warmest Miocene deep ocean isotopic paleotemperatures 15–18 Ma. This species was not common in the deep Pacific Ocean after 14 Ma, but continued on in shallower (<2.5 km) ocean margin environments after 14 Ma suggesting that a deep ocean change occurred which was not well tolerated by this species. Part of the early Miocene uniformity may be due to poorer early Miocene sample preservation, the result of lower

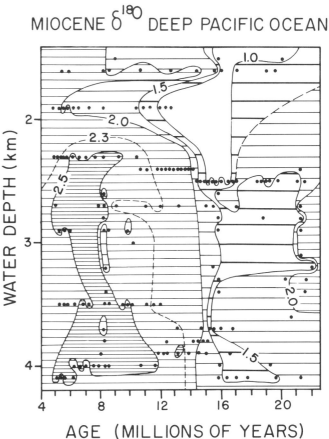

Figure 26. $\delta^{18}O$ values for most of the faunal samples in this study. The "warmest" isotopic paleotemperatures occur approximately 16 Ma, "cooling after 14.5 Ma and "coldest" after 8-10 Ma, exclusive of ice volume effects. Most of the data have been published (Keigwin, 1977; Loutit, 1981; Savin et al., 1981; Savin ct al., unpublishcd). All data are from monospecific samples of benthic foraminifera calibrated to *Cibicidoides* (Graham et al., 1981) and to PDB (subtract 0.06 per mil from B-1 standard; L. Keigwin Jr., personal communication). The equivalent diagram containing the $\delta^{13}C$ values is in Woodruff and Savin, 1985.

carbonate supply rates (van Andel et al., 1975) and increasing diagenesis with increasing burial depth.

Benthic foraminiferal species associated in the current literature with relatively low oxygen and/or intermediate water (Table 4, Factors 2, 3, 5) weakly suggest that the early Miocene deep-ocean was less well-oxygenated than the mid-late Miocene Pacific Ocean. Assemblages associated with low oxygen water were more prevalent in deeper water in early Miocene times, particularly the early Miocene intermediate water assemblage dominated by *Bulimina* species (Factor 3, Fig. 17). It is not suggested that intermediate water was found at deeper water depths before 15 Ma but rather that there existed a less well-defined interface between deep and intermediate water. Pacific intermediate water must be found above approximately 2.5 km, as it is the return flow of newly generated dense Southern Ocean water which sank and flowed northward below 2.5 km. As the water ages its organic matter oxidizes so that the return flow is relatively low in oxygen. Longer ocean residence times, a lower vertical thermal gradient, and a less well-stratified ocean would promote low oxygen water at greater depths.

Middle Miocene

Modern Pacific Ocean intermediate and bottom water assemblages (Burke, 1981) replaced early Miocene assemblages 14 Ma (Figs. 17, 19). A faunal turnover also occurred in the bathyal and abyssal Atlantic Ocean at approximately the same time and involved many of the same species (Berggren, 1972). The early Miocene intermediate water assemblage was characterized by *Buliminella grata, Bulimina jarvisi, B. glomarchallengeri, B. tessellata, Stilostomella spinata* and *S. annulifera*. Early Miocene bottom water assemblages were characterized by *Epistominella umbonifera, Cibicidoides grimsdalei, C. havanensis, C. bradyi, Laticarinina pauperata, Melonis barleeanum,* and *Bolivinopsis cubensis* (Table 6). Both early and mid-late Miocene intermediate water assemblages were dominated by *Uvigerina, Bulimina* and *Stilostomella* species with *Rotorbinella lobatula, Ehrenbergina trigona, Osangularia culter, Bulimina rostrata* and *Uvigerina proboscidea* becoming dominant in mid-late Miocene time. Both early and mid-late Miocene bottom water assemblages are characterized by *E. umbonifera* with *E. exigua* and the miliolids becoming more prevalent in mid-Late Miocene time.

During the periods of accelerated evolution 14–16 Ma, 11 Paleogene species from both intermediate and bottom water depths became extinct, at least 3 new species evolved (Tables 2, 3), and many species changed their depth of maximum distribution by moving from lower abyssal to upper abyssal realms and from upper abyssal to lower bathyal realms. The change is characterized by the origination of *Cibicidoides cenop* n. sp. which replaced *C. kullenbergi* in dominance below but not above 2.5 km water depth, 14 Ma. Since little is known about modern species-environment correlations in the Pacific one can only speculate that the major deep-ocean change, perhaps a cooling as suggested by the rapid increase in $\delta^{18}O$ 14 Ma, caused the deep ocean to become uninhabitable for many species which either became extinct, less common, or moved out of the deep ocean to shallower depths. The species which filled the void either evolved or moved up from greater depths becoming more common at upper abyssal and lower bathal depths after 14 Ma. This suggests that the source of the change was in the deep and bottom water masses. A major cooling of the deep ocean would affect the fauna at all depths and could be responsible for extinctions in all water masses.

An assemblage characteristic of the Pacific bottom water corridor during the Miocene became characteristic of the whole Pacific Basin during the period of maximum isotopic deep ocean "cooling" 15–8 Ma (Table 5, positive assemblage factor 5). *Cibicidoides* sp. cf. *C. pseudoungerianus* characterized the corridor throughout Miocene time; *C. cenop* n. sp. characterized the corridor and Pacific Basin in mid-late Miocene (see plates 1-3). The sites at which the assemblage was dominant (317, 167, 289, 310)

are those where modern bottom water temperatures are colder than 1.1°C (Fig. 25). It is clear that the isotopic "cooling" and expansion of the bottom water corridor assemblage were related (Figures 22-26). Even in the earliest Miocene (19–21 Ma) the assemblage correlated with "cool" isotopic values (greater than 2 per mil, Figures 23, 26) below 2.4 km. Also, the sites which show bottom water corridor assemblage affinities 21 Ma (292, 55, 448, 167) are all located in the western Pacific north of New Guinea and have "cold" oxygen values (average 2.1 per mil), whereas those which are negatively correlated 21 Ma (208, 171, 289, 317, 296, 2066, 159, 320, 495, 77) are located throughout the rest of the Pacific and have "warm" isotopic values (average 1.7 per mil). This suggests that a bottom water corridor existed across the Indo-Pacific (Tethys) at least until 21 Ma. It seems most likely that the expansion of the bottom water corridor assemblage 15 Ma was related to a major influx of cold dense water from the Antarctic region.

However, although this assemblage contains *Epistominella exiqua*, associated with oxygenated young water in the Atlantic) (Schnitker, 1979; Streeter and Shackleton, 1979), it is not as heavily weighted toward species preferring highly-oxygenated water as was the mid-late Miocene assemblage which became most prevalent after 8–10 Ma (Figure 13, Tables 5 and 8). The bottom water corridor assemblage appears to have dominated during the transition between establishment of the first permanent Antarctic ice sheet 14.5 Ma and major expansion of the Antarctic glacier after 8–10 Ma. Small ice shelves began to form along the margins of the Ross Sea at about 16 Ma, and the Ross Ice Shelf (located on the southern Pacific Ocean in the Antarctic region) had formed by 10–8.5 Ma, leading to bottom water intensification that created disconformities from 14.7 to 0.6 Ma in the Ross Sea (Savage and Ciesielski, 1983). Nevertheless, the possibility remains that the assemblage expansion reflects an infusion of old, saline, North Atlantic deep water into the Pacific through the circumpolar 13 to 7 Ma, or dense saline waters formed in the North Pacific terminating 7 Ma (Schnitker, 1980; Blanc and Duplessy, 1982).

After 15 Ma the distribution of the mid-late Miocene assemblage composed primarily of *Melonis, Uvigerina* and miliolid species suggest that there was a general and progressive increase in specimen preservation and organic nutrient availability due to an increase in carbonate surface productivity especially along the equator (Figure 14). This is reflected in the mid-late Miocene higher benthic sample diversity, the higher percentages of less well-preserved specimens (miliolid and fragile species) and the lower percentages of dissolution-resistant species (robust agglutinated forms, *Globocassidulina subglobosa*). A general increase in sediment accumulation rates occurred in the Pacific after the early Miocene (Fig. 2). The prevalence of the mid-late Miocene assemblage beneath the equatorial high-productivity zone suggests that the increase in sample preservation was partially due to an increase in surface productivity raining more dissolution-prone planktonic calcite to the seafloor which reduced the chance for benthic dissolution and accelerated burial rates. The increase in the mid-late Miocene assemblage first occurred in the eastern equatorial Pacific (even at deep sites) where upwelling is prevalent today. It is seen progressively later in the western Pacific (Fig. 14). Surface productivity probably increased along the whole equator simultaneously but water near the bottom water corridor to the west may have been more corrosive to carbonate while upwelling regimes to the east may have dispensed more carbonate, resulting in a pattern progressing from east to west. Rates of biogenic carbonate supply from highly productive surface regions steadily increase along the equator throughout the Miocene but were balanced by steadily increasing rates of dissolution from bottom waters (van Andel et al., 1975). An increase in the latitudinal thermal gradient and intensified atmospheric-oceanic circulation (Savin, 1977) probably brought food to the surface for planktonic organisms along the equator. Closure of the Indo-Pacific passage in the middle Miocene (Edwards, 1975) could also have enhanced surface productivity by diverting large volumes of water into the western boundary currents which would return as intensified equatorial circulation.

An ocean margin assemblage became important between 16 and 12 Ma (Factor 4, negative assemblage, Figs. 20, 21). Its distribution (sites 158, 470, 173, 310, 289) suggests a preference for areas beneath highly productive, particularly siliceous, regions such as the California and Kuroshio current systems and the equator away from areas of the central gyres and those of low sediment accumulation (less than 10 m/Ma). The assemblage is attributed to invigorated atmospheric-oceanic circulation around the Pacific margin where siliceous surface productivity is associated with invigorated current activity, mixing, upwelling and tectonic activity (Ingle, 1973, 1981). Sites 173 and 470, under the California current system, and Site 158, located in the eastern equatorial upwelling zone, are all in known areas of high siliceous sedimentation. Increased silica is found at Site 289 when this assemblage became prevalent (Romine, this volume) and Site 310 contains the first diatoms when it emerged from under the central gyre to beneath the Kuroshio current 7 Ma (Burckle, personal communication). The assemblage includes *Uvigerina multicostata* and *Cibicidoides bradyi*. Species of *Uvigerina* have a preference for organic-rich sediments in the California Borderland (Douglas, 1981) as would be found below productive regions. Burke (1981) finds *C. bradyi* associated with upwelling (ridging) in the Recent western equatorial Pacific. This dissolution-resistant assemblage may be the result of the combined effects of increased organic food on the ocean floor and the increased dissolution at the sediment interface related to increased oxidation of organic matter (Berger and Winterer, 1974). This assemblage appears to reflect an increase in surface productivity around the Pacific ocean margins related to intensified atmospheric-oceanic circulation after 16 Ma.

The very detailed record at Site 289 shows that the planktonic isotopic paleotemperatures started to "warm" and the benthic ones started to "cool" 16 Ma (Savin et al., 1981; Shackleton, 1982). The planktonic isotopic paleotemperatures did not "warm" in the subpolar surface waters (Shackleton and Kennett,

1975b). This suggests that the latitudinal thermal gradient began to increase 16 Ma, before the sharp isotopic deep ocean "cooling" of 14.5 Ma. Therefore, it is likely that atmospheric-oceanic circulation intensified before deep ocean "cooling" occurred. This could be the reason increased ocean margin productivity preceded expansion of the "cold" bottom water corridor assemblage by approximately 1 Ma.

Late Miocene

A second period of faunal change occurred in late Miocene time (9-5 Ma) involving increases in *Uvigerina* and *E. umbonifera* abundance. These changes appear to be primarily related to changes in water mass quality and increases in sediment organic carbon.

The greater abundances of *Uvigerina* in the late Miocene may be related to the effects of approaching major Antarctic glaciation. *Uvigerina* species have been associated with glacial periods (Schnitker, 1974; Corliss, 1982; Lutze, 1978) and increased organic carbon in the sediments and low oxygen water (Douglas, 1981). Steep thermal gradients intensify atmospheric-oceanic circulation during glacial periods, leading to increased surface productivity and the raining of more organic carbon to the seafloor (Arrhenius, 1952; Pederson, 1983). There is abundant evident for late Miocene intensified circulation resulting from steepening latitudinal thermal gradients (Lohmann and Carlson, 1981) and increased wind velocities, compatible with late Miocene increased silica productivity in the Southern Ocean, upwelling in the equatorial Pacific, Indonesian area and around the Pacific rim (Kennett, 1982, 1977; Brewster, 1980; Leinen, 1979; Van Gorsel and Troelstra, 1981; Ingle, 1981; Gilbert and Summerhayes, 1981). Climatic fluctuations associated with latest Miocene glaciation (Adams et al., 1977; Hamilton, 1972; Shackleton and Kennett 1975a) caused increased rates of upwelling, associated surface productivity intensifying the oxygen minima around the Pacific rim (Summerhayes, 1981; Gilbert and Summerhayes, 1981; Birch, 1979; Siesser, 1980). These conditions appear ideal for *Uvigerina* species (see section on species distribution).

The clearly defined shallowing of *E. umbonifera* after 9 Ma (Fig. 7) may have been related to the effects of increased AABW in latest Miocene time, a period of maximum Miocene glaciation (Shackleton and Kennett, 1975b). This species has been associated with AABW in all modern ocean basins. Prior to 15 Ma the species is commonly found below 4 km but it appears intermittently at shallower depths (2.7 km) between 15 and 8 Ma at Site 317 in the bottom water corridor south of the equator. It alternates in abundance with *E. exiqua*, a species associated with Pacific Bottom Water in and above the *E. umbonifera* realm (Burke, 1981). The intermittant appearances of *E. umbonifera* could be due to occasional influxes of AABW following permanent establishment of the Antarctic ice sheet 15 Ma. The abundant appearances after 9 Ma could be due to major Antarctic ice expansion. This is compatible with the oxygen isotope data (Fig. 26).

The shoaling of abundant specimens of *E. umbonifera* 8-5 Ma throughout the Pacific Ocean (Fig. 7) could also have been caused by a shoaling of the Pacific calcite compensation depth (CCD). Bremer and Lohmann (1982) have shown that *E. umbonifera* is associated with waters undersaturated in carbonate. However, most estimates of the Pacific CCD place its shoalest Miocene extent at 10 Ma, not 5 Ma (van Andel et al. 1975; Berger, 1973). Further, the alternation of *E. umbonifera* and *E. exigua* abundances in the bottom water corridor at 2.7 km suggest that the fluctuations are related to changes in water mass quality well above the Pacific CCD of 4-4.8 km.

If *E. umbonifera* and *E. exiqua* (Fig. 7) are used as indicators of bottom water masses of somewhat different qualities (Antarctic Bottom Water and Pacific Bottom Water, respectively) their areal distributions (not depicted) suggest the following: Between 23 and 16.2 Ma *E. umbonifera* was generally rare (1 to 3%), was virtually nonexistent (0 to 1%) outside the equatorial area and was common (4 to 9%) only at the deepest sites. *E. exiqua*, also rare, was found at slightly shallower sites. Between 15 and 16 Ma both species increased in abundance, with *E. umbonifera* becoming common in the deep central Pacific and *E. exiqua* in the east and west Pacific. Between 15 and 9 Ma the *E. umbonifera* range decreased somewhat while *E. exiqua* became common after 13 Ma in the central deep ocean (particularly in the North Pacific) and became slightly more abundant throughout the whole Pacific but interestingly, was not found at the coastal equatorial sites (158, 55, 62). Suddenly, 8.5–5.5 Ma *E. umbonifera* became abundant (10 to 43%) in the central deep ocean and by 5.5–4.5 Ma became abundant at all sites below 2.4 km (well above the lysocline) while *E. exiqua* became rare almost everywhere. *Epistominella umbonifera* may be reflecting the influence of fluctuating amounts of glacially-derived water after 16 Ma and major glacial conditions and AABW when it became abundant 8.5 Ma. *Epistominella exiqua* which became common 13 to 8 Ma, well after the period of major isotopic "cooling," may reflect Pacific bottom water not dominated by AABW in an already cool ocean, or interglacial conditions, or NADW influences. *Epistominella umbonifera* may reflect the influence of fluctuating amounts of glacially derived water after 16 Ma and major glacial conditions and AABW when it became abundant after 8.5 Ma.

By 8 Ma the benthic foraminiferal assemblages were similar to Quaternary assemblages. Most of the species identified by Burke (1981) in the Recent Pacific were found in the late Miocene assemblages. A Quaternary Antarctic circumpolar glacial assemblage (Corliss, 1982) correlates with high scores of the late Miocene species prevalent after 14 Ma and particularly after 8 Ma (Factor 1, Table 5). Corliss' "glacial assemblage" includes *Melonis barleeanum* (+0.49), *Melonis pompilioides* (+0.45) and *Uvigerina peregrina* (related to *U. hispido-costata*, +0.45). The late Miocene assemblage is less well-correlated to Corliss' "interglacial assemblage": *Cibicidoides wuellerstorfi* (+0.47), *Ehrenbergina trigona* (+0.40), *Gyroidina soldanii* (–0.06) and *Globocassidulina subqlobosa* (–0.15). Apparently the late Miocene ocean resembled Quaternary and particularly the glacial Quaternary ocean.

The simplest explanation for the numerous benthic foraminiferal changes in the intermediate, deep and bottom waters in both the ocean margins and the deep sea 14–16 Ma was a gradual adjustment to paleoceanographic changes as the ocean adjusted to the effects of climatic deterioration asssociated with the development of a permanent Antarctic ice sheet which reached its maximum Miocene extent in latest Miocene time. Most of the species which became extinct 14 Ma were Paleogene in origin, whereas most of the species which dominate the late Miocene are common in the Pacific today.

ACKNOWLEDGMENTS

I wish to thank every member of the Cenozoic paleoceanography team (CENOP) for much shared information and many fascinating discussions. In addition, Dr. Burton Jones, of the University of Southern California, Department of Biological Sciences, provided the computer guidance; Christopher Morris helped with sample preparation; John Quinn provided benthic data for DSDP Site 159; R. Commeau of the U.S. Geological Survey (Woods Hole) photographed the SEM plates. Dr. Gerta Keller of the U.S. Geological Survey (Menlo Park), is responsible for most of the planktonic biostratigraphic correlations and Professor Samuel Savin and Linda Abel handled the stable isotope analyses at Case Western Reserve University. Many useful discussions, comments and suggestions provided by reviewers R. G. Douglas, G. P. Lohmann, D. Schnitker, Z. Reiss, were incorporated into this manuscript. This study was supported by the National Science Foundation, Oceanography Section, submarine geology and geophysics program OCE-7919093 (CENOP).

APPENDIX 1. NEW SPECIES

Cibicidoides cenop Woodruff, n. sp. Holotype, Plate I, Figs. 1-3; paratypes Plate II, 7-12.

Description: The test is biconvex with pores on both the dorsal and ventral sides. The pores become more evident and numerous on the ventral side in the larger specimens. In specimens smaller than 200 microns the ventral pores are often difficult to see with the light microscope. The species is often large, greater than 250 microns. Many of the pores on the larger specimens have become enlarged due to dissoultion. There are usually 10-12 chambers in the last whorl, all visible dorsally but with only those of the final whorl visible on the ventral side. The aperature is interiomarginal, extraumbilical-equatorial and extends along the spiral suture in the last 1 to 2 chambers. The species ranges from the middle Miocene to the Recent.

Type locality: Ontong-Java Plateau, Pacific Ocean, DSDP Site 289, 0o28'S, 158o30'E.

Type sample: DSDP leg 30, Site 289, core 37, section 4, 93-100 cm.

Type level: Middle Miocene (Foraminiferal Planktonic Zone N14).

Comments: This species appears to be the one figured by Brady (1884) at 3.6 km challenger 296 in the South Pacific which he referred to Truncatulina sp. near T. haidingerii (d'Orbigny). Barker (1960) in his revision of the Brady 1884 plates ascribes the spcies to Cibicidoides mundulus (Brady et al., 1888), a species described from 0.5 km, Plumber Station 4, Abroholos Bank in the South Atlantic. The lectotype of Cibiciodoides mundula from Plumber Station 4, designated by Loeblich and Tappan (1964) bears little resemblance to Brady's (1884) species which has no ventral pores, is more compressed, has fewer chambers and a raised unbilical nob.

APPENDIX 2. CHANGES IN MIOCENE DEEP-SEA BENTHIC FORAMINIFERAL DISTRIBUTION IN THE PACIFIC OCEAN: RELATIONSHIP TO PALEOCEANOGRAPHY

(See microfiche in pocket inside back cover)

APPENDIX 3. CHANGES IN MIOCENE DEEP-SEA BENTHIC FORAMINIFERAL DISTRIBUTION IN THE PACIFIC OCEAN; RELATIONSHIP TO PALEOCEANOGRAPHY

(See microfiche in pocket inside back cover)

APPENDIX 4. LISTING OF SPECIES IDENTIFIED

(The number to the left is the faunal number used in the data compliation and factor analysis.)

____ Allomorphina pacifica Cushman and Todd, 1949, Cushman Lab. Foram. Res., Contr., v. 25, pt. 3, p. 68, pl. 12, figs. 6-9.

1 Anomalinoides sp 1; see Douglas, 1973, Init. Repts. Deep Sea Drill. Proj., v. 17, p. 607-671.

5 Anomalinoides alazanesis (Nuttall), Anomalina alazanensis, Nuttall 1932, J. Paleont., v. 6, p. 32, pl. 8, figs. 5-7.

4 Anomalinoides cicatricosa (Schwager). Anomalina cicatricosus Schwager, 1866, Novara-Exped., Geol. Theil., v. 2, p. 260, pl. 7, figs. 108, 4.

3 Anomalinoides globulosa (Chapman and Parr). Anomalina globulosa Chapman and Par, 1937, foraminifera: Australasian Antarct. Exped. 1911-1914, Sci. Repts., Ser. C (Zool. Bot.), v. 1, pt. 2, p.1-190, pl. 7-10.

2 Anomalinoides semicribrata (Beckmann). Anomalina pompilioides Galloway and Hemminway var. semicribrata Beckmann 1953, Ec. Geol. Helv., v. 10, p. 400-401, pl. 27, fig. 3.

6 Astrononion guadalupae (Parker), Melonis guadalupae Parker; 1964, J. Paleont. v. 38, n. 4, p. 626, pl. 100, figs. 13, 14.

8 Bolivina sp cf B. punctata d'Orbigny, 1939, Voyage dans l'Amerique Meridionale; Foraminiferes. Strasbourg, France; Levrault, 1939, tome 5, pt. 5, p. 63, pl. 8, figs. 10-12.

APPENDIX 4. LISTING OF SPECIES IDENTIFIED (continued)

7   Bolivina optima (Cushman). Bolivina pisciformis Galloway and Morrey
     var. optima Cushman, 1943, Cushman Lab. Foram. Res. Contr. v. 19, p.
     91, pl. 16, fig. 2.

104 Bolivinopsis cubensis (Cushman and Bermudez). Spiroplectoides
     cubensis,1 Cushman and Bermudez, 1937, Cushman Lab. Foram. Res. Contr.,
     v. 13, pl. 1, figs. 44, 45.

16   "Bulava indica" Boltovskoy, 1976, Rev. Espanola Micropaleont., v. 8, N.
     2, p. 301-303.

13   Buliminella grata Parker and Bermudez, 1937, J. Paleont., v. 11, p.
     515, pl. 59, fig. 6.

14   Bulimina inflata Sequenza, 1862, Accad. Gioenica Sci. Nat. Atti., Ser.
     2, v. 18, p. 25, pl. 1, fig. 10.

___ Bulimina inflata Sequenza, 1862, Accad. Gioenica Sci. Nat. Atti., Ser
     2, v. 18, p. 25, pl. 1, fig. 10

15   Bluimina glomarchallengeri, Tjalsma and Lohmann, 1983, Micropaleont.
     Spec. Plub. N. 4, p. 25, pl. 13, figs. 8-12c.

10   Bulimina jarvisi Cushman and Parker, 1936, Cushman Lab. Foram. Res.
     Contr., v. 12, p. 34, pl. 7, fig. 1.

12   Bulimina miolaevis Finlay, 1940, Trans. Roy. Soc. New Zealand, v. 69,
     p. 454, pl. 64, fig. 70, 71.

9   Bluimina rostrata,Brady, 1884, Challenger Repts., Zool. v. 9, p. 408,
     pl. 51, figs, 14, 15.

11   Bulimina tessellata, Cushman and Todd; Bulimina marginata d'Orbigny
     var. tessellata Cushman and Todd, 1945, Cushman Lab. Foram. Res., Spc.
     Publ. 15, p. 39, pl. 6, fig. 9.

17   Cassidella bradyi (Cushman), Virgulina bradyi Cushman, 1922, p.
     115, pl. 24, fig. 1.

20   Cassidulina angulosa Cushman, 1933,Cushman Lab. Foram. Res. Contr., v.
     9, p. 93, pl. 10, fig. 6.

18   Cassidulina galvinensis Cushman and Frizzell, 1940, Cushman Lab. Foram.
     Res. Contr., v. 16, p. 43, pl. 8, fig. 10.

21   Cassidulina reflexa Galloway and Wissler, 1927, J. Paleont., v. 1, n.
     1, p. 80, pl. 12, fig. 13.

19   Cassidulina tricamerata Galloway and Heminway, 1941, Sci. Surv. Porto
     Rico, Virgin Islands, v. 3, pt. 4, p. 425, pl. 32, fig. 3.

___ Chilostomella oolina Schwager, 1878, R. Com. Geol. Ital., Boll., v. 9,
     p. 527, p. 1, fig. 16.

23   Chrysalogonium lanceolum Cushman and Jarvis, 1934, Cushman Lab. Foram.
     Res. Contr., v. 10, p. 74, pl. 10, fig. 16.

22a Chrysalogonium longicostatum Cushman and Jarvis, 1934, Cushman Lab.
     Foram. Res. Contr., v. 10, p. 74, pl. 10, fig. 12.

22b Chrysalogonium tenuicostatum Cushman and Bermudez, 1936, Cushman Lab.
     Foram. Res. Contr., v. 12, p. 27, pl. 5, figs. 3-5.

36   Cibicidoides trincherasensis (Bermudez) Cibicides trincherasensis
     Bermudez, 1949, Cushman Lab. Foram. Res. Spec. Publ. 25, p. 307, pl. 25,
     figs. 1-3.

28   Cibicidoides cenop Woodruff n. sp., This paper.

34   Cibicidoides dohmi (Bermudez), Cibicides dohmi Bermudez, 1949;
     Tertiary smaller Foraminifera of the Dominican Republic, Cush. Lab.
     Foram. Res., v 25, p. 297, pl. 24, figs, 25-27.

30   Cibicidoides sp. cf C. pseudoungerianus (Cushman). Truncatulina
     pseudoungerianus, Cushman, 1922, U.S. Geol. Surv. Prof. Pap., No
     129-E, p. 97, pl. 20, fig. 9.

35   Cibicidoides bradyi (Trauth). Truncatulina bradyi Trauth, 1918, K.
     Akad. Wiss. Wein, Math-Naturw. Kl., Denkschr., v. 95, p. 235.

APPENDIX 4. LISTING OF SPECIES IDENTIFIED (continued)

33   Cibicidoides grimsdalei (Nuttall). Cibicides grimsdalei Nuttall, 1930, J. Paleont., v. 4, p. 291, pl. 25, figs. 7, 8, 11.

32   Cibicidoides havanensis (Cushman and Bermudez). Cibicides havanensis Cushman and Bermudez, 1937, Cushman Lab. Foram. Res. Contr., v. 13, p. 28, pl. 3, figs. 1-3.

24   Cibicidoides kullenbergi (Parker). Cibicides kullenbergi Parker 1953 in Phleger, Parker and Peirson, 1953, Repts. Swedish Deep Sea Exped., v. 7, n. 1, p. 49, pl. 11, figs. 7, 8.

___   Cibicidoides subhaidingerii Parr, 1950, B.A.N.Z. Antarctic Res. Exped. 1929-1931, Repts., Adelaide, Ser. B., v. 5, pt. 6, p. 364, pl. 15. fig. 7.

37   Cibicidoides trinitatensis (Nuttall). Truncatulina trinitatensis Nuttall, 1928, Geol. Soc. London Quart. Jour., v. 84, p. 97, pl. 7, figs 3, 5, 6.

25   Cibiciodoides wuellerstorfi (Schwager). Anomalina wuellerstorfi Schwager, 1866, Novara-Exped. Geol. Theil., v. 2, p. 258, pl. 7, figs. 105-107.

26   Cibicidoides sp. B.

27   Cibicidoides sp. C.

29   Cibicidoides sp. N.

31   Cibicidoides sp. R.

105   Dorothea brevis Cushman and Stainforth, 1945, Cushman Lab. Foram. Res. Spec. Publ. 14, p. 18, pl. 2, fig. 5.

107   Eggerella sp. L.

106   Eggerella bradyi (Cushman). Verneuilina bradyi Cushman, 1911, U.S. Nat. Mus. Bull. 71, pt. 2, p. 54, tf. 87, p. 55.

42   Ehrenbergina spp

40   Ehrenbergina bosoensis, Takayangai, 1951, Paleo. Soc. Japan, Trans. Proc., Tokyo, n.s. N. 3, p. 87, p. 88, tf 8.

38   Ehrenbergina caribbea Galloway and Hemingway, 1941, Acad. Sci., Sci. Surv. Puerto Rico, Virgin Islands, v. 3, p. 427, pl. 32, fig. 4.

41   Ehrenbergina spct E. hystrix1 Brady, 1884, Repts. Voy. Challenger, v. 9, p. 434, pl. 55, figs. 8-11.

39   Ehrenbergina trigona (Goes). Ehrenbergina serrata Reuss var. trigona Goes, 1896, Harvard College, Mus. Comp. Zool., Bull., v. 29, p. 49, pl. 6, figs. 183-184.

___   Ellipsoglandulina multicostata (Galloway and Morrey). Daucina multicosta Galloway and Morrey, 1929, Bulls. American Paleont. v. 15, n. 55, p. 42, pl. 6, fig. 13.

___   Ellipsodimorphina robusta (Cushman). Nodosarella robusta Cushman, 1943, Cushman Lab. Foram. Res. Contr., v. 19, pt. 4, p. 92, pl. 16, fig. 8.

43   Epistominella exigua (Brady). Pulvinulina exigua Brady, 1884, Rept. Voy. Challenger, Zool., v. 9, p. 696.

44   Epistominella umbonifera (Cushman). Pulvinulinella umbonifera Cushman, 1933, Cushman Lab. Foram. Res. Contr., v. 9, pt. 4, p. 90, pl. 9, fig. 9.

45   Eponides sp. E.

47   Favocassidulina australis Eade, 1967, New Zealand J. Mar. Fresh-water Res., Wellington, New Zealand, v. 1, n. 4, p. 425, 428.

49   Favocassidulina elegantissima (Cushman), Cassidulina elegantissima, Cushman, 1925, Contr. Cushman Lab. Foram. Res., Sharon, Mass, U.S.A., vol. 1, n. 10, p. 37, pl. 7, fig. 5.

46   Favocassidulina favus (Brady), Pulvinulina favus Brady, 1877, Geol. Mag., v. 4, p. 535.

APPENDIX 4. LISTING OF SPECIES IDENTIFIED (continued)

48 <u>Favocassidulina</u> <u>sp</u> <u>cf</u>. <u>V</u>. <u>favus</u>

50 <u>Francisita</u> <u>advena</u> (Cushman). <u>Virgulina</u> (?) <u>advena</u> Cushman, 1922,
       U.S. Nat. Mus. Bull. 104, pt. 3, p. 120, pl. 25, figs. 1-3.

___ Gaudryina trinitatensis Nuttall, 1928, Geol. Soc. London, Quart. Jour.,
       v. 84, p. 76, pl. 3, figs. 15-16.

51 <u>Globocassidulina</u> <u>subglobosa</u> (Brady). <u>Cassidulina</u> <u>subglobosa</u> Brady,
       1881, Quart. Micr. Sci., v. 21, p. 60, Repts. Voy. Challenger, v. 9, pl.
       54, fig. 17.

___ <u>Globocassidulina</u> <u>ornata</u> (Cushman). <u>Cassidulina</u> <u>subglobosa</u> Brady,
       ornata Cushman, 1927, Scripps Inst. Oceanography Bull., Tech. Ser., v.
       1, p. 167.

___ <u>Gyroidina</u> <u>soldani</u> d'Orbigny var. <u>altiformis</u> Stewart and Stewart,
       1930, J. Paleont., v. 4, n. 1, p. 67, pl. 9, fig. 2.

___ <u>Gyroidina</u> <u>broeckhiana</u> Karrer, 1878, Fragmente zu Einer Geologie der
       Insel Luzon. K. Gerold's Shon, p. 98, pl. 5, fig. 26.

52 <u>Gyroidina</u> <u>io</u> Resig, 1958, Micropaleontology, v. 4, n. 3, p. 304, tf.
       15a-c.

53 <u>Gyroidina</u> <u>sp.cf</u> <u>G</u>. <u>10</u>

54 <u>Gyroidina</u> <u>neosoldani</u> Brotzen var. <u>acuta</u> Boomgaart, 1949, smaller
       foraminifera from Bodjonegoro (Java) (Utecht, Univ., doct. diss.)
       Utrecht: the author, 1949, p. 125, pl. 14, fig. 1.

55 <u>Gyroidina</u> <u>neosoldanii</u> Brotzen, 1936, Schonen. Sweden, Sver. Geol.
       Unders. Avh., Stockholm, Sverige, Ser. C., N. 396, p. 158, pl. 107,
       figs. 6, 7.

57 <u>Gyroidina</u> <u>torulus</u>, Belford, 1966, Miocene and Pliocene Smaller
       Foraminifera from Papua and New Guinea, Commonwealth of Australia Dept.
       of Nat. Developm., Bur. of Mineral Res., Geol., Geophysics, Bull. 79, p.
       168, pl. 28, fig. 10-20.

58 <u>Gyroidina</u> <u>girardana</u> (Reuss). Rotalina girardina Reuss, 1851, A.
       Deutsch, Geol. Gesell., v. 3, p. 73, pl. 5, fig. 34.

___ <u>Gyroidina</u> <u>lamarkiana</u> (d'Orbigny). <u>Rotalina</u> <u>lamarckiana</u> d'Orbigny,
       1839, Foraminiferes des Iles Canaries par. M.M.P. Barker-webb et Sabin
       Berthelot. Bethune, Paris, France, tome 2, p. 2, Zool., p. 131, pl. 2,
       figs. 13-15.

___ <u>Gyroidina</u> <u>perampla</u> Cushman and Stainforth. <u>Gyroidina</u> <u>girardana</u>
       (Reuss) var. <u>perampla</u>Cushman and Stainforth, 1945, Cushman Lab. Foram.
       Res. Spec. Publ., v. 14, p. 61, pl. 10, fig. 19.

___ <u>Gyroidina</u> <u>quinqueloba</u> Uchio, 1960, Cushman Found. Foram. Res. Spec.
       Publ. n. 5, p. 66, pl. 8, figs. 22-25.

56 <u>Gyroidina</u> <u>planulata</u> Cushman and Renz, 1941, Cushman Lab. Foram. Res.
       Contr., v. 17, p. 23, pl. 4, fig. 1.

___ <u>Gyroidina</u> <u>zealandica</u> Finlay, 1939, Roy. Soc. New Zealand Trans. Proc.,
       v. 67, p. 434, pl. 28, figs. 138-140.

59 <u>Hanzawaia</u> <u>cushmani</u> (Nuttall). <u>Cibicides</u> <u>cushmani</u> Nuttall, 1930, J.
       Paleont., v. 4, p. 291, pl. 25, figs. 3, 5, 6.

60 <u>Heterolepa</u> <u>spp</u>

61 <u>Herronallenia</u> <u>Kempii</u> (Heron-Allen and Erland). <u>Discorbis</u> <u>kempii</u>
       Heron-Allen and Erland, 1929, Roy. Micros. Soc. Jour., ser. 3, v. 49,
       pt. 4, art. 27, p. 332, pl. 4, figs. 40-48.

108 <u>Karreriella</u> <u>bradyi</u> (Cushman). <u>Gaudryina</u> <u>bradyi</u> Cushman, 1911, U.S.
       Nat. Mus. Bull. 71, pt. 2, p. 67, fig. 107.

___ <u>Karreriella</u> <u>novangliae</u>(Cushman). <u>Gaudryina</u> <u>baccata</u> Schwager var.
       novangliae Cushman, 1922, U.S. Nat. Mus. Bull. 104, pt. 3, p. 76, pl.
       13, fig. 4.

62 <u>Laticarinina</u> <u>pauperata</u> (Parker and Jones). <u>Pulvinulina</u> <u>repanda</u>
       menardii pauperata Parker and Jones, 1865, Phil. Trans., v. 155, p.
       395, pl. 16, figs. 50, 51.

APPENDIX 4. LISTING OF SPECIES IDENTIFIED (continued)

63   Lenticulina clericci (Fornasini). Cristellari clericii Fornasini,
     1901, Accad. Sci. 1st. Bologna Mem., Ser. 5, v. 9, p. 65, fig. 17.

___ Martinotiella levis (Finlay). Listerella levis Finlay, 1939, Roy.
     Soc. New Zealand, Trans. Proc., v. 69, pt. 1, p. 97, pl. 14, fig. 79.

___ Martinotiella suteri Cushman and Stainforth, 1945, Cushman Lab. Foram.
     Res. spec. Publ. 14, p. 19, pl. 2, fig. 26.

65   Melonis soldanii (d'Orbigny), Nonionina soldanii d'Orbigny, 1848,
     Foraminiferes fossiles du bassin tertiare de Vienne. Gide et Comp.,
     Paris, France, p. 109, p. 5, fig. 15-16.

66   Melonis affinis (Reuss). Nonionina affinis Reuss, 1851, Deutsch.
     Geol. Gesell, Zeitschr., Bd. 3, p. 72, pl. 5, fig. 32.

67   Melonis barleeanum (Williamson). Nonionina barleeana Williamson,
     1858, On the Recent foraminifera of Great Brittain. Roy. Soc. London, p.
     32, pl 3, figs. 68-69.

64   Melonis pompilioides (Fichtel and Moll). Nautilus pompilioides
     Fichtel and Moll. 1798, Test. Micr., Wien, p. 31, pl. 2, figs. a-c.

68   Melonis sp cf. regina (Martin). Anomalina Renina Martin, 1943,
     Stanford Univ. Publ., Univ. Ser., Geol. Sci., v. 3, n. 3, p. 118, pl. 9,
     fig. 3.

69   Nonion havanensis Cushman and Bermudez, 1937, Cushman Lab. Foram. Res.
     Contr., v. 13, pt. 1, pl. 2, figs. 13-14.

70   Oridorsalis umbonatus (Reuss). Rotalina umbonata Reuss, 1851,
     Deutsch. Geol. Gesell., v. 3, p. 75, pl. 5, fig. 35.

71a Orthomorphina havanensis (Cushman and Bermudez). Nodogenerina
     havanensis Cushman and Bermudez, 1937, Cushman Lab. Foram. Res. Contr.,
     v. 13, pt. 1, p. 14, pl. 1, figs. 47-48.

71b Orthomorphina jedlitschkal (Thalmann). Nodogenerina jedlitschkai
     Thalmann, 1937, Eclogae Geol. Helv. Lausanne, Suisse, vol. 30, p. 341,
     fig. 16.

72   Osangularia culter (Parker and Jones). Planorbulina farcta (Fichtel
     and Moll) var. ungeriana (d'Orbigny) subvar. culter Parker and
     Jones, 1865, Roy. Soc. London, Philos. Trans., v. 155, p. 382, pl. 19,
     fig. 1a-b.

___ Osangularia interupta Cushman, 1927, Cushman Lab. Foram. Res. Contr.,
     v. 3, p. 115, pl. 22, fig. 10.

73   Planulinagigas Keijzer. Planulina marialana Hadley var. gigas
     Keijzer, 1945, Utrecht, Univ. Georgr. Geol. Meded., Physiogr.-Geol.
     Recks, Utrecht, 1945, Ser. 1, n. 6, p. 206, pl. 5, fig. 77.

77   Pleurostomella acuminata Cushman, 1922, U.S. Nat. Mus. Bull. 104, p.
     50, pl. 19, fig. 6.

75   Pleurostomella alternans Schwager, 1866, Fossile Foraminiferen Von Kar
     Nikobar, Novara Exped. 1857-1859, Geol. Theil, Bd., Abt. 2, p. 238, pl.
     6, figs. 79-80.

___ Pleurostomella dominicana Bermudez. Pleurostomella schuberti
     Cushman and Harris var. dominicana Bermudez, 1949, kCushman Lab.
     Foram. Res. Spc. Publ., n. 25, p. 230, pl. 14, figs. 67-70.

74   Pleurostomella elliptica Galloway and Heminway, 1941, Sci. Surv. Puerto
     Rico, Virgin Islands, v. 111, pt. 4, p. 438, pl. 35, fig. 3.

___ Pleurostomella bierigi Palmer and Bermudez var. hebeta Cushman and
     Stainforth, 1945, Cushman Lab. Foram. Res. Spec. Publ. n. 13, p. 52, pl.
     8, fig. 16.

___ Pleurostomella alternanas Schwager var. hians Schubert, 1899, Lotos,
     N.F., Bd. 19, p. 223, p. 5, fig. 4a-b.

76   Pleurostomella naranjoensis Cushman and Bermudez, 1937, Cushman Lab.
     Foram. Res. Contr., v. 13, p. 16, pl. 1, figs. 59, 60.

79   Pleurostomella praegerontica Cushman and Stainforth, 1945, Cushman Lab.
     Foram. Res., Spec. Publ. n. 14, p. 52, pl. 8, figs. 13-14.

APPENDIX 4. LISTING OF SPECIES IDENTIFIED (continued)

78   Pleurostomella recens Dervieuz. Pleurostomella rapa gumbel var.
      recens Dervieux, 1899, Soc. Geol. Ital., Boll., v. 18, p. 76.

___   Pleurostomella schuberti Cushman and Harris, 1927, Cushman Lab. Foram.
      Res. Contr., v. 3, pt. 2, p. 1233, pl. 25, fig. 19a-b.

___   Pleurostomella subglobosa Rey, 1955, Soc. Geol. France, Bull., Ser. 6.
      v. 4, fasc. 4-6, p. 211, pl. 12, figs. 6a-c.

84   Pullenia alazanensis Cushman, 1927, J. Paleont. v. 1, p. 168, pl. 26,
      figs. 14, 15.

81   Pullenia angusta Cushman and Todd. Pullenia quinqueloba Reuss var.
      angusta Cushman and Todd, 1943, Cushman Lab. Foram. Res. Contr., v.
      19, pt. 1, p. 10, pl. 2, fig. 3.

80a   Pullenia bulloides (d'Orbigny). Nonionina bulloides d'Orbigny,
      1846, Foram. Foss. Bass. Tert. Vienne, p. 107, pl. 5, figs. 9, 10.

80b   Pullenia duplicata Stainforth, 1949, J. Paleonto., v. 23, no. 4, p.
      436.

83   Pullenia quadriloba Reuss. Pullenia compressiuseula Reuss
      quadriloba Reuss, 1867, Sitz Akad. Wiss. Wein, v. 55, pl. 3, fig. 8.

82   Pullenia quinqueloba (Reuss). Nonionina quinqueloba Reuss, 1851,
      Zeitschr. deutsch. geol. Ges., v. 3, p. 71, pl. 5, fig. 31.

101a Pyrgo depressa (d'Orbigny). Biloculina depressa d'Orbigny, 1826,
      Ann. Sci. Nat., v. 7, p. 298, modele 90.

101b Pyrgo murrhina (Schwager). Biloculina murrhina Schwager, 1866,
      Nova-Exped. 1857-59, Geol. Theil., v. 2, p. 303, pl. 4. figs. 15a,c.

___   Pyrulina extensa (Cushman). Polymorphina extensa Cushman, 1923, U.S.
      Nat. Mus. Bull. 104, pt. 4, p. 156, pl. 41, figs. 7,8.

___   Pyrulinoides cylindroides (Roemer) Polymorphina cylindroides Roemer,
      1938, Neves Jahrb. Min. Geogn. Geol. Petref.-Kunde, Stuttgart, Germany,
      p. 385, pl. 3, fig. 26a-b.

102a Quinqueloculina Spp.

102b Triloculina spp.

95   Rotorbinella lobatula (Parr). Discorbis lobatulus Parr, 1950,
      Foraminifera B.A.N.Z. Antarctic Research Exped. 1929-1931, Repts.,
      Adelaide, Ser. B., v. 5, pt. 6, p. 354.

103   Sigmoilina tenuis (Czjzek). Quinqueloculina tenuis Czjzek, 1848.
      Naturw. Abh. Wien, Bd. 2, Abth. 1, p. 149, pl. 13, figs. 31-34.

85   Siphogenerina pohana Finlay, 1939. Roy. Soc. New Zealand Trans., v. 69,
      p. 109, pl. 13, fig. 44, 45.

___   Siphotextularia rolshauseni Phleger and Parker, 1951, Geol. Soc. America
      Mem. 46, pt. 2, p. 4, pl. 1, figs. 23, 24.

86   Sphaeroidina bulloides d'Orbigny, 1826, Ann. Sci. Nat., Paris, France,
      Ser. 1, tome 7, p. 267, (Modeles, no. 65, 3me liveaison).

94   Stetsonia sp.

87   Stilostomella abyssorum (Brady). Nodosaria abyssorum Brady, 1881,
      Quat. Jour. Micr. Sci., v. 21, p. 31-71; Repts. Voy. Challenger 1873-1876,
      v. 9, p. 584, pl. 63, figs. 8, 9.

88   Stilostomella annulifera (Cushman and Bermudez). Elliposonodosaria
      annulifera Cushman and Bermudez, 1936, Cushman Lab. Foram. Res. Contr.,
      v. 12, pt. 3, p. 28, pl. 5, figs. 8-9.

89   Stilostomella curvatura (Cushman). Ellipsonodosaria curvatura Cushman
      1939, Cushman Lab. Foram. Res. Contr. v. 15, pt. 71, pl. 12, fig. 6

93   Stilostomella mappa (Cushman and Jarvis). Ellipsonodosari mappa
      Cushman and Jarvis, 1934, Cushman Lab. Foram. Res. Contr., v. 10, pt. 3,
      p. 73, pl. 10, fig. 8.

92   Stilostomella modesta (Bermudez). Ellipsonodosaria modesta Bermudez,
      1937, Mem. Soc. Cubana Hist. Nat., v. 11, p. 238, pl. 20, Fig. 3.

APPENDIX 4. LISTING OF SPECIES IDENTIFIED (continued)

____ Stilostomella recta (Palmer and Bermudez). Ellipsonodosaria Recta
Palmr and Bermudez, 1936, pl. 18, fig. 6, 7.

91   Stilostomella spinata (Cushman). Nodogenerina spinata Cushman, 1934,
Bernice P. Bishop. Mus., Bull., Honolulu, Hawaii, v. 119, P. 123, pl.
14, fig. 14.

90   Stilostomella subspinosa (Cushman). Ellipsonodosaria subspinosa
Cushman, 1943, Cushman Lab. Foram. Res. Contr., v. 19, p. 92, pl., 16,
fig. 7a-b.

____ Textularia pseudocollinsi (Cushman and Stainforth) Gaudryina
pseudocollinsi Cushman and Stainforth, 1945,Cushman Lab. Foram. Res. Spec.
Publ., v. 14, pl. 2, figs. 1-3.

99   Uvigerina hispido-costata Cushman and Todd, 1945, Cushman Lab. Foram.
Res., Spec. Publ., N. 15, p. 51, pl. 7, figs. 27, 31.

110 Uvigerina hispida Schwager,1866, Fossile Foraminiferen von Kar Nikobar.
Novara Exped. 1857-1859, Wien, Osterreich, Geol. Theil, Bd. 2, Abt. 2,
p. 249, pl. 7, fig. 95.

98   Uvigerina multicostata Le Roy, 1939. Natuurk. Tijdschr. Nederli-Indie,
Batavia, Java, 1939, dl. 99, afl. 6, p. 251, pl. 2, figs. 4-5.

____ Uvigerina interrupta Brady, 1884, Challenger Repts. Zool., n. 9,
p. 580, pl. 75, figs. 2-4.

97   Uvigerina proboscidea Schwager, 1866, Novara-Exped. Geol. Theil., v. 2,
p. 250, pl. 7, fig. 96.

96   Uvigerina vadescens(Cushman). Uvigerina proboscidea Schwager var.
vadescens Cushman, 1933, Cushman Lab. Foram. Res. Contr., v. 9, p. 85,
pl. 8, figs. 14, 15.

100 Vaginulina pseudoclavata Colom, 1943, Roy. Soc. Espanola Hist. Nat.,
Bol., Madrid, v. 41, p. 319, pl. 21, figs. 6-13, 16-17.

____ Vulvulina leuzengeri (Cushman and Renz). Textularia leuzengeri Cushman
and Renz, 1941, Cushman Lab. Foram. Res. Contr., v. 17, p. 3, pl. 1, fig. 2.

109 Vulvulina miocenica Cushman and Renz. Vulvulina spinosa Cushman var.
Miocenica Cushman, 1932, Cushman Lab. Foram. Res. Contr., v. 8, pt. 4,
n. 123, p. 80.

APPENDIX 5. CHANGES IN MIOCENE DEEP-SEA BENTHIC FORAMINIFERAL DISTRIBUTION IN
THE PACIFIC OCEAN: RELATIONSHIP TO PALEOCEANOGRAPHY

(See microfiche in pocket inside back cover)

APPENDIX 6. STATION FACTOR SCORES

(See microfiche in pocket inside back cover)

REFERENCES

Adams, C. G., Benson, R. H., Kidd, R. B., Ryan, W.B.F., and Wright, R. C.,
1977, The Messinian Salinity Crisis and evidence of late Miocene eustatic
changes in the World Ocean: Nature, 269; 383–386.

Andrews, J. E., Packham, G., Eade, J. V., Holdsworth, B. K., Jones, D. L., Klein,
G., Kroenke, L. W., Saito, T., Shafik, S., Stoeser, D. B., van der Lingen,
G. J., 1975, Site 289: Initial Reports of the Deep Sea Drilling Project, 30,
231–398.

Arrhenius, G. O., 1952, Sediment cores from the East Pacific, *in* Petterson, H.,
ed., Reports of the Swedish Deep-Sea Expedition, 5 (fasc. 1), 117 p.

Bandy, O. L., 1960, International Geology Congress, 21st, Copenhagen 1960,
Report pt. 22, p. 7–18.

Bandy, O. L., 1971, Recognition of Upper Miocene Neogene Zone 18, Experimental Mohole, Guadalupe Site: Nature, 233, 476–478.

Barker, R. W., 1960, Taxonomic notes on the species figures by H. B. Brady in his
report on the foraminifera dredged by H.M.S. Challenger during the years
1873–1876: Society Economic Paleontologists and Mineralogists Spec. Publication 9, 1–238.

Barrera, E., Keller, G., Savin, S. M., this volume, Evolution of the Miocene Ocean
in the Eastern North Pacific as inferred from oxygen and carbon isotope
ratios of foraminifera.

Barron, J. A., 1980, Lower Miocene to Quaternary diatom biostratigraphy of Leg
57 off Northeastern Japan, Deep Sea Drilling Project: Initial Reports of the
Deep Sea Drilling Project, 57, 507–538.

Barron, J. A., Keller, G., Dunn, D. A., This volume. A multiple microfossil
biochronology for the Miocene.

Beckmann, J., 1953, Die Foraminifern der oceanic Formation (Eocaen-
Oligocenaen) von Barbadoes: Eclogae Geologische Helvetia, 46, 301–412.

Belford, D. J., 1966, Miocene and Pliocene smaller foraminifera from Papua and

New Guinea: Department of National Development Bureau of Mineral Resources Bull., 79, 1–306.

Bender, M. L., and Keigwin, L. D. Jr., 1979, Speculations about the Upper Miocene change in abyssal Pacific dissolved bicarbonate, ^{13}C: Earth and Planetary Science Letters, 45, 383–393.

Berger, W. H., 1970, Biogenous deep-sea sediments: Fractionation by deep-sea circulation: Bulletin of Geological Society of America, 81, 1385–1402.

Berger, W. H., 1973, Cenozoic sedimentation in the eastern tropical Pacific: Bulletin Geological Society of America, 84, 1941–1954.

Berger, W. H., and Johnson, T. C., 1976, Deep-sea carbonates: Dissolution and mass wasting on Ontong-Java Plateau: Science, 192, 785–787.

Berger, W. H., and Winterer, E. L., 1974, Plate stratigraphy and the fluctuating carbonate lines, *in* Hsu, K. J., and Jenkins, H. C., eds., Pelagic sediments on land and under the sea: International Association Sedimentologists Special Publication, 1, 11–48.

Berggren, W. A., 1972, Cenozoic biostratigraphy and paleogeography of the North Atlantic: Initial Reports of the Deep Sea Drilling Project, 12, 965–1001.

Bermudez, P. J., 1949, Tertiary smaller foraminifera of the Dominican Republic: Cushman Laboratory for Foraminiferal Research Special Publication 25, 1–322.

Birch, G. F., 1979, Phosphatic rocks on the western margin of South Africa: Journal of Sedimentary Petrology, 49, 93–110.

Blanc, P-L., and Duplessy, J-C., 1982, The deep-water circulation during the Neogene and the impact of the messinian salinity crisis: Deep-Sea Research, 29 (12A), 1391–1414.

Blanc, P-L., Rabussier, D., Vergnand-Grazzini, C., Duplessy, J-C., 1980, North Atlantic deep water formed by the late middle Miocene: Nature, 283, 553–555.

Blasco, D., Estrada, M., Jones, B. H., 1980, Relationship between phytoplankton distribution and composition and the hydrography in the Northwest African upwelling region near Cabo Corbeiro: Deep-Sea Research, 27 (10A), 799–823.

Boltovskoy, E., 1978, Late Cenozoic foraminifera of the Ninetyeast Ridge (Indian Ocean): Marine Geology, 26, 139–175.

Boltovskoy, E., 1980, Benthonic foraminifera of the bathyal zone from Oligocene through Quaternary: Revista Espanola de Micropaleontologie 12, 283–304.

Brady, H. B., 1884, Report on the foraminifera dredged by H.M.S. Challenger, during the Years 1873–1876, *in* Reports of the Scientific Results of the Voyage of H.M.S. Challenger, London, England, v. 9, 1–814.

Brady, H. B., Parker, W. K., Jones, T. R., 1888, On some foraminifera from Abroholos Bank: Zoological Society London, Transactions, 12 (7), 211–239.

Bremer, M. L., and Lohmann, G. P., 1982, Evidence for primary control of the distribution of certain Atlantic Ocean benthonic foraminifera by degree of carbonate saturation: Deep-Sea Research 29 (8A), 987–998.

Brewster, N. A., 1980, Cenozoic biogenic silica sedimentation in the Antarctic Ocean, based on two Deep Sea Drilling Sites: Geological Society of America Bulletin, 91, 337–347.

Burke, S. C., 1981, Recent benthic foraminifera of the Ontong-Java Plateau: Journal Foraminiferal Research, 11, 1–19.

Ciesielski, P. F., and Weaver, F. M., 1983, Neogene and Quaternary paleo-environmental history of Deep Sea Drilling Project Leg 71 sediments, Southwest Atlantic Ocean: Initial Reports of the Deep Sea Drilling Project, 71, 461–477.

Corliss, B. H., 1979, Recent deep-sea benthonic foraminiferal distributions in the Southwest Indian Ocean: Inferred bottom-water routes and ecological implications: Marine Geology 31, 115–138.

Corliss, B. H., 1982, Linkage of North Atlantic and Southern Ocean deep-water circulation during glacial intervals: Nature, 298(5873), 458–460.

Corliss, B. H., and Honjo, S., 1981, Dissolution of deep sea benthonic foraminifera: Micropaleontology, 27(4), 356–378.

Cushman, J. A., and Parker, F. L., 1947, Bulimina and related foraminiferal genera: United States Geological Society Professional Paper 210-D, 1–176.

Cushman, J. A., and Stainforth, R. M., 1945, The foraminifera of the Cipero Marl Formation of Trinidad, B.W.I. Cushman Laboratory for Foraminiferal Research Special Publication 14, 1–74.

Donnelly, T. W., 1982, Worldwide continental denudation and climatic deterioration during the Late Tertiary: Evidence from deep-sea sediments: Geology, 10, 451–454.

Douglas, R. G., 1973, Benthonic foraminiferal biostratigraphy in the Central North Pacific, Leg 17, Deep Sea Drilling Project, Initial Reports of the Deep Sea Drilling Project, 17, 607–671.

Douglas, R. G., 1981, Paleoecology of continental margin basins: A modern case history from the borderland of Southern California, *in* Douglas, R. G., Colburn, I. P., and Gorsline, D. S., eds., Depositional systems of active continental margin basins: Short Course Notes, Society Economic Paleontologists and Mineralogists (Pacific Section), Los Angeles, California, 121–156.

Douglas, R. G. and Woodruff, F., 1981, Deep sea benthic foraminifera, *in* C. Emiliani, ed., The oceanic lithosphere, The Sea, v. 7: Wiley-Interscience, N.Y., 1233–1327.

Edwards, A. R., 1975, Southwest Pacific Cenozoic paleoceanography and an integrated Neogene paleocirculation model: Initial Reports of the Deep Sea Drilling Project, 30, 667–684.

Finlay, H. J., 1939a, New Zealand foraminifera: Key Species in Stratigraphy-N.1: Royal Society New Zealand, Transactions Proceedings 68, pt. 4, 504–569.

Finlay, H. J., 1939b, New Zealand foraminifera: Key species in stratigraphy-N.2: Royal Society New Zealand, Transactions Proceedings, 69, pt. 1, 89–128.

Finley, H. J., 1939c, New Zealand foraminifera: Key species in stratigraphy-N.3: Royal Society New Zealand, Transactions Proceedings, 69, pt. 3, 309–329.

Finley, H. J., 1940, New Zealand foraminifera: Key species in stratigraphy-N.4: Royal Society New Zealand, Transactions Proceedings, 69, pt. 4, 448–472.

Galloway, J. J., and Heminway, C. E., 1941, The Tertiary foraminifera of Puerto Rico: Scientific Survey Puerto Rico, Virgin Islands, 3, p. 4, 275–491.

Galloway, J. J., and Morrey, M., 1929, A Lower Tertiary foraminiferal fauna from Manta Ecuador: Bulletin American Paleontologists, 15(55), 1–57.

Gilbert, D., and Summerhayes, C. P., 1981, Distribution of organic matter in sediments along California Continental Margins: Initial Reports of the Deep Sea Drilling Project, 63, 757–761.

Gofas, S., 1978, Une approche du paleoenvironnement oceanique: les foraminiferes benthiques, calcaires, traceurs de la circulation abyssale [Doctoral Dissertation]: l'Universite de Bretagne Occidentale, Brest, France, 1–149 p.

Graham, D. W., Corliss, B. H., Bender, M. L., Keigwin, L. D., Jr., 1981, Carbon and Oxygen Disequilibria of Recent Deep-Sea Benthic Foraminifera: Marine Micropaleontology, 6, 483–497.

Hamilton, W., 1972, Hallett volcanic province, Antarctica: United States Geological Survey, Professional Paper 456-C, 1–62.

Haq, B. V., Worsley, J. R., Burckle, L. H., Douglas, R. G., Keigwin, L. D., Opdyke, N. D., Savin, S. M., Sommer, M. A., Vincent, E., and Woodruff, F., 1980, The Late Miocene marine carbon-isotopic shift and the synchroneity of some phytoplanktonic biostratigraphic datums: Geology, 8, 427–431.

Heath, G. R., Moore, T. C., Dauphin, J. P., 1977, Organic carbon in deep sea sediments of fossil fuel CO_2 in the oceans, proceedings from Symposium, Anderson, N. R. and Malahoff, A., Eds.: Plenum Press, N.Y., p. 605–626.

Hessler, R. P. and Jumars, P., 1974, Abyssal community analysis from replicate box cores in the Central North Pacific: Deep-Sea Research, 21, 185–209.

Hodell, D. A., and Kennett, J. P., 1982, Abyssal circulation in the Western South Atlantic during the late Miocene and Pliocene: Geological Society of America Abstract, New Orleans meeting, p. 516.

Ingle, J. C., 1973, Summary comments on Neogene biostratigraphy, physical stratigraphy and paleo-oceanography in the marginal Northeastern Pacific Oceans, Initial Reports of the Deep Sea Drilling Project, 18, 949–960.

Ingle, James C., 1981, Origin of Neogene diatomites around the North Pacific Rim, *in* Garrison, R. E., Douglas, R. G., Pisciotto, K. E., Isaacs, C. M., and Ingle, J. C., eds., The Monterey Formation and related siliceous rocks of California: Society of Economic Paleontologists and Mineralogists, Special Publication, 159–180.

Janecek, T. R., and Rea, D. K., 1983, Eolian deposition in the Northeast Pacific Ocean: Cenozoic history of atmospheric circulation: Geological Society of America Bulletin, 94, 730–738.

Keigwin, L. D., Jr., 1979, Late Cenozoic stable isotope stratigraphy and paleoceanography of DSDP Sites from the east equatorial and central north Pacific Ocean: Earth and Planetary Science Letters, 45, 361–382.

Keller, G., 1980a, Early to middle Miocene planktonic foraminiferal datum levels of the equatorial and subtropical Pacific: Micropaleontology, 26, 372–391.

Keller, G., 1980b, Middle to late Miocene planktonic foraminiferal datum levels and paleoceanography of the North and Southeastern Pacific Ocean: Marine Micropaleontology, 249–281.

Keller, G., 1981a, Miocene biochronology and paleoceanography of the North Pacific: Marine Micropaleontology, 6, 535–551.

Keller, G., 1981b, Planktonic foraminiferal faunas of the Equatorial Pacific suggest early Miocene origin of present ocean circulation: Marine Micropaleontology, 6, 269–295.

Keller, G., and Barron, J. A., 1981, Integrated planktic foraminiferal and diatom biochronology for the Northeast Pacific and the Monterey Formation, *in* Garrison, R. E., Douglas, R. G., Pisciotto, K. E., Isaacs, C. M., and Ingle, J. C., eds., The Monterey Formation and related siliceous rocks of California: Society Economic Paleontologists and Mineralogists Special Publication, 43–54.

Keller, G., and Barron, J. A., 1983, Paleoceanographic implications of Miocene deep-sea hiatuses: Geological Society America Bulletin, 94, 590–613.

Keller, G., Barron, J. A., Burckle, L., 1982, North Pacific late Miocene correlations using microfossils, stable isotopes, percent CaCO$_3$ and magnetostratigraphy: Marine Micropaleontology 7, 327–357.

Kennett, J. P., 1977, Cenozoic evolution of Antarctic glaciations, the Circum-Antarctic Ocean and their impact on global paleoceanography: Journal of Geophysical Research, 82, 3843–3860.

Kennett, J. P., 1982, Marine geology: Prentice-Hall, Inc., N.J., 813 p.

Kroopnick, 1974, Dissolved O$_2$-CO$_2$-^{13}C system in the Equatorial Pacific: Deep-Sea Research, 21, 211–217.

Legendre, L., and Legendre, P., 1979, Ecologic numerique: Masson, Paris, 254 pp.

Leinen, M., 1979, Biogenic silica accumulation in the central equatorial Pacific and its implications for Cenozoic paleoceanography: Summary: Geological Society America Bulletin 90: 801–803.

Loeblich, A. R., and Tappan, H., 1964, Sarcodina, chiefly "Thecamoebians" and Foraminiferida, Treatise on Invertebrate Paleontology, Part C, Protista 2, R. C. Moore, Ed.: Geological Society of America and University of Kansas Press, 1-2, 1–900.

Lohmann, G. P., 1978, Abyssal Benthonic Foraminifera as Hydrographic Indicators in the Western South Atlantic Ocean: Journal of Foraminiferal Research, v. 8, (1), 6–34.

Lohmann, G. P., and Carlson, J. J., 1981, Oceanographic significance of Pacific late Miocene calcareous nannoplankton: Marine Micropaleontology, 6, 553–579.

Loutit, T. S., 1981, Late Miocene paleoclimatology: Subantarctic water mass, Southwest Pacific: Marine Micropaleontology, 6, 1–27.

Lutze, G. F., 1978, Neogene benthonic foraminifera from Site 369, Leg 41, Deep Sea Drilling Projects: Initial Reports of the Deep Sea Drilling Project, 41, 659–666.

Lutze, G. F., 1979, Benthic foraminifers at Site 397: Faunal fluctuations and ranges in the Quaternary: Initial Reports of the Deep Sea Drilling Project, 42, 1, 419–431.

Lutze, G. F., and Coulbourn, W. T., 1984, Recent benthic foraminifera from the continental margin of Northeast Africa: Community Structure and Distribution: Marine Micropaleontology, 8, 361–401.

Miller, K. G., and Lohmann, G. P., 1982, Environmental distribution of Recent benthic foraminifera on the Northeast United States continental slope: Geological Society of America Bulletin 93: 200–206.

Miller, K. G., Curry, W. B., Ostermann, D. R., 1985, Late Paleogene (Eocene to Oligocene) benthic foraminiferal paleoceanography of the Goban Spur Region, DSDP Leg 80; Initial Reports of the Deep Sea Drilling Project, 80.

Moore, T. C., Jr., van Andel, T. H., Sancetta, C., Pisias, N., 1978, Cenozoic hiatuses in pelagic sediments: Micropaleontology, 24(2), 113–138.

Nuttall, W.L.F., 1932, Lower Oligocene foraminifera from Mexico: Journal of Paleontology, 6, 3–35.

Parker, F. L., 1964, Foraminifera from the experimental Mohole Drilling near Guadalupe Island, Mexico: Journal of Paleontology, 38(4), 617–636.

Parker, F. L. and Berger, W. H., 1971, Faunal and solution patterns of planktonic foraminifera in surface sediments of the South Pacific: Deep-Sea Research, 18, 73–107.

Paul, A., and Menzies, R., 1973, Benthic ecology of the high arctic deep-sea, ONR Final Report from Contract NO. 0014-67-A-0235-0005.

Pederson, T. F., 1983, Increased productivity in the eastern equatorial Pacific during the last glacial maximum (19,000 to 14,000 yr. B.P.): Geology, 11, 16–19.

Phleger, F. B., Parker, F. L., Peirson, J. F., 1953, North Atlantic foraminifera repts.: Swedish Deep-Sea Expedition, 7, 1, 3–121.

Quinn, J., 1982, Early Miocene rates of faunal change, and foraminiferal preservational facies, Equatorial Pacific Ocean [Unpubl. M.S. Thesis]: Univ. Southern California, Los Angeles, California, 98 p.

Resig, J., 1976, Benthic foraminiferal stratigraphy, Eastern Margin, Nazca Plate: *in* Initial Reports of the Deep Sea Drilling Project, 37, 743–760.

Resig, J. M., 1981, Biogeography of benthic foraminifera of the Northern Nazca Plate and adjacent Continental Margin: Geological Society of America Memoir, 154, 619–665.

Romine, K., This volume, Evolution of Pacific circulation in the Miocene: Radiolarian Evidence from Site 289.

Savage, M. L., and Ciesielski, P. F., 1983, A revised history of glacial sedimentation in the Ross Sea region, *in* Antarctic earth science, Oliver, R. L., James, P. R., Jago, J. B., eds., Australian Academy of Science, Canberra, 555–559.

Savin, S. M., 1977, The history of the earth's surface temperature during the past 100 million years: Annual Review Earth and Planetary Science, 5, 319–355.

Savin, S. M., Douglas, R. G., Keller, G., Killingley, J. S., Shaughnessy, L., Sommer, M. A., Vincent, E. and Woodruff, F., 1981, Miocene benthic foraminiferal isotope record: A synthesis: Marine Micropaleontology 6, 423–450.

Savin, S. M., Douglas, R. G., and Stehli, F. G., 1975, Tertiary marine paleotemperatures: Geological Society America Bulletin, 86, 1499–1510.

Schlanger, S. O., and Douglas, R. G., 1974, The Pelagic ooze-chalk-limestone transition and its implications for marine stratigraphy: Special Publications International Association Sedimentologists, 1, 117–148.

Schnitker, D., 1974, West Atlantic abyssal circulation during the past 120,000 years: Nature, 248, 385–387.

Schnitker, D., 1979, Cenozoic deep water benthonic foraminifers, Bay of Biscay: Initial Reports of the Deep Sea Drilling Project, 48, 377–413.

Schnitker, D., 1980, Quaternary deep-sea benthic foraminifers and bottom water masses: Annual Review Earth and Planetary Science, 48, 343–370.

Sclater, J. G., Anderson, R. N., and Bell, M. L., 1971, Elevation of ridges and evolution of the central eastern Pacific: Journal Geophysical Research, 76, 7888–7893.

Sclater, J. G., Meinke, L., Bennett, A., and Murphy, C., this volume, The depth of the ocean through the Neogene.

Shackleton, N. J., 1982, The deep-sea record of climate variability: Progress Oceanography, 11, 199–218.

Shackleton, N. J., and Kennett, J. P., 1975a, Late Cenozoic oxygen and carbon isotopic changes at DSDP Site 284: Implications for glacial history of the northern hemisphere: Initial Reports of the Deep Sea Drilling Project, 29, 801–807.

Shackleton, N. J., and Kennett, J. P., 1975b, Paleotemperature history of the Cenozoic and the initiation of Antarctic glaciation: oxygen and carbon isotope analyses in DSDP Sites 277, 279, and 281: Initial Reports of the Deep Sea Drilling Project, 29, 743–755.

Siesser, W. G., 1980, Late Miocene origin of the upwelling system off Northern Namibia: Science, 208, 283–285.

Srinivasan, M. S. and Kennett, J. P., 1981, A review of Neogene planktonic foraminiferal biostratigraphy: Applications in the Equatorial and South Pacific, *in* Warme, J. E., Douglas, R. G., and Winterer, E. L., eds., A decade of ocean drilling: Society Economic Paleontologists and Mineralogists Special Publication 32, 395–432.

Streeter, S. S., 1972, Living benthonic foraminifera of the Gulf of California, a factor analysis of Phleger's (1964) data: Micropaleontology, 18(1), 64–73.

Streeter, S. S., 1973, Bottom water and benthonic foraminifera in the North Atlantic—Glacial-Interglacial Contrasts: Quaternary Research, 3, 131–141.

Streeter, S. S., and Shackleton, N. J., 1979, Paleocirculation of the deep North Atlantic: 150,000 year round record of benthic foraminifera and 0-18: Science, 203, 168–170.

Summerhayes, C. P., 1981, Oceanographic controls on organic matter in the Miocene Monterey Formation, Offshore California, *in* Garrison, R. E., Douglas, R. G., Pisciotto, K. E., Isaacs, C. M., and Ingle, J. C., The Monterey Formation and related siliceous rocks of California: Society Economic Paleontologists and Mineralogists (Pacific Sec.), Special Publication, Los Angeles, California 213–220.

Thompson, P. R., 1982, Foraminifers of the Middle America Trench: Initial Reports of the Deep Sea Drilling Project, 67, 351–380.

Thierstein, H. R., 1979, Paleoceanographic implications of organic carbon and carbonate distribution in Mesozoic deep sea sediments, *in* Talwani, M., Hay, W., and Ryan, W.B.F., eds., Deep sea drilling results in the Atlantic Ocean: Continental margins and paleoenvironment: Maurice Ewing Series 3, American AGeophysical Union, (Washington, D.C.), 249–274.

Tjalsma, R. C., and Lohmann, G. P., 1983, Paleocene-Eocene bathyal and abyssal benthic foraminifera from the Atlantic Ocean: Micropaleontology, Special Publication 4, 1–90.

Vail, P. R., and Hardenbol, J., 1979, Sea-level changes during the Tertiary: Oceanus, 22, 71–79.

Valencia, M. J., 1973, Calcium carbonate and gross-size analysis of surface sediments, Western Equatorial Pacific: Pacific Science, 27, 290–303.

van Andel, T. H., Heath, G. R., Moore, T. C. Jr., 1975, Cenozoic history and paleoceanography of the Central Equatorial Pacific Ocean: Geological Society of America Memoir, 14, 1–134.

Van Couvering, J. A., Berggren, W. A., Drake, R. E., Aguirre, E., and Curtis, G. H., 1976, The terminal Miocene event: Marine Micropaleontology, 1, 263–286.

Van Gorsel, J. T., and Troelstra, S. R., 1981, Late Neogene planktonic foraminiferal biostratigraphy and climatostratigraphy of the Solo River Section (Java, Indonesia): Marine Micropaleontology, 6, 183–209.

Vincent, E., 1981, Neogene carbonate stratigraphy of Hess Rise (Central North Pacific) and paleoceanographic implications: Initial Reports of the Deep Sea Drilling Project, 62, 571–606.

Vincent, E., Killingley, J. S. and Berger, W. H., 1980, The magnetic Epoch-6 carbon shift: Change in the Oceans C^{13}/C^{12} ratio 6.2 million years ago: Marine Micropaleontology 5, 185–203.

Walch, C., 1978, Recent abyssal benthic foraminifera from the Eastern Equatorial Pacific [Unpubl. M.S. Thesis]: Univ. So. Calif., Los Angeles, California, 117 p.

Weyl, R.,1980, Geology of Central America: Berlin, Gebruder Born-Traeger, 371 p.

Woodruff, F., and Douglas, R. G., 1981, Response of deep-sea benthic foraminifera to Miocene paleoclimatic events: DSDP Site 289: Marine Micropaleontology, 6, 617–632.

Woodruff, F., and Savin, S. M., 1985, $\delta^{13}C$ values of Miocene Pacific benthic foraminifera: Correlations with sea level and biological productivity: Geology, 13, 119–122.

Woodruff, F., Savin, S. M., and Douglas, R. G. , 1981, Miocene stable isotope record: A detailed deep Pacific Ocean study and its paleoclimatic implications: Science, 212, 665–668.

Wyrtki, K., 1962, The subsurface water masses in the western South Pacific Ocean: Australian Journal of Marine and Freshwater Research, 131, 18–47.

Wyrtki, K., 1966, Oceanography of the Eastern Equatorial Pacific Ocean: Oceanogr. Marine Biology Annual Review, 4, 33–68.

MANUSCRIPT ACCEPTED BY THE SOCIETY DECEMBER 17, 1984

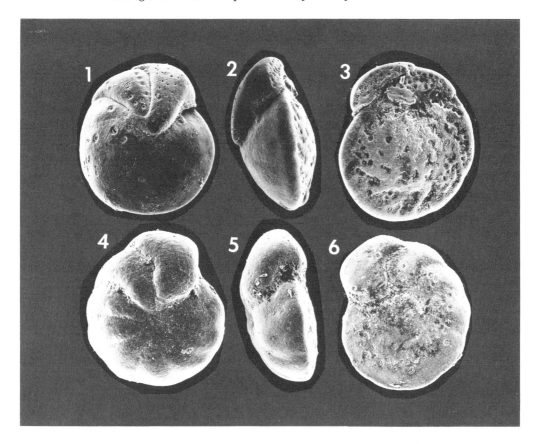

PLATE 1

1-3   *Cibicidoides cenop* Woodruff, n. sp. holotype.
        DSDP Site 289-37-4-93 cm; middle Miocene, Planktonic Foraminiferal Zone
        N14; (×78); 1 dorsal views; 2 side views; 3 ventral view; medium size. This species
        is found throughout the deep Pacific Ocean after 15 Ma.

4-6   *Cibicidoides* sp cf C. *pseudoungerianus* (Cushman.
        DSDP Site 289-37-4-93 cm; middle Miocene, Planktonic Foraminiferal Zone
        N14; (×82); 1 dorsal view; 2 side view; 3 ventral view. This species is most
        common in the Pacific bottom water corridor throughout Miocene time.

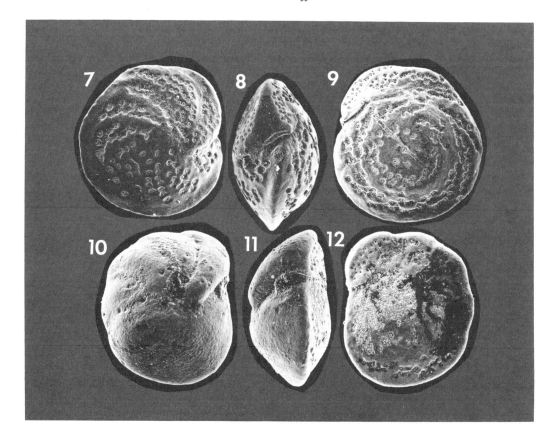

PLATE 2

7-9   *Cibicidoides cenop* Woodruff, n. sp. paratype, large specimen.
        DSDP Site 289-37-4-93 cm; middle Miocene, Planktonic Foraminiferal Zone
        N14; (×55); 1 dorsal view; 2 side view; 3 ventral view. This is a large specimen and
        shows the effects of dissolution causing enlarging and coalescing of pores.

10-12   *Cibicidoides cenop* Woodruff n. sp. paratype, small specimen.
        DSDP Site 289-37-4-93 cm; middle Miocene, Planktonic Foraminiferal Zone
        N14; (×85); 1 dorsal view; 2 side views; 3 ventral views. This is a small specimen
        showing that the dorsal and ventral pores are rare and difficult to see in the small
        specimens.

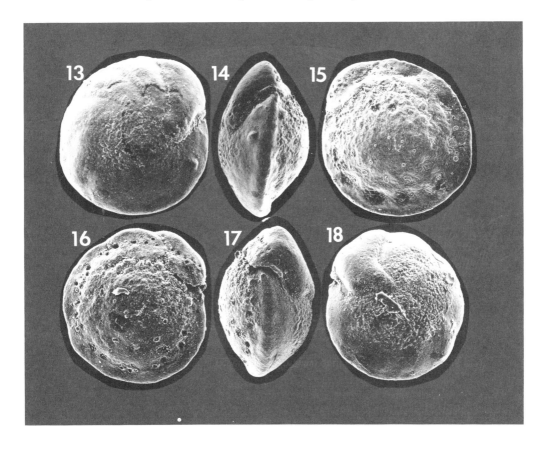

PLATE 3

13-15   *Cibicidoides kullenbergi* (Parker); large specimen.
         DSDP Site 289-37-4-93 cm; middle Miocene, Planktonic Foraminiferal Zone
         N14; (×70); 1 dorsal view; 2 side views, 3 ventral views. This species is not
         common in the deep Pacific Ocean, below 2500 m, after 14 Ma.

16-18   *Cibicidoides kullenbergi* (Parker); small specimen.
         DSDP Site 289-37-4-93 cm cm; middle Miocene, Planktonic Foraminiferal Zone
         N14; (×100); 1 ventral view, 2 side views, 3 dorsal views.

Printed in U.S.A.

Geological Society of America
Memoir 163
1985

Depth stratification of planktonic foraminifers in the Miocene ocean

Gerta Keller*
U.S. Geological Survey
345 Middlefield Road
Menlo Park, California 94025
and
Department of Geology
Stanford University
Stanford, California 94305

ABSTRACT

A depth stratification of planktonic foraminifers based on oxygen isotopic ranking is proposed for the Miocene. Species are grouped into surface, intermediate, and deep dwellers based upon oxygen isotopic composition of individual species. The depth stratification is applied to planktonic foraminiferal populations in three Miocene time-slices (21 Ma, 16 Ma, and 8 Ma) in the equatorial, north, west, and east Pacific. The late Miocene time-slice is compared with modern Pacific GEOSECS transect water-mass profiles of temperature and salinity in order to illustrate the similarities between the depth ranking of planktonic foraminifers and temperature and salinity conditions. The geographic distribution of inferred surface, intermediate, and deep water dwellers was found to be very similar to modern temperature profiles: surface dwellers appear to be associated with warmest temperatures ($>20^{\circ}$C), upper intermediate water dwellers with temperatures between 10 and 20°C, and lower intermediate and deep water dwellers with temperatures below 10°C. Tropical high-salinity water appears to be associated with the upper intermediate *Globorotalia menardii* group in the modern ocean.

Depth stratification applied to two Miocene time-series analyses in the equatorial Pacific (Sites 77B and 289) indicates increased vertical and latitudinal provincialism between early, middle, and late Miocene time. The early and middle Miocene equatorial Pacific was dominated by the warm surface water group, which shows distinct east-west provincialism. This provincialism is interpreted as the periodic strengthening of the equatorial surface circulation during polar cooling phases. During the late Miocene the upper intermediate group increased and the surface group declines. At the same time the east-west provincialism disappeared. This faunal change may have been associated with the major Antarctic glaciation and resultant strengthening of the general gyral circulation and the strengthening of the Equatorial Countercurrent due to the closing of the Indonesian Seaway at that time.

*Present address: Department of Geological and Geophysical Science, Princeton University, Princeton, New Jersey 08544

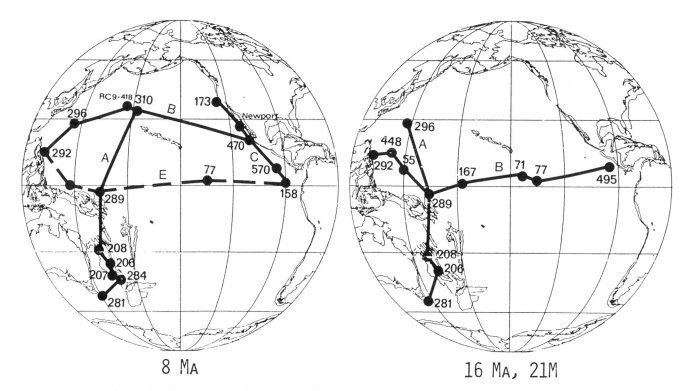

Figure 1. a. Miocene data points in the Pacific Ocean. Figure 1a, late Miocene (8 Ma) data points: (A) West Pacific, (B) North Pacific, (C) Northeast Pacific, (D) Equatorial Pacific. b. Early Miocene data points (21 Ma, 16 Ma): (A) West Pacific, (B) Equatorial Pacific.

INTRODUCTION

The stable isotope composition of planktonic foraminifers has become an increasingly valuable tool for paleoclimatic and paleoceanographic studies. Many workers have reported a good agreement between depth stratification of species from plankton tows (Williams et al. 1979; Bé et al. 1971) and inferred depth from oxygen isotopic data although some species appear to be in disequilibrium (Emiliani 1954; Shackleton and Vincent 1978; Douglas and Savin 1978; Boersma and Shackleton 1978; Williams and Williams 1980; Vincent et al. 1981; Keller 1983; Savin et al. 1985; Poore and Matthews 1984, a,b). Although the species-specific differences in the $^{18}O/^{16}O$ ratio of the calcite shells of planktonic foraminifers from deep-sea cores is still debated, the calcite shells of many ecologically important species appear to calcify in equilibrium with seawater (Williams and Williams 1980; Fairbanks et al. 1980, 1982). Fairbanks et al. (1980) noted that non-spinose species appear to calcify in oxygen isotope equilibrium whereas spinose species appear to calcify out of equilibrium by –0.3 to –0.4 $^o/_{oo}$ in $\delta^{18}O$ values. They also observed that in the Gulf Stream species appear to calcify in significantly narrower depth zones than their actual vertical distribution would imply. Specifically, they found that species appear to calcify in the photic zone, or upper 100 m of the water column, and that they are stratified within this zone. This implies that species calcify at specific temperature and density conditions within the photic zone and then migrate to greater depths, as

opposed to calcifying within the upper 400 m of the water column as previously assumed. Poore and Matthews (1984b) suggest that the two hypotheses can be reconciled, if we consider that in most oceanographic situations the temperature and salinity conditions of the upper 400 m of the water column can be found at times in the photic zone. In either case, depth stratified planktonic foraminiferal data should provide information of the temperature conditions in the surface water-masses.

The purpose of this report is to discuss and illustrate the Miocene quantitative foraminiferal data of three time-slice intervals (21 Ma, 16 Ma, and 8 Ma) and two time-series analyses (Sites 77B and 289) in terms of (1) groups inferred to dwell in relative surface, intermediate, and deep waters based on oxygen isotope ranking of individual species, (2) the relationships of planktonic foraminifers to water-mass properties such as temperature and salinity, based on comparison of late Miocene faunal transects with modern water-mass profiles of the Pacific GEO-SECS transect profiles (Craig et al. 1981) and (3) water-mass stratification changes in the Miocene ocean and inferred paleoclimatic changes.

Four late Miocene (8 Ma) faunal transects have been chosen to illustrate faunal depth stratification in various regions of the Pacific: West Pacific, North Pacific, Equatorial Pacific, and northeast Pacific (California Current province) (Figure 1). In addition, West Pacific and Equatorial Pacific data points are also

TABLE 1. DEPTH RANKING OF PLANKTONIC FORAMINIFERS IN THREE MIOCENE TIME-SLICES
AT 21 Ma, 16 Ma, AND 8 Ma*

	8 Ma	16 Ma	21 Ma
Surface	Globigerinoides mixed spp. G. trilobus-G. sacculifer Globoquadrina altispira +Pulleniatina obliquiloculata *Orbulina universa	Globigerinoides mixed spp. G. trilobus-G. sacculifer G. subquadratus Gl. siakensis-G. mayeri Globoquadrina altispira	Globorotalia kugleri Globigerinoides mixed spp. G. trilobus-G. sacculifer Gl. siakensis-Gl. mayeri Globigerina angustiumbilicata Globoquadrina altispira
Upper Intermediate	Globorotalia menardii group Sphaeroidinella seminulina +Globigerina nephenthes-G. drury +Globorotalia continuosa +Globorotalia acostaensis +Globoquadrina dehiscens	Globorotalia peripheroronda Gl. fohsi group +Globorotalia continuosa *Globoquadrina dehiscens	*Globoquadrina dehiscens
Lower Intermediate	Globorotalia conoidea +Globigerina woodi Globigerina bulloides Neogloboquadrina pachyderma	Globorotalia miozea +Globigerina woodi Globigerina bulloides	Globorotalia miozea +Globigerina woodi Globigerina bulloides
Deep	Globoquadrina venezuelana	Globoquadrina venezuelana	Globoquadrina venezuelana Catapsydrax ssp. Globoquadrina tripartita

*Note: Taxa are grouped into surface, upper and lower intermediate, and deep dwellers based on $\delta^{18}O$ ranking. Asterisk marks species with variable $\delta^{18}O$ values. Species needing further study to rank them confidently are marked with +.

illustrated for late early Miocene (16 Ma) and early Miocene (21 Ma) time-slices in order to illustrate changes in depth stratification and relative abundances of species during the Miocene (Figure 1). Changes in Miocene depth stratification are also illustrated in two time-series analyses (Sites 77B and 289) and climatic oscillations are interpreted from abundance fluctuations of temperature sensitive species. The three time-slice intervals and age determinations are discussed in Barron et al. (1985).

DATA ANALYSIS

The relative depth stratification of planktonic foraminifers is based on isotopic analyses of various species in three Miocene time-slices at 21 Ma, 16 Ma, and 8 Ma. Isotopic ranking of species is based on Deep Sea Drilling Project (DSDP) sites with multiple species analyses for each time-slice interval. Isotopic data are published in Tables 1–4 of Savin et al. (1985). Quantitative data on Miocene planktonic foraminifers have been generated by CENOP workers and are published in Keller (1980a, b; 1981a, b), Srinivasan and Kennett (1981a, b; Kennett and Srinivasan, 1985), Barron and Keller (1983), and Barrera et al. (1985). Relative abundance data of species is based on representative sample splits (using an Otto microsplitter) of 300–500 specimens per sample in the size fraction greater than 150 microns. Three to eight samples were analyzed for each time-slice interval in each deep-sea core, and samples were averaged to obtain a representative faunal assemblage.

ISOTOPIC RANKING

To establish isotopic depth ranking of taxa, oxygen isotope

values have been plotted against latitudes for each site analyzed in the three time-slice intervals (see Figures 2–4). Consistency in the relative isotopic ranking among the various taxa is interpreted to reflect depth ranking in the water column. Table 1 suggests the relative depth ranking of Miocene species for each time-slice, and Plates 1 and 2 illustrate the depth ranking of species in the early and middle to late Miocene. No absolute oxygen isotope values are proposed in Table 1 because isotopic values change across latitudes and are related to water-mass conditions at each locality.

Late Miocene: 8 Ma

A variety of planktonic foraminiferal taxa have been analyzed in the late Miocene time-slice interval in 18 DSDP sites across latitudes, as illustrated in Figure 2. In low to middle latitudes, *Globoquadrina venezuelana* consistently has the heaviest oxygen isotope values and is interpreted as a deep dweller living near the lower part of the thermocline.

Globigerinoides trilobus-G. sacculifer appear to be the shallowest, or lightest species in the Pacific and equally light, or slightly heavier than *Globoquadrina altispira* in the Indian Ocean and South Pacific (Figure 2). Mixed species of *Globigerinoides* retain very light oxygen isotope values. This suggests that species of the genus *Globigerinoides* are shallow, or surface dwellers along with *G. altispira* and *Pulleniatina obliquiloculata.* Oxygen isotope values of *Orbulina universa* are variable and may range from heavier than *Globorotalia menardii* to lighter than *Globigerinoides trilobus-G. sacculifer* (Figure 2). This species is tentatively grouped as a surface dweller.

Among isotopically intermediate taxa in low latitudes, the *Globorotalia menardii* group (includes *Globorotalia limbata*) and

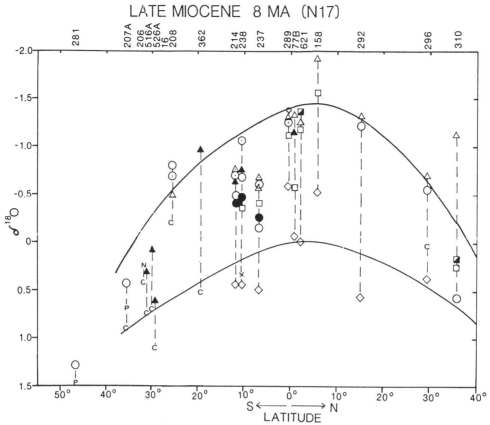

Figure 2. Oxygen isotope values of planktonic foraminiferal taxa from DSDP sites of the late Miocene time-slice (8 Ma) in the Pacific, Atlantic and Indian Oceans plotted against latitude north and south of the equator. Oxygen isotope data in Savin et al. (1985, Table 2). Upper line indicates surface temperature gradient and the lower line indicates thermocline position across latitudes.

Sphaeroidinella seminulina are consistently 0.1 to 0.5 ⁰/₀₀ heavier than the *Globigerinoides* group or *Globoquadrina altispira*, but considerably lighter (0.5 to 1.0 ⁰/₀₀) than *Globoquadrina venezuelana* (Figure 2). These species are therefore interpreted as upper intermediate dwellers (UIW) (Table 1). The *Globigerina nepenthes-G. druryi* group, *Globorotalia continuosa-Gl. acostaensis* group and *Globoquadrina dehiscens* have also been tentatively placed in this group, although further multiple species isotopic analyses are necessary to confirm their depth ranking.

In temperate middle latitudes *Globorotalia conoidea* consistently has heavy oxygen isotope values whereas *Neogloboquadrina pachyderma* appears lighter than *Globorotalia conoidea*, but heavier than *Orbulina*. Both *Gl. conoidea* and *N. pachyderma* are interpreted as lower intermediate (LIW) species (Table 1). *Globigerina bulloides* and *G. woodi* are also tentatively placed in this group. Figure 2 illustrates the isotopic values of species across latitudes with the upper line indicating the surface temperature gradients as inferred from the isotopically lightest species and the lower line indicating the thermocline position as inferred from the isotopically heaviest species.

Late Early Miocene: 16 Ma

Multiple species oxygen isotope analyses of 24 DSDP sites across latitudes are illustrated in Figure 3. *Globoquadrina venezuelana*, analyzed in ten sites, is consistently the heaviest planktonic foraminifer. The isotopically lightest species, or surface dwellers, include *Globorotalia siakensis*, *Globigerinoides trilobus-G. sacculifer*, *G. subquadratus*, *Globigerinoides* mixed species, and *Globoquadrina altispira*, similar to the late Miocene time slice (Table 1).

The intermediate group is less well defined. *Globorotalia peripheroronda* is 0.2 to 0.5 ⁰/₀₀ heavier than *G. trilobus-G. sacculifer* and *Globigerinoides* mixed species, except at Site 366A where the oxygen isotope values are similarly light perhaps due to diagenesis (Figure 3). *Globorotalia continuosa* also appears to fall into the intermediate group, although this species has not been analyzed from the same sample as other species in Figure 3. *Globoquadrina dehiscens* analyzed in 10 sites ranks 0.4 to 1.0 ⁰/₀₀ heavier than the surface group and *Globorotalia peripheroronda*, except at the high latitude Site 408 where it is anomalously light

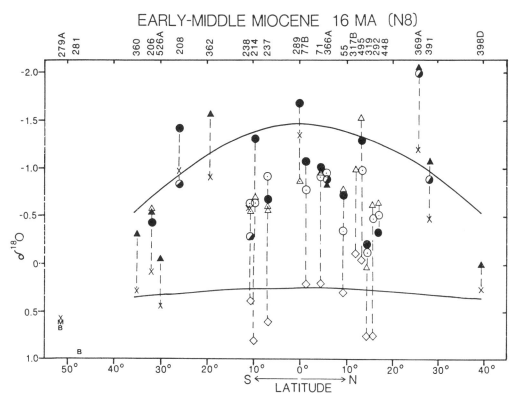

Figure 3. Oxygen isotope values of planktonic foraminiferal taxa from DSDP sites of the late early Miocene time-slice (16 Ma) in the Pacific, Atlantic, and Indian Oceans plotted against latitude north and south of the equator. Oxygen isotope data in Savin et al. (1985, Table 2). Upper line indicates surface temperature gradient and lower line indicates thermocline position across latitudes.

similar to *Globigerinoides* mixed species. This suggests that *Gq. dehiscens* ranks as intermediate species in low latitudes, but ranks as a surface dweller in higher latitudes. This may be a general problem with species which range from high to low latitudes; such species tend to occupy deeper positions in low latitudes and may become surface dwellers in high latitudes.

The lower intermediate group in middle to high latitudes comprises cool to temperate taxa and includes *Globorotalia miozea*, *Globigerina bulloides* and *G. woodi* (Table 1).

Early Miocene: 21 Ma

Multiple species oxygen isotope analyses of 19 DSDP sites across latitudes are illustrated for the early Miocene time-slice in Figure 4. Among the heaviest taxa are *Catapsydrax* spp., *Globoquadrina tripartita*, and *Gq. venezuelana* (Table 1). *Globoquadrina venezuelana* has anomalously enriched oxygen isotope values at Sites 71 and 55 perhaps due to diagenesis.

Surface dwellers comprise the largest group in the early Miocene (Table 1) *Globorotalia kugleri* is consistently the lightest species (10 sites). *Globorotalia siakensis* (7 sites) is heavier than *Gl. kugleri* and *Globigerina angustiumbilicata* except at Site 55, and lighter than *Globigerinoides trilobus–G. sacculifer* or *Globigerinoides* mixed species (Figure 4. *Globigerinoides trilobus-G.*

sacculifer has an unusually heavy oxygen isotope value at Site 292 (heavier than *Gq. venezuelana*) which may be due to dissolution. There is little change in the position of the surface temperature gradient or the thermocline across latitudes between early and middle Miocene time as inferred from isotopically lightest and heaviest species (Figures 3 and 4).

There are few isotopically intermediate species in the early Miocene time-slice. The lower intermediate group is composed of the cool to temperate species *Globorotalia miozea*, *Globigerina bulloides*, and *G. woodi*, similar to the middle Miocene time-slice. *Globoquadrina dehiscens* has been tentatively placed into the upper intermediate group although isotope values of this species are highly variable ranging from 0.3 to 1.2 °/₀₀ below *Globigerinoides* species to values as low as *Globoquadrina venezuelana*. This high variability may be in part a taxonomic problem because *Gq. dehiscens* evolved from *Gq. tripartita* (a deep dweller) in the early Miocene and transitional forms may have been included in the analyses. It is also possible that *Gq. dehiscens* adapted to a different water-mass after its evolution from *Gq. tripartita*.

MIOCENE FAUNAL TRANSECTS

In Figures 5 to 7 planktonic foraminifers have been grouped

EARLY MIOCENE 21MA (N4B)

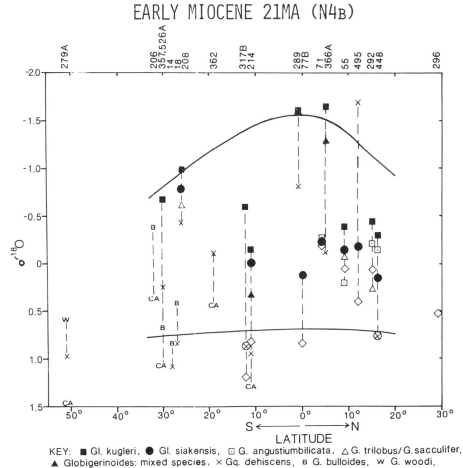

Figure 4. Oxygen isotope values of planktonic foraminiferal taxa from DSDP sites of the earliest Miocene time-slice (21 Ma) in the Pacific, Atlantic, and Indian Oceans plotted against latitude north and south of the equator. Oxygen isotope data in Savin et al. (1985, Table 2). Upper line indicates surface temperature gradient and lower line indicates thermocline position across latitudes.

into surface, upper intermediate (UIW), lower intermediate (LIW) and deep water dwellers for the three Miocene time-slices and plotted in four faunal transects (West Pacific, North Pacific, Northeast Pacific, and Equatorial Pacific) in order to illustrate faunal changes through the Miocene and across latitudes. The late Miocene (8 Ma) faunal transects are compared with modern ocean GEOSECS water-mass profiles of temperature and salinity in order to illustrate the similarities between the depth ranking of planktonic foraminifers as inferred from oxygen isotope ranking, and temperature and salinity conditions across latitudes. Ideally, modern water-mass profiles should be compared with modern faunal transect data; unfortunately, no Recent quantitative faunal data are available for the Pacific. One might argue that a comparison of modern ocean temperature and salinity profiles with late Miocene faunal data are invalid, or at least suspect, since oceanographic and paleoclimatic conditions have changed considerably over the last 8 million years. However, large scale oceanographic conditions such as the major gyral circulation in the Pacific may not have been very different from the present as indicated by

faunal studies that suggest that the late Miocene marks the onset of paleoceanographic conditions similar to the present (Lohmann and Carlson 1981, Burckle et al. 1982; Barron and Keller 1983). In this study it is assumed that a consistent similarity between faunal groups and the temperature or salinity profiles in each transect implies that there may be a relationship between the faunal groups (surface, intermediate, or deep water dwellers), and temperature (or density) and salinity conditions. However, ultimately this assumption will have to be tested on modern faunal data.

Late Miocene: 8 Ma

The west Pacific faunal transect (Figure 5) illustrates the bipolarity of fauna in the northern and southern hemispheres, which has also been shown for Pliocene and Pleistocene fauna (Cifelli 1969; Kennett 1978; Keller 1978). Nearly identical faunal assemblages are found at Site 310 (43°N) and Site 208 (31°S) with the exception that the UIW species *Globigerina nepenthes,*

GEOSECS WEST PACIFIC TRANSECT 1973-1974

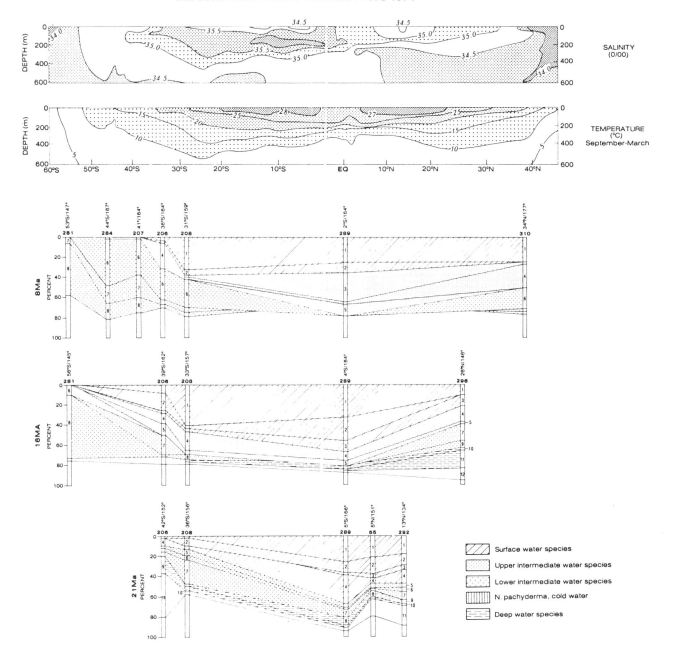

Figure 5. GEOSECS West Pacific transect profiles of temperature and salinity of the modern ocean (Craig et al., 1981) and depth stratification of planktonic foraminifers in the West Pacific transect at 8 Ma, 16 Ma and 21 Ma. Depth ranking of species is based on $\delta^{18}O$ ranking. Species are grouped into surface, upper intermediate, lower intermediate, and deep water groups. Key: 8 Ma time-slice: (1) *Globigerinoides*, (2) *Globoquadrina altispira*, (3) *Globorotalia menardii*, (4) *Globigerina nepenthes-G. druryi*, (5) *Gl. acostaensis*, (6) *Gl. conoidea*, (7) *G. woodi*, (8) *G. bulloides*, (10) *Gq. venezuelana*, (11) *Globigerinita glutinata*.

16 Ma time-slice: (1) *Globigerinoides*, (2) *Globorotalia siakensis-Gl mayeri*, (3) *Globoquadrina altispira*, (4) *Gq. dehiscens*, (5) *Globorotalia peripheroronda-Gl. fohsi*, (6) *Gl. continuosa*, (7) *Gl. miozea*, (8) *Globigerina woodi*, (9) *G. bulloides*, (10) *Globoqudrina venezuelana*, (11) *Sphaeroidinella disjuncta*, (12) *Globigerinita glutinata*.

21 Ma time-slice: (1) *Globigerina angustiumbilicata*, (2) *Globigerinoides*, (3) *Globoquadrina altispira*, (4) *Globorotalia kugleri*, (5) *Gl. siakensis-Gl. mayeri*, (6) *Globigerina woodi*, (7) *G. bulloides-G. praebulloides*, (9) *Globoquadrina venezuelana*, (10) *Catapsydrax*, (11) *Globigerinita glutinata*.

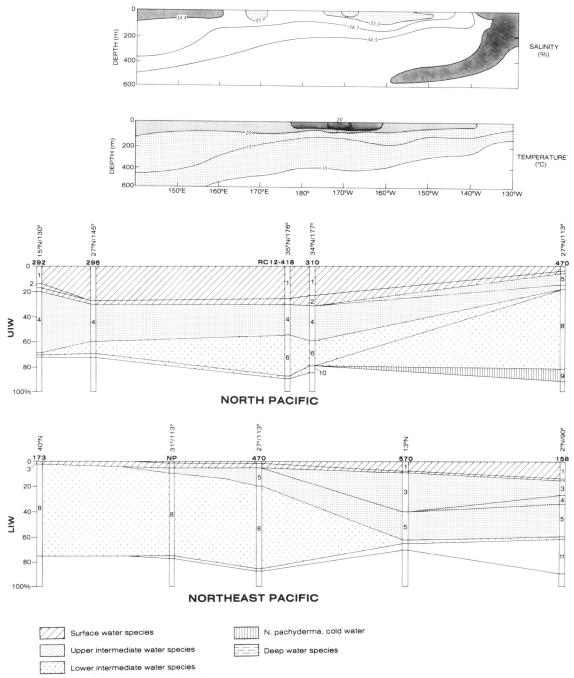

Figure 6. GEOSECS North Pacific transect profiles of temperature and salinity of the modern ocean and depth stratification of planktonic foraminifers in the north Pacific and northeast Pacific at 8 Ma. See Figure 5 caption for key to taxa.

which is abundant in Site 310 and Site 206 (36°S) but not at Site 208. The LIW group, the temperate species *Globorotalia conoidea,* and cool water species *Globigerina woodi* and *G. bulloides,* are rare or absent in equatorial water, but are common in middle latitudes and dominate in high latitudes.

Comparison of inferred late Miocene faunal depth stratification with modern GEOSECS water-mass profiles (Figure 5) reveals similarities, especially between temperature and depth stratification of species. Surface dwellers appear to be restricted to waters of greater than 20°C. In the GEOSECS west Pacific temperature profile, the 20°C isotherm outcrops at about 35°N and 35°S, similar to the latitudinal restrictions of the inferred late Miocene surface dwelling fauna in the South Pacific. The temperate UIW dweller *Globigerina nepenthes* and temperate LIW dweller *Globorotalia conoidea* disappear between 44°S and 53°S latitude near the outcrop location of the 10°C isotherm (Figure

5). The cool water component of the LIW group, *Globigerina woodi* and *G. bulloides,* dominate faunal assemblages south of 44°S, which compares to water temperatures cooler than 10°C in the modern ocean.

The similarity between temperature profiles and faunal depth stratification is also apparent in the North Pacific and northeast Pacific transects (Figure 6). The northeast Pacific transect illustrates a north-south transect through the California Current province and to the equator. Faunal changes in this region are especially pronounced and show the southward transport of a cool high latitude fauna with the California Current. Inferred faunal depth stratification suggests that the cool LIW group dominates north of 27°N latitude (Site 470) whereas the warm UIW group dominates to the south. Unfortunately, no GEOSECS temperature data are available for this region.

In the North Pacific transect the surface group comprises only 6% of the total fauna in the cool water California Current province (Site 470, 27°N lat.), but increases to 20–32 percent in the warmer central and western Pacific. The distribution of the surface group is similar to the 20°C isotherm which outcrops in the eastern Pacific and deepens westward (Figure 6). The UIW group also increases in abundance westward from 8% in the eastern Pacific to 24% and 45% in the central and western Pacific respectively. Similarly, the 10°C isotherm deepens westward.

In the equatorial Pacific the GEOSECS transect profiles illustrate east-west differences in the modern ocean near Sites 77B and 289 respectively (Figure 7). For instance, significantly higher surface water temperatures (~3°C) prevail in the west equatorial Pacific, the 20°C isotherm is at about 200 m depth as compared to 50–100 m depth in the upwelling region of the east equatorial Pacific, and the 10°C isotherm lies between 300–400 m in the west, but varies between 200–400 m in the east equatorial Pacific (Figure 7). These east-west temperature differences appear to be reflected in the faunal depth stratification. For example, the shallowing of the 20°C isotherm in the east equatorial Pacific is reflected in the lower abundance of surface dwelling species. The increased abundance in UIW dwellers, which coincides with temperatures between 10°C and 20°C, may be due to the increased upwelling in the eastern equatorial Pacific, but may also reflect higher salinity waters. Salinity profiles (Figures 5 and 7) show high salinity tongues between 150–200 m depth south of the equator in both west and east equatorial Pacific. *Globorotalia menardii,* which may prefer high salinity water (Poore 1981), is equally abundant in both regions (Sites 77B and 289).

The similarities between both the outcroppings of the 10°C and 20°C isotherms, and the general deepening of the isotherms, to the abundance distribution of the inferred surface, UIW and LIW groups suggests that the surface group may occupy waters warmer than 20°C, the UIW may occupy waters between 10°C and 20°C, and the LIW group waters cooler than 10°C. However, since late Miocene faunal groups are compared with modern temperature profiles, it can only be assumed that the temperature gradients of the late Miocene and the modern ocean were similar.

There is some evidence to support an interpretation of late Miocene faunal stratification based on Recent water-mass profiles from a study by Bé et al. (1971) based on plankton tow data (between 0–200 m) and the temperature profile (upper 700 m) in the Atlantic. This study shows a close correlation in both the geographic distribution and abundance of *Globigerinoides* (inferred surface dweller) with surface water temperatures greater than 20°C. The greatest abundance of cool water species *Globigerina bulloides* and *Neogloboquadrina pachyderma* (inferred LIW group) is found in temperatures of less than 10°C. However, many species occupying the intermediate temperature range, between 10–20°C in the modern ocean (*Globorotalia inflata, Gl. crassaformis, Gl. truncatulinoides*) evolved during Pliocene–Pleistocene times.

Early Miocene: 21 Ma, 16 Ma

Inferred depth stratification of planktonic foraminifers of the earliest Miocene (21 Ma) and late early Miocene (16 Ma) time-slices are also illustrated for the equatorial and west Pacific transects (Figures 5, 7). Comparison of the early, late early, and late Miocene time-slices reveals marked changes in the depth stratification of species through the Miocene.

In the southwest Pacific the surface group increases from 12% in the earliest Miocene (Site 208) to 45% in the late early Miocene and decreases to 30% in the late Miocene. A sharp drop to lower surface values occurs between 31° and 36°S latitude (Sites 208, 206) in the late Miocene. At these latitudes the earliest and late Miocene are dominated by the cool LIW group, whereas in the late early Miocene the warmer surface and UIW group dominates (Figure 5).

In the equatorial Pacific the surface group gradually decreases from a high of 65–80% in the earliest Miocene, to 55–70% in the late early Miocene and to 25–35% in the late Miocene. At the same time there is a decrease in the deep dwelling group (extinction of *Catapsydrax*) and near disappearance of cold water species *Globigerina bulloides* and *G. woodi* between the earliest and late early Miocene (Figure 7). In the late early Miocene, species which inhabit the UIW appear in the equatorial Pacific and in the mid-latitude west Pacific (*Globorotalia peripheroronda-Gl. fohsi* group, *Gl. continuosa*). This group becomes dominant in the late Miocene with the evolution of *Gl menardii* in the tropics and *Gl. acostaensis* in tropical to subtropical waters. Hence, major changes in the depth stratification are the two-thirds decrease in the late Miocene surface group associated with an expansion of the UIW group and differentiation into tropical and temperate components.

The changing depth stratification in the three Miocene time-slices implies that the upper water-mass (0–500 m) has gradually evolved from a relatively simple to a highly complex stratified ocean by late Miocene time associated with an increased faunal provincialism and steepening of the thermal gradients between equator and poles as observed by many previous workers (Kennett 1978; Lohmann and Carlson 1981; Keller 1981a; Moore and

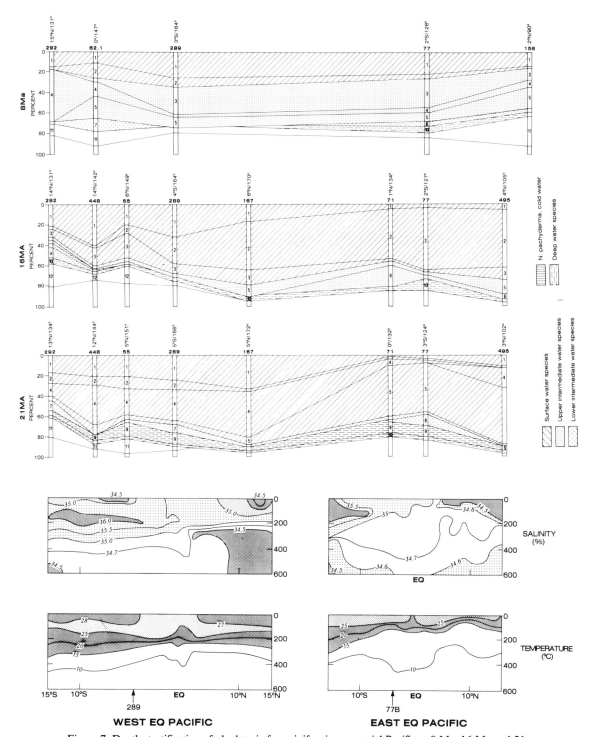

Figure 7. Depth stratification of planktonic foraminifers in equatorial Pacific at 8 Ma, 16 Ma and 21 Ma. (See Figure 5 caption for key to taxa) and GEOSECS west and east equatorial Pacific transects of temperature and salinity. Relative location of the GEOSECS transects near Sites 289 and 77B.

Lombari 1981; Thunell and Belyea 1982; Barron and Keller 1983). This faunal provincialism implies generally cooler surface water temperatures in middle latitudes during early and late Miocene, whereas surface water temperatures increased or remained unchanged in the tropics throughout the Miocene (suggested by the presence of long ranging forms, such as *Globigerinoides*). A general warming of surface waters in the tropics between the early and late Miocene is also observed in the oxygen isotope data as discussed by Savin et al. (1985).

DEPTH STRATIFICATION: TIME—SERIES ANALYSIS

Depth stratification of planktonic foraminifers during the Miocene is illustrated from the west and east equatorial Pacific (Sites 289 and 77B, Figures 8 and 9) in order to: (a) analyze the nature of water-mass stratification through the Miocene as implied by relative abundances of surface, intermediate, and deep dwelling species; (b) identify water-mass changes that may have caused or contributed to the major faunal changes in the early, middle, and late Miocene; and (c) identify species abundance changes that correlate with cold-warm oscillations in the oxygen isotope record. Quantitative faunal analysis of Site 289 was done by Srinivasan and Kennett (1981a, b; Kennett and Srinivasan 1985). They also provided the data for this study. Faunal data for Site 77B were reported earlier in Keller (1980a, 1981b).

Depth stratification of planktonic foraminifers at Site 77B and 289 imply major changes in water-mass stratification through the Miocene. The earliest Miocene ocean (Zones N4–N5) was dominated by the warm water surface group (70–80%) marked by pronounced east-west faunal provincialism and minor intermediate and deep water groups (Figures 8 and 9). This suggests that water temperatures remained warm (>20°C?) through most of the living habitat of planktonic foraminifers at this time. It is noteworthy that the N4/N5 Zone boundary is characterized by a major faunal turnover, primarily in the surface group, resulting in the decline and eventual extinction of the Paleogene fauna and evolution of the Neogene fauna (Kennett 1978; Keller 1981a). A gradual change in the water-mass stratification is also implied at this time by a permanent decline in the deep water group (extinction of *Catapsydrax*), associated with a major cool event in the $\delta^{18}O$ record (Boersma and Shackleton 1978; Savin et al. 1985), widespread deep-sea unconformity and sea level lowstand (Vail and Hardenbol 1979; Keller and Barron 1983).

The surface group dominates through the late early and middle Miocene with *Globorotalia siakensis–Gl. mayeri* dominant in Site 77B and *Globigerinoides* dominant in Site 289 (Figures 8 and 9). The early middle Miocene is marked by an increase followed by a gradual decline in the intermediate group associated with generally cooler $\delta^{18}O$ temperatures. The evolution and range of the intermediate dwelling *G. fohsi* plexus characterizes this interval. Deep-water dwellers are nearly absent. The continued dominance of the surface group suggests that in the tropics the living habitat of this group remained essentially un-

changed during the major phase of polar glaciation in the Neogene. However, cooling is indicated by the increasing abundance of the cooler intermediate group, indicating increased stratification of the upper water-mass.

The most dramatic change in water-mass stratification is implied during the late Miocene. Beginning at the middle/late Miocene boundary (~12 Ma) there is an abrupt decline in the surface group (extinction of *Gl. siakensis–Gl. mayeri*) and increase in the intermediate group (evolution of *Gl. menardii–Gl. tumida, G. nepenthes–G. druryi* groups), coincident with significantly lower $\delta^{18}O$ temperatures. No east-west faunal provincialism is apparent at this time. This faunal change suggests the establishment of a distinct broad intermediate water layer at the expense of a reduced surface water layer (Figures 8 and 9). Lower abundances in the surface group and presence of a deep water group at Site 77B suggests that in the eastern equatorial Pacific water was significantly cooler, perhaps due to increased upwelling, and more stratified than in the west equatorial region.

FAUNAL CLIMATIC CURVE

Early and Middle Miocene

The early and middle Miocene equatorial Pacific is dominated by a surface group that exhibits definite east-west faunal provinces. This provincialism is expressed in a western equatorial fauna (Site 289) dominated by *Globigerinoides* spp. *Globoquadrina altispira, G. angustiumbilicata,* and *Gl. kugleri,* and an east equatorial fauna (Site 77B) dominated by the *Gl. siakensis–Gl. mayeri* group. Such faunal differences are presumably related to water-mass properties and therefore may reveal oscillating climatic conditions and fluctuations in the intensity of the equatorial current circulation. A faunal climatic curve has been constructed based on this east-west provincialism. In this curve the western fauna is represented as the proportion of the total eastern (E) and western (W) fauna W/W+E, whereas in the east equatorial Pacific the eastern fauna is proportional E/E+W. Hence, the faunal climatic curves in Figures 8 and 9 illustrate incursions of the migrating eastern and western faunal components during cool-warm oscillations.

In the west Pacific Site 289 examination of the surface dwelling group, the faunal climatic curve and the benthic $\delta^{18}O$ curve suggest a close relationship between these groups during the early and middle Miocene. For paired samples, there is a one-to-one correspondence (solid lines, Figure 9) between high abundance peaks in the surface fauna, high proportion of the west equatorial faunal component (climatic curve), and lighter $\delta^{18}O$ values, or climatic warming. (No isotope data are available for the lower part of the early Miocene at Sites 289 and 77B). Incursions of the east equatorial faunal component generally correlate with heavier $\delta^{18}O$ values, or climatic cooling. Exceptions to this correlation are one point in the faunal climatic curve at 15.9 Ma and two points in the $\delta^{18}O$ curve at 12.3 Ma and 12.4 Ma that correlate with heavier $\delta^{18}O$ values indicating cooling, not warm-

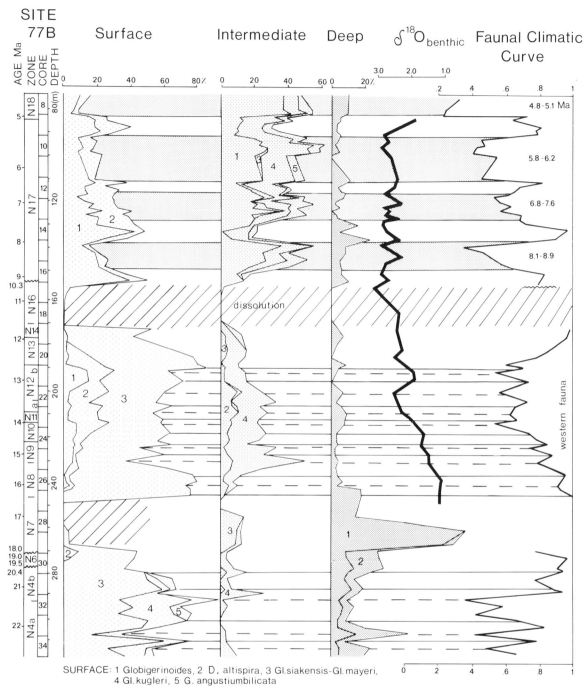

SURFACE: 1 Globigerinoides, 2 D. altispira, 3 Gl.siakensis-Gl.mayeri,
　　　　　4 Gl.kugleri, 5 G. angustiumbilicata

INTERMEDIATE: 1 Gl.menardii-Gl.tumida, 2 Gl.peripheroronda-Gl.fohsi, 3 Gq.dehiscens,
　　　　　　　4 Gl.continuosa-Gl.acostaensis, 5 G. nepenthes-G.druryi

DEEP: 1Gq.venezuelana, 2 Catapsydrax

Figure 8. Planktonic foraminiferal fauna of Site 77B grouped into surface, intermediate, and deep water
dwellers, benthic oxygen isotope curve of Savin et al. (1981), and faunal climatic curve based on
east-west provincialism in surface dwellers in the early and late early Miocene, and based on surface
dwellers versus *Globorotalia menardii* in the late Miocene. The time scale is after Berggren et al. (in
press) and ages for faunal climatic events have been extrapolated from sediment rate curves. Solid lines
in early and middle Miocene mark correlation between high percent surface dwellers and high percent
eastern faunal component, dashed lines correlate low percent surface dwellers with high western faunal
components indicating global cooling and warming respectively. Late Miocene gray intervals show
correlation between low percent surface dwellers and high percent *Gl menardii* assumed to indicate
polar cooling. Zig-zag lines mark hiatuses.

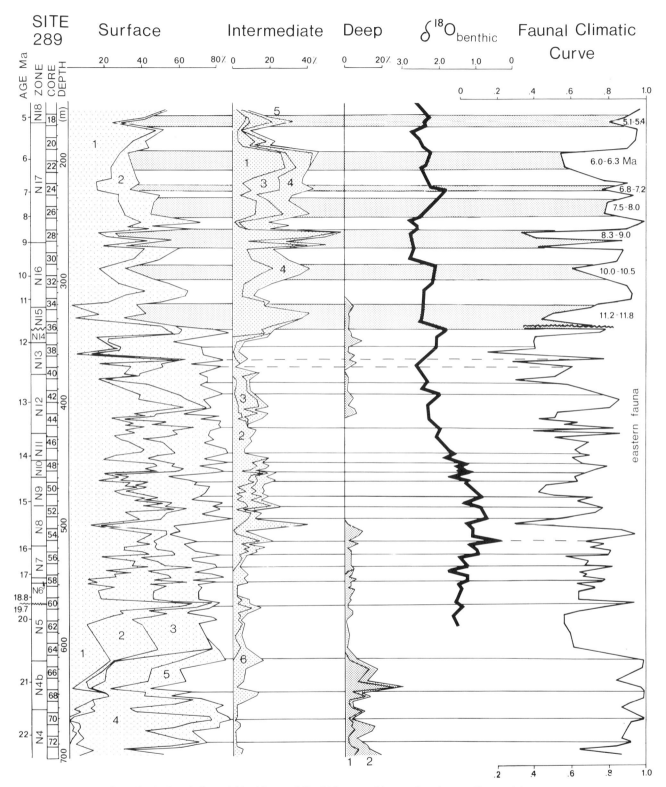

Figure 9. Planktonic foraminiferal fauna of Site 289 grouped into surface, intermediate, and deep water dwellers, benthic oxygen isotope curve of Woodruff et al. (1981), and faunal climatic curve. Solid lines in early and middle Miocene show correlation between high peaks in surface dwellers, high peaks in western faunal components, and polar warming as indicated by light $\delta^{18}O$ values. For complete explanation see Figure 8 caption.

ing (dashed lines, Figure 9). Nevertheless, this excellent correlation indicates that in the west equatorial Pacific high percent abundance in the surface group and dominance of the west equatorial fauna record warm climatic events.

The polar cooling trend observed in the $\delta^{18}O$ curve between 16 and 13 Ma (Woodruff et al., 1981) is not reflected in the surface group (Figure 9), except for an increase in the cooler intermediate group. This suggests that overall surface conditions in the west equatorial Pacific remained warm. There is, however, a slight increase in the eastern faunal component at Site 289 at this time suggesting intensified circulation and equatorial upwelling, as would be expected during periods of polar cooling and ice build-up.

Figure 10 illustrates the early and middle Miocene benthic isotope record and faunal climatic curve at Site 289. Cool climatic phases predicted from the faunal curve and benthic isotope data are shaded. In the faunal climatic curve, cooling episodes are identified by incursions of the eastern faunal component into the west equatorial Pacific, presumably due to intensified equatorial current circulation. Ages for cooling events have been calculated from sediment accumulation curves based on datum levels of planktonic foraminifers, nannofossils, diatoms, and radiolarians tied to the revised time scale of Berggren et al. (in press; also see Barron et al. 1985).

In the east equatorial Pacific Site 77B there is a good correlation between high abundance peaks in the surface group and high proportion of the eastern faunal component (solid lines, Figure 8), and low abundance of surface dwellers with incursions of the western faunal component (dashed lines, Figure 8).

However, there is only a general correspondence between faunal fluctuations and the benthic isotope curve (Savin et al. 1981; 1985). The disparity is partly due to fewer data points and because the benthic isotope values record bottom water temperatures. No continuous isotope record of planktonic foraminifers is available at this time. Nevertheless, there appears to be a general correlation between heavier isotope values, or cooling, and increased eastern faunal component. By analogy with the west equatorial Pacific Site 289, it is assumed that the high eastern faunal component indicates climatic cooling, whereas incursions of the western faunal component at Site 77B indicate climatic warming.

The warm western faunal component generally increases in Site 77B between 16 and 13 Ma, correlating with the polar cooling trend (Figure 8). This suggests that at the time of middle Miocene high latitude cooling, there is climatic warming of surface waters in the equatorial Pacific. Planktonic isotope data of the time-slice intervals corroborate this observation (Savin et al. 1985). Similar observations based on $\delta^{18}O$ data have been made by earlier workers (Savin et al. 1975; Shackleton and Kennett 1975).

Late Miocene

During the late Miocene the faunal differences between the

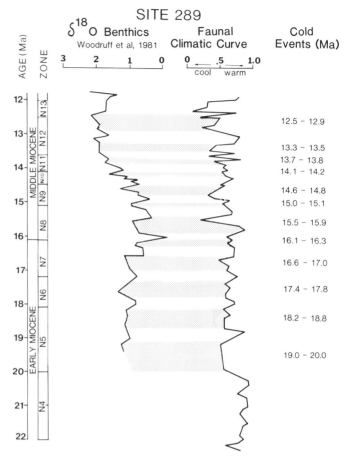

Figure 10. Early and middle Miocene benthic oxygen isotope curve and faunal climatic curve based on east-west faunal provincialism at Site 289, plotted against time. Time scale of Berggren et al. (in press) and Barron et al. (1985). Dotted intervals mark cold events observed in both benthic $\delta^{18}O$ data and the faunal climatic curve (incursions of eastern faunal component). Ages of cold events extrapolated from sediment accumulation curve.

east and west equatorial Pacific disappeared (Figures 8 and 9) along with the extinction of the surface *Globorotalia siakensis-Gl. mayeri* group and the *Gl. fohsi* group of the intermediate water dwellers. No significant new surface dwelling species evolved to replace the *Gl. siakensis-Gl. mayeri* group. The western surface dwellers (primarily *Globigerinoides* spp.) established themselves in the east equatorial Pacific instead (Figure 8). A major expansion occurred in the intermediate group with the evolution of the *Globorotalia menardii-Gl. tumida* groups. The expansion appears to be associated with the high salinity, high density Equatorial Countercurrent, and the more temperate *Globigerina nepenthes-G. druryi* group. In addition, *Gl. acostaensis* evolved from the early and middle Miocene *Gl continuosa*. The deep water group remained an insignificant part of the faunal assemblage in Site 77B (<10%) and is nearly absent in Site 289.

A reciprocal relationship exists between warm surface water dwellers (S) and the cooler intermediate dweller *Gl. menardii-Gl. tumida* group (Im) and this relationship S/S+Im has been used to develop the late Miocene faunal climatic curve (Figures 8 and 9). Early and middle Miocene correlations between abundance peaks in the surface group and light $\delta^{18}O$ values have indicated that the surface group expands during climatic warming. I assume here that the *Gl. menardii–Gl. tumida* group expands during polar cooling phases. Based on this assumption, cooling phases are identified in Site 289 at 11.2–11.8 Ma, and 10.0–10.5 Ma; sediments of these intervals are removed by a hiatus in Site 77B. The Site 289 cool event at 8.3–9.0 Ma is present in Site 77B at about 8.1–8.9 Ma; cool events 7.5–8.0 Ma and 6.8–7.2 Ma appear to be combined into one cooling phase between 6.8–7.6 Ma in Site 77B. Latest Miocene cool events are present in Site 289 at 6.0–6.3 Ma and in Site 77B at 5.8–6.2 Ma, whereas the cool event at 5.1–5.4 Ma is not present in Site 77B, probably due to a short hiatus. However, there is only a general correlation between inferred faunal cooling phases and the benthic isotope curve, and no time-series isotope data are available for planktonic foraminifers at this time. Isotope analyses of planktonic foraminifers will be necessary to corroborate the cooling phases predicted from faunal data.

PALEOCLIMATIC IMPLICATIONS

Most of the faunal parameters discussed, the major evolutionary assemblage changes, and the increasing vertical and latitudinal stratification between early, middle, and late Miocene as well as east-west equatorial provincialism can be explained by paleoclimatic and paleoceanographic changes. Both oxygen isotope studies (Shackleton and Kennett 1975; Savin et al. 1975, 1985) and faunal climatic indices indicate a successive warming in the equatorial Pacific between early, middle, and late Miocene that correlates with increased polar cooling and hence resulted in increased thermal gradients between equator and poles. Many of the faunal assemblage changes appear to be related to these paleoclimatic fluctuations. For instance, the major faunal change in the early Miocene surface and deep water groups at about 21 Ma that marks the extinction of the late Oligocene fauna (e.g. *Gl kugleri, Catapsydrax* spp.) and evolution of the Neogene fauna may be linked to both a major cooling phase and the development of a deep circum-Antarctic circulation (Kennett et al. 1975; Keller 1981a). East-west faunal provincialism during the early and late early Miocene may be due to intensified equatorial surface circulation during polar cooling (transport of eastern fauna westward) and more sluggish surface circulation during warm events. The most significant faunal change during the middle Miocene is the evolution of the *Globorotalia fohsi* group, an intermediate dwelling species group that indicates increased stratification of the water column in the tropics. This group became extinct at the end of the middle Miocene and was replaced by the late Miocene *Globorotalia menardii* group which appears to prefer high salinity water.

Equatorial warming continued into the late Miocene, as indicated by the increased abundance of surface dwellers in the east equatorial Pacific and abundance of the intermediate water

Gl. menardii–Gl. tumida group. However, the abrupt faunal change between middle and late Miocene, the disappearance of significant east-west provincialism, and most of all, the evolution and dominance of the intermediate water group are difficult to explain by paleoclimatic changes alone. These faunal changes are interpreted to reflect the development or strengthening of the Equatorial Countercurrent system during the middle Miocene when the Indonesian Seaway effectively closed, and the general strengthening of the gyral circulation that resulted from increased Antarctic glaciation at that time (see also Kennett et al. 1985).

SUMMARY AND CONCLUSIONS

Planktonic foraminifers have been grouped into surface, intermediate, and deep dwellers based on oxygen isotope ranking of individual species. This depth ranking should be considered as preliminary because (a) relatively few oxygen isotope analyses have been done on individual species and unusual values due to diagenesis and dissolution often can not be recognized and (b) the depth habitat of planktonic foraminifers is still not well understood in modern assemblages.

The Miocene depth stratification inferred from oxygen isotope ranking of species has been compared with Pacific GEOSECS water-mass profiles and a consistent relationship is observed between temperature and the faunal groups. The distribution of the surface group (lightest oxygen isotope values) corresponds to water temperatures greater than 20°C, the distribution of the upper intermediate group (intermediate oxygen isotope values) corresponds to water temperatures between 10°–20°C, and the distribution of the lower intermediate and deep water groups (heavy oxygen isotope values) correspond to waters cooler than 10°C. However, since modern ocean temperature profiles are compared with late Miocene faunal assemblages the actual temperature values are suspect. Quantitative faunal analyses of Recent assemblages will be necessary to corroborate these observations. Nonetheless, the consistent correlation of these temperatures with the depth ranked faunal groups makes depth stratification of species a potentially useful tool in paleoceanographic interpretations.

Miocene depth stratification of species indicates increasing vertical as well as latitudinal stratification between the early, middle, and late Miocene. The increasing latitudinal stratification is well known from the geographic distribution of several microfossil groups (foraminifers, diatoms, nannofossils) and lends support to the isotopic depth ranking of species. The vertical water-mass stratification information provides a potentially new tool to study changes in the surface waters (0–500 m). This study indicates that the early Miocene equatorial Pacific has a dominant surface water group and very minor deep and intermediate water groups. During the middle Miocene a significant upper intermediate water fauna develops and during the late Miocene the upper intermediate group expands and dominates the faunal assemblages during warm climatic intervals whereas surface dwellers dominate during cool events. Hence, the Miocene surface ocean evolved from a relatively simple, uniform, warm early Miocene ocean dominated by isotopically light species (surface) to a more stratified middle Miocene ocean with the development of an intermediate surface water layer (represented by isotopically heavier species), and to a complex stratified late Miocene ocean with a decreasing warm surface layer and increasing intermediate layer.

Depth stratification of species has been applied to two time-series analyses of Miocene sequences in the east and west equatorial Pacific (Sites 289 and 77B). These time-series analyses show the dominance of the surface group during the early and middle Miocene and their decline during the late middle Miocene (extinction of *Globorotalia siakensis*) and increase in the intermediate group (evolution of *Globorotalia menardii-Gl. tumida* group). A distinct east-west faunal provincialism is apparent in the surface water group during the early and middle Miocene with the *Globigerinoides* group dominant in the western region and the *Gl. siakensis-Gl. mayeri* dominant in the eastern region. During the late Miocene this provincialism disappears. This pro-

vincialism is interpreted as the result of intensified equatorial surface circulation during polar cooling phases and more sluggish circulation during warm events. The disappearance of the provincialism during the late middle Miocene and development of the upper intermediate faunal group may be associated with the major Antarctic glaciation and resultant strengthening of the general gyral circulation and the strengthening of the Equatorial Countercurrent due to the closing of the Indonesian Seaway at that time.

ACKNOWLEDGMENTS

I would like to thank John A. Barron and James P. Kennett for comments, and James P. Kennett and S. Srinivasan for providing faunal abundance data of the southwest Pacific sites and Site 289. DSDP samples were provided by the National Science Foundation through the Deep Sea Drilling Project. This study was supported in part by NSF Grant OCE 20-008879.00 and OCE 29-18285 (CENOP) to Stanford University.

REFERENCES CITED

Barrera, E., Keller, G., and Savin, S. M., 1985. The Evolution of the Miocene ocean in the eastern North Pacific Ocean as inferred from stable isotope ratios of foraminifera, *in* Kennett, J. P., ed., The Miocene ocean: Paleoceanography and biogeography: Geological Society of America Memoir 163.

Barron, J. A. and Keller, G., 1983, Paleotemperature oscillations in the middle and late Miocene: Micropaleontology, v. 29(2), p. 150–181.

Barron, J. A., Keller, G., and Dunn, D. A., 1985, A Multiple Microfossil Biochronology for the Miocene, *in* Kennett, J. P., ed., The Miocene Ocean: Paleoceanography and biogeography: Geological Society of America Memoir 163.

Bé, A.W.H., Vilks, G., and Lott, L., 1971, Winter distribution of planktonic foraminifera between the Grand Banks and the Caribbean: Micropaleontology 17(1), p. 31–42.

Berggren, W. A., Kent, D. V., and Flynn, J. J., in press, Paleogene geochronology and chronostratigraphy. *in* N. J. Snelling, ed., Geochronology and the Geological Record, Geological Society of London, Special Paper.

Boersma, A., and Shackleton, N. J., 1978, Oxygen and carbon isotope record through the Oligocene, DSDP Site 366, Equatorial Atlantic; *in* Lancelot, Y. and Siebold, E., eds., Initial Reports of the Deep Sea Drilling Project, v. 41: Washington, D.C., U.S. Government Printing Office, p. 957–962.

Burckle, L. H., Keigwin, L. D., and Opdyke, N. D., 1982, Middle and late Miocene stable isotope stratigraphy: Correlation to the paleomagnetic reversal record: Micropaleontology, v. 28(4), p. 329–334.

Cifelli, R., 1969, Radiation of the Cenozoic planktonic foraminifera. Systemic? Zoology, 18: 154–168.

Craig, H., Broecker, W. S., and Spencer, D., 1981, GEOSECS Pacific Expedition Volume 4, Sections and Profiles; Washington, D.C., U.S. Government Printing Office.

Douglas, R. G., and Savin, S. M., 1978, Oxygen and isotope evidence for the depth stratification of Tertiary and Cretaceous planktic Foraminifera: Marine Micropaleontology, v. 3, p. 175–196.

Emiliani, C., 1954. Temperatures of Pacific bottom waters and polar superficial waters during the Tertiary: Science, v. 119, p. 853.

Fairbanks, R. G., Wiebe, P. H., and Bé, A.H.W., 1980. Vertical Distribution and isotopic composition of Living planktonic Foraminifera in the Western North Atlantic: Science, v. 207, p. 61–63.

Fairbanks, R. G., Sverdlove, M., Free, R., Wiebe, P. H., and Bé, A.W.H., 1982.

Vertical distribution and isotopic fractionation of living planktonic Foraminifera from the Panama Basin: Nature, v. 298, p. 841–844.

Keller, G., 1978, Late Neogene Biostratigraphy and Paleoceanography of DSDP Site 310 Central North Pacific and correlation with the southwest Pacific: Marine Micropaleontology, v. 3, p. 97–119.

—— 1980a, Middle to late Miocene planktonic foraminiferal datum levels and paleoceanography of the north and southeastern Pacific Ocean: Marine Micropaleontology, v. 5, p. 249–281.

—— 1980b, Early to Middle Miocene planktonic foraminiferal datum levels of the equatorial and subtropical Pacific: Micropaleontology, v. 26, p. 372–319.

—— 1981a, Planktonic foraminiferal faunas of the equatorial Pacific suggest early Miocene origin of present oceanic circulation: Marine Micropaleontology, v. 6, p. 269–295.

—— 1981b, Miocene biochronology and paleoceanography of the North Pacific: Marine Micropaleontology, v. 6, p. 535–551.

—— 1983, Paleoclimatic analyses of middle Eocene through Oligocene planktonic foraminiferal faunas: Paleoclimatology, Paleoecology and Paleoceanography, v. 43, p. 73–94.

Keller, G., and Barron, T. A., 1983, Paleoclimatic Implications of Miocene deep-sea Hiatuses: Geological Society of America Bulletin, v. 94, p. 590–613.

Kennett, J. P., 1978, The development of planktonic biogeography in the southern ocean during the Cenozoic: Marine Micropaleontology, v. 3, p. 301–345.

Kennett, J. P., Houtz, R. E., Andrews, P. B., Edwards, A. R., Gostin, V. A., Hajos, M., Hampton, M. A., Jenkins, D. G., Margolis, S. V., Ovenshine, A. T., and Perch-Nielson, K., 1975, Cenozoic paleoceanography in the southwest Pacific ocean, Antarctic glaciation, and the development of the circum-Antarctic current: *in* Kennett, J. P. and Houtz, R. E., eds., Initial Reports of the Deep Sea Drilling Project, Washington, D.C., U.S. Government Printing Office, v. 29, p. 1155–1170.

Kennett, J. P., Keller, G., and Srinivasan, M. S., 1985, Miocene planktonic foraminiferal biogeography and paleoceanographic development of the Indo-Pacific region, *in* Kennett, J. P., ed., The Miocene ocean: Paleoceanography and biogeography: Geological Society of America Memoir 163.

Lohmann, G. P., and Carlson, J. J., 1981, Oceanographic significance of Pacific late Miocene calcareous nannoplankton: Marine Micropaleontology, v. 6, nos. 5/6, p. 553–580.

Moore, T. C., Jr., and Lombari, G., 1981, Sea-Surface temperature changes in the North Pacific during the late Miocene: Marine Micropaleontology, v. 6, nos. 5/6, p. 581–598.

Poore, R. Z., 1981, Late Miocene biogeography and paleoclimatology of the central north Atlantic: Marine Micropaleontology, v. 6, p. 599–616.

Poore, R. Z., and Matthews, R. K., 1984a, Late Eocene-Oligocene oxygen and carbon isotope record from south Atlantic Ocean DSDP Site 522: *in* Hsü, K. J. and LaBrecque, J. L., eds., Initial Reports of the Deep Sea Drilling Project, v. 73, Washington, D.C., U.S. Government Printing Office, p. 725–735.

——1984b, Oxygen isotope ranking of Late Eocene and Oligocene planktonic foraminifers: Implications for Oligocene sea surface temperatures and global ice volume. Marine Micropaleontology, v. 9, p. 111–134.

Savin, S. M., Abel, L., Barrera, E., Bender, M., Hodell, D., Keller, G., Kennett, J. P., Killingley, J., Murphy, M., and Vincent, E., 1985, The Evolution of Miocene surface and near-surface Oceanography: Oxygen and Isotope evidence, *in* Kennett, J. P., ed., The Miocene ocean: Paleoceanography and biogeography: Geological Society of America Memoir 163.

Savin, S. M., Douglas, R. G., and Stehli, F. G., 1975, Tertiary marine paleotemperatures: Geological Society of America Bulletin, v. 86, p. 1499–1510.

Savin, S. M., Douglas, R. G., Keller, G., Killingley, J. S., Shaughnessy, L., Sommer, M. A., Vincent, E., and Woodruff, F., 1981, Miocene benthic foraminiferal isotope records: A synthesis: Marine Micropaleontology, v. 6, p. 423–450.

Shackleton, N. J., and Kennett, J. P., 1975, Paleotemperature history of the Cenozoic and the initiation of Antarctic glaciation: oxygen and carbon isotope analyses in DSDP Sites 277, 279, and 281; *in* Kennett, J. P. and Houtz, R. E., eds., Initial Reports of the Deep Sea Drilling Project, v. 29, Washington, D.C., U.S. Government Printing Office, p. 743–755.

Shackleton, N. J., and Vincent, E., 1978, Oxygen and carbon isotope studies in Recent foraminifera from the Southwest Indian Ocean: Marine Micropaleontology, v. 3, p. 1–13.

Srinivasan, M. S., and Kennett, J. P., 1981a, A review of Neogene planktonic foraminiferal biostratigraphy: Application in the Equatorial and South Pacific; *in* Warme, J. E., Douglas, R. G., and Winterer, E. L., eds., Deep Sea Drilling Project: A decade of Progress: Society of Economic Paleontologists and Mineralogists Special Publication no. 32, p. 395–432.

——1981b, Neogene planktonic foraminiferal biostratigraphy and evolution: equatorial and subantarctic South Pacific: Marine Micropaleontology, v. 6, p. 499–533.

Thunell, R., and Belyea, P., 1982, Neogene planktonic foraminiferal biogeography of the Atlantic Ocean: Micropaleontology, v. 28(4): p. 381–398.

Vail, P. R., and Hardenbol, J., 1979, Sea level changes during the Tertiary: Oceanus, v. 22(3), p. 71–80.

Vincent, E., Killingley, J., and Berger, W. H., 1981. Stable isotope composition of benthic foraminifera from the equatorial Pacific: Nature, v. 289 (5799), p. 639–643.

Woodruff, F., Savin, S. M., and Douglas, R. G., 1981, Miocene stable isotope record: A detailed deep Pacific Ocean study and its paleoclimatic implications: Science, v. 212, p. 665–668.

Williams, D. R., and Healy-Williams, N., 1980, Oxygen isotopic-hydrographic relationships among recent planktonic foraminifera from the Indian Ocean: Nature, v. 283, p. 848–852.

Williams, D. F., Bé, A.W.H., and Fairbanks, R., 1979, Seasonal oxygen isotopic variations in living planktonic foraminifera off Bermuda: Science, v. 206, p. 447–449.

Manuscript Accepted by the Society December 17, 1984

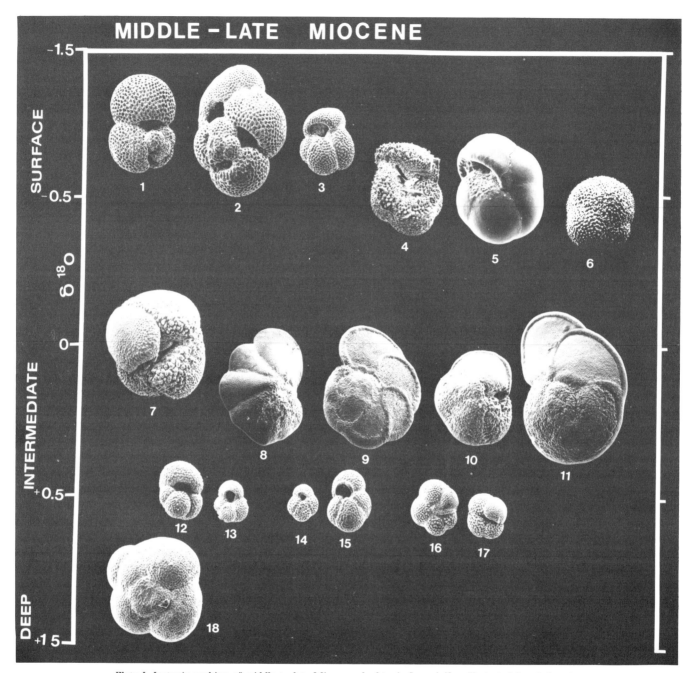

Plate I. Isotopic ranking of middle to late Miocene planktonic foraminifers illustrated in relation to general low latitude δ¹⁸O values. All species are illustrated at the same magnification to show relative size differences. Key to taxa: Surface: (1) *Globigerinoides trilobus;* (2) *G. sacculifer;* (3) *G. obliquus;* (4) *Globoquadrina altispira;* (5) *Pulleniatina primalis;* (6) *Orbulina.* Intermediate: (7) *Globoquadrina dehiscens;* (8) *Globorotalia fohsi;* (9) *Gl. menardii;* (10) *Gl conomiozea;* (11) *Gl. tumida;* (12) *Globigerina bulloides;* (13) *G. woodi;* (14) *G. druryi;* (15) *G. nepenthes;* (16) *Gl. acostaensis.* Deep: (17) *Globoquadrina venezuelana.*

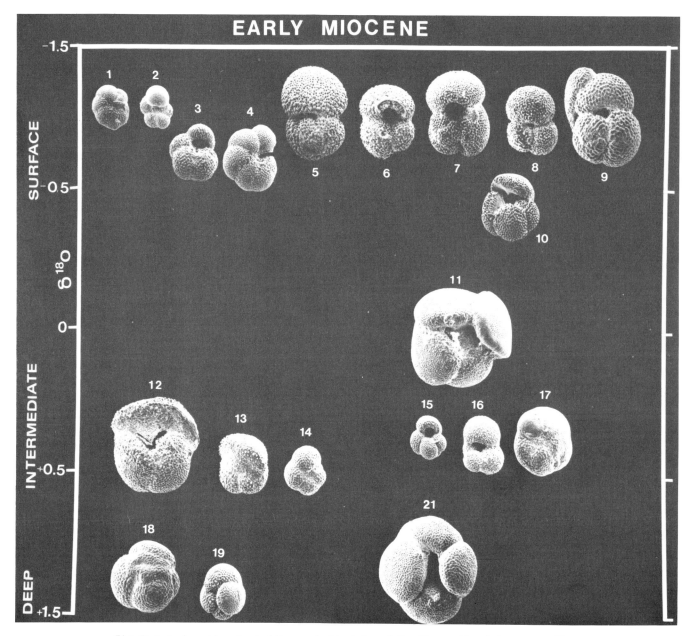

Plate II. Isotopic ranking of early Miocene planktonic foraminifers illustrated in relation to general low latitude δ^{18} values. All species are illustrated at the same magnification to show relative size differences. Key to taxa: Surface: (1) *Globorotalia kugleri;* (2) *Globigerina angustiumbilicata;* (3) *G. ciperoensis;* (4) *Gl. siakensis;* (5) *Globigerinoides trilobus;* (6) *G. subquadratus;* (7) *G. altiaperturus;* (8) *Globigerinoides parawoodi;* (9) *G. sacculifer;* (10) *Globoqudrina altispira.* Intermediate: (11) *Globoquadrina dehiscens;* (12) *Gq. praedehiscens;* (13) *Globigerina binaiensis;* (14) *Globorotalia continuosa;* (15) *Globigerina woodi;* (16) *G. bulloides praebulloides;* (17) *Gl miozea.* Deep: (18) *Catapsydrax dissimilis;* (19) *C. unicava;* (20) *Globoquadrina tripartita;* (21) *Gq. venezuelana.*

Geological Society of America
Memoir 163
1985

Miocene planktonic foraminiferal biogeography and paleoceanographic development of the Indo-Pacific region

James P. Kennett
Graduate School of Oceanography
University of Rhode Island
Narragansett, Rhode Island 02882

Gerta Keller*
U.S. Geological Survey
345 Middlefield Road
Menlo Park, California 94025

M. S. Srinivasan
Banaras Hindu University
Varanasi 221005
India

ABSTRACT

Biogeographic patterns of Pacific planktonic foraminifera have been quantitatively mapped for two time-slices in the early Miocene (22 and 16 Ma) and one in the late Miocene (8 Ma). Important differences are apparent between the early and late Miocene that resulted from changes in surface water circulation within the Pacific Ocean and between the tropical Pacific and Indian Oceans.

In the early Miocene, tropical Pacific planktonic foraminifera were dominated by different taxa in the eastern and western areas, but by the late Miocene the assemblages were similar across the entire tropical Pacific. East to west faunal differences were probably due to differences in the surficial water-mass structure and temperature. It is likely that a deeper thermocline existed in the west favoring shallow water dwellers such as *Globigerinoides* and *Globigerina angustiumbilicata*, and a shallower thermocline in the east favoring slightly deeper-dwelling forms, especially *Globorotalia siakensis* and *G. mayeri*. During the late Miocene a trans-equatorial assemblage developed, dominated by *Globorotalia menardii–G. limbata* and *Globigerinoides* groups. These faunal changes are interpreted to reflect both the development, during the middle Miocene, of the Equatorial Undercurrent system when the Indonesian Seaway effectively closed and the general strengthening of the gyral circulation and Equatorial Countercurrent that resulted from increased Antarctic glaciation and high-latitude cooling during the middle Miocene.

The trans-equatorial planktonic foraminiferal distribution patterns typical of the late Miocene did not persist to the present-day oceans when east-west differences are again evident. However, these differences in modern assemblages are exhibited within forms that usually inhabit deeper waters. There is a successive changing dominance from west to east of *Pulleniatina obliquiloculata* to *Globorotalia tumida* to *Neogloboquadrina dutertrei*. The modern west to east differences in these deeper-dwelling forms reflect an intensification of the Equatorial Undercurrent system and its shallowing towards the east to depths well within the photic zone. Shallow-water forms, such as *Globigerinoides*, maintain trans-tropical distribution patterns in the modern ocean un-

---

*Present address: Department of Geological and Geophysical Science, Princeton University, Princeton, New Jersey 08544.

like the early Miocene that lacked an effective equatorial countercurrent system in the Pacific.

The distribution of faunas in the North Pacific indicates that the gyral circulation system was only weakly developed in the early Miocene, but was strong by the late Miocene. In the northwest Pacific, temperate faunas were displaced northward as the Kuroshio Current intensified in the late Miocene. In the South Pacific, more distinct latitudinal faunal provinces appeared during the middle to late Miocene along with a northward expansion of the polar-subpolar provinces and contraction of the tropical province. These faunal changes resulted from the continued areal expansion of the polar and subpolar water masses as Australia drifted northward from Antarctica and from the steepening of pole to equator thermal gradients related to increased Antarctic glaciation.

INTRODUCTION

A major objective of paleoceanography is to better understand patterns of surface-water circulation and the character of the upper part of the water column in ancient oceans. There are two principal approaches: determination of regional gradients in the oxygen isotopic composition of planktonic foraminiferal tests which, in part, reflect changes in temperature and salinity related to paleocirculation; and changes in the distribution and character of planktonic microfossil assemblages. Modern planktonic microfossil groups represent sensitive tracers of surface and near-surface water masses (Bradshaw 1959; Bé 1977; Vincent and Berger 1981; Kennett 1982). In particular, planktonic foraminifera have long been used to study oceanic surface-water paleocirculation in many regions (Ingle 1967; Kennett 1967; Bandy 1968; Keller 1981a, b, c). A major objective of the Cenozoic Paleoceanography (CENOP) project (see Kennett 1981) has been to quantitatively map biogeographic distribution patterns for a number of planktonic microfossil groups on a broad scale to better understand the history of surface-water circulation of the Miocene ocean. CENOP has attempted to produce a synoptic picture of biogeographic patterns from selected intervals of time in the Miocene and represents the first large-scale biogeographic mapping project of the global ocean prior to the Pleistocene Period (CLIMAP 1976; 1981).

The purpose of this contribution is to present quantitative biogeographic maps of planktonic foraminifers in the Indo-Pacific region for three intervals of time during the Miocene that have been used in all of the CENOP time-slice studies: the earliest Miocene (22 Ma, Zone N4B); the late early Miocene (16 Ma, Zone N8) and the late Miocene (8 Ma, Zone N16/N17). Hodell and Kennett (1985) discuss similar data from the South Atlantic that are also shown on our species distribution maps. The biogeographic patterns are employed to define, in broad terms, the nature of surface-water circulation during each of these intervals. It is of particular interest to examine changes in biogeographic patterns that might exist between the three time-slices. This in turn will help in understanding the evolution of surface-water circulation and its structure during the Miocene. It is well known that Oligocene and Neogene planktonic foraminiferal faunas exhibit a continuing trend towards latitudinal differentiation (Kennett 1977; Berggren 1984). Oligocene assemblages exhibit a marked uniformity over a large latitudinal extent, a pattern that was replaced during the Neogene by numerous distinct latitudinal provinces.

The three time-slices were chosen for two main reasons. First, they generally represent periods of deep-sea sediment accumulation rather than erosion and unconformities (Keller and Barron 1983). Second, these periods were chosen to characterize biogeographic patterns at times when the oceans were potentially quite different. Thus an interval was selected to include the earliest Miocene at a time of Oligocene to Miocene biogeographic transitions and to include the beginning of the well-known Neogene evolutionary radiations in marine microfossil groups (Cifelli 1969; Berggren 1969; Lipps 1970; Kennett 1977, 1983; Thunell 1981; Keller 1981a). Another interval was selected in the late early Miocene during a distinct climatic warming following the major early Miocene evolutionary radiation and preceding the distinct oxygen isotopic shift of the middle Miocene interpreted by many (e.g. Shackleton and Kennett 1975; Savin et al. 1975; Kennett 1977; Woodruff et al. 1981) but not all workers (Matthews and Poore 1980) to represent a major growth phase of the east Antarctic ice sheet. A third interval was selected from within the late Miocene, following this major global climatic/glacial event and before the distinct $\delta^{13}C$ shift (Keigwin 1979; Haq et al. 1980) and further major climatic deterioration during the latest Miocene generally associated with the Messinian salinity crisis (Kennett 1967; Adams et al. 1977). Questions addressed in this investigation include: Were there significant changes in biogeographic patterns during the Miocene?: If so when did the changes occur?: What do the changes indicate about the development of surface-water circulation and what may have caused the changes?

Almost all of the stratigraphic sections used in this mapping experiment (Figures 1, 2, and 3) were drilled as part of the Deep Sea Drilling Project (DSDP). Also used in the 8 Ma time-slice are a piston core from the North Pacific (RC12-418) and the Newport Beach section, in southern California. The modern position, water depth, and stratigraphic intervals encompassed in the time-slices and the number of samples counted for each slice are listed in Table 1. In general, core coverage is poor and is perhaps still the most severe restriction upon such ocean mapping experiments. For the Pacific Ocean, there are slightly fewer available

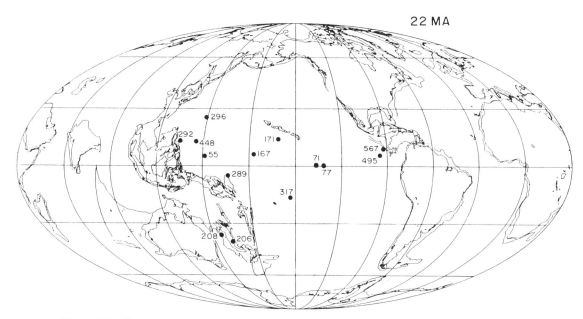

Figure 1. Earliest Miocene global paleogeography (22 Ma) (Sclater et al. 1985) showing location of DSDP sites used for mapping paleobiogeography of planktonic foraminifera within this time-slice.

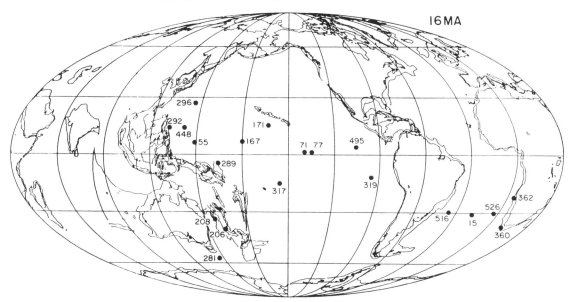

Figure 2. Latest early Miocene (16 Ma) global paleogeography (Sclater et al. 1985) showing location of DSDP sites used for mapping paleobiogeography of planktonic foraminifera within this time-slice.

sections of appropriate age in the early Miocene, as expected from the incomplete deep sea sedimentary record (Moore et al. 1978); 14 sections for the 22 Ma time-slice; 15 for 16 Ma; and 19 for 8 Ma. Large areas have no drilled sections; this is especially true of the southeast and south central Pacific and the higher latitudes of the North Pacific. Core coverage is best in the tropical to subtropical latitudes of the Pacific Ocean. Although cores of suitable ages are few in number, broad trends can be defined. Future drilling should provide needed additional materials to enhance mapping studies of the ancient ocean.

Biogeographic patterns are plotted on global paleogeogra-

phic maps of Sclater et al. (1985). (Figures 1–3) and individual sites have been accordingly backtracked. The most significant changes in the ocean basins potentially effecting Indo-Pacific circulation during the Miocene were the steady constriction and final effective closure to surface circulation of the Indonesian Seaway; the constriction of deep water flow through the central American Seaway and the expansion of the circum-Antarctic Current and polar-subpolar surface waters as Australia drifted northward from Antarctica. Tectonic changes at each of these interocean gateways almost certainly would have had major effects upon surface-water circulation. Tectonic changes that may

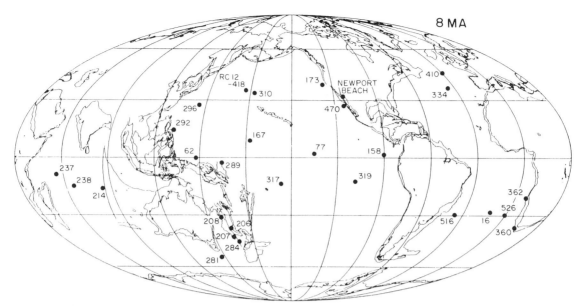

Figure 3. Late Miocene (8 Ma) global paleobiogeography (Sclater et al., 1985) showing location of sections used for mapping paleobiogeography of planktonic foraminifera within this time-slice.

have affected the character of deep and intermediate water circulation during the Miocene are not considered here.

METHODS

Quantitative counts were made of planktonic foraminifera in the >150 μm size fraction. Samples were oven-dried at 50°C, weighed, disaggregated in hot Calgon solution, washed over a 63 μm Tyler screen, dried, and reweighed. Samples were microsplit to obtain an average of 300 specimens per sample. Samples were counted and identified following the taxonomy of Kennett and Srinivasan (1983). Up to 20 samples were analyzed for each time-slice interval in each deep sea core, with an average of 6 or 7 samples per core for each time-slice interval. Samples were averaged to obtain a representative faunal assemblage and plotted on maps to show individual species distributions. Assemblages were not included in the data set when they clearly showed alteration resulting from calcium carbonate dissolution. Nevertheless, assemblages from deeper waters must have been affected to some extent by dissolution, although this was not quantitatively analyzed.

The species percent data for the Pacific were analyzed for each time slice using Q-mode factor analysis, a multivariate statistical technique (Imbrie and Kipp 1971). The Atlantic data were not included in this analysis. This technique examines the relative proportions of species within each sample in a data set, groups the data into a predetermined number of factors (in this study, five factors have been employed to define assemblages; Table 2) and ranks each sample by providing a factor value within each assemblage indicating its compositional similarity to that assemblage. Thus, biogeographic patterns in the oceans are revealed by the changes in the factor values.

PHYSICAL OCEANOGRAPHY

The most conspicuous features of modern ocean, surface-water circulation in the Pacific (Figure 4) are the large anticyclonic gyres (clockwise in the North Pacific; counterclockwise in the Southern Hemisphere) in the tropical and subtropical regions of each ocean. The North Pacific gyre is separated from the South Pacific gyre by a well-defined series of eastward and westward zonal surface currents, including an eastward-flowing undercurrent near the equator (Reid 1962; Wyrtki 1967). In each gyre, the flow is most narrow and intense on the western side of the ocean (such as the Kuroshio Current), and more weak and diffuse in the eastern regions. The western boundary currents (Kuroshio and East Australian currents) flow in the direction of the poles. Conversely, the eastern boundary currents (California and Peru-Chiole currents) flow back toward the equator as cold currents, thus completing the system of gyral circulation.

The major ocean currents near the equator are more complex and are a response to the wind system. The equatorial circulation consists of three primary components: the westward-flowing north and south equatorial currents (Figure 4) that lie beneath the trade winds; a relatively narrow eastward-flowing Countercurrent that occurs between the trade wind belts at the zone of minimum wind stress; and the eastward flowing Equatorial Undercurrent (Cromwell Current), found directly on the equator beneath the sea surface (Wyrtki 1967). The Equatorial Countercurrent returns warm water to the eastern Pacific. The equatorial wind system produces some upwelling along the equator and along the northern fringes of the surface Countercurrent. The Pacific Equatorial Undercurrent is narrow (300 km) and thin (200 m) and its depth of flow is centered at about 200 m in the west and rises to less than 50 m in the eastern Pacific where it

TABLE 1. SAMPLE LOCATION DATA FOR EACH OF THE THREE MIOCENE TIME-SLICES EMPLOYED IN THIS STUDY
(22 MA, 16 MA, AND 8 MA)

Time Slice	Site	Position	Present-Day Water Depth (M)	Paleo Water Depth (M)	Interval Sampled	Depth Below Surface (M)	Interval Thickness (M)	Number of Samples
22 MA	55	9°18. 1'N;142°32. 1'E	2850	1998	12-2, 52- 58 cm 13-2, 75- 80 cm	105.32- 114.75	9.43	7
	71	4°28.28'N;140°18.91'W	4419	4051	30-6, 87- 92 cm 33-6, 15- 19 cm	270.37- 297.65	27.60	7
	77(B)	0°28.90'N;133°13.70'W	4291	3732	30-5,103-104 cm 31-6,104-105 cm	275.33- 289.04	13.71	5
	167	7°04. 1'N;176°49. 5'W	3176	3161	9 cc	232.00	0.00	1
	171	169°27. 6'W; 19°07. 9'N	2290	2055	3-6,147-150 cm	54.97	0.00	1
	206(C)	32°00.75'S;165°27.15'E	3196	3119	2-2, 48- 50 cm 4 cc	414.98- 440.00	25.02	9
	208	26°06.61'S;161°13.27'E	1545	1439	23-3,100-102 cm 24-4,100-102 cm	375.00- 404.50	29.50	14
	289	0°29.92'S;158°30.69'E	2206	2398	65-3, 28- 42 cm 69-5, 38- 42 cm	611.38- 652.38	41.00	10
	292	15° 9.11'N;124°39.05'E	2943	2366	13-6,140-144 cm 15-3, 90- 94 cm	119.90- 133.90	14.00	10
	296	29°20.41'N;133°31.52'E	2920	2308	34-1, 45- 56 cm 34-5, 52- 60 cm	310.95- 317.02	6.07	5
	317(B)	11°00.09'S;162°15.78'W	2598	2520	25-1,142-143 cm 25-4,142-143 cm	226.42- 230.92	4.50	2
	448	16°20.46'N;134°52.45'E	3483	2725	6-1, 78- 82 cm 6-1,140-144 cm	43.78- 45.90	2.12	2
	495	12°29.78'N; 91°02.26'W	4140	2764	38-1, 80- 82 cm 39-1,140-144 cm	352.30- 362.40	10.10	2
	567	12°42.96'N; 90°55.99'W	5529	5339	12-1,101-105 cm 13 cc	5827.00- 5845.20	17.80	6
16 MA	15	30°53.38'S; 17°58.99'W	3938	3286	6-6, 46- 50 cm 7-4, 48- 57 cm	112.96- 118.98	4.02	2
	55	9°18. 1'N;142°32. 1'E	2850	2287	10-1, 42- 50 cm 10-6, 42- 50 cm	82.72- 90.22	7.50	6
	71	4°28.28'N;140°18.91'W	4419	4133	19-2,140-144 cm 22-6,100-102 cm	163.90- 197.50	33.60	10
	77(B)	0°28.90'N;133°13.70'W	4291	3929	26-2, 92- 94 cm 27-2, 92- 94 cm	237.22- 246.32	9.10	5
	167	7°04. 1'N;176°49. 5'W	3176	3163	7-5, 70- 72 cm 7 cc	154.70- 158.00	3.30	2
	171	169°27. 6'W; 19°07. 9'N	2290	2125	2-6,147-150 cm	36.97	0.00	1
	206	32°00.75'S;165°27.15'E	3196	3136	31-1, 90- 92 cm 32 cc	277.90- 295.00	17.10	20
	208	26°06.61'S;161°13.27'E	1545	1492	21-3,100-102 cm 21-6,105-107 cm	319.00- 322.55	3.55	7
	281	47°59.84'S;147°45.85'E	1591		10-3, 53- 54 cm 10-6, 90- 91 cm	87.03- 91.90	4.87	7
	289	0°29.92'S;158°30.69'E	2206	2373	52-2, 82- 90 cm 55-3, 88- 92 cm	486.82- 516.88	30.06	13
	292	15°49.11'N;124°39.05'E	2943	2552	12-2,140-144 cm 12-4,102-103 cm	104.40- 107.02	2.62	3
	296	29°20.41'N;133°31.52'E	2920	2543	28 cc 30-1,118-119 cm	263.00- 273.68	10.68	6
	317(B)	11°00.09'S;162°15.78'W	2598	2551	18-1, 86- 90 cm 18-3, 54- 58 cm	159.36- 162.04	2.68	2
	319	13°01.04'S;101°31.46'W	4296	3492	11-4, 50- 52 cm 12-2, 48- 50 cm	100.00- 106.48	6.48	6
	360	35°50.75'S; 18°05.79'E	2949	3021	22-2, 74- 79 cm 22-6,146-150 cm	319.24- 325.96	6.72	2
	362	19°45.45'S; 10°31.95'E	1325	1580	36 cc 37-2,145-159 cm	606.00- 618.45	12.45	2
	448	16°20.46'N;134°52.45'E	3483	2963	3-2, 40- 42 cm 3-2, 94- 98 cm	16.40- 16.94	0.54	2
	495	12°29.78'N; 91°02.26'W	4140	3521	26-1, 75- 79 cm 26-6,103-107 cm	238.50- 247.00	8.50	15
	516	30°16.59'S; 35°17.10'W	1313	1227	21-1, 51- 55 cm 22-2, 50- 54 cm	86.61- 92.50	5.89	2
	526(A)	30°07.36'S; 03°08.28'E	1054	758	21-1, 50- 54 cm 21-4, 11- 16 cm	116.50- 120.61	4.11	2

TABLE 1. SAMPLE LOCATION DATA FOR EACH OF THE THREE MIOCENE TIME-SLICES EMPLOYED IN THIS STUDY
(22 MA, 16 MA, AND 8 MA) (continued)

Time Slice	Site	Position	Present-Day Water Depth (M)	Paleo Water Depth (M)	Interval Sampled	Depth Below Surface (M)	Interval Thickness (M)	Number of Samples
8 MA	16	30°20.15'S; 15°42.79'W	3526	3061	9-1, 58- 65 cm / 10-5, 50- 55 cm	135.08- / 150.20	15.12	2
	62(.1)	1°52. 2'N;141°56. 3'E	2591	2452	23-5, 47- 51 cm / 24-2,143-147 cm	222.47- / 227.93	5.46	6
	77(B)	0°28.90'N;133°13.70'W	4291	4127	15-4, 30- 34 cm / 16-2,100-104 cm	142.00- / 148.80	6.80	8
	158	6°37.36'N; 85°14.16'W	1953	1592	19-6, 40- 44 cm / 21-1, 38- 42 cm	169.90- / 180.38	10.48	9
	167	7°04. 1'N;176°49. 5'W	3176	3169	4 cc	75.00	0.00	1
	173	39°57.71'N;125°27.12'W	2927		16-2, 01- 02 cm / 17 cc	140.00- / 157.50	17.50	5
	206	32°00.75'S;165°27.15'E	3196	3205	24-1,106-107 cm / 24 cc	211.06- / 219.00	7.94	7
	207(A)	36°57.75'S;165°26.06'E	1389	1330	6-3, 43- 48 cm / 6-6,100-104 cm	95.43- / 100.50	5.07	8
	208	26°06.61'S;161°13.27'E	1545	1525	16-1, 50- 51 cm / 16-5, 50- 51 cm	194.50- / 200.50	6.00	6
	214	11°20.21'S; 88°43.08'E	1671	1578	14-1, 40- 42 cm / 15-2, 10- 12 cm	123.90- / 134.60	10.70	10
	237	7°04.99'S; 58°07.48'E	1623	1533	12-6, 01- 03 cm / 13-2, 01- 03 cm	109.00- / 112.50	3.50	5
	238	11°09.21'S; 70°31.56'E	2832	2707	24-1, 01- 03 cm / 27-5, 50- 52 cm	215.00 / 250.00	35	12
	281	47°59.84'S;147°45.85'E	1591		6-4, 45- 46 cm / 7-4,105-106 cm	50.45- / 60.55	10.10	13
	284	40°30.48'S;167°40.81'E	1078		20-1, 81- 82 cm / 21-6, 40- 41 cm	180.31- / 196.90	16.59	13
	289	0°29.92'S;158°30.69'E	2206	2283	17-5,142-150 cm / 29-3, 38- 46 cm	254.42- / 269.38	14.96	9
	292	15°49.11'N;124°39.05'E	2943	2761	9-1, 93- 97 cm / 9 cc	73.93- / 82.50	8.57	11
	296	29°20.41'N;133°31.52'E	2920	2782	21-6, 60- 62 cm / 23 cc	195.10- / 215.50	20.40	14
	310	36°52.11'N;176°54.09'E	3516	3474	8-5, 50- 52 cm / 8-6,147-150 cm	68.50- / 70.97	2.47	11
	317(B)	11°00.09'S;162°15.78'W	2598	2580	9-5,143-144 cm / 10-2, 66- 67 cm	80.43- / 84.66	4.26	7
	319	13°01.04'S;101°31.46'W	4296	3955	3-2,118-120 cm / 3-3,118-120 cm	21.68- / 23.18	1.50	3
	334	37°02.13'N; 34°24.87'W	2619	2185	8-2,112-114 cm / 14-1,108-110 cm	189.12- / 244.58	55.46	9
	360	35°50.75'S; 18°05.79'E	2949	2984	8-2, 50- 55 cm / 11-6, 50- 55 cm	148.00- / 182.50	34.50	2
	362	19°45.45'S; 10°31.95'E	1325	1460	24-1, 52- 56 cm / 27-6, 52- 57 cm	350.02- / 414.52	64.50	2
	410	45°30.51'N; 29°28.56'W	2975	2584	28-1, 94- 96 cm / 31-2, 54- 56 cm	255.44- / 285.04	29.60	8
	470	28°54.46'N;117°31.11'W	3549	3193	9-1, 54- 59 cm / 9 cc	76.54 / 85.50	8.96	4
	516	30°16.59'S; 35°17.10'W	1313	1276	13-1, 70- 72 cm / 14-3, 70- 72 cm	51.60 / 59.00	7.40	2
	526(A)	30°07.36'S; 03°08.28'E	1054	940	9-1, 50- 54 cm / 11-3, 55- 59 cm	63.70- / 75.55	11.85	2
	Newport Beach	22°38'N;177°53'W	onshore		N7A / N5	256.00- / 235.00	21.00	5
	RC12-418	38°06'N;170°01'E	3842		452 cm / 636 cm	4.52- / 6.36	1.81	8

TABLE 2. SCALED VARIMAX FACTOR SCORES (Q-MODE FACTOR ANALYSIS) FOR
PLANKTONIC FORAMINIFERAL ASSEMBLAGES AT 22 MA, 16 MA, AND 8 MA

	Factor 1	Factor 2	Factor 3	Factor 4	Factor 5
Earliest Miocene (22 Ma)					
Globigerina angustiumbilicata	1.342	-0.020	-0.084	1.704	-0.571
Globigerina praebulloides	0.030	-0.159	0.343	-0.276	3.200
Globigerina woodi and woodi connecta	0.016	-0.029	0.088	0.205	0.261
Globigerinita glutinata	-0.170	-0.055	2.934	0.740	-0.408
Globigerinoides spp.	-0.341	-0.260	-0.588	2.594	0.959
Dentoglobigerina altispira	-0.078	-0.004	-0.115	0.889	-0.175
Globoquadrina praedehiscens	-0.153	0.181	0.090	0.774	-0.093
Globoquadrina venezuelana	0.037	0.221	0.332	0.306	0.001
Globorotalia opima nana	-0.071	0.273	0.324	0.015	0.052
Globorotlia siakensis and mayeri	-0.114	3.419	-0.134	0.174	0.185
Globorotalia kugleri	3.161	0.109	0.081	-0.332	0.282
Catapsydrax dissimilis	0.089	0.206	1.628	-0.277	0.348
Early Miocene (16 Ma)					
Globigerina praebulloides	0.193	0.027	3.252	-0.008	-0.014
Globigerina woodi	-0.034	0.049	0.009	0.436	0.107
Globigerinita glutinata	-0.132	0.167	0.271	0.959	1.395
Globigerinoides spp.	3.268	-0.051	-0.178	-0.021	-0.126
Dentoglobigerina altispira	0.207	-0.282	-0.177	0.042	2.959
Globoquadrina dehiscens	0.239	0.175	-0.066	2.452	-0.395
Globoquadrina venezuelana	0.047	-0.295	-0.011	0.042	0.007
Globorotalia continuosa	0.048	-0.254	0.548	-0.056	-0.075
Globorotalia conoidea and miozea	-0.302	0.030	-0.024	1.819	-0.265
Globorotalia peripheroronda	0.236	-0.474	-0.002	0.744	-0.063
Globorotalia siakensis and mayeri	-0.108	-3.237	0.009	0.094	-0.192
Late Miocene (8 Ma)					
Globigerina nepenthes and druryi	3.046	0.018	-0.069	-0.087	0.091
Globigerina bulloides	0.003	3.146	0.009	0.291	-0.055
Globigerina woodi	-0.344	0.019	-0.198	1.000	0.808
Globigerinita glutinata	0.330	0.055	-0.446	-0.077	2.479
Globigerinoides spp.	0.346	-0.100	1.958	0.540	0.423
Dentoglobigerina altispira	-0.064	-0.018	0.443	0.058	0.672
Globoquadrina venezuelana	0.259	0.038	0.634	-0.108	-0.445
Neogloboquadrina acostaensis	-0.520	-0.000	0.812	0.102	1.440
Globorotalia conoidea	0.179	-0.292	-0.127	2.918	-0.394
Globorotalia menardii-limbata	-0.038	0.067	2.157	-0.265	-0.366

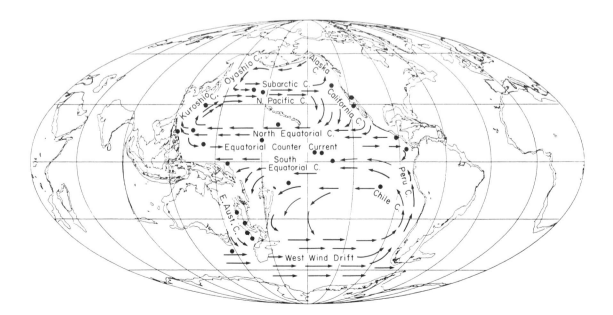

Figure 4. Major components of Modern Pacific Ocean surface circulation.

becomes involved with upwelling created by wind induced divergence (Pak and Zaneveld 1974; Wyrtki 1967; Reid 1962). The origin of the Pacific Equatorial Undercurrent seems to be linked with the balancing of geostrophic flow near the equator as a result of an increasing pressure gradient towards the west.

A weak cyclonic gyre (counterclockwise) occurs in the subpolar area and the North Pacific (Figure 4) with the Alaskan and Oyashio Currents as the main components. An equatorial Southern Hemisphere gyre is lacking because mid-latitude zonal flow is not obstructed by continental barriers. Instead, the Antarctic Circumpolar Current circles Antarctica and flows eastward at all depths.

As the major gyres circulate around the oceans, they either receive or give up heat to the atmosphere or to adjacent currents. Surface-water temperatures thus change around the circuit. Near the equator, the westward-flowing portion of the gyre is heated by high insolation and heat transfer from the warm atmosphere and adjacent warm parts of the central gyres, whereas at high latitudes the eastward-flowing part of the circuit loses heat to the atmosphere and cool subpolar waters. Thus the poleward-flowing western boundary currents are considerably warmer than the eastern boundary currents with consequent displacement of associated planktonic biota along the ocean margins.

Also of biogeographic importance are the division of surface waters into distinct water masses separated by oceanic fronts. Of particular importance in this study are the Subtropical Convergence in the South Pacific that separates Subantarctic from temperate (cool subtropical) surface waters and the Subtropical Divergence (Tasman Front) that separates temperate from warm subtropical surface waters. In the modern ocean each of these fronts is known to have an important effect upon the distribution of planktonic microfossils forming sharp transition zones (Parker 1971; Eade 1973; Bé 1977; Kennett 1978).

Modern ocean circulation would not be expected to be completely analogous to that of the Miocene Ocean. The changing configuration of continents forming the boundaries of the ocean basins created different circulation patterns in key areas, and changes in the latitudinal temperature gradient affected the strength of the circulatory system. There should have been large differences in tropical Pacific circulation patterns during the early Miocene when the Indonesian and Central American seaways were open to surface circulation. During the early Miocene, largely uninterrupted westward equatorial surface currents flowed from the eastern Pacific (or perhaps even the equatorial Atlantic) through the Indonesian Seaway north of Australia-New Guinea (Audley-Charles et al. 1972; Hamilton 1979) and into the Indian Ocean (Edwards 1975). Return-flow in the form of the Equatorial Undercurrent probably did not originate in the west Pacific as today, but in the Indian Ocean upon contact with Africa. The Equatorial Countercurrent system was weaker due to lower equator to pole thermal gradients and reduced wind strengths in the tropics. Largely uninterrupted westward flow of tropical currents into the Indian Ocean also reduced the volume and strength of the tropical waters directed into the western boundary currents in

the North and perhaps the South Pacific. One of the most important oceanic changes during the Miocene was the closure of the Indonesian Seaway by the middle Miocene (Moberly 1972; Hamilton 1979), which should have resulted in important circulatory changes and affected the biogeographic distribution of plankton (Edwards 1975; Kennett 1977, 1978).

STRATIGRAPHY

CENOP workers chose three time-slice intervals for detailed study close to major Miocene paleoceanographic events so as to describe the evolving Miocene Ocean and its paleobiogeographic evolution. The three time-slice intervals are the earliest Miocene at 22 Ma, corresponding to Zone N4b of Srinivasan and Kennett (1981); late early Miocene at 16 Ma, Zone N8 which straddles the early-middle Miocene boundary; and the late Miocene at 8 Ma, corresponding to the boundary between Zones N16 and N17. These three time-slice intervals are biostratigraphically defined and discussed in detail by Barron et al. (1985). Uncertainties of correlation are minimized by study of multiple samples from each of the time-slice intervals in various stratigraphic sections. The multidisciplinary time-scale developed by Barron et al. (this volume) and employed by CENOP Project workers allows resolution of Miocene time approaching 100,000 years. Correlations between temperate and tropical sections approach a time resolution of between 200,000 and 500,000 years although in a few sections this was expanded to up to one million years to alleviate problems of nonrecovery during parts of the interval. Such lengthy intervals sometimes contained large quantitative faunal oscillations, but the counting of an average of 6 or 7 samples per time-slice in each core provided average values that produce coherent biogeographic patterns in the oceans when plotted on maps.

RESULTS

General Trends

It is well established that modern planktonic foraminiferal species are limited in their distribution to certain water masses and latitudinal ranges (Bé 1977; Ruddiman et al. 1970; Parker 1971). Similarly the distribution of fossil planktonic foraminifera in stratigraphic sections shows that they were also latitudinally restricted in the past ocean. The latitudinal ranges of most Neogene planktonic foraminifera are now fairly well known because of investigations by large numbers of workers on DSDP sections from a wide range of latitudes (for summary of literature see Srinivasan and Kennett 1981; Kennett and Srinivasan 1983; Barron and Keller, 1983). General microfossil assemblages have been mapped for particular intervals during the Neogene (Sancetta 1978; Thunell and Belyea 1982). However, this is the first time that planktonic foraminiferal species patterns have been quantitatively mapped on a broad scale. The centers of evolution and biogeographic distribution of most species lie either in tropical or temperate water masses, although the maximum extent of individual species often includes several water masses.

Table 3 lists the most important planktonic foraminiferal species making up assemblages associated with different water-masses during seven intervals from the latest Oligocene and Miocene. Four broad assemblages are shown in Table 2: Subantarctic; temperate-transitional; warm-subtropical; and tropical. An Antarctic assemblage also occurred but it is not included in the table. Equatorial assemblages, distinct from more broadly distributed assemblages, also occurred at times. In the modern ocean, the Subantarctic water mass lies between the Antarctic Convergence (Polar Front) and the Subtropical Convergence, generally between 60° and 45° south of the equator. Temperate water masses are those that occur at latitudes of about 35° to 45° south or north of the equator. Transitional assemblages are most similar to the temperate, but also contain significant proportions of subtropical elements. The warm subtropical area is immediately adjacent to the tropics between about latitudes 20° to 30° north and south of the equator. Tropical assemblages are those that are largely restricted to areas within the present day tropics between 20° north and south of the equator.

Biogeographic knowledge of modern planktonic foraminifera is of limited value in the study of Miocene assemblages because of the different taxonomic composition of the two assemblages due to evolution and extinction. The species distribution patterns now mapped for each time-slice provide the basis for our paleoceanographic interpretations. During the Miocene, a number of species or lineages adjusted their environmental preferences. For example, *Globigerina praebulloides* exhibited the highest frequencies in warm subtropical-transitional areas in the early Miocene but later developed a preference for the subpolar-temperate regions, which its descendant *Globigerina bulloides* still maintains. *Globorotalia menardii* was a warm subtropical form upon its evolution from *G. praemenardii* 12 m.y. ago and became a tropical form about 10 Ma. Also *Globoquadrina praedehiscens* was a tropical form, while its ancestor *G. dehiscens* is a transitional-temperate form. Nevertheless, the switching of water-mass preferences through time seems to be exceptional. Most evolutionary lineages have remained closely associated with the same water masses during the Neogene (Kennett and Srinivasan 1983).

For each of the three time-slices, distribution maps are presented for individual species as well as for five factors resulting from Q-mode factor analysis. Species were selected because of their quantitative importance in some assemblages and because of the coherent patterns displayed. Quantitatively unimportant species are not shown, although many of these also exhibit clear associations with certain water masses. Only a few species were found to exhibit irregular patterns and these are mostly quantitatively unimportant elements within the assemblages.

Earliest Miocene: 22 Ma

Earliest Miocene assemblages still contained the last rem-

nants of typical Oligocene taxa such as *Catapsydrax dissimilis, Globorotalia kugleri* and *Globigerina angustiumbilicata*. Evolutionary radiation had already added the earliest representatives of a number of typical Neogene lineages such as *Globigerinoides, Globoquadrina dehiscens* and *Dentoglobigerina altispira*. For this time-slice, distribution patterns are shown for eleven taxa. Temperate-transitional asscmblages were characterized by high frequencies of *Globigerinita glutinata* (Figure 5). *Catapsydrax dissimilis* clearly exhibited highest frequencies in temperate areas (Figure 6), but extended its geographic range into the transitional and eastern tropical Pacific region. This species exhibited low frequencies in the western tropical Pacific. Oxygen isotopic data (Biolzi 1983; Savin et al.; 1985) indicates that this was a deep-dwelling species.

The warm subtropical areas were marked by high frequencies of *Globigerina praebulloides* (Figure 7) and the *Globigerina woodi–G. woodi connecta* complex (Figure 8).

Tropical faunal distributions included two major categories; species that were distributed relatively evenly across the tropical Pacific and those that seemed to be restricted to either the western or eastern sectors.

Globorotalia kugleri (Figure 9), distinctly exhibited highest frequencies in the western tropical Pacific region. Other trans-Pacific tropical elements were *Dentoglobigerina altispira* (Figure 10) and *Globoquadrina praedehiscens* (Figure 11).

Three forms exhibited distinctly higher frequencies in the western tropical Pacific. These include *Globigerinoides* that broadly encompassed the tropical-warm subtropical western Pacific (Figure 12); *Globigerina angustiumbilicata* (Figure 13) and *Globoquadrina venezuelana* (Figure 14). Keller (1981a, b) previously recognized the western provincialism of *G. angustiumbilicata.*

The eastern tropical Pacific was dominated by a single complex; *Globorotalia siakensis* and *mayeri* (Figure 15), forming a distinct biogeographic province. This species complex extended in much reduced frequencies across the tropical-warm subtropical Pacific.

The factor analysis, employing 5 factors, incorporates 96 percent of the faunal variance within the earliest Miocene (22 Ma) data set. Five distinct faunal assemblages are recognizable, as follows (Figure 16).

Factor 1. Western tropical-Subtropical Assemblage (32% of the faunal variance). This assemblage (factor) is dominated by *Globorotalia kugleri* that tended to be more dominant in the western tropics, and *Globigerina angustiumbilicata* that was distributed broadly in the western subtropics and tropics but did not extend in abundance toward the eastern tropical Pacific.

Factor 2. Eastern Tropical Assemblage (23% of the faunal variance). This assemblage is dominated by the *Globorotalia siakensis–mayeri* complex that clearly defined a biogeographic province in the eastern tropical Pacific region.

Factor 3. Temperate Assemblage (16% of the faunal variance). This assemblage is dominated by *Globigerinita glutinata* and *Catapsydrax dissimilis* occurring in the temperate regions of

TABLE 3. DOMINANT PLANKTONIC FORAMINIFERAL SPECIES DURING SPECIFIC TIME INTERVALS
OF THE MIOCENE AS RECORDED IN TROPICAL, WARM SUBTROPICAL, TRANSITIONAL-TEMPERATE,
AND SUBANTARCTIC PROVINCES*

Epoch	Ma	Tropical	Warm-Subtropical	Transitional-Temperate	Subantarctic
Early Miocene	17.5-16.5	Globigerinoides spp. Gr. siakensis Gq. dehiscens D. altispira Gr. peripheroronda	Globigerinoides spp. Gq. dehiscens Gr. siakensis D. altispira Gg. praebulloides Gr. peripheroronda	Gr. periphero-ronda Gq. dehiscens Globigerinoides spp. Gr. miozea Gg. woodi Gg. praebulloides	Gr. miozea Gg. woodi
Early Miocene	20-20.5	Gr. kugleri Gg. angustiumbili-cata Gq. venezuelana Globigerinoides spp. Gr. siakensis-mayeri D. altispira Gq. praedehiscens	Gq. dehiscens Gg. praebulloides Globigerinoides spp. Gr. kugleri	Catapsydrax spp. Gq. dehiscens Gr. incognita Gg. woodi Gr. kugleri Ga. glutinata	Gg. woodi
Late Oligocene	24-25	Gr. kugleri Gg. angustiumbili-cata Catapsydrax spp. Gr. siakensis Gq. venezuelana	Catapsydrax spp. Gq praebulloides Gr. kugleri Gr. siakensis Gg. woodi	Catapsydrax spp. Gg. praebulloides Gr. kugleri Gq. venezuelana Gg. woodi	
Middle Miocene	12.5-11	Globigerinoides spp. Gr. siakensis Gr. menardii D. altispira Gq. venezuelana	Globigerinoides spp. Gr. menardii Gr. conoidea Gq. dehiscens D. altispira	Gr. conoidea Gq. dehiscens Gr. mayeri Gg. decoraperta Gg. praebulloides Gg. falconensis Gg. woodi	Gg. prae-bulloides Gr. challen-geri N. continuosa Gr. panda Gg. woodi
Middle Miocene	14-15.5	Globigerinoides spp. Gr. siakensis-mayeri D. altispira Gr. peripheroacuta Gq. venezuelana	Globigerinoides spp. Gr. siakensis-mayeri D. altispira Gq. dehiscens Gr. peripheroacuta	Gr. mayeri Gr. conoidea/miozea Gq. dehiscens Gg. druryi/decoraperta Gg. praebulloides Gg. woodi	Gr. mayeri/challengeri Gg. praebull-oides Gr. panda Gr. conoidea Gg. woodi
Late Miocene	5-6	Globigerinoides spp. Gr. menardii Gq. dehiscens Gg. decoraperata D. altispira N. acostaensis P. primalis	Globigerinoides spp. Gr. conomiozea Gg. decoraperta Gg. woodi Gr. menardii N. acostaensis	Gr. conomiozea N. pachyderma Gg. bulloides Gg. woodi	N. pachyderma Gg. bulloides Gr. cibaoen-sis/crassula
Late Miocene	9-10	Globigerinoides spp. Gr. menardii D. altispira N. acostaensis Gg. nepenthes-druryi Gq. venezuelana	Gr. conoidea Globigerinoides spp. Gr. menardii Gg. nepenthes-druryi Gg. decoraperta D. altispira Gg. woodi	Gr. conoidea Gg. bulloides Gg. woodi N. pachyderma	N. pachyderma Gg. bulloides

*Note: Gg = Globigerina; Gr = Globorotalia; Gq = Globoquadrina; D = Dentoglobigerina; N = Neogloboquadrina; P = Pulleniatina; Ga = Globigerinita.

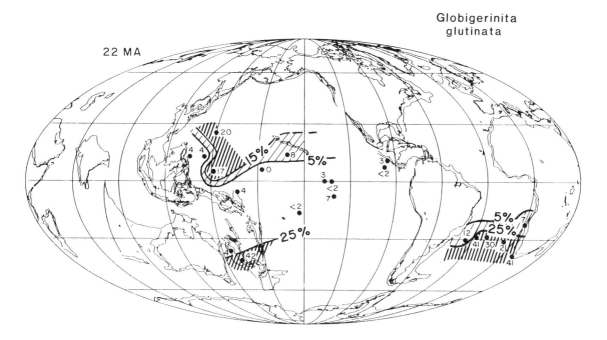

Figure 5. Percent distribution of *Globigerinita glutinata* during the earliest Miocene (22 Ma).

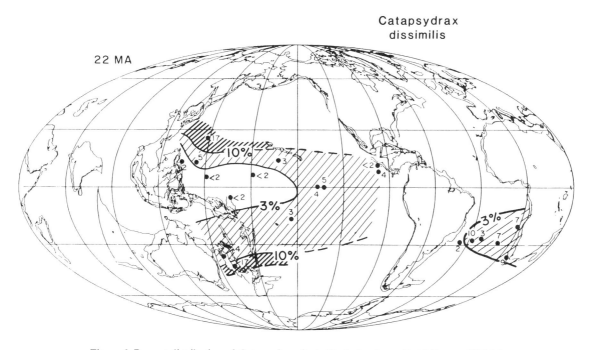

Figure 6. Percent distribution of *Catapsydrax dissimilis* during the earliest Miocene (22 Ma).

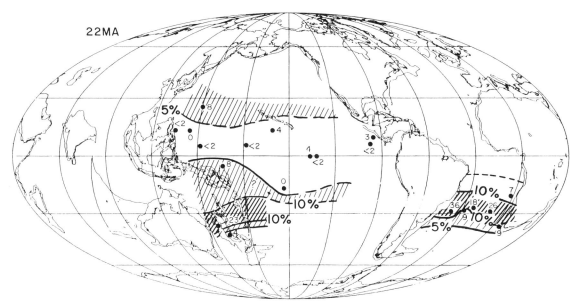

Figure 7. Percent distribution of *Globigerina praebulloides* during the earliest Miocene (22 Ma).

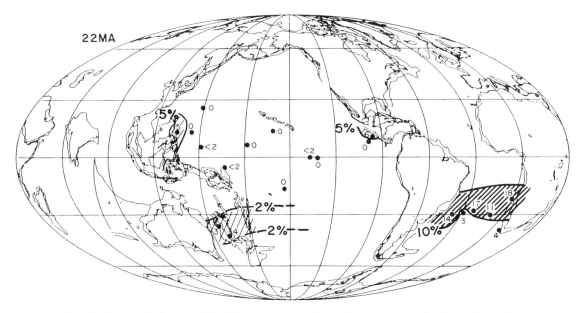

Figure 8. Percent distribution of the *Globigerina woodi–G. woodi connecta* complex during the earliest Miocene (22 Ma).

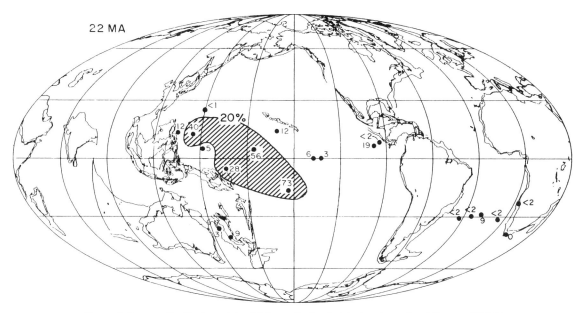

Figure 9. Percent distribution of the *Globorotalia kugleri* during the earliest Miocene (22 Ma).

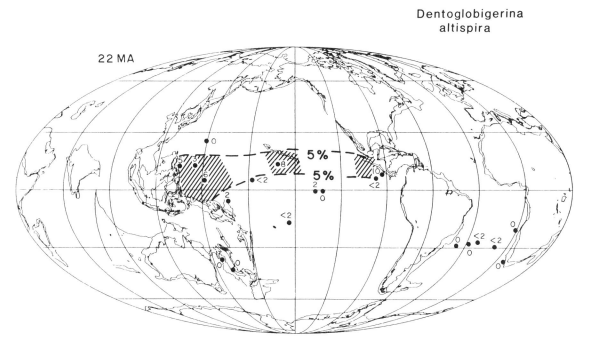

Figure 10. Percent distribution of *Dentoglobigerina altispira* during the earliest Miocene (22 Ma).

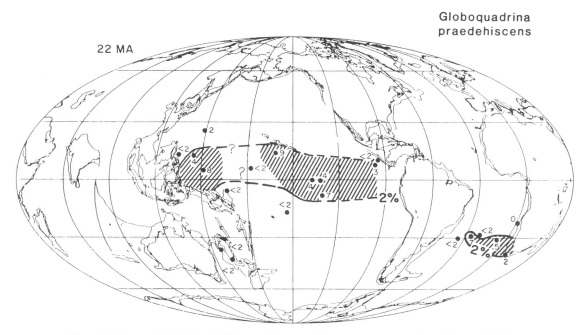

Figure 11. Percent distribution of *Globoquadrina praedehiscens* during the earliest Miocene (22 Ma).

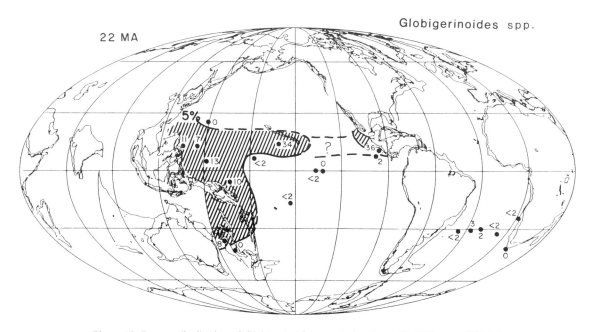

Figure 12. Percent distribution of *Globigerinoides* spp. during the earliest Miocene (22 Ma).

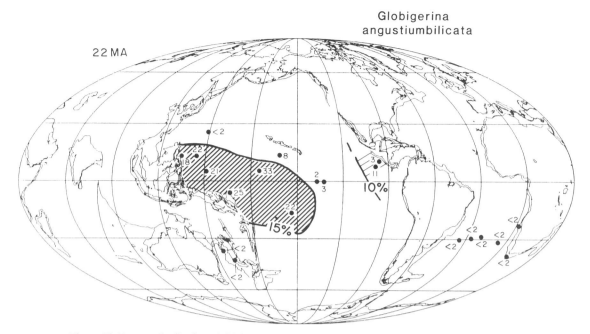

Figure 13. Percent distribution of *Globigerina angustiumbilicata* during the earliest Miocene (22 Ma).

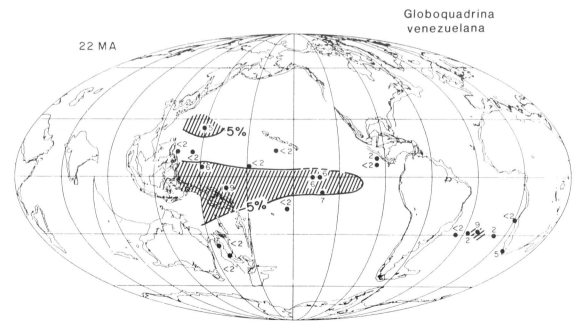

Figure 14. Percent distribution of *Globoquadrina venezuelana* during the earliest Miocene (22 Ma).

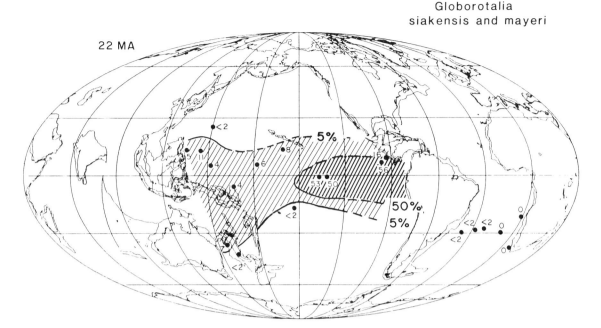

Figure 15. Percent distribution of the *Globorotalia siakensis-G. mayeri* complex during the earliest Miocene (22 Ma).

the Pacific. In the South Pacific this assemblage would almost certainly be latitudinally distributed across the temperate latitudes but there is insufficient sample coverage to define this pattern. It is also significant that the Temperate Assemblage occurred in relatively low latitudes in the northwestern Pacific, an area presently under the strong influence of the tropical-warm subtropical western boundary current.

Factor 4. Western Tropical (18% of the faunal variance). This assemblage is dominated by *Globigerinoides* and *Globigerina angustiumbilicata.* This factor was restricted to the low latitudes of the western Pacific. It broadly overlaps with Factor 1 but was more restricted to north central regions within the tropics.

Factor 5. Southwest warm subtropical (8% of the faunal variance). This assemblage is dominated by *Globigerina praebulloides* occurring in the warm subtropics of the southwest Pacific. This assemblage was almost certainly distributed latitudinally across the South Pacific but cannot be identified because of insufficient sample coverage.

Late Early Miocene: 16 Ma

For this time-slice, distributions are shown for eight taxa. By the late early Miocene subantarctic assemblages were dominated by *Globigerina praebulloides* (Figure 17) which also extends its range at much lower frequencies into temperate areas. During most of the early Miocene, this species was largely a warm subtropical species (Figure 7) but by the latest early to early middle Miocene it had switched its environmental tolerance to high latitudes, at least in the Pacific region.

Faunal elements clearly restricted to temperate areas were not apparent in the late early Miocene except for the *Globorotalia miozea-G. conoidea* complex. Species tended to exhibit broader distribution patterns and range into the warm subtropics. Important temperate-subtropical faunal elements during 16 Ma included *Globigerinita glutinata,* which also ranged into the warm subtropics and even into the tropics of the western Pacific (Figure 18); *Globoquadrina dehiscens* (Figure 19); and *Globigerina woodi* (Figure 20).

As during the earliest Miocene, tropical faunal elements tended to be distributed either in the western or eastern sectors of the Pacific. Faunal elements important in the west were *Globigerinoides* (Figure 21) which, similar to the earliest Miocene (Figure 12), were broadly distributed in both the tropics and subtropics; and *Dentoglobigerina altispira* (Figure 22). Compared with its distribution in the earliest Miocene (Figure 10), *Dentoglobigerina altispira* distinctly increased its frequencies in the western tropical Pacific, although it still extended to the east.

As in the earliest Miocene, the *Globorotalia siakensis-G. mayeri* complex dominated the eastern tropical Pacific (Figure 23), but higher frequencies ranged further westward than in the early Miocene. Similarly *Globoquadrina venezuelana* exhibited slightly higher frequencies in the eastern tropical Pacific (Figure 24).

Factor analysis of the 16 Ma data (5-factors; 96% of the faunal variance) shows the distribution of 5 biogeographic provinces as follows (Figure 25).

Factor 1. Western Tropical-Subtropical Assemblage (33% of the faunal variance). This assemblage, dominated by *Globige-*

22 MA

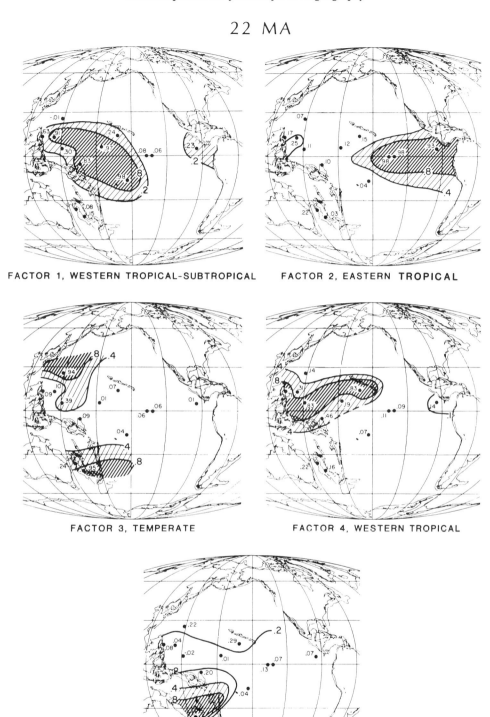

FACTOR 1, WESTERN TROPICAL-SUBTROPICAL

FACTOR 2, EASTERN TROPICAL

FACTOR 3, TEMPERATE

FACTOR 4, WESTERN TROPICAL

FACTOR 5, SOUTHWEST WARM SUBTROPICAL

Figure 16. Distribution of planktonic foraminiferal assemblages in the Pacific at 22 Ma (earliest Miocene). Time-slice cores are plotted with indicated factor scores (from Q-mode factor analysis). Contoured and shaded areas delineate regions dominated by each assemblage: Factor 1, Western Tropical-Subtropical Assemblage; Factor 2, Eastern Tropical Assemblage; Factor 3, Temperate Assemblage; Factor 4, Western Tropical Assemblage; and Factor 5, Southwest Warm Subtropical Assemblage.

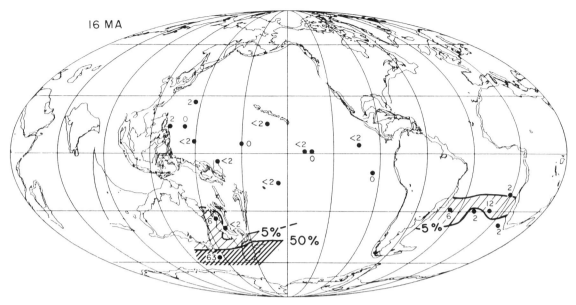

Figure 17. Percent distribution of *Globigerina praebulloides* during the latest early Miocene (16 Ma).

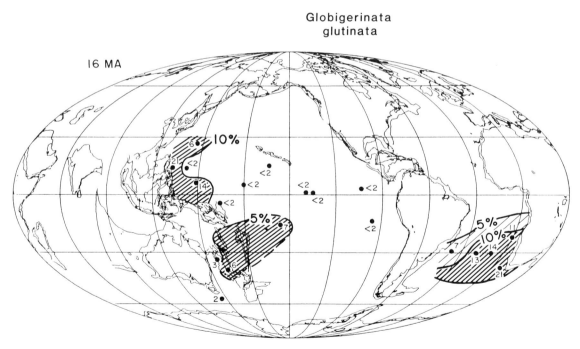

Figure 18. Percent distribution of *Globigerinita glutinata* during the latest early Miocene (16 Ma).

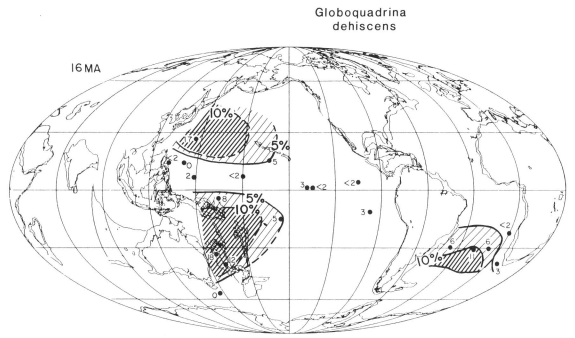

Figure 19. Percent distribution of *Globoquadrina dehiscens* during the latest early Miocene (16 Ma).

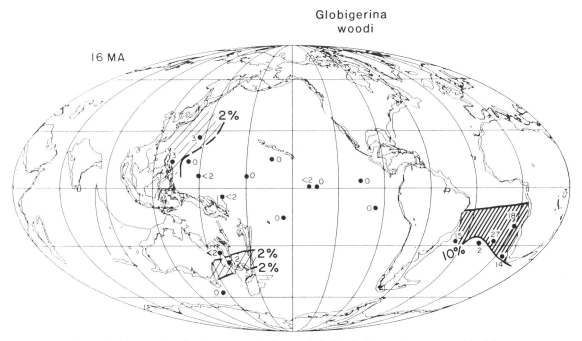

Figure 20. Percent distribution of *Globigerina woodi* during the latest early Miocene (16 Ma).

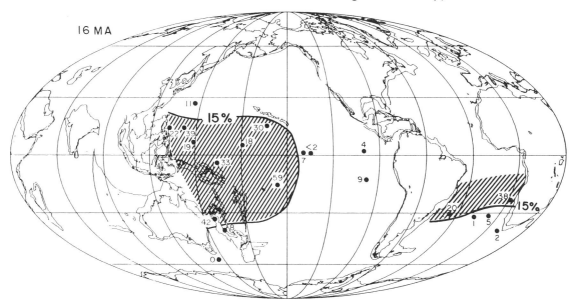

Figure 21. Percent distribution of *Globigerinoides* spp. during the latest early Miocene (16 Ma).

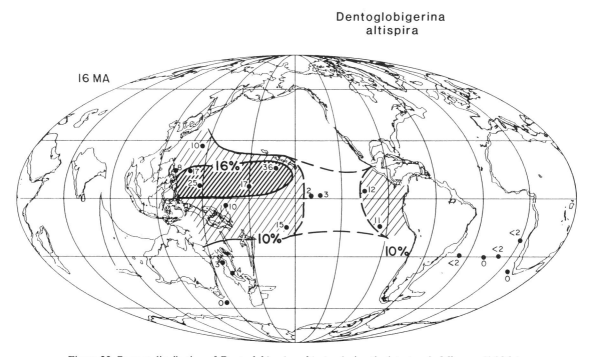

Figure 22. Percent distribution of *Dentoglobigerina altispira* during the latest early Miocene (16 Ma).

Globorotalia
siakensis and mayeri

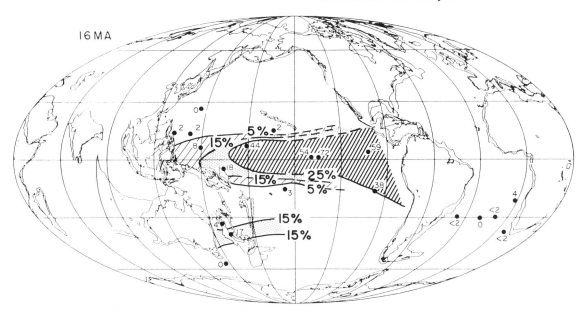

Figure 23. Percent distribution of the *Globorotalia siakensis–G. mayeri* complex during the latest early Miocene (16 Ma).

Globoquadrina
venezuelana

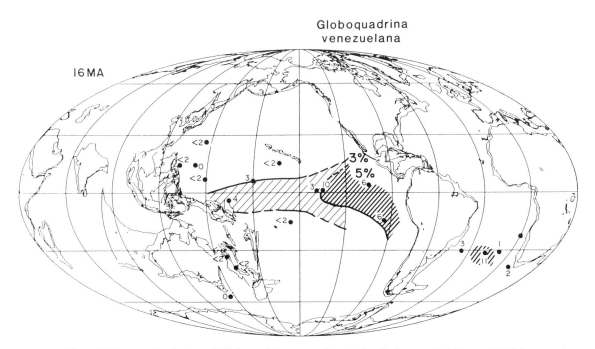

Figure 24. Percent distribution of *Globoquadrina venezuelana* during the latest early Miocene (16 Ma).

16 MA

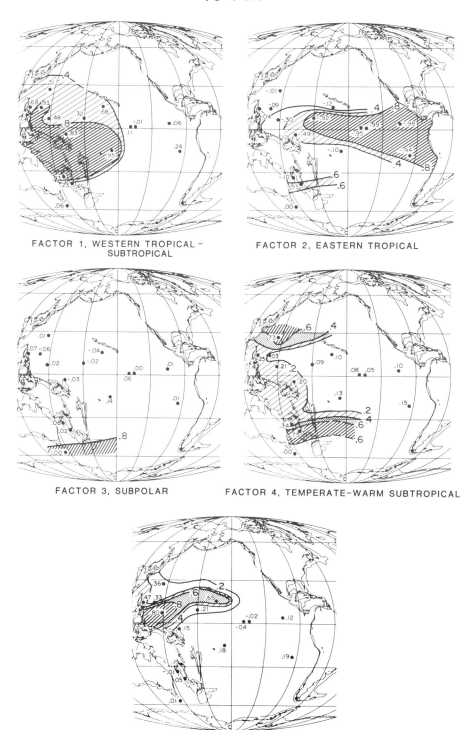

Figure 25. Distribution of planktonic foraminiferal assemblages in the Pacific at 16 Ma (latest early Miocene). Time-slice cores are plotted with indicated factor scores (from Q-mode factor analysis). Contoured and shaded areas delineate regions dominated by each assemblage: Factor 1, Western Tropical-Subtropical Assemblage; Factor 2, Eastern Tropical Assemblage; Factor 3, Subpolar Assemblage; Factor 4, Temperate-Warm Subtropical Assemblage; and Factor 5, Northwestern-Subtropical-Tropical Assemblage.

rinoides, was distributed broadly over the western tropical-subtropical Pacific, but did not extend toward the eastern tropical Pacific.

Factor 2. Eastern Tropical Assemblage (35% of the faunal variance). This assemblage is dominated, as in the earliest Miocene (Figure 16), by the *Globorotalia siakensis-G. mayeri* complex, that clearly defined an eastern tropical biogeographic province but extended further to the west than in the earliest Miocene (Figure 16).

Factor 3. Subpolar Assemblage (7% of the faunal variance). This assemblage is dominated by *Globigerina praebulloides* that clearly marked the subantarctic region in the South Pacific. This assemblage seems to have been latitudinally distributed around the Antarctic (including the South Atlantic), but insufficient samples were available to define this pattern.

Factor 4. Temperate-Warm Subtropical Assemblage (10% of the faunal variance). This assemblage is dominated by *Globoquadrina dehiscens* and the *Globorotalia miozea-G. conoidea* complex. This assemblage was most strongly developed in the temperate areas of the North and South Pacific but extended into warm subtropical and even tropical areas of the western Pacific. It was inferred that this assemblage was latitudinally distributed in temperate water masses, but not in the tropics where there was a relatively greater importance of this assemblage in the west compared to the east.

Factor 5. Northwestern Subtropical–Tropical Assemblage (11% of the faunal variance). This assemblage is dominated by *Dentoglobigerina altispira* and *Globigerinita glutinata* and was distributed in tropical-subtropical latitudes in the northwest Pacific.

Late Miocene: 8 Ma

For this time-slice, the distribution is shown for ten taxa. As with the late early Miocene time-slice, the subpolar assemblage was clearly dominated by *Globigerina bulloides* (Figure 26) (The ancestral form of *G. bulloides* is *G. praebulloides.)* This species also extended into temperate areas as an important element. It showed up clearly as the dominant element associated with the California Current; and was an important cool-water element in both hemispheres. It was also an important element in temperate latitudes of the southeast Atlantic adjacent to South Africa associated with regions of upwelling (Hodell and Kennett 1985).

The temperate regions were marked by high frequencies of *Globigerina woodi* (Figure 27) and *Globorotalia conoidea* (Figure 28), although *G. woodi* was an unimportant component of north Pacific assemblages.

In this time-slice, the distribution of most tropical taxa were unlike that of the two early Miocene time-slices. Instead of exhibiting high frequencies in the western or eastern tropics, the species were generally distributed across the tropics. The patterns are most different between the 8 and 22 Ma maps, while the 16 Ma maps tend to be intermediate in character. Such patterns are shown by the *Globorotalia menardii-limbata* complex that was

almost equatorial in its distribution (Figure 29); *Globigerinoides* was more broadly tropical and warm subtropical (Figure 30); the *Globigerina nepenthes-G. druryi* complex (Figure 31) although widespread (temperate to tropics) favored tropical to warm subtropical areas away from the equator; and *Globoquadrina venezuelana* that was closely associated with the trans-equatorial Pacific region (Figure 32).

Another important change in the distribution of planktonic foraminifera between the 8 Ma and the earlier time-slices was the strong northward distributional extension into the northwest Pacific of certain tropical taxa; in particular *Globigerinoides* and the *Globigerina nepenthes-G. druryi* complex (Figures 30 and 31). In the early Miocene no tropical species were found to extend their ranges so far north in such high frequencies.

There are two tropical species that may not have been distributed entirely across the tropical Pacific: *Dentoglobigerina altispira* (Figure 33) and *Neogloboquadrina acostaensis* (Figure 34). The latter, however, showed relatively high frequencies in both west and eastern sectors, separated by low frequencies in only a single mid-ocean sample (Figure 34).

The distribution of *Globigerinita glutinata* was unusually disjointed over a wide range of latitudes (Figure 35), with highest frequencies in the far eastern equatorial Pacific. Tolderlund and Bé (1971) have shown that *G. glutinata,* a surface dweller, is the most nearly ubiquitous planktonic foraminiferal species in the Modern ocean (Bé and Hamlin 1967).

Factor analysis of the 8 Ma data (5 factors; 93% of the faunal variance) shows the distribution of four distinct faunal assemblages and one less distinctive grouping as follows (Figure 36).

Factor 1. Tropical–Trans-Pacific Assemblage (23% of the faunal variance). This assemblage is dominated by the *Globigerina nepenthes-G. druryi* complex, which was distributed across the tropical Pacific and swept northward into the northwest Pacific as far as Japan in association with the western boundary current.

Factor 2. Subpolar-Temperate Assemblage (21% of the faunal variance). This assemblage is dominated by *Globigerina bulloides* and was distributed in polar to temperate latitudes in both hemispheres.

Factor 3. Trans-Equatorial Assemblage (18% of the faunal variance). This assemblage is dominated by the *Globorotalia menardii-G. limbata* complex and *Globigerinoides* and was distributed across the Pacific Ocean in close association with the equatorial and warmer tropics. High abundance of this assemblage was more closely associated with and restricted to latitudes close to the equator compared with the tropical-transpacific assemblage. Highest frequencies were closely associated with the equatorial region as in the modern ocean (Tolderlund and Bé 1971).

Factor 4. Temperate-Transitional Assemblage (20% of the faunal variance). This assemblage is dominated by *Globorotalia conoidea* and *Globigerina woodi* and almost certainly formed a latitudinally distributed zone in the Southern Hemisphere, where

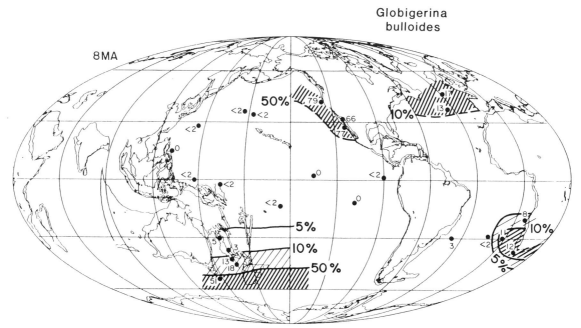

Figure 26. Percent distribution of *Globigerina bulloides* during the late Miocene (8 Ma).

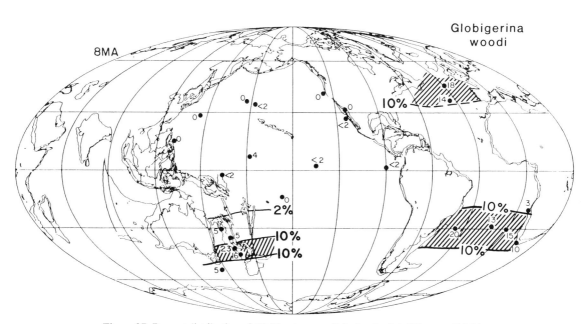

Figure 27. Percent distribution of *Globigerina woodi* during the late Miocene (8 Ma).

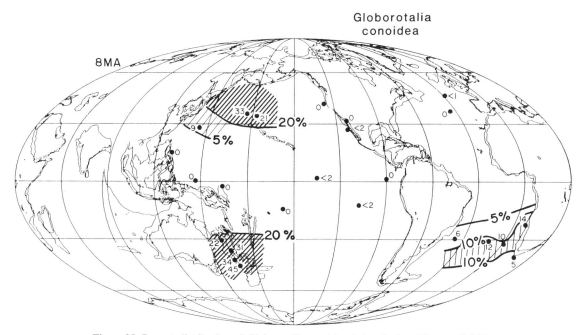

Figure 28. Percent distribution of *Globorotalia conoidea* during the late Miocene (8 Ma).

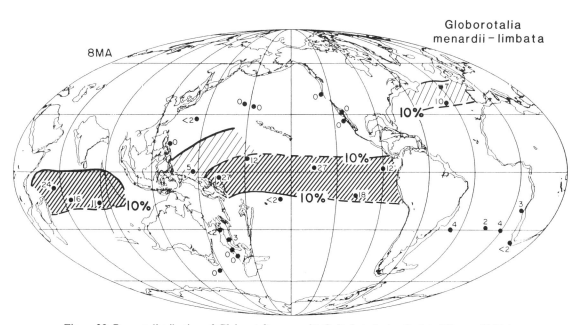

Figure 29. Percent distribution of *Globorotalia menardii–G. limbata* during the late Miocene (8 Ma).

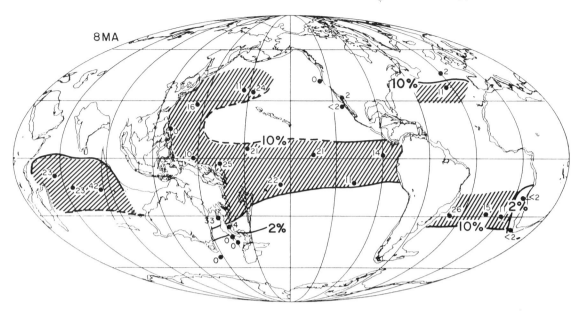

Figure 30. Percent distribution of *Globigerinoides* spp. during the late Miocene (8 Ma).

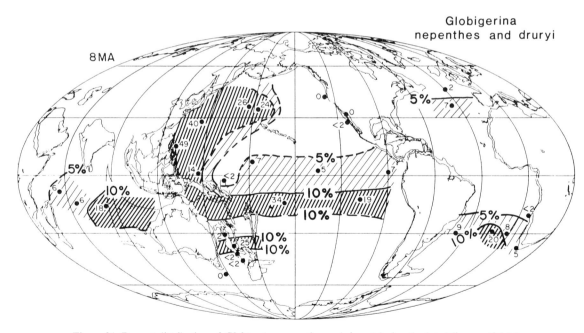

Figure 31. Percent distribution of *Globigerina nepenthes* and *druryi* during the late Miocene (8 Ma).

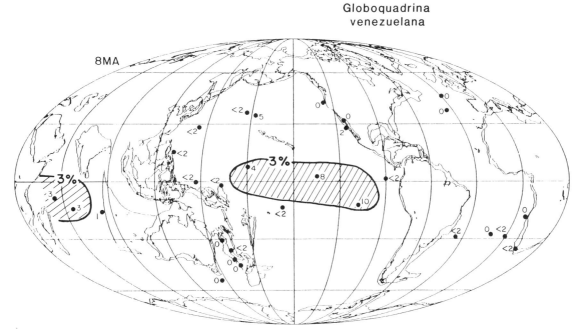

Figure 32. Percent distribution of *Globoquadrina venezuelana* during the late Miocene (8 Ma).

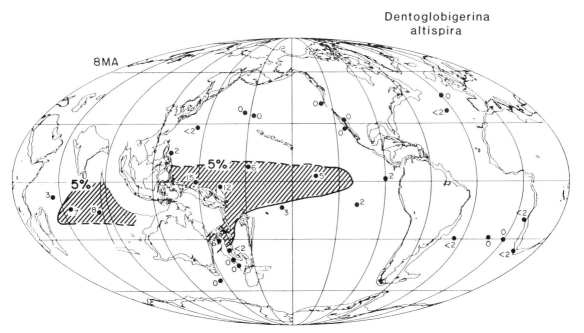

Figure 33. Percent distribution of *Dentoglobigerina altispira* during the late Miocene (8 Ma).

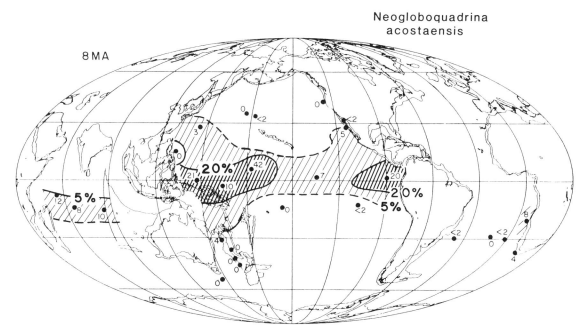

Figure 34. Percent distribution of *Neogloboquadrina acostaensis* during the late Miocene (8 Ma).

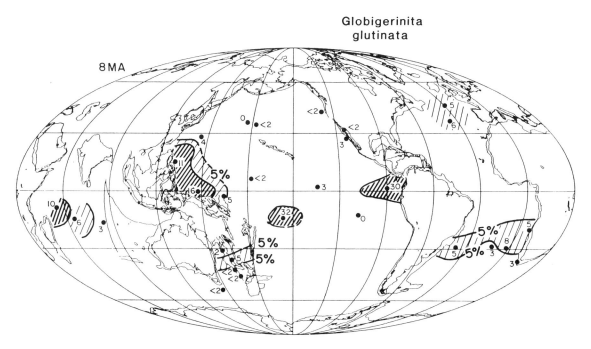

Figure 35. Percent distribution of *Globigerinita glutinata* during the late Miocene (8 Ma).

8 MA

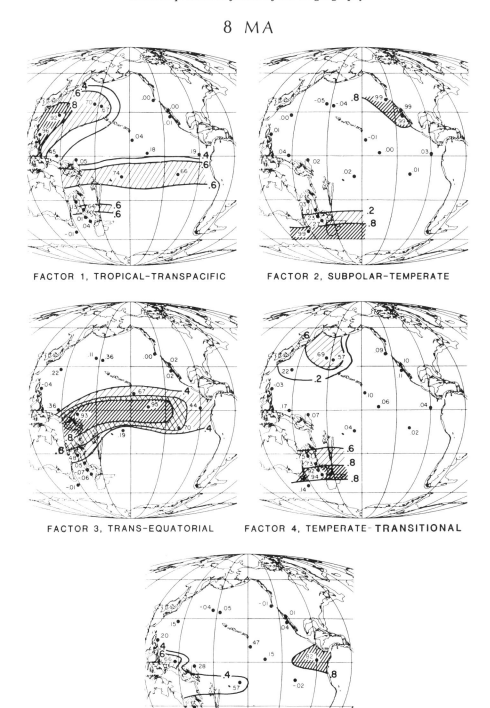

FACTOR 1, TROPICAL–TRANSPACIFIC

FACTOR 2, SUBPOLAR–TEMPERATE

FACTOR 3, TRANS–EQUATORIAL

FACTOR 4, TEMPERATE–**TRANSITIONAL**

FACTOR 5, TROPICAL–MARGINAL

Figure 36. Distribution of planktonic foraminiferal assemblages in the Pacific at 8 Ma (late Miocene). Time-slice cores are plotted with indicated factor scores (from Q-mode factor analysis). Contoured and shaded areas delineate regions dominated by each assemblage: Factor 1, Tropical-Transpacific Assemblage; Factor 2, Subpolar-Temperate Assemblage; Factor 3, Trans-Equatorial Assemblage; Factor 4, Temperate-Transitional Assemblage; and Factor 5, Tropical-Marginal Assemblage.

it was most strongly developed. In the Northern Hemisphere it appeared to form an extension of the tongue of the warm-water assemblage represented by Factor 1.

Factor 5. Tropical-Marginal Assemblage (10% of the faunal variance). This assemblage is dominated by *Globigerinita gluti-nata* and *Neogloboquadrina acostaensis*. This assemblage was most strongly developed in the eastern equatorial Pacific, but occurred to a lesser extent in the western tropical Pacific, relatively close to the continental areas.

DISCUSSION

The primary objective of this mapping experiment was to determine if fundamental differences existed in the distribution patterns of planktonic foraminiferal species and assemblages between each of the three time-slice intervals. Differences might provide useful information about changes in surface-water circulation during the Miocene. Changes in Miocene biogeographic patterns should have occurred as a result of major changes in the boundary conditions of the oceans and of global climates. During each of the Miocene time-slices the Central American Seaway was open. However, during the middle to early late Miocene three fundamental changes occurred that should have affected Indo-Pacific surface-water circulation. First, continued northward expansion occurred of the Antarctic-Subantarctic water masses as Australia continued its drift northwards and as the polar glacial regime developed (Kennett 1977). This also would have contributed to the increase in temperature gradients between equator and poles. Loutit et al. (1983b) calculated that the late Miocene latitudinal temperature gradient in the Southern Hemisphere was about ¾ of the present surface ocean and that it doubled during the Miocene from about 6° to about 12°C. This was in response to the thermal isolation of Antarctica. Second, there occurred major accumulation of Antarctic ice in the form of an ice sheet from about 15 to 13 m.y. ago (Shackleton and Kennett 1975; Savin et al 1975) that resulted from the thermal isolation and cooling of the Antarctic region. The development of this ice sheet would have created positive climatic feedback and further cooling of the Antarctic region, and led to further steepening of the equator to pole temperature gradient. Third, the Indonesian Seaway became increasingly constricted, so that by the middle Miocene–early late Miocene it ceased to be an effective gateway for surface-water transportation between the tropical Pacific and Indian Oceans (van Andel et al., 1975; Edwards 1975; Kennett 1977; Sclater et al., 1985).

Our descriptions of the biogeographic maps demonstrate that important changes did occur in the biogeographic patterns of planktonic foraminifera during the Miocene and that the most important of these occurred between 16 and 8 m.y. ago (late-early and late Miocene) (See Keller 1985). Patterns are relatively similar between the two early Miocene time-slices. Thunell and Belyea (1982) have previously shown that one of the major changes in planktonic foraminiferal biogeography of the Atlantic occurred during the middle Miocene as a result of steepening of

the latitudinal temperature gradients. The primary biogeographic differences between the early and late Miocene require summary before discussion of their implications. They are as follows.

1. During the early Miocene there existed distinct faunal assemblages (provinces) in the western and eastern sectors of the tropical Pacific (Figure 37). By the late Miocene these differences had largely disappeared and trans-tropical distributions began to prevail.

2. During the early Miocene, planktonic foraminiferal patterns in the northwest and western Pacific suggest that the gyre system was more weakly developed than during the late Miocene, although greater sample coverage is needed in the North Pacific to confirm this.

3. During the early Miocene, temperate assemblages extended much further south in the northwest Pacific. By the late Miocene these were displaced northward by subtropical-tropical assemblages as the Kuroshio Current intensified.

4. During the early Miocene, there was less distinct latitudinal zonality in the South Pacific and tropical North Pacific, although this was already developed in temperate and warm subtropical areas. By the late Miocene, these zones became narrower and more distinctly latitudinal in distribution (Kennett 1983). Increased provincialism of faunas during the Miocene has been recorded previously by several workers (Berggren and Hollister 1974, 1977; Sancetta 1978; Haq 1980; Lohmann and Carlson 1981). The provincialism apparently occurred in response to changes in the thermal structure of water masses and steepening gradients.

5. During the early Miocene, there were no distinctly different equatorial and tropical assemblages. By the late Miocene, separate assemblages seemed to mark the equatorial and tropical regions, as distinct from a broader tropical assemblage.

6. During the early Miocene in the southwest Pacific, the tropical biogeographic province was broader than during the late Miocene. Keller (1981a) noted that tropical assemblages were broader during the early Miocene. From studies in the Atlantic, Thunell and Belyea (1982) described a contraction of the tropical-subtropical assemblages during the Neogene. In the early Miocene, warmer-water assemblages extended to higher latitudes in the Atlantic, with the North Atlantic warmer than the South Atlantic.

What created the changes in tropical planktonic foraminiferal distribution patterns during the middle Miocene to early late Miocene? Selective calcium carbonate dissolution of planktonic foraminiferal assemblages can potentially produce much geographic variation during any time interval. We do not consider that the major biogeographic patterns described were caused by selective foraminiferal dissolution that concentrated certain species for the following reasons.

1. Samples containing very high frequencies of *G. siakensis* and *G. mayeri* do not exhibit clear evidence of having undergone strong dissolution, as would be shown by common specimen breakage. Also the walls of *G. siakensis* and *G. mayeri* are not noticeably thickened to suggest specimen concentration as a re-

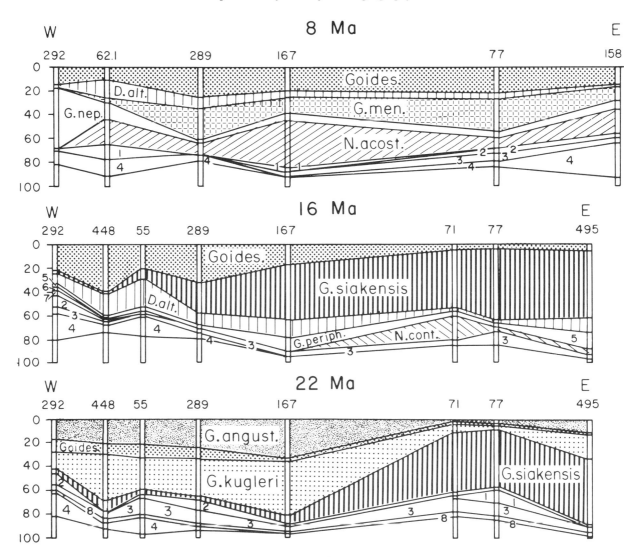

Figure 37. West to east trans-tropical Pacific traverses for three time-slices (22 Ma; 16 Ma; 8 Ma) to show changes in percent frequency distribution in planktonic foraminiferal assemblages (See Figures 1 to 3 for DSDP site locations. Abbreviations for taxa are as follows: 8 Ma: Goides: *Globigerinoides*; D. alt.: *Dentoglobigerina altispira*; G. men.: *Globorotalia menardii limbata*; G. nep.: *Globigerina nepenthes-druryi*; N. acost.: *Neogloboquadrina acostaensis*; 1: *Globigerina woodi*; 2: *Globigerina bulloides*; 3: *Globoquadrina venezuelana*; 4: *Globigerinita glutinata*. 16 Ma: Goides: *Globigerinoides*; G. siakensis also includes *Globorotalia mayeri*; D. alt.: *Dentoglobigerina altispira*; G. periph. and 5: *Globorotalia peripheroronda*; N. cont. and 6: *Neogloboquadrina continuosa*; 2: *Globigerina bulloides*; 3: *Globoquadrina venezuelana*; 4: *Globigerinita glutinata*. 22 Ma: G. angust.: *Globigerina angustiumbilicata*; Goides: *Globigerinoides*; G. kugleri: *Globorotalia kugleri*; G. siakensis also includes *Globorotalia mayeri*; 1: *Globigerina woodi*; 2: *Globigerina praebulloides*; 3: *Globoquadrina venezuelana*; 4: *Globigerinita glutinata*; 8: *Catapsydrax dissimilis*.

sult of severe dissolution that would be required to eliminate nearly all specimens of *Globigerinoides*. In other words, the large differences between eastern and western tropical Pacific assemblages during the early Miocene cannot be explained by differences in preservation, because there is no clear evidence for this. Such evidence would include common specimen breakage and concentration of specimens with relatively thickened tests.

2. The faunal differences between eastern and western tropical sectors (22 and 16 Ma) had largely disappeared by the late

Miocene (8 Ma) when the paleodepths of individual sites were deeper than during the early Miocene (Sclater et al. 1985). This, by itself, should have increased dissolution of assemblages, and hence exaggerated east-west differences. The calcium carbonate compensation depth (CCD) in the Pacific shallowed by about 250 m between the earliest Miocene and the late Miocene (8 Ma) (van Andel et al. 1975). This should have increased dissolution over a wider area in the late Miocene relative to the early Miocene.

Instead, unlike the early Miocene, late Miocene assemblages contain abundant (11 to 21%) *Globigerinoides* spp., a solution susceptible form, across the equatorial Pacific region. Indeed, the deepest site in the late Miocene tropical traverse (Site 77; 4127 m paleodepth) has among the highest frequencies (21%) of *Globigerinoides*. Altogether, these environmental changes would have increased dissolution during the late Miocene compared to the early Miocene and enhanced rather than decreased east-west faunal differences as the data shows.

3. In general, the deepest sites studied are in the eastern tropical Pacific, which by itself suggests a greater potential effect of dissolution upon the assemblages in this area. However, the detailed faunal-paleodepth relations do not support this. For example, in the 22 Ma time-slice, the deepest site (Site 567; 5339 m paleodepth) contains high frequencies of *Globigerinoides*. In contrast, the 16 Ma time-slice, a relatively shallow site (Site 167; 3163 m paleodepth) containing high frequencies (44%) of *G. siakensis* and *G. mayeri* also contains high frequencies (18%) of *Globigerinoides*. This association indicates that *G. siakensis* and *G. mayeri* were not concentrated by dissolution.

We believe that the faunal differences are valid biogeographic differences and that these are largely related to paleoceanographic changes. The changes were almost certainly caused, in part, by the development during the middle Miocene–early late Miocene of an effective barrier to Indo-Pacific tropical surface-water circulation. Paleoreconstructions have been made by Hamilton (1979) of the Banda Sea region between Indonesia and Australia-New Guinea (Figure 38). These show that the Indonesian Seaway would effectively have been closed during the middle Miocene, preventing any further surface-water circulation between the tropical Pacific and Indian Oceans. During the early and middle Miocene, surface-water transport would have been enhanced through the seaway (up to 1300 km wide) by the submergence of much or all of Java, Sulawesi, and Sumatra. All of these areas record regional subsidence and marine transgression during the middle Tertiary, followed by uplift, emergence, and related regression during the late Neogene, leading to the present, largely continental setting (Hamilton 1979). Audley-Charles et al. (1972) showed that the first major middle to late Cenozoic orogenic phase to affect the Banda Arc was during the middle Miocene. On the other hand, Borneo behaved more or less as a craton in middle and late Cenozoic times and had a major effect on diverting Pacific equatorial currents to the north and south. The position of Borneo relative to the equator during the early and middle Miocene is important (and not known with any accuracy; E. Silva, personal communication), since it affected the relative strengths of the western boundary current (Kuroshio Current) and the equatorial currents flowing through the Indonesian Seaway. New Guinea represents the northern leading edge of the northward moving Australian craton (Figure 38). During the early and middle Miocene, it was probably emergent enough to form the southern continental margin to the Indonesian Seaway. On the other hand, major orogeny first occurred in eastern New Guinea during the middle Miocene (Audley-Charles et al., 1972),

about 11 to 12 Ma, as a result of plate collision as Australia-New Guinea moved to the north. This resulted in initiation of turbidite deposition in the Coral Sea (Burns, et al. 1973). By the late middle to early late Miocene, the Indonesian and New Guinea land masses were sufficiently "interlocked" to prevent any further Pacific equatorial current transport into the Indian Ocean.

The development of this barrier created the Equatorial Undercurrent system that would have been previously undeveloped or nonexistent across the equatorial Pacific. The barrier would have created a pile-up of surface waters in the western equatorial Pacific and hence the development of an easterly flowing Equatorial Undercurrent (Knauss 1963). The barrier also may have helped strengthen the Equatorial Countercurrent, which results largely from the strength of winds and their differences in the equatorial region. A strong undercurrent system should have existed in the Indian Ocean because of the barrier formed by Africa to further westward surface-water transport. One of the principal effects of the modern Pacific countercurrent system is to return warm water to the eastern tropical Pacific (Figure 4). Van Andel et al. (1975) suggested that the Equatorial Undercurrent first developed about 12–11 Ma, a conclusion based upon observed large increases in biogenic sedimentation rates along the equator that they assumed resulted from increased upwelling. Leinen (1979) found no evidence of strong east-to-west gradients in opal accumulation characteristic of the Equatorial Undercurrent upwelling, before the middle Miocene. Siliceous biogenic productivity increased during the middle Miocene and peaked in the late Miocene about 8 Ma. This increase was interpreted to reflect the strengthening of the Equatorial Undercurrent system. Loutit et al., (1983b) suggest that equatorial Pacific sea-surface temperatures increased during the middle Miocene, at the time when surface waters cooled at higher latitudes.

In the late Miocene, the trans-tropical distribution pattern of *Globigerinoides*, which are warm-water, surface dwellers, is considered to record this paleocirculation change. Furthermore, the development of a distinct trans-equatorial province (as distinct from a broader, undifferentiated tropical province) marked by the *Globorotalia menardii–G. limbata* complex, may have been in response to the development of the Undercurrent system. These forms are non-spinose and in the Modern ocean are generally deeper-dwelling forms (except in areas of a shallow thermocline) living below 50 m and probably deeper during late ontogeny (Bé and Tolderlund 1971; Keller 1985). In the Atlantic, Jones (1967) observed that *G. menardii* in abundance is closely associated with the Equatorial Undercurrent with maximum frequencies of more than 25 percent. In the Pacific, Bradshaw (1959) showed highest abundances of *G. menardii* from plankton tows closely associated with the equator in the western Pacific but with a wider tropical distribution in the eastern Pacific. Curry et al. (1983) inferred that *G. menardii* calcifies on the thermocline in the Panama Basin. At the time of the year when *G. menardii* is most abundant in the Panama Basin, the thermocline is at a depth of only 10 m.

During the first two million years of its range, following its evolution from *G. praemenardii* 12 m.y. ago, *G. menardii* was

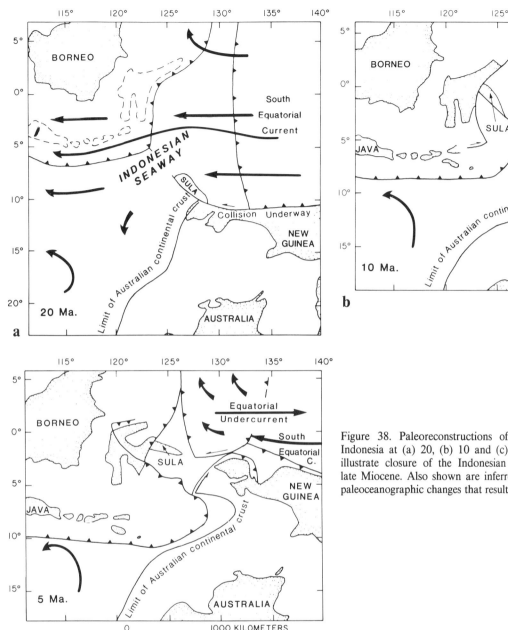

Figure 38. Paleoreconstructions of region between Australia and Indonesia at (a) 20, (b) 10 and (c) 5 Ma (after Hamilton 1979) to illustrate closure of the Indonesian Seaway between the early and late Miocene. Also shown are inferred surface-water and near-surface paleoceanographic changes that resulted from closure of seaway.

largely a warm-subtropical rather than equatorial form. However, about 10 Ma this species became more abundant close to the equator, as it is today (Tolderlund and Bé 1971), and seems to have replaced the important equatorial lineage—*Globorotalia (Fohsella)* which became extinct about 12 Ma, and perhaps even the tropical lineage *Globorotalia (Jenkinsella)* which became extinct about 11 Ma. The extinction of major Neogene lineages with no ancestral forms are unusual. These extinctions during the late middle Miocene may have been in response to the inferred large-scale paleocirculation changes in the tropical-equatorial regions.

Thus, the increased development of trans-tropical faunal distributions during or following the middle Miocene, and the con-

spicuous reduction of east-west biogeographic differences is considered to be partly in response to the development of this countercurrent system. These currents assisted in reducing inferred stronger environmental gradients that had existed between the eastern and western tropical Pacific during the early Miocene.

Another change that may have helped reduce east-west tropical biogeographic differences in the early Miocene would have been a general strengthening of the oceanic gyral systems resulting from a steepening of global latitudinal temperature gradients. An intensification of latitudinal temperature gradients almost certainly occurred during the middle Miocene when much of the east Antarctic ice sheet is believed to have accumulated (Shackleton & Kennett 1975; Savin et al. 1975). In the North Pacific, faunal evidence indicates that such a strengthening did take place. A more sluggish oceanic circulation system might weaken east-west surface water transport in the tropics and in turn would favor the maintenance of the faunal differences that we have observed in the early Miocene. Conversely, an increase in gyre strength should help reduce the east-west faunal differences leading to a late Miocene pattern. For the Quaternary ocean, Moore (1978) observed that during the last ice age tropical radiolarian faunal distributions are much more trans-Pacific compared with the modern ocean where there are distinct west to east faunal differences. These changes were interpreted to reflect differences in strength of the circulation.

Gyral circulation in the North Pacific was intensified by the closing of the Indonesian Seaway during the middle Miocene. This should have strengthened the Kuroshio Current and led to the northward expansion of tropical assemblages into the northwest Pacific. The southeastern areas of the Japanese islands may have exhibited increased warming during the late middle Miocene (as expressed in subtropical species distributions along the marginal northwest Pacific) at a time when global climates were cooling. However, there is a widespread hiatus in Neogene sections on the Pacific coast of Japan, in the late middle Miocene (12–14 Ma) (Ujiie 1984) that will make such evidence difficult to obtain. However, in the northern half of Japan, tropical faunas were replaced by temperate faunas after about 13 Ma (Saito 1963; Berggren 1984) as the Oyashhio Current intensified. Thus a very large latitudinal thermal gradient developed in the ocean east of Japan.

An important corollary to these findings is that an open equatorial-tropical seaway system between several oceans does not, by itself, produce uniform surface-water paleoenvironmental conditions girdling the tropics and hence uniformity of biogeographic patterns. With the absence of a relatively strong global circulation and an absence of a strong countercurrent system, the open central American and Indonesian Seaways seemed to have had minimal effect in producing a trans-tropical faunal province. Less mixing took place between the east and western assemblages and faunas of the North Pacific gyre may have been largely isolated from the influence of tropical circulation.

It is more difficult to determine the actual environmental changes in the surface water masses that lead to the changes in Miocene biogeographic patterns. Specifically, what changes occurred in water-mass structure that led to such a dominance of the *G. siakensis–G. mayeri* plexus in the eastern tropical Pacific during the early Miocene (Figure 37) and different forms in the western tropical Pacific (22 Ma time-slice—*Globigerina angusti-umbilicata, Globigerinoides* and *Globorotalia kugleri*; 16 Ma time-slice *Globigerinoides* and *Globorotalia kugleri*; 16 Ma time-slice *Globigerinoides* and *Dentoglobigerina altispira*)? If we knew the preferred depth habitat of these taxa, the problem would be easier to solve, but most of these species are now extinct and several, for example, the *G. siakensis–G. mayeri* plexus, do not even have modern living descendents.

The oxygen isotopic composition of the foraminiferal tests can serve as a general guide to the depth of calcification of foraminiferal tests in the upper part of the water column (Douglas and Savin 1978); but this, by itself, can be unreliable because several modern species, especially those spinose forms with symbiotic zooxanthellae, may fractionate oxygen isotopes out of equilibrium with temperature and the oxygen isotopic composition of the sea water in which they live (Fairbanks et al. 1980; Fairbanks et al. 1982). Modern forms calcify out of oxygen isotopic equilibrium by approximately –0.3 to 0.4 per mil in δO^{18} values (Fairbanks et al. 1982; Curry et al. 1983). *Globigerinoides* are extant forms in which all species are spinose surface dwellers that prefer tropical-warm subtropical water masses. Most modern spinose species of planktonic foraminifera exhibit maximum concentrations in the upper 10 m of the water column (Bé and Tolderlund 1971). All species of *Globigerinoides* may have had symbiotic zooxanthellae (Bé 1977; Bé and Hutson 1977) and thus have calcified out of oxygen isotopic equilibrium. Oxygen isotopic measurements of Miocene *Globigerinoides* (Biolzi 1983; Savin et al. 1984; Keller 1985) show that this taxa almost always provides the lightest measured values of several taxa within an assemblage. Also, neither have the basic morphology of *Globigerinoides* species, nor have the latitudinal distribution patterns of these forms changed much during the Neogene, which suggests that they have not changed their depth preference in the uppermost part of the water column.

The preferred depth habits of other taxa are more difficult to interpret. Deeper-dwelling non-spinose, modern species may be at, or close to oxygen isotopic equilibrium (Fairbanks et al. 1982; Curry et al. 1983). Therefore, the oxygen isotopic values of inferred deeper-dwelling taxa in Miocene assemblages are probably more reliable. But what, for instance, was the preferred depth habitat of the *Globorotalia siakensis–G. mayeri* complex? Oxygen isotopic values of these forms are almost always relatively light compared with most species, suggesting a surface-water preference (Biolzi 1983; Keller 1985). In general, these forms have slightly heavier values than the *Globigerinoides,* (Biolzi 1983, shows heavier values of between 0.2 and 0.8 per mil in δO^{18} values), but then *Globigerinoides,* a known surface-dweller, fractionates to provide lighter values than the water in which it lives. More important is that *G. mayeri* is consistently heavier by .07 to .83 per mil in δO^{18} values than *Globoquadrina dehiscens* (Biolzi 1983), a non-spinose form that almost certainly did not live in the upper few meters of the ocean. The surface ultrastructure (Kennett and Srinivasan 1983) of the *G. siakensis–G. mayeri* complex indicates a lack of spines. Furthermore, general morphological considerations suggest that this plexus may have preferred a slightly deeper habitat than *Globigerinoides.* The morphology of this plexus is generally similar to *Neogloboquadrina* that includes extant, known deeper-dwelling species that calcify on the thermocline (Curry et al. 1983), but there is no close phylogenetic relationship between these forms. The morphology of *G. siakensis* is similar to *N. dutertrei* while that of *G. mayeri* is generally similar to *N. pachyderma.*

Our preferred interpretation is that the *G. siakensis–G.*

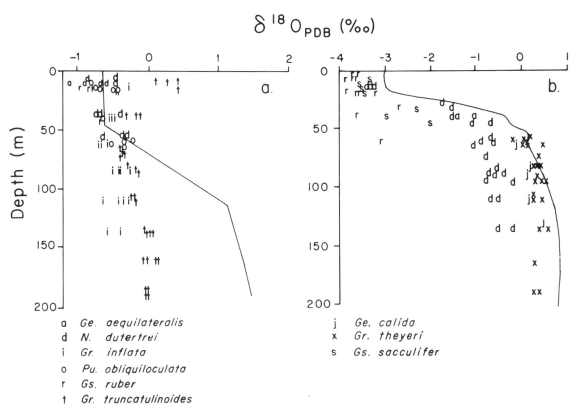

Figure 39. Oxygen isotopic composition of the tests of various living planktonic foraminiferal species and size fractions (from planktonic tows) as a function of water depth in (a) region of deep thermocline; and (b) shallow thermocline. Solid lines in (a) and (b) are the calculated equilibrium $CaCo_3-\delta^{18}O$ variation with depth. Note similarity of oxygen isotopic composition within and between species in area of deep thermocline (a) compared with changing values with increasing water depth associated with shallow thermocline (b). (a) is after Fairbanks et al. (1980) for the western North Atlantic; (b) is after Fairbanks et al. (1982) for the Panama Basin.

mayeri plexus lived at slightly deeper depths than *Globigeri-noides,* closer to the thermocline. Other evidence suggests that deeper-dwelling species were quantitatively favored in early Miocene assemblages in the eastern tropics. Higher frequencies of *Catapsydrax dissimilis* occur in the eastern tropics, but not to the west (Figure 6). This taxa is heavily calcified, provides by far the heaviest δO^{18} values and light δO^{13} values typical of a deep pelagic habitat, and hence was a deep-dwelling form probably living below the thermocline (Biolzi 1983; Keller 1985).

The relative quantitative importance of species in assemblages may be determined by the position of the thermocline within the upper part of the photic zone where more food is available (i.e., upper 100 m of the water column). Some species prefer to live within the thermocline, some above and some below. If a species prefers to live and calcify within the thermocline, it will adjust its depth habitat to follow any movement of the thermocline. If the thermocline is shallow, deeper-water species will be quantitatively favored in the assemblage. Fairbanks et al. (1982) have found that most modern foraminiferal species precipitate most of the calcium carbonate in their tests in the upper 100 m of the water column (upper photic zone), even though a species may be found in abundance below 100 m.

Figure 39 shows two contrasting models of planktonic foraminiferal growth and oxygen isotopic composition dependent upon a deeper (Figure 39a) or shallower (Figure 39b) thermocline (Fairbanks et al. 1980; Fairbanks et al. 1982). Where the thermocline is deeper relative to the photic zone, most foraminiferal productivity and calcification occurs in the photic zone above the thermocline, and thus between-species $\delta^{18}O$ values are similar because of the relatively uniform water-mass temperatures in which calcification takes place. Where the thermocline is deep, shallower-dwelling forms such as *Globigerinoides* are quantitatively more important in assemblages (Thunell et al. 1983). In contrast, where a shallow thermocline exists within the photic zone (Figure 39b) most calcification and growth of foraminifera occurs within and below the thermocline in association with the chlorophyl maximum where food is more plentiful. This leads to greater between-species differences in $\delta^{18}O$ values and a higher quantitative importance of deeper-dwelling taxa (Thunell et al. 1983).

There seem to be two possible hypotheses to explain east-west faunal differences during the early Miocene. The first hypothesis is that surface waters were generally cooler in the eastern tropical Pacific, giving rise to the biogeographic differences

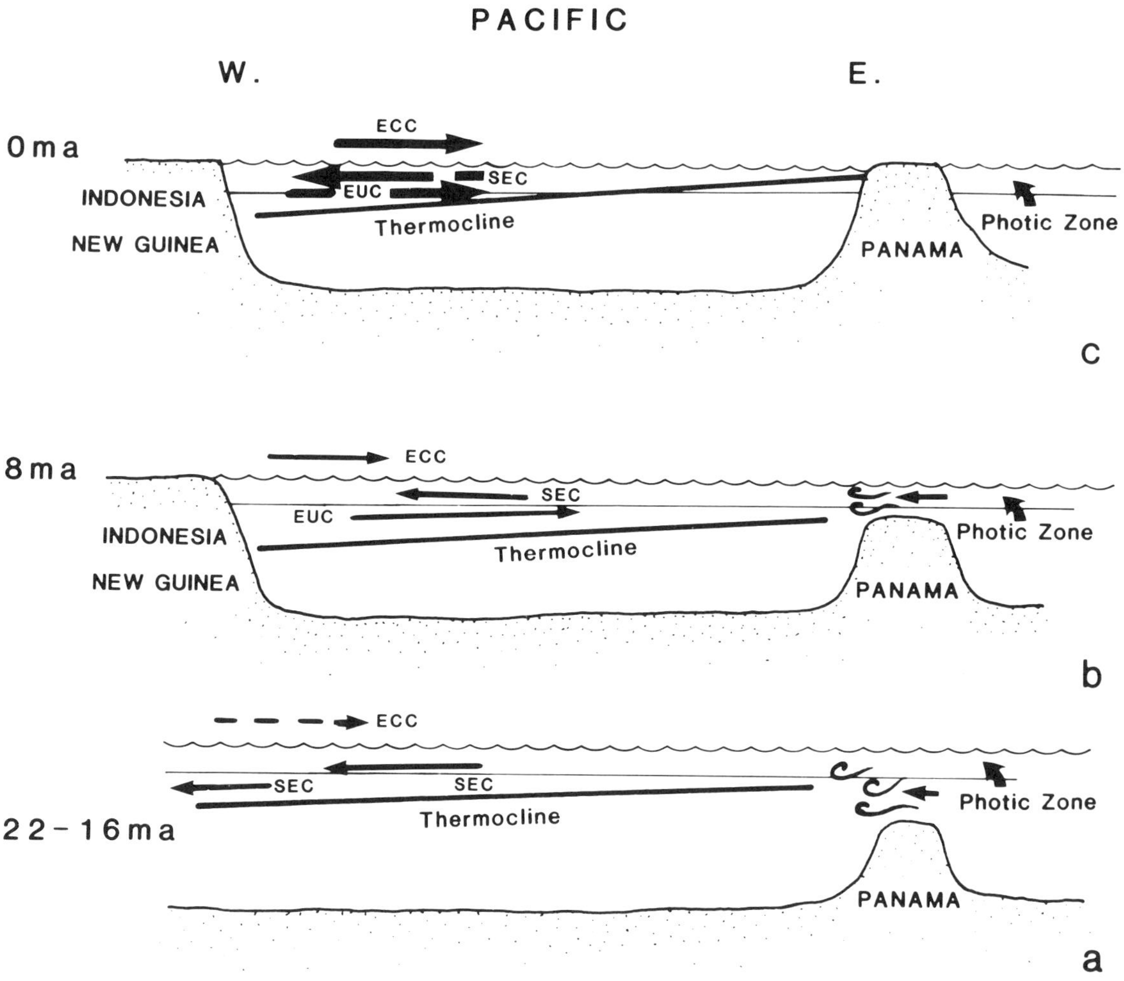

PACIFIC

W.   E.

Figure 40. Suggested Neogene changes in surface water-mass structure in the equatorial Pacific Ocean between early Miocene (a); late Miocene (b); and the Modern ocean (c). Shown in each case is the position of the thermocline relative to the photic zone and current directions of major currents as follows: SEC: South Equatorial Current; ECC: Equatorial Countercurrent; EUC: Equatorial Undercurrent. a. Early Miocene—22–16 Ma. Both Indonesian and Central American Seaways open to surface waters. Thermocline is relatively deep across the equatorial Pacific in absence of an Equatorial Undercurrent. Equatorial Countercurrent (dotted line) is weak. b. Late Miocene—8 Ma. Indonesian Seaway is now closed, the barrier forming the Equatorial Undercurrent. Thermocline is shallower in the photic zone. Equatorial Countercurrent is moderately strong. c. Modern Ocean—0 Ma. Central American Seaway is now closed. All surface water circulation is more vigorous. Stronger Equatorial Undercurrent further raises thermocline within photic zone especially in eastern Pacific. Equatorial Countercurrent is more vigorous transporting warm water to east.

among surface or near-surface forms. In the western tropical Pacific, surface waters were not only distinctly warmer, but tropical conditions were also broader in the South Pacific. The relatively sluggish equatorial circulation, the absence of an Equatorial Undercurrent, and a weak Equatorial Countercurrent allowed distinct biogeographic differences to develop between the east and the west.

The alternate hypothesis states that the east-west differences resulted from a relatively shallow thermocline over broad areas of the eastern tropical Pacific, and a deeper thermocline in the western tropical Pacific (Figure 40). Nevertheless, the thermocline

across the tropical Pacific was somewhat deeper perhaps than that of the modern ocean and located everywhere in the deeper part of, or below the photic zone (Figure 40a). Shallow-water warmer taxa (such as *Globigerinoides* spp.) were favored in the west and slightly deeper, cooler species (especially *G. siakensis-G. mayeri*) were favored in the east.

With a strengthening of the Equatorial Countercurrent by the late Miocene (Figure 40b), warm, shallow-water dwellers such as *Globigerinoides* spread across almost the entire tropical Pacific. Also, the development of the Equatorial Undercurrent system (Figure 40b) returned relatively warmer, deeper waters to the eastern tropics. Deeper-dwelling forms such as the *Globorotalia menardii–G. limbata* plexus and *N. acostaensis* were able to exploit this water mass across much of the tropical-equatorial Pacific (Figure 37). Thus, by the late Miocene, both shallow (*Globigerinoides*) and deeper forms (*G. menardii–G. limbata; N. acostaensis*) occurred in faunas across the equatorial Pacific as a result of seasonal changes in the depth of the thermocline and of upwelling that must have been much stronger than during the early Miocene.

Late Miocene equatorial patterns of planktonic foraminiferal distributions did not persist into the Modern ocean. As in the early Miocene, modern assemblages again exhibit important east-west differences across the tropical Pacific. These differences, however, are largely restricted to taxa that normally live deeper within the water column. The eastern province, marked by strong seasonal upwelling, is dominated by *N. dutertrei,* followed by *G. tumida* and then *P. obliquiloculata* along a westward path (Parker and Berger 1971). The Panama Basin exhibits a strong dominance of *N. dutertrei* and *Globorotalia theyeri* as a result of strong seasonal upwelling (Thunell et al. 1983). In contrast, typically shallow-dwelling forms (such as *G. ruber* and *G. sacculifer*) occur at important elements from west to east across the equatorial region (Parker 1960; Bé 1977; Bradshaw 1959; Coulbourn et al. 1980; Thompson 1981), as in the late Miocene, but unlike the early Miocene. Also, *G. menardii* is distributed across the tropical Pacific with highest frequencies occurring in close association with the equator.

The modern pattern is thus different from both early and late Miocene patterns, by exhibiting east-west differences in deeper-dwelling forms but at the same time maintaining important populations of shallow-water taxa across the region. The pattern is promoted by a relatively shallow thermocline (which seasonally becomes extremely shallow in the east) maintained by a strong Equatorial Undercurrent. This favors production of the "deeper-dwelling" taxa (*N. dutertrei, G. tumida* and *P. obliquiloculata*). There is also a strong Equatorial Countercurrent that returns warmer waters to the east and favors seasonal production of "surface-dwelling" taxa (*G. ruber, G. sacculifera* and *G. aequilateralis*). It seems that the Modern equatorial Pacific differs from the early Miocene (Figure 40A) in having a shallower thermocline located well within the photic zone (Figure 40C), especially in the east due to more vigorous circulation and a strong Equatorial Undercurrent, and a strong return flow of warm surface waters and associated faunas as a result of a stronger Equatorial Countercurrent.

CONCLUSIONS

1. Biogeographic patterns of Pacific planktonic foraminifera have been quantitatively mapped for two time-slices in the early Miocene (22 Ma and 16 Ma) and one in the late Miocene (8 Ma).

2. Important changes occurred in the biogeographic patterns of planktonic foraminifera, especially between the early (16 Ma) and late Miocene (8 Ma), which are interpreted as

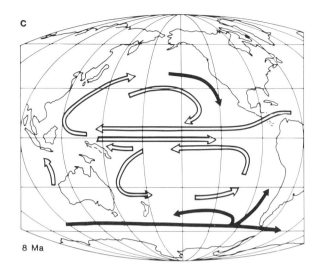

c

8 Ma

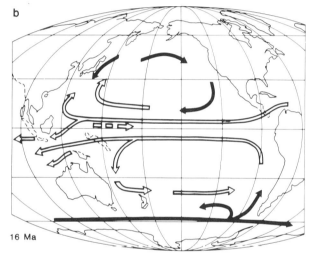

b

16 Ma

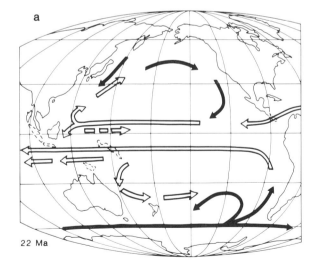

a

22 Ma

Figure 41. Inferred circulation patterns of surface and near surface waters in the Pacific Ocean at (a) 22 Ma, (b) 16 Ma, and (c) 8 Ma. Closed arrows indicate cold currents; open arrows indicate warm currents. Paleoreconstructions after Sclater et al. (1985).

reflecting major changes in surface-water circulation of the Indo-Pacific region (Figure 41).

3. In the tropical Pacific during the early Miocene, the planktonic foraminiferal assemblages were dominated by different taxa in the eastern and western areas. This changed by the late Miocene when the assemblages became more similar across the entire expanse of the tropical Pacific. These changes are interpreted to reflect both the development, during the middle Miocene, of the Equatorial Undercurrent system when the Indonesian Seaway effectively closed and the general intensification of tropical and gyral surface-water circulation, including a strengthening of the Equatorial Countercurrent, that resulted from a steepening of latitudinal temperature gradients.

4. The east to west faunal differences across the tropical Pacific in the early Miocene were due to differences in the surficial water-mass structure. Either generally warmer surface waters or the presence of a deeper thermocline in the west quantitatively favored shallow-water dwellers such as *Globigerinoides,* while a shallower thermocline in the east favored slightly "deeper-dwelling" forms. These differences across the equatorial Pacific were maintained because the Equatorial Countercurrent system was only weakly developed (Figure 41a, b).

5. A separate equatorially distributed assemblage (*Globorotalia menardii–G. limbata* and *Globigerinoides*) had developed by the late Miocene, perhaps in response to the development of the Equatorial Undercurrent system and the strengthening of the Equatorial Countercurrent (Figure 41c). Two important tropical Neogene lineages became entirely extinct about this time, perhaps in response to these water-mass changes.

6. Modern distribution patterns in the equatorial Pacific differ from those of both the early and late Miocene. East-west differences occur in "deeper-dwelling" forms while "shallow-dwelling" forms occur in seasonal abundance across the region. This pattern resulted from the intensification of the Countercurrent system that had developed by the late Miocene. This had the effect of shallowing the thermocline in the equatorial Pacific and strengthening the return flow of warm waters to the east within the Equatorial Countercurrent.

7. In the northwest and western Pacific, the distribution of faunas indicates that the gyral circulation system was only weakly developed in the early Miocene, but was strong by the late Miocene. This resulted from the closure of the Indonesian Seaway during the middle Miocene and the general intensification of ocean circulation as the latitudinal temperature gradient steepened.

8. Other changes that occurred between the early and late Miocene were the northward displacement of temperate faunas in the northwest Pacific as the Kuroshio Current intensified; the development of more distinct latitudinally distributed, narrower provinces in the south and tropical Pacific; and in the south Pacific, the northward expansion of the polar-subpolar province and the latitudinal contraction of the tropical provinces. These changes resulted from the continued areal expansion of the polar and subpolar water masses as Australia drifted northward from Antarctica and from the steepening of pole to equator thermal gradients related to increased Antarctic glaciation.

9. The early Miocene tropical distribution patterns indicate that an open equatorial-tropical seaway system between several oceans has not, by itself, led to the development of uniform surface-water paleoenvironmental conditions girdling the tropical regions and producing a uniform biogeographic province.

10. Several planktonic foraminiferal lineages and species changed their surface-water preferences during the Miocene, although most lineages remain closely associated with particular water masses.

ACKNOWLEDGMENTS

We are most grateful to the other workers in the CENOP project for their valuable discussions and their diverse contributions to the project. Indian Ocean data for several maps was provided by E. Vincent; Atlantic data by D. Hodell and R. Poore. We appreciate the detailed, constructive discussions with K. Romine, D. Hodell, K.-Y. Wei, and S. Savin. This manuscript benefited greatly from suggestions made by K. Romine, T. Moore, R. Thunell, and J. Ingle. N. Meader typed the manuscript. Drafting was done by B. Watkins, L. Defusco and M. McHugh and photography by S. Silvia. Research was supported by the U.S. National Science Foundation, grants OCE79-14594 (CENOP) and OCE82-14937 (Geological Oceanography) to the University of Rhode Island and, in part, by NSF Grant OCE20-00887900 and OCE29-18285 (CENOP) to Stanford University.

REFERENCES CITED

Adams, C. G., Benson, R. H., Kidd, R. B., Ryan, W.B.F., and Wright, R. C., 1977, The Messinian salinity crisis and evidence of Late Miocene eustatic changes in the world ocean: Nature, v. 269, p. 383–386.

Audley-Charles, M. G., Carter, D. J., and Milsom, J. S., 1972, Tectonic development of Eastern Indonesia in relation to Gondwanaland dispersal: Nature Physical Science, v. 239, p. 35–39.

Bandy, O. L., 1968, Cycles in Neogene paleoceanography and eustatic changes: Palaeogeography, palaeoclimatology, and palaeoecology, v. 5, p. 63–73.

Barron, J. A., and Keller, G., 1983, Paleotemperature oscillations in the Middle and Late Miocene of the northeastern Pacific: Micropaleontology, v. 29, n. 2, p. 150–181.

Barron, J. A., Keller, G., and Dunn, D. A., 1985, A multiple microfossil bio-chronology for the Miocene, *in* Kennett, J. P., ed., The Miocene Ocean: Paleoceanography and biogeography: Geological Society of America Memoir 163, this volume.

Bé, A.W.H., 1977, An ecological, zoogeographic and taxonomic review of recent planktonic foraminifera; *in* Ramsay, A.T.S., ed., Oceanic Micropaleontology: London, Academic Press, v. 1, p. 1–100.

Bé, A.W.H., and Hamlin, W. H., 1967, Ecology of Recent planktonic foraminifera: Part 3—Distribution in the North Atlantic during the summer of 1962: Micropaleontology, v. 13, n. 1, p. 87–106.

Bé, A.W.H., and Hutson, W. H., 1977, Ecology of planktonic foraminifera and biogeographic patterns of life and fossil assemblages in the Indian Ocean: Micropaleontology, v. 23, n. 4, p. 369–414.

Bé, A.W.H., and Tolderlund, D. S., 1971, Distribution and ecology of living planktonic foraminifera in surface waters of the Atlantic and Indian Oceans; *in,* Funnell, B. M., and Riedel, W. R., eds., The Micropaleontology of Oceans: London, Cambridge University Press, p. 105–149.

Berggren, W. A., 1969, Rates of evolution in some Cenozoic planktonic foraminifera: Micropaleontology, v. 15, p. 351–365.

—— 1984, Neogene Planktonic Foraminiferal Biostratigraphy and Biogeography: Atlantic, Mediterranean, and Indo-Pacific Regions, *in* Ikebe, N., and Tsuchi, R., eds., Pacific Neogene Datum Planes: Tokyo, University of Tokyo Press, p. 111–161.

Berggren, W. A., and Hollister, C. D., 1974, Paleogeography, paleobiogeography and the history of circulation in the Atlantic, *in* Hay, W. W., ed., Studies in Paleo-Oceanography: Society of Economic Paleontologists and Mineralogists Special Publication 20, p. 126–186.

—— 1977, Plate tectonics and paleocirculation-commotion in the ocean: Tectonophysics, v. 38, p. 11–48.

Biolzi, M., 1983, Stable isotopic study of Oligocene-Miocene sediments from DSDP Site 354, equatorial Atlantic: Marine Micropaleontology, v. 8, p. 121–139.

Bradshaw, J. S., 1959, Ecology of living planktonic foraminifera in the north and equatorial Pacific Ocean: Contributions Cushman Foundation for Foraminiferal Research, v. 10, part 2, p. 25–64.

Burns, R. E., et al., 1973, Initial Reports of the Deep Sea Drilling Project, Volume 21, Washington, D.C., U.S. Government Printing Office, 931 p.

Cifelli, R., 1969, Radiation of Cenozoic planktonic Foraminifera: Systemic Zoology, v. 18, p. 154–168.

CLIMAP, 1976, The surface of the ice-age earth: Science, v. 1919, p. 1131–1137.

—— 1981, Seasonal Reconstructions of the Earth's surface at the last glacial maximum: Geological Society of America Map and Chart Series, MC-36.

Coulbourn, W. T., Parker, F. L., and Berger, W. H., 1980, Faunal and solution patterns of planktonic foraminifera in surface sediments of the North Pacific: Marine Micropaleontology, v. 5, p. 329–399.

Curry, W. B., Thunell, R. C., and Honjo, S., 1983, Seasonal changes in the isotopic composition of planktonic foraminifera collected in Panama Basin sediment traps: Earth and Planetary Science Letters, v. 64, p. 33–43.

Douglas, R. G., and Savin, S. M., 1978, Oxygen isotopic evidence for the depth stratification of Tertiary and Cretaceous planktic foraminifera: Marine Micropaleontology, v. 3, p. 175–196.

Eade, J. V., 1973, Geographical distribution of living planktonic foraminifera in the Southwest Pacific; *in* Fraser, R., compiler, Oceanography of the South Pacific: Wellington, New Zealand, National Commission for UNESCO, p. 249–256.

Edwards, A. R., 1975, Southwest Pacific Cenozoic Paleogeography and an Integrated Neogene Paleocirculation Model; *in* Andrews, J. E., et al., Initial Reports of the Deep Sea Drilling Project, Volume 30: Washington, D.C., U.S. Government Printing Office, p. 667–684.

Fairbanks, R. G., Sverdlove, M., Free, R., Wiebe, P. H., and Bé, A.W.H., 1982, Vertical distribution and isotopic fractionation of living planktonic foraminifera from the Panama Basin: Nature, v. 298, p. 841–846.

Fairbanks, R. G., Wiebe, P. H., and Bé, A.W.H., 1980, Vertical distribution and isotopic composition of living planktonic foraminifera in the western North Atlantic: Science, v. 207, p. 61–63.

Hamilton, W., 1979, Tectonics of the Indonesian Region: Geological Society of America Professional Paper No. 1078, 345 p.

Haq, B. U., 1980, Biogeographic history of Miocene calcareous nannoplankton and paleoceanography of the Atlantic Ocean: Micropaleontology, v. 26, n. 4, p. 414–443.

Haq, B. U., Worsley, T. R., Burckle, L. H., Douglas, R. G., Keigwin, L. D., Jr., Opdyke, N. D., Savin, S. M., Sommer, M. A., III, Vincent, E., and Woodruff, F., 1980, Late Miocene marine carbon-isotopic shift and synchroneity of some phytoplanktonic biostratigraphic events: Geology, v. 8, p. 427–431.

Hodell, D. A., and Kennett, J. P., 1985, Miocene Paleoceanography of the South Atlantic Ocean, *in* Kennett, J. P., ed., The Miocene Ocean: Paleoceanography and biogeography: Geological Society of America Memoir 163, this volume.

Imbrie, J., and Kipp, N. G., 1971, A new micropaleontological method for quantitative paleoclimatology: Application to a Late Pleistocene Caribbean core; *in* Turekian, K. K., ed., The Late Cenozoic Glacial Ages: New Haven, Yale University Press, p. 71–181.

Ingle, J. C., Jr., 1967, Foraminiferal biofacies variation and the Miocene-Pliocene boundary in Southern California: Bulletin of American Paleontology, v. 52, n. 236, p. 217–394.

Jones, J. I., 1967, Significance of distribution of planktonic foraminifera in the Equatorial Atlantic Undercurrent: Micropaleontology, v. 13, n. 4, p. 489–501.

Keigwin, L. D., Jr., 1979, Late Cenozoic stable isotope stratigraphy and paleoceanography of Deep Sea Drilling Project sites from the East Equatorial and Central North Pacific Ocean: Earth and Planetary Science Letters, v. 45, p. 361–382.

Keller, G., 1981a, Planktonic foraminiferal faunas of the equatorial Pacific suggest Early Miocene origin of Present ocean circulation: Marine Micropaleontology, v. 6, p. 269–295.

—— 1981b, Miocene biochronology and paleoceanography of the North Pacific: Marine Micropaleontology, v. 6, p. 535–551.

—— 1981c, The genus *Globorotalia* in the Early Miocene of the equatorial and northwestern Pacific: Journal of Foraminiferal Research, v. 11, n. 2, p. 118–132.

—— 1985, Depth Stratification of planktonic foraminifers *in* Kennett, J. P., ed. The Miocene ocean: Paleoceanography and biogeography: Geological Society of America Memoir 163, this volume.

Keller, G., and Barron, J. A., 1983, Paleoceanographic implications of Miocene deep-sea hiatuses: Geological Society of America Bulletin, v. 94, n. 5, p. 590–613.

Kennett, J. P., 1967, Recognition and correlation of the Kapitean Stage (Upper Miocene, New Zealand): New Zealand Journal of Geology and Geophysics, v. 10, p. 1051–1063.

—— 1977, Cenozoic evolution of Antarctic glaciation, the Circum-Antarctic Ocean, and their impact on global paleoceanography: Journal of Geophysical Research, v. 82, n. 27, p. 3843–3860.

—— 1978, The development of planktonic biogeography in the southern ocean during the Cenozoic: Marine Micropaleontology, v. 3, p. 301–345.

—— 1982, Marine Geology: Englewood Cliffs, N.J.: Prentice-Hall, 813 p.

—— 1983, Paleo-oceanography: Global ocean evolution, *in* U.S. National report to International Union of Geodesy and Geophysics 1979–1982, Reviews of Geophysics and Space Physics, v. 21, n. 5: American Geophysical Union, p. 12581–1274.

Kennett, J. P., ed., 1981, Cenozoic paleoceanography: Marine Micropaleontology, v. 6, p. V–VI.

Kennett, J. P., and Srinivasan, M. S., 1983, Neogene planktonic foraminifera: New York, Hutchinson Ross, 263 p.

Knauss, J. A., 1963, Equatorial Current Systems, *in* Hill, M. W., and others, eds., The Sea: New York, Wiley Interscience Publications, v. 2, p. 235–252.

Leinen, M., 1979, Biogenic silica accumulation in the central equatorial Pacific and its implications for Cenozoic paleoceanography: Geological Society of America Bulletin, v. 90, p. 1310–1376.

Lipps, J. H., 1970, Plankton evolution: Evolution, v. 24, p. 1–22.

Lohmann, G. P., and Carlson, J. J., 1981, Oceanographic significance of Pacific Late Miocene calcareous nannoplankton: Marine Micropaleontology, v. 6, p. 553–579.

Loutit, T. S., 1981, Late Miocene paleoclimatology: Subantarctic water mass, Southwest Pacific: Marine Micropaleontology, v. 6, p. 1–27.

Loutit, T. S., Pisias, N. G., and Kennett, J. P., 1983a, Pacific Miocene carbon isotope stratigraphy using benthic foraminifera: Earth and Planetary Science Letters, v. 66, p. 48–62.

Loutit, T. S., Kennett, J. P., and Savin, S. M., 1983b, Miocene equatorial and Southwest Pacific paleoceanography from stable isotope evidence: Marine Micropaleontology, v. 8, n. 3, p. 215–233.

Matthews, R. K., and Poore, R. Z., 1980, Tertiary $\delta^{18}O$ record and glacioeustatic sea-level fluctuations: Geology, v. 8, p. 501–504.

Moberly, R., 1972, Origin of Lithosphere behind Island Arcs, with Reference to the Western Pacific: Geological Society of America Memoir No. 132, p. 35–55.

Moore, T. C., Jr., 1978, The distribution of radiolarian assemblages in the modern and ice-age Pacific: Marine Micropaleontology, v. 3, p. 229–266.

Moore, T. C., Jr., Burckle, L. H., Geitzenauer, K., Luz, B., Molina-Cruz, A., Robertson, J. H., Sach, H., Sancetta, C., Thiede, J., Thompson, P., and Wenkam, C., 1980, The reconstruction of sea surface temperatures in the Pacific Ocean of 18,000 B.P.: Marine Micropaleontology, v. 5, p. 215–247.

Moore, T. C., Jr., van Andel, T. H., Sancetta, C., and Pisias, N., 1978, Cenozoic hiatuses in pelagic sediments: Micropaleontology, v. 24, p. 113–138.

Pak, H., and Zaneveld, J. R., 1974, Equatorial front in the eastern Pacific Ocean: Journal of Physical Oceanography, v. 4, p. 570–578.

Parker, F. L., 1960, Living planktonic foraminifera from the equatorial and southeast Pacific: Science Reports of the Tohoku University, Sendai, Japan, Second Series (Geology), Special Volume, No. 4, (Hanzawa Memorial Volume), p. 71–82.

—— 1971, Distribution of planktonic foraminifera in Recent deep-sea sediments, *in* Funnell, B. M., and Riedel, W. R., eds., The Micropaleontology of Oceans: London, Cambridge University Press, p. 289–307.

Parker, F. L., and Berger, W. H., 1971, Faunal and solution patterns of planktonic foraminifera in surface sediments of the South Pacific: Deep-Sea Research, v. 18, p. 73–107.

Reid, J. L., 1962, On the circulation, phosphate-phosphorus content and zooplankton volumes in the upper part of the Pacific Ocean: Limnology and Oceanography, v. 7, p. 287–306.

Ruddiman, W. F., Tolderlund, D. S., and Bé, A.W.H., 1970, Foraminiferal evidence of a modern warming of the North Atlantic Ocean: Deep-Sea Research, v. 17, p. 141–155.

Saito, T., 1963, Miocene planktonic foraminifera from Honshu; Japan: Tohoku University Science Report, 2nd Series (Geology), v. 35, n. 2, p. 123–209.

Sancetta, C. A., 1978, Neogene Pacific Microfossils and Paleoceanography: Marine Micropaleontology, v. 3, p. 347–376.

Savin, S. M., 1977, The history of the earth's surface temperature during the past 100 million years: Annual Review of Earth and Planetary Science Letters, v. 5, p. 319–355.

Savin, S. M., Abel, L., Barrera, E., Bender, M., Hodell, D., Keller, G., Kennett, J. P., Killingley, J., Murphy, M., and Vincent, E., 1985, Evolution of Miocene surface and near surface waters: Oxygen isotope evidence, *in* The Miocene ocean: Paleogeography and biogeography: Geological Society of America Memoir 163, this volume.

Savin, S. M., Douglas, R. G., and Stehli, F. G., 1975, Tertiary marine paleotemperatures: Geological Society of America Bulletin, v. 86, p. 1499–1510.

Sclater, J. G., Meinke, L., Bennett, A., and Murphy, C., 1985, The depth of the ocean through the Neogene, *in* The Miocene ocean: Paleogeography and biogeography: Geological Society of America Memoir 163, this volume.

Shackleton, N. J., and Kennett, J. P., 1975, Paleotemperature history of the Cenozoic and the initiation of Antarctic glaciation: Oxygen and carbon analyses in DSDP Sites 277, 279, and 281, *in* Kennett, J. P., Houtz, R. E., *et al.,* Initial Reports of the Deep Sea Drilling Project, Volume 29: Washington, D.C., U.S. Government Printing Office, p. 743–755.

Srinivasan, M. S., and Kennett, J. P., 1981, A review of Neogene planktonic foraminiferal biostratigraphy: applications in the Equatorial and South Pacific, *in* Warme, J. E., Douglas, R. G., and Winterer, E. L., eds., The Deep Sea Drilling Project: A Decade of Progress, Society of Economic Paleontologists and Mineralogists Special Publication 32, p. 395–432.

—— 1981, Neogene planktonic foraminiferal biostratigraphy and evolution: equatorial to subantarctic, South Pacific: Marine Micropaleontology, v. 6, p. 499–533.

Thompson, P. R., 1981, Planktonic foraminifera in the western North Pacific during the past 150,000 years: comparison of modern and fossil assemblages: Palaeogeography, Palaeoclimatology, and Palaeoecology, v. 35, p. 241–279.

Thunell, R. C., 1981, Cenozoic palaeotemperature changes and planktonic foraminiferal speciation: Nature, v. 289, p. 670–672.

Thunell, R. C., and Belyea, P., 1982, Neogene planktonic foraminiferal biogeography of the Atlantic Ocean: Micropaleontology, v. 28, n. 4, p. 381–398.

Thunell, R. C., Curry, W. B., and Honjo, S., 1983, Seasonal variation in the flux of planktonic foraminifera: time series sediment trap results from the Panama Basin: Earth and Planetary Science Letters, v. 64, p. 44–56.

Tolderlund, D. S., and Bé, A.W.H., 1971, Seasonal distribution of planktonic foraminifera in the western North Atlantic: Micropaleontology, v. 17, p. 3, p. 297–329.

Ujiié, H., 1984, A Middle Miocene hiatus in the Pacific region: its stratigraphic and paleoceanographic significance: Palaeogeography, Palaeoclimatology, and Palaeoecology, v. 46, p. 143–164.

van Andel, Tj. H., Heath, G. R., Moore, T.C., Jr., 1975, Cenozoic history and paleoceanography of the Central Equatorial Pacific Ocean: Geological Society of America Memoir 143, 134 p.

Vincent, E., and Berger, W. H., 1981, Planktonic foraminifera and their use in paleoceanography; *in* Emiliani, C., ed., The Oceanic Lithosphere, Volume 7: New York, Wiley Interscience, p. 1025–1120.

Woodruff, F., Savin, S. M., and Douglas, R. G., 1981, Miocene stable isotope record: a detailed deep Pacific Ocean study and its paleoclimatic implications: Science, v. 212, p. 665–668.

Wyrtki, K., 1967, Circulation and water masses in the eastern equatorial Pacific Ocean: International Journal of Oceanology and Limnology, v. 1, p. 117–147.

MANUSCRIPT ACCEPTED BY THE SOCIETY DECEMBER 17, 1984

Geological Society of America
Memoir 163
1985

Radiolarian biogeography and paleoceanography of the North Pacific at 8 Ma

Karen Romine*
Graduate School of Oceanography
University of Rhode Island
Narragansett, Rhode Island 02882-1197

ABSTRACT

The paleoceanography of the North Pacific Ocean at 8 Ma has been determined using radiolarian biogeographic patterns that were defined through Q-mode factor analysis. Forty-four late Miocene radiolarian species categories were grouped into assemblages exhibiting latitudinal zonality and distinct ecological preferences. The assemblages represent three major water mass types: (1) tropical, shallow water; (2) subarctic, deep to intermediate water; and (3) transitional, intermediate water. Several extant species that characterize the subarctic and transitional regions have survived for the past 8 Ma without apparent morphologic change. These species may have developed the ability to withstand large changes in depth habitat, allowing them to live deeper in water masses at lower latitudes when climatic variations made it necessary.

Comparison of late Miocene and Recent biogeographic patterns using extant species indicates important differences between the 8 Ma time-slice and the Recent, which can be interpreted as changes in surface-water circulation. At 8 Ma, flow of Atlantic water into the equatorial Pacific (through the Central American Seaway) disrupted the flow pattern of the Equatorial Countercurrent, causing warm water, which is usually returned to the eastern tropical Pacific, to be diverted into the northward-flowing western limb of the subtropical gyre. This resulted in increased transport of warmer waters by the subtropical gyre and warmer sea-surface temperatures in the North Pacific.

As the Panama Isthmus became a barrier to surface water flow between the Atlantic and Pacific, transport of warm tropical water into the middle and high latitudes of the Pacific decreased, resulting in progressive cooling of the subpolar regions. The absence of a late Miocene equivalent to the Modern Eastern Tropical assemblage suggests that changes in the character of water masses in the eastern equatorial Pacific occurred at some time after 8 Ma. Decreased equatorial sea-surface temperatures in response to increased advection of cold subpolar water (via eastern boundary currents), an increase in colder, upwelled waters at the equatorial divergence, and an increased importance in surfacing of the Equatorial Undercurrent in the eastern tropical Pacific are likely oceanographic responses to the creation of a geographic barrier to inter-ocean exchange. The gradual cessation of Atlantic-Pacific surface water communication accelerated the global cooling trend that began in the middle Miocene with the development of the East Antarctic ice sheet. Termination of this inter-ocean flow in the middle Pliocene marked the beginning of a new chapter in Earth history in which climate was characterized by glacial-interglacial extremes due to cyclic variations in continental ice volume.

---

*Present address: Exxon Production Research Company, P.O. Box 2189,
Houston, Texas 77001.

INTRODUCTION

As a field, paleoceanography has grown remarkably in the last decade as a result of the Deep Sea Drilling Project (DSDP) and multi-institutional programs such as Climate/Long range Investigation, Mapping, and Prediction (CLIMAP). Using various stratigraphic tools and multivariate analysis of quantitative data, CLIMAP was able to produce a detailed, high resolution history of Late Quaternary paleoclimate and paleocirculation (CLIMAP 1976; 1982). The next logical step is to apply these techniques to other times in geologic history. The Cenozoic Paleoceanography (CENOP) project, for example, set out to look at several time-slices within the Miocene. My study examines the distribution of radiolarians in the Pacific during the late Miocene time-slice chosen by the CENOP project, compares them to Modern biogeographic patterns and uses the comparison to make deductions about late Miocene paleoceanography.

The development of a major continental ice sheet on East Antarctica during the middle Miocene was associated with a trend toward increased thermal gradients between the equator and the poles as the polar regions became more thermally isolated by circumpolar currents and upwelled deep waters whose ultimate sources were the polar regions. By late Miocene time, oscillations in the climate of Antarctica (linked to fluctuations in the amount of continental ice) were associated with changes in the spatial patterns and intensity of both atmospheric and oceanic circulation. These changes included shifts in the latitudinal position of water mass boundaries as well as variations in intensity of upwelling.

Modern and Quaternary studies have shown that biogeographic patterns of plankton reflect surface water mass distributions in the oceans (see, for example, Bé and Tolderlund 1971; McGowan 1971; Casey 1971a, b; Kipp 1976; and Moore, 1978). CENOP has attempted to produce a synoptic picture of biogeographic patterns from selected times in the Miocene in order to deduce circulation patterns at those times. As a contribution to this effort, this study examines a time-slice at approximately 8 Ma using Radiolaria in samples from Deep Sea Drilling sites in the Pacific Ocean. The time investigated represents the stage in oceanographic evolution that followed the development of the Southern Hemisphere ice sheet in middle Miocene and preceded ice sheet formation in the northern Hemisphere. Paleobiogeographic maps have been developed that provide information on North Pacific paleocirculation during the late Miocene and can be contrasted with similarly constructed maps from studies of glacial and interglacial intervals of the Pleistocene (see, for example Moore 1978; Nigrini 1970; Robertson 1975; Romine 1982; Sachs 1973; and Sancetta 1979).

METHODS

The strategy for selection of this time-slice was to look for an interval in the late Miocene during which the oxygen isotope record was relatively stable, (i.e. no large oscillation from warm to cold or interglacial to glacial conditions), so that high stratigraphic precision was not required to define the average conditions for this late Miocene interval. The period between approximately 7.5 and 8.5 Ma was selected, and determination of the correct position of the interval in a number of Deep Sea Drilling sites and long piston cores followed (see Stratigraphy section below). For this study 15 sites were chosen. Each site contained both the time-slice interval and sufficient radiolarians for quantitative studies (Figure 1, Table 1). The time-slice interval was sampled at 50,000 to 100,000 year intervals.

Each sample was prepared in a standard manner: calcium carbonate was dissolved using 50 percent hydrochloric acid; hydrogen peroxide was added to eliminate organic matter; samples were washed and random-mounts were prepared according to a technique developed by Moore (1973) with modifications by Molina-Cruz (1977).

Approximately one hundred species categories were tabulated, with counts of between 600–1000 individuals recorded per sample (slide). There are a large number of rare species and the abundances of others varies widely. Those that were rare (<0.5%) in all samples were eliminated from subsequent data treatment. In addition, species within an evolutionary lineage but whose geologic range overlapped at 8 Ma were counted as a single species category if their geographic distributions were the same. The remaining species (see Appendix for taxonomic list) were analyzed using a multivariate mathematical technique, Q-mode factor analysis (Imbrie and Kipp 1971). This technique compares all the samples in a data set, examines the relative proportion of species within each sample, splits the data into a predetermined number of factors (assemblages) and ranks each sample (gives it a factor value) within each assemblage indicating its compositional similarity to that assemblage. Each factor or assemblage can be mapped using the assigned ranking of the samples. Thus, this technique is useful for determining microfossil biogeographic patterns.

OCEANOGRAPHIC SETTING

The continental boundaries of the late Miocene Pacific Ocean (Figure 2) were similar to those of Modern times. Therefore, the general circulation pattern was probably also similar, with a large subtropical and smaller subpolar gyre in the Northern Hemisphere and with an Antarctic circulation marked by a circum-Antarctic current and distinct separation from the subtropical gyre in the Southern Hemisphere. Western boundary currents would have been warm, narrow, and relatively high-velocity currents, balanced on the eastern boundary by cold, broad, slow-moving currents with upwelling waters near the continental margins. Figures 2a and 2b show the more important elements of Modern Pacific circulation and water mass distribution.

The most notable difference between the Modern Pacific and that of 8 Ma is in the eastern equatorial Pacific where communication between the Atlantic and Pacific Oceans was possible

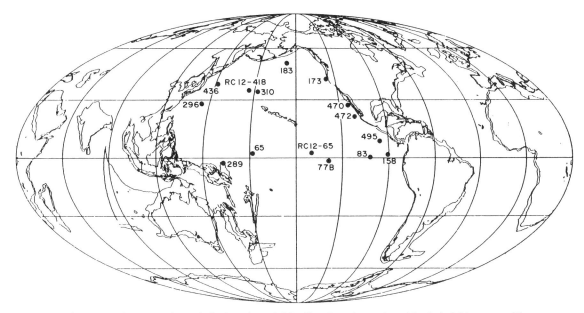

Figure 1. Paleogeography and site locations, 8 Ma. Sites have been plotted in their 8 Ma geographic position. (See Sclater et al. 1985, for details.) (Note: All maps except for figures 5a, 6a, 7a, 8a, 9a, 10a, and 11 will use the 8 Ma reconstruction as a base map.)

TABLE 1. CORE LOCATIONS*

Site	Recent (0 Ma) Latitude (N+,S-)	Recent (0 Ma) Longitude (E+,W-)	Late Miocene (8 Ma) Latitude (N+,S-)	Late Miocene (8 Ma) Longitude (E+,W-)
183	52.57	-161.20	49.91	-156.10
173	39.96	-125.45	40.54	-126.20
310	36.87	176.90	34.32	-177.17
RC12-418	38.10	170.02	35.65	176.19
436	39.93	145.56	38.13	152.53
296	29.34	138.52	27.80	145.08
470	28.91	-117.52	27.13	-113.20
472	213.01	-113.99	21.36	-109.47
495	12.50	-91.03	8.20	94.76
158	6.62	-85.24	1.59	-90.01
83	4.05	-95.74	-0.13	-99.97
77B	0.48	-133.23	-1.76	-127.84
RC12-65	4.65	-144.97	2.16	-139.65
65	4.35	176.98	1.80	-177.60
289	-0.50	158.51	-2.67	163.84

*From Sclater and Meinke (1985)

through the Central American Seaway. The Tethys as a current system was no longer in existence by the middle Miocene due to the development of various geographic barriers to its flow during the Tertiary, that is, isolation of the Mediterranean Basin, northward movement of the Indian continent, and tectonic development of the Indonesian region. Circulation in the equatorial Atlantic and Pacific today is characterized by tradewind-driven, westward-flowing equatorial currents. This pattern probably existed in the late Miocene as well. However, Luyendyk et al. (1972) demonstrated experimentally that some eastward flow from the Pacific into the Atlantic occurs under "nonglacial" wind-conditions (westerlies centered at 50°N) with a geographic barrier to flow in the western Pacific and an open Central American Seaway. They do not present results using these paleogeographic constraints and "glacial" wind conditions (westerlies

centered at 40°N, which is the average position in the Modern ocean). In the absence of any geological evidence supporting the results of the "nonglacial" experiment, certain assumptions have been made for the purposes of this study: (1) the location of major zonal components of atmospheric and oceanic circulation in the North Pacific at 8 Ma were not radically different from today's (westerlies are centered about 40°N); (2) surface circulation through the Central American Seaway was ordinarily westward except possibly during times of El Niño-like conditions when the southeast tradewinds were abnormally weak; (in the Modern ocean; Firing and Lukas 1983); (3) westward surface inflow through the Central American Seaway may have been balanced by subsurface outflow to the east. This study will address the possible effects of this inter-ocean equatorial circulation on the general circulation of the tropical Pacific.

STRATIGRAPHY

The position of the one million-year study interval (7.5–8.5 Ma) was estimated using Shaw diagrams (Shaw 1964; e.g. Figure 3) based on the published biostratigraphic data for various sites. This estimate was refined using diatom and radiolarian datums (Barron 1980, 1981, and personal communication; Westberg and Riedel 1978; Keller et al. 1982) as well as calcium carbonate stratigraphy where available. Variation in calcium carbonate concentration in these sections primarily reflects the degree of corrosiveness of the deeper Pacific waters. Deep waters have a similar chemistry throughout the Modern Pacific, and within the mixing time of the oceans (approximately 1600 years), tend to respond to changes as a single water mass. Therefore, it is assumed that temporal changes in deep-water chemistry and the ability to dis-

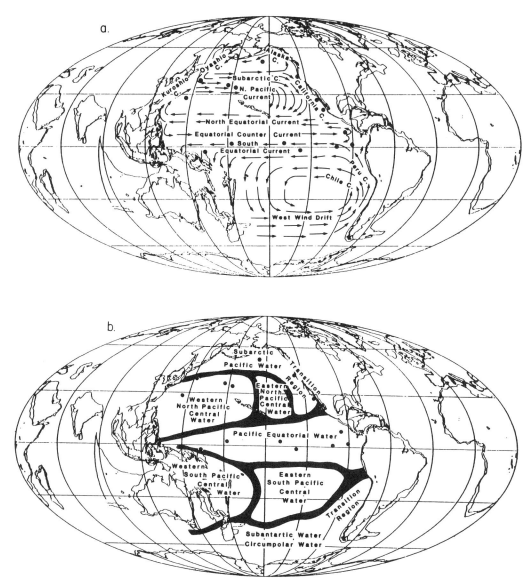

Figure 2. Generalized circulation of the modern Pacific Ocean. a: Surface currents; b: Water masses. (Adapted from Sverdrup et al. 1942.)

solve calcium carbonate are recorded synchronously in deep-sea sediments and revealed by examining calcium carbonate content of these sediments.

Age control on the sampled interval was provided by a diatom, *Thalassiosira burckliana,* whose first and last appearances have been tied to the magnetostratigraphic record and span the interval from 8.7 to 7.6 Ma (Barron 1981). In addition, there is a maximum in calcium carbonate concentration within magnetic epoch 8 at approximately 8 Ma (peak 8b, Dunn and Moore 1981). Two radiolarian datums, the last appearance of *Diartus hughesi* and the first appearance of *Acrobotrys tritubus* were used to correlate between sites that had no diatoms and/or no calcium carbonate and those with diatoms and/or calcium carbonate. Figure 4 shows the relationship of the datums within these sites.

Rather than take an average faunal assemblage from the

entire one million-year long interval, the 8 Ma level was determined in each site by interpolation and the sample or an average of two samples nearest that determination was selected to represent the 8 Ma level. An estimate of the error associated with this choice of sample(s) representing the 8 Ma level has been made and ranges from an average of ±80,000 years at best to ±160,000 years.

Two-piston cores, RC12-65 and RC12-418, were used in this study. Both have published magnetic polarity reversal records and microfossil stratigraphies. Together these data allowed for more precise correlation between low (RC12-65) and high (RC12-418) latitude sites (Embley and Johnson 1980, RC12-65; Haq, et al. 1980, RC12-418). In addition, the calcium carbonate stratigraphy of sites 77B, 158, and 310 has been tied to magnetostratigraphy using various combinations of biostratigraphic

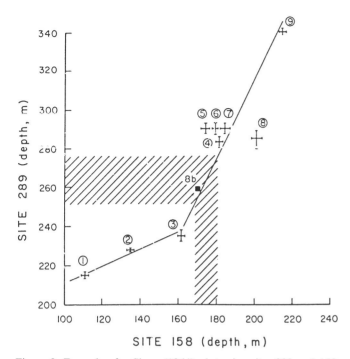

Figure 3. Example of a Shaw (1964) plot using sites 289 and 158. Datums used are: (1) First Appearance Datum (FAD) *A. amplificus* (nannofossil); (2) Last Appearance Datum (LAD) *Didymocyrtis antepenultimus* (R = radiolaria); (3) LAD *Diartus hughesi* (R); (4) FAD *Acrobotrys tritubus* (R); (5) FAD *Didymocyrtis penultimus* (R); (6) LAD *Diartus petterssoni* (R); (7) LAD *Didymocyrtis laticonus* (R); (8) FAD *Globorotalia merotumida* (foraminifera); (9) FAD *Didymocyrtis antepenultimus* (R); and the carbonate maximum, 8b (Dunn and Moore, 1981). Shaded regions indicate sampled interval.

datums from foraminifera, radiolaria, diatoms, and calcareous nannofossils (Dunn and Moore 1981, sites 158 and 77B; Vincent 1981, site 310). A recent publication by Keller et al. (1982) reviews the late Miocene correlations of several sites used in this study (158, 77B, 310 and 296).

Once tied to the magnetic polarity time scale (Barron 1980, 1981) the diatom datums were used independently to estimate the 8 Ma level in each site. These estimates were then compared to the depth of the calcium carbonate datum level, 8b. In all cases the estimated 8 Ma datum depth agreed with the calcium carbonate datum depth and was well within the maximum average error of ±160,000 years.

Correlation of DSDP sites 158, 310, 77B, and 173 and piston core RC12-65 was based mainly on the diatom datums and calcium carbonate records. Comparison of these data in the high and low latitude sites indicates that the microfossil datums are approximately synchronous in the interval studied.

DSDP sites 470, 472, and 495 and piston core RC12-418 have diatom datums but no calcium carbonate records available. The 8 Ma level was estimated using only the diatom datum ages. DSDP site 436 does not have the *T. burckliana* datums and was correlated instead to DSDP site 438A nearby (which does have *T. burckliana*), using both diatom and radiolarian datums.

Site 289 and 65 had Radiolaria but no diatoms. Two radiolarian datums, the last appearance of *Diartus hughesi* and the first appearance of *Acrobotrys tritubus*, were used to estimate the position of the 8 Ma level after establishing their relationship to other datums in sites 77B and 158. Site 289 has a calcium carbonate record that allowed confirmation of the accuracy of the estimate in that site (D. Dunn, personal communication).

DSDP site 83 has a quite detailed calcium carbonate record in the Initial Reports of Leg 9. The position of the 8b carbonate peak at site 83 was determined by correlating the record of calcium carbonate and several foraminiferal and radiolarian datums to sites 77B and 158 as shown in Dunn and Moore (1981).

The stratigraphic estimation of the 8 Ma level at site 183 is based on unpublished work on diatom stratigraphy at this site by J. Barron (personal communication). Core 18 of site 183 is bounded by hiatuses above and below, and by using the occurrences of two diatoms, *Thalassiosira hirosakiensis* and *Nitzschia pliocenica,* an 8 Ma level was determined (see Figure 4). The time-slice interval within core 18 is very condensed because of a low accumulation rate. Therefore, this site represents the worst case in terms of error associated with the selection of the 8 Ma level. However, using a single sample estimated to be approximately 8 Ma appears to be satisfactory since an average of several samples around the selected level produces no significantly different values in species percentages.

Radiolaria in the samples within the time-slice interval in site 296 were poorly preserved and only one sample was considered usable. Fortunately, the stratigraphy of site 296 as revied in Keller, et al. (1982) indicated that this sample was approximately equivalent to the 8 Ma level.

Special Problems

The samples from DSDP site 310 originally obtained by CENOP did not sample deep enough to reach the 8 Ma level. However, the deepest sample is probably ca. 7.8–7.9 Ma and, therefore, has been used in the data set. Results from this site are generally consistent with those of RC12-418 which is located nearby (Figure 1; see Figure 6b, 7b, and 8b for comparison).

The latest appearance (LA) of *Diartus hughesi* in the sites studied here usually occurs higher in the section than the 8 Ma datum, between 7.5 and 7.8 Ma. In the piston core where the datum was reported by Theyer et al. (1978, Fig. 5), it occurs at the top of magnetic epoch 9 which is approximately 8.7–8.8 Ma. This author believes that the numerous occurrences of the datum at levels younger than 8 Ma in the sites used in this study outweigh the single reported occurrence of the datum at 8.7–8.8 Ma in one piston core in Theyer et al. (1978) and the younger age between 7.5 and 7.8 Ma has been assumed correct.

RESULTS AND DISCUSSION

The following results and discussion sections are in two parts. Part One presents the results of a comparison of the bio-

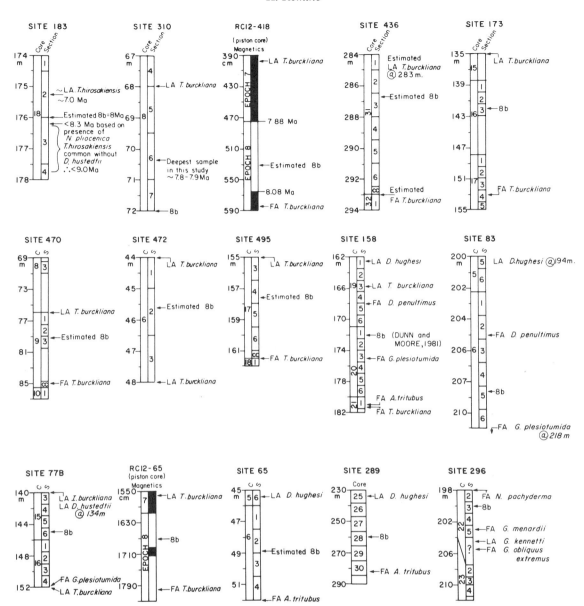

Figure 4. Stratigraphy of the sites used in this study. Subbottom depth, core, and section are plotted within the sampled time-slice interval in each site. Microfossil datums and the actual or estimated calcium carbonate peak 8b (Dunn and Moore 1981) are plotted against subbottom depth to demonstrate correlation. Sources for datum levels include Keller, et al. 1982; Haq, et al. 1980; Barron, 1980, 1981; Westberg and Riedel, 1978; Embley and Johnson, 1980; and unpublished data from J. Barron and D. Dunn (personal communications).

geographic patterns of a group of radiolarians which are present in both the Modern and late Miocene samples. Part two examines the distribution of late Miocene radiolarian assemblages that not only include extant species but also those forms that are typically abundant in the late Miocene.

Part 1: Mathematical comparison of late Miocene biogeography with the Present, 8 Ma vs 0 Ma.

Results. Moore and Lombari (1981) generated maps of Modern assemblage distributions based on a factor analysis of 27

extant species (Table 2). The proportional species composition of each late Miocene sample (this study) was mathematically compared to the species composition of these Modern factors or assemblages (Moore and Lombari 1981). The late Miocene samples were then assigned factor values that ranked them as to their degree of similarity to each Modern assemblage (Klovan and Imbrie 1971). In this analysis the late Miocene species percentage data was compared, sample by sample, to the Modern assemblages used by Moore and Lombari (1981). Thus, a comparison of the habitats of the late Miocene radiolarians with respect to their Modern distributions can be made in order to

COMMUNALITIES

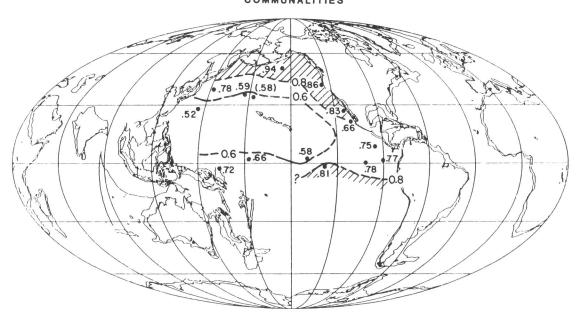

Figure 5. Communalities of 8 Ma samples. Each sample location has been plotted with its corresponding communality. Contours are dashed where sample control allows only an estimation of the contour. Question marks indicate areas where an estimate of the contour is not possible based on the sample control points. Note that parentheses are used around values for site 310; see *Special Problems* in Stratigraphy section for explanation. (Diagonal pattern delineates regions of values greater than 0.8).

TABLE 2. EXTANT SPECIES

Species categories used in the comparison of modern and late Miocene paleobiogeographic patterns.

Cenosphaera cristata	Pylospira octopyle
Spongurus (?) sp.	*Zygocircus spp.
Styptosphaera spumacea	*Giraffospyris angulata
Heliodiscus asteriscus	Eucyrtidium hexagonatum
*Didymocyrtis tetrathalamus	Eucyrtidium acuminatum
Hymeniastrum euclidis	Phormospyris stabilis
	scaphipes
Stylodictya validispina	Theocalyptra bicornis
Stylochlamidium asteriscus	Theocalyptra bicornis var.
**Spongopyle osculosa	Theocalyptra davisiana
**Spongotrochus glacialis	*Anthocyrtidium ophirense
**Spongotrochus (?) venustrum	Lamprocyrtis maritalis
Tetrapyle octacantha	*Botryostrobus aquilonaris
Larcopyle butschlii	*Phormostichoartus corbula
*Larcospira quadrangula	Pterocorys zanceleus
Lithelius minor	

*See Appendix for listing of species included in these categories.
**These species were combined in the analysis to form a single species category.

determine whether changes in ecologic tolerances and/or migrations have occurred that may indicate differences in circulation patterns of the late Miocene relative to those of the Modern Pacific Ocean.

Communalities (Figure 5). A sample's communality is the sum of squares of the factor values from each of its factors. The resulting numerical value is a quantitative measure of the proportion of variance in the species percentage data of that sample accounted for by the mathematic data treatment. Ideally, the

value is equal to 1.0 if all information contained in the sample is described by the factor model. In this analysis the communality of a late Miocene sample measures how well that sample compares to the Modern or how good an analog the Modern assemblage is for the late Miocene. Most of the samples have communalities of 0.6 or greater (Figure 5). The highest communalities (>0.8) occur where Modern water masses are cold and the assemblages represent either cold surface waters or cold water advected equatorward in eastern boundary currents (off California and the South American coast). The lowest communalities occur on the inner margins of the anticyclonic subtropical gyre of the North Pacific. The late Miocene species compositions of these particular samples have been altered somewhat by dissolution of silica which decreased their similarity to the Modern assemblages and resulted in lowered communalities. In general, however, the majority of late Miocene samples compare well with the Modern data, indicating that the Modern assemblages are good analogs for those of the late Miocene.

Factor 1: Eastern tropical assemblage (40% of the variance in Modern data; Figure 6). The Modern eastern tropical assemblage delineates two areas of divergence, upwelling, and mixing of water masses: the eastern to central tropical Pacific; and the northwestern Pacific off Japan. The first region is associated with the permanent shallow thermocline in the eastern tropical Pacific that is maintained by advection of cold water from eastern boundary currents (e.g. Peru Current), equatorial divergence, and surfacing of the Equatorial Undercurrent. This area is also a

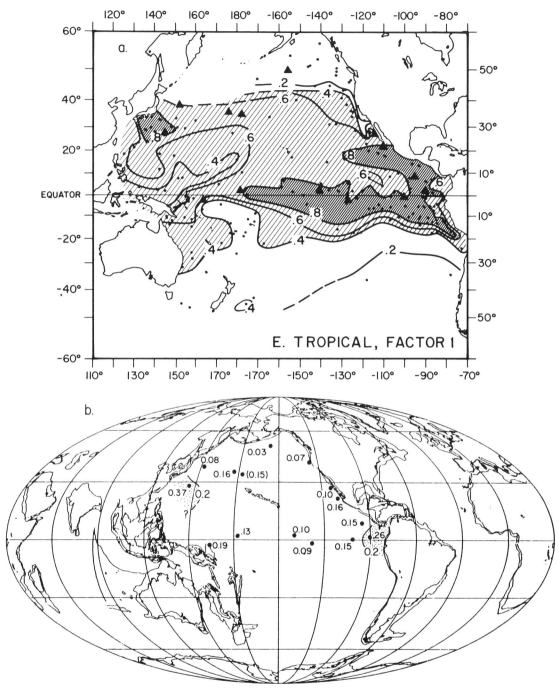

Figure 6. Distribution of the Eastern Tropical assemblage in the (a) modern and (b) 8 Ma North Pacific. The modern map is from Moore and Lombari (1981). Small dots represent the surface sediment samples whose factor values have been contoured. The heavy diagonal pattern delineates regions of values greater than 0.8; lighter diagonal pattern, regions of values greater than 0.8–0.4. Large triangles show the locations of the 8 Ma sites used in the late Miocene analysis and whose factor values are plotted and contoured in (b); shading patterns follow that of (a). (Note: In certain cases where highest factor values are 0.4–0.2, a dotted pattern is used.)

region of mixing of surface water masses where warm North Equatorial Countercurrent water converges and mixes with colder water masses—the California Current to the north and the South Equatorial Current to the south (Wyrtki 1964, 1965, 1966, 1967). The second area is a convergence zone where the cold Oyashio Current flowing south meets the warmer Kuroshio Current flowing north (Figures 6a and 2a). These water masses meet, mix, and flow eastward along 30° to 40° N latitude as the Subarctic and North Pacific Currents.

The Modern species dominating this assemblage is *Tetrapyle octacantha.* This species is very scarce in the late Miocene at 8 Ma, which explains the restricted and minor presence of the factor (Figure 6b). Of the abundant late Miocene species considered, there is no other which shows such a strong eastern tropical preference suggesting that the development of this assemblage occurred after the 8 Ma time-slice.

Factor 2: Subpolar assemblage (22% of the variance in Modern data; Figure 7). A Modern assemblage of cold-water species defines the subpolar and transitional regions of the Pacific exhibiting highest factor values within the subarctic and subantarctic water masses (Figure 7a). The dominant species group consists of *Spongopyle osculosa, Spongotrochus glacialis, S. venustum* (all lumped into one species category) with some lesser contribution from *Lithelius minor* and *Botryostrobus aquilonaris.* These species existed in similar relative abundances in the same geographic region in the late Miocene at 8 Ma and can be seen in Figure 7b. The Modern subpolar assemblage is a good analog for the late Miocene, suggesting persistence of these high-latitude water masses for at least the last 8 Ma.

Factor 3: S.W. Tropical assemblage (21% of the variance in Modern data; Figure 8). An assemblage whose dominant member is a species group, *Zygocircus* spp., occurs in a band across the southern tropical Pacific with highest values in the southwest where core-top control is best (Figure 8a). This factor is most likely indicative of silica-poor sediments in the south Pacific where dissolution rates are high and productivity low. The absence of core tops for control points in the central south Pacific is also related to the absence of silica in sediments in the region. Lombari and Boden (1982) mapped regions of good and poor radiolarian preservation in the Pacific Ocean and showed a distribution pattern of poorly preserved material in the South Pacific similar to the biogeographic pattern of this assemblage, suggesting that dissolution of the silica exerts some control on the distribution of this factor.

The late Miocene data (Figure 8b) show a band of highest factor values (0.27–0.35) across the equatorial region, whereas the Modern distribution shows a band of highest values at 10°S. The factor values are quite low in the late Miocene samples, however, indicating that the assemblage is not an important component in the area and that *Zygocircus* spp. is not very abundant. If samples south of the equator could have been included in this analysis, it is possible that the distribution of high factor values may have been more similar to the Modern data that had somewhat better sample coverage.

Factor 4: N.W. Tropical assemblage (7% of the variance in Modern data; Figure 9). This assemblage basically delineates the regions of the Modern subtropical gyres or central water masses in both the northern and southern hemispheres with highest factor values occurring in the northwest Pacific (Figure 9a). The dominant species is *Stylochlamydium asteriscus,* with a substantial contribution from *Heliodiscus asteriscus.* The late Miocene data (Figure 9b) exhibit the highest factor values in the equatorial sites (0.6–0.8) indicating that this assemblage is much more important there than the S.W. Tropical. In fact, the N.W. Tropical assemblage appears to have higher factor values relative to the Modern in all the sites, suggesting an expansion of the warm, subtropical gyre.

Factor 5: Eastern Subpolar assemblage (2% of the variance in Modern data; Figure 10). The Eastern Subpolar assemblage is restricted to a very small region of the northeastern Pacific where today the Subarctic Current splits into the southward-flowing California Current and the northward-flowing Alaska Current (Figure 2a and 10a). The fauna of this region is dominated by an association of cold and/or deep dwelling radiolarians, *Botryostrobus aquilonaris, Pterocorys zancleus,* and *Cenosphaera cristata.*

The late Miocene distribution of this assemblage is quite different and low factor values indicate lesser relative abundances of the species. Highest factor values occur in a small region right along the Equator (Figure 10b), perhaps reflecting the occurrence of upwelling at the equatorial divergence in the late Miocene.

Factor 6: Eastern Temperate assemblage (2% of the variance in Modern data; Figure 11). The Eastern Temperate assemblage in the Northern Hemisphere today occupies an area between about 25° and 40° N latitude where water temperatures are between 15° and 20°C, coincident with the Eastern North Pacific Central Water (Figures 2a and 11a). A single species, *Lithelius minor,* dominates the assemblage. In the late Miocene this assemblage follows more closely the limits of the transition zone and appears to have moved "offshore" since 8 Ma (Figure 11b).

Discussion. The purpose of the above analysis was to determine the degree of similarity between the character of late Miocene and Modern Pacific ocean surface circulation through comparison of radiolarian biogeographic patterns. From the paleobiogeography it appears that the subpolar species group has the most similar distribution between 8 Ma and the Present. This strong similarity indicates the presence of a distinct subarctic water mass in the late Miocene. The dominant species group is composed of three species, *Spongopyle osculosa, Spongotrochus glacialis,* and *S. venustum* that have undergone no visually detectable morphologic evolution in the past 8 Ma. This could suggest a stable or slowly changing environment with little if any stressing of ecologic tolerances. On the other hand, these species may have broad ecologic tolerances. In Modern plankton tows these species are associated with upwelling in the California Current and with intermediate to deep water masses in the Sargasso Sea (Casey et al. 1979a; Spaw 1979). The distribution of *S. glacialis* (includes *S. osculosa*) in Modern sediments in high

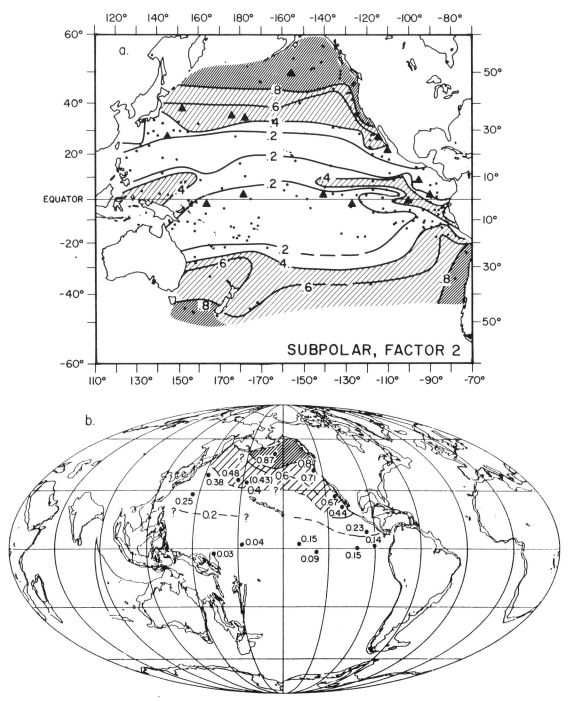

Figure 7. Distribution of the Subpolar assemblage in the (a) modern and (b) 8 Ma North Pacific. See Figure 6 for explanation.

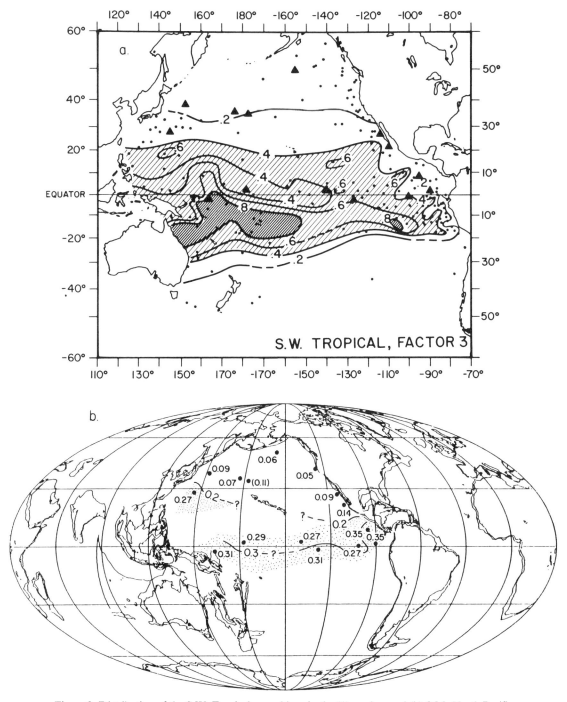

Figure 8. Distribution of the S.W. Tropical assemblage in the (1) modern and (b) 8 Ma North Pacific. See Figure 6 for explanation.

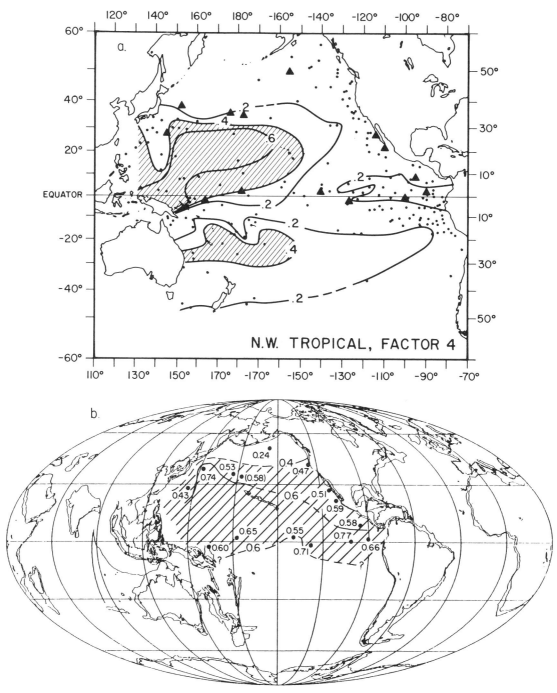

Figure 9. Distribution of the N.W. Tropical assemblage in the (a) modern and (b) 8 Ma North Pacific. See Figure 6 for explanation.

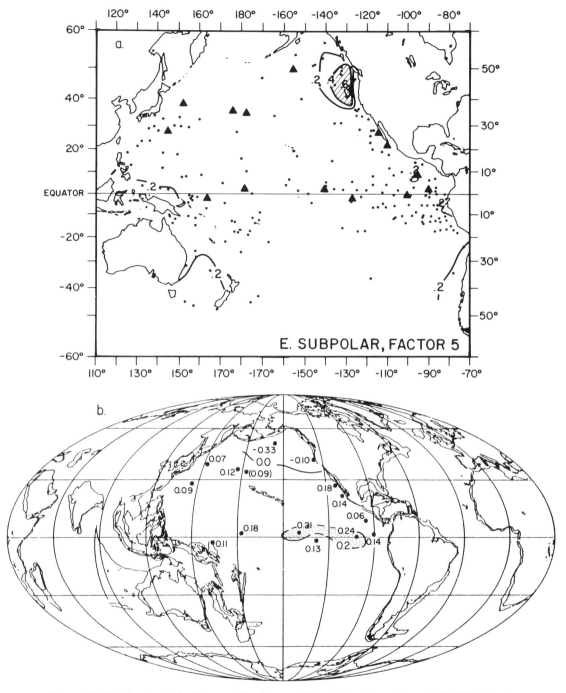

Figure 10. Distribution of the Eastern Subpolar assemblage in the (a) modern and (b) 8 Ma North Pacific. See Figure 6 for explanation.

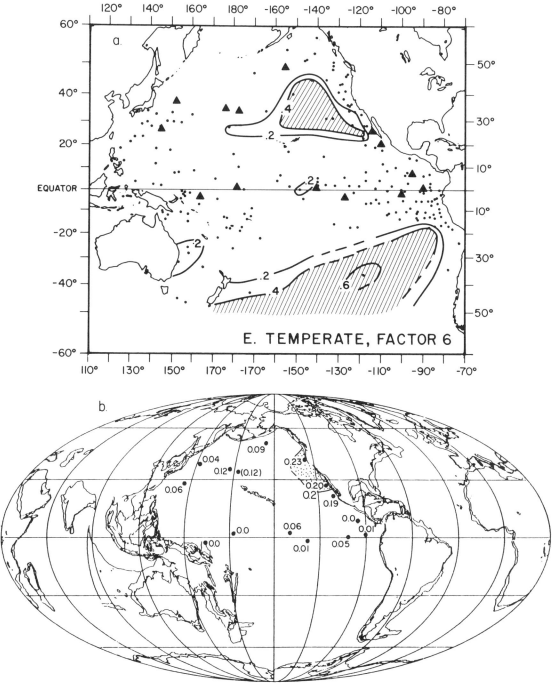

Figure 11. Distribution of the Eastern Temperate assemblage in the (a) modern and (b) 8 Ma North Pacific. See Figure 6 for explanation.

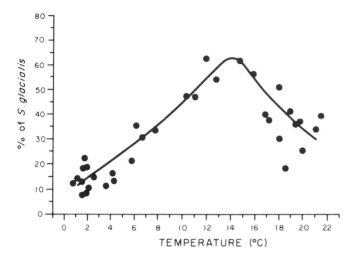

Figure 12. Percent abundance of *Spongotrochus glacialis* versus sea-surface temperature in the high latitudes of the Southern Hemisphere (Lozano and Hays 1976).

southern latitudes is associated with a wide range of surface-water temperatures (Figure 12; Lozano and Hays 1976), with maximum abundances occurring between the 12° and 16°C isotherms and just north of the subtropical convergence. This temperature range is the same as that associated with transitional and subarctic waters of the North Pacific. The wide range of sea-surface temperatures associated with the seafloor distribution of these subpolar species may indicate that they are able to change the depth limits of their habitats in order to maintain a relative position within a specific water mass, for example, living at relatively shallow depths within the subpolar water mass and then adapting to deeper depths as they follow the subpolar water mass as it dives below subtropical waters south of the transition (Casey 1982; Casey et al. 1982). An adaptability to depth changes would be a key factor in their persistence through such a geologically long period of time.

The remaining paleobiogeographic patterns show some important differences between the late Miocene and Modern ocean. The Eastern Tropical assemblage is undeveloped at 8 Ma, in part because the Modern species dominating the assemblage, *Tetrapyle octacantha,* is scarce in the 8 Ma time-slice. At 8 Ma the Modern S.W. Tropical and N.W. Tropical assemblages both show maximum factor values in the equatorial Pacific with the N.W. Tropical assemblage being the more important component. Comparison of the Modern and late Miocene distributions of the N.W. Tropical assemblage that occupies the region of the western North Pacific gyre today (Figure 2b) indicates that the subtropical gyre was much expanded to the north and east in the late Miocene data (Figure 9a,b). As a consequence, the subarctic gyre must have been somewhat compressed and warmer subtropical water replaced much of the usual flow of subarctic water southward along the California coast. At 8 Ma the 15° and 20°C isotherms were located further north along the California coast (Ingle 1973; Moore and Lombari 1981) and the Eastern Temperate assemblage then occupied a coastal position east of its Modern location (Figure 11a,b). The Eastern Subpolar assemblage did not occupy its Modern position in the late Miocene either. Since the isotherms were shifted northward, this assemblage may have occupied a position within the Alaskan gyre to the north, but out of the range of sample control in the late Miocene data. However, the data indicate that the assemblage was located along the equatorial divergence at 8 Ma (Figure 10b). Since it is located today in the region where the Subarctic Current diverges and becomes two currents, one flowing north (Alaskan Current) and one flowing south (California Current), it is possible that the assemblage may have migrated between 8 Ma and the Present, perhaps moving from deeper waters beneath the late Miocene equatorial divergence to higher latitude, shallower waters but still associated with divergence.

These late Miocene assemblage distributions indicate a circulation regime with some important differences from the Modern patterns. The most well-documented change in circulation since 8 Ma was the termination in the early-middle Pliocene of Atlantic-Pacific communication through the Central American Seaway (Saito 1976; Keigwin 1978). Flow of Atlantic equatorial water into the equatorial Pacific may have been sufficient to cause a reduced Equatorial Countercurrent (Figure 13). Surface water flow through the proto-Isthmus region could have suppressed the normal eastward return-flow of warm surface water via the Countercurrent which instead would have been diverted northward within the late Miocene equivalent to the Kuroshio Current, increasing the volume and sea-surface temperature of water entering the temperate and subarctic North Pacific. Evidence for increased sea-surface temperatures in five of the sites used in this study has been produced by Moore and Lombari (1981; Figure 14) and supports this view. As the Isthmus later began to bar the flow of Atlantic surface water into the Pacific, the character of eastern equatorial Pacific water masses would have been influenced. The creation of a barrier to flow between the Atlantic and Pacific Oceans may have resulted in shallowing of the eastern Pacific thermocline and the actual *surfacing* of the Equatorial Undercurrent as observed in the Modern Pacific (Pak and Zaneveld 1973). An increase in equatorial divergence and advection of cooler water by eastern boundary currents that help to maintain the permanent shallow thermocline in the eastern Pacific may also have occurred as an indirect result of the closure of the Central American Seaway. As warmer waters that normally flowed eastward as the Equatorial Countercurrent were no longer diverted into the higher latitudes, cooling in high-latitudes of the North Pacific brought about increased vigor in atmospheric circulation (Leinen and Heath 1981; Rea and Janacek 1982; and Janacek and Rea 1983) that enhanced both equatorial divergence and advection by eastern boundary currents. These changes in the eastern equatorial Pacific were probably responsible for the displacement of the late Miocene N.W. Tropical and S.W. Tropical assemblages to the westward and away from the Equator to their Modern northwestern and southwestern locations. The displacement and migration of the Eastern Subpolar assemblage to its Modern location may have also been a result. Further, as the

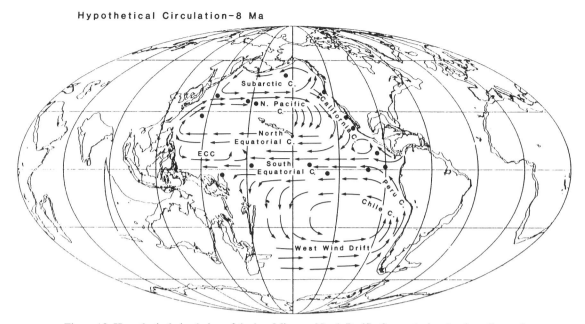

Figure 13. Hypothetical circulation of the late Miocene North Pacific Ocean during the time-slice at 8 Ma (ECC: Equatorial Countercurrent).

California Current became a colder current supplied with sub-arctic water, the Eastern Temperate assemblage would have moved offshore as the 15° and 20°C isotherms began to bend southward, at the same time allowing the Eastern Subpolar assemblage to move into its Modern habitat at approximately 40–50° N.

Part 2: Late Miocene radiolarian assemblage distributions

Results. The previous section has dealt with the biogeographic patterns of 27 species/species groups and lineages that were found in late Miocene North Pacific sediments at 8 Ma and still can be found in surface sediments of the North Pacific today. This section is intended to show late Miocene biogeography when more of the forms endemic to and abundant in this time period are included in the analysis. Q-mode factor analysis grouped 44 late Miocene species (Table 3) into 5 factors (assemblages). The five assemblages accounted for more than 90 percent of the variance in the data.

Factor 1: Tropical assemblage (53% of the variance; Figure 15). Once again the assemblage accounting for most of the variance is a tropical assemblage. The highest factor values occur in the equatorial sites and in the northwest subtropical Pacific where the warm late Miocene equivalent of the Kuroshio Current exerts its influence. Relatively high factor values occur up to 30° N latitude in the northeast and to about 40° N latitude in the northwest Pacific. The dominant species in the late Miocene data, *Phorticium pylonium,* has a broad latitudinal distribution (Figure 16). The remaining species making up the assemblage are responsible for the high factor values in the equatorial region: *Didymocyrtis* spp. (Figure 17), *Disolenia* spp. (Figure 18; includes Otosphaerids and Solenosphaerids), *Heliodiscus asteriscus*

TABLE 3. LATE MIOCENE SPECIES

Species categories used in the 8 Ma paleobiogeographic reconstruction.

*Disolenia spp.	Pylospira octopyle
*Actinomma spp.	*Acrosphaera spp.
Hexacontium spp.	*Antarctissa spp.
Stylatractus universus	Ceratocyrtis histricosa
Spongurus (?) sp.	*Giraffospyris angulata
Spongurus sp. B	Lophospyris pentagona pentagona
Heliodiscus asteriscus	Phormospyris stabilis scaphipes
*Diartus spp.	Phormospyris stabilis stabilis
*Didymocyrtis spp.	Dendrospyris bursa
Stylodictya validispina	Cornutella profunda
Circodiscus microporus group	Stichocorys delmontensis
Stylochlamidium asteriscus	Stichocorys peregrina
Spongopyle osculosa	Theocalyptra bicornis
Spongotrochus glacialis	Theocalyptra davisiana
Spongotrochus venustum	Theocorys redondoensis
Phorticium polycladum	Anthocyrtidium ophirense
Phorticium pylonium	Calocycletta caepa
Tetrapyle octacantha	Lamprocyrtis maritalis
Larcopyle butschlii	Pterocorys zancleus
Larcospira quadrangula	*Botryostrobus aquilonaris
Lithelius minor	*Phormostichoartus corbula
Lithelius sp.	Siphostichartus corona

*See Appendix for listing of species included in these categories.

(Figure 19), *Larcospira quadranqula* (Figure 20), and *Diartus* spp. (Figure 21; includes *D. hughesi* and *D. petterssoni*). As was suggested by the previous analysis using only extant species, there is no segregation of the eastern tropical region as delineated in the Modern distributions; however, low sample resolution may be partly responsible for this pattern.

Factor 2: Subarctic assemblage (15% of the variance; Figure 22). The biogeographic pattern presented in Figure 22 is dis-

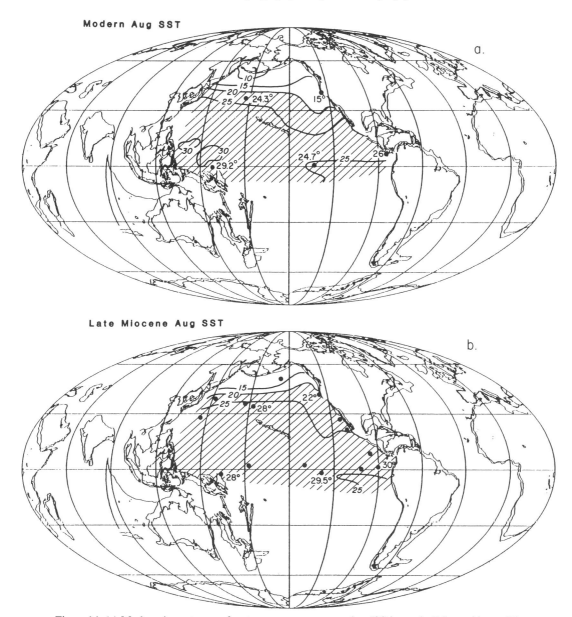

Figure 14. (a) Modern August sea-surface temperatures contoured at 5°C intervals. Paleopositions of 5 late Miocene sites (Sites 310, 173, 158, 77B, and 289) are shown with their associated modern August temperatures. (b) Sea-surface temperatures at 8 Ma for these 5 sites. Isotherms are drawn to be consistent with these temperatures and modern isotherm patterns. Regions with temperatures greater than 20°C are shaded. (Data from Moore and Lombari 1981).

tinctly subarctic with an assemblage composed of *Stichocorys delmontensis, S. peregrina,* and *Cornutella profunda* (Figures 23, 24 and 25). Site 183, the northernmost of the late Miocene sites, has the highest factor value and greatest abundance of the *Stichocorys* species. Relatively high factor values also occur in the eastern equatorial Pacific and probably indicate the position of equatorial divergence and upwelling. The Subarctic assemblage was most important in the northeast Pacific region within a cyclonic gyre (equivalent to the Modern Alaskan gyre; Figure 2a) where upwelling would have occurred. Diatoms overwhelmed

the radiolarians in samples from site 183, evidence of the high productivity of the region. The *Stichocorys* species reported here were thickly silicified so it is possible that the high abundances of these two species is in part due to the preservation quality of the samples. However, these species do represent a subpolar endmember in this analysis, so their abundance in site 183 is consistent with the general biogeographic pattern of the North Pacific and suggests that even under ideal preservation conditions, the *Stichocorys* group would still make up the larger proportion of this subarctic assemblage. Both *Stichocorys delmontensis* and *S.*

TROPICAL ASSEMBLAGE

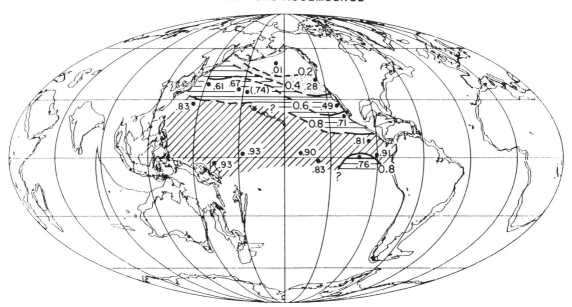

Figure 15. Distribution of the Tropical assemblage at 8 Ma. Contours are dashed where sample control is limited. Question marks indicate uncertainty with regard to the continuation of a particular contour. Diagonal pattern delineates regions of factor values greater than 0.8; horizontal pattern, regions of 0.8–0.4; dotted pattern, regions of 0.4–0.2.

Phorticium pylonium

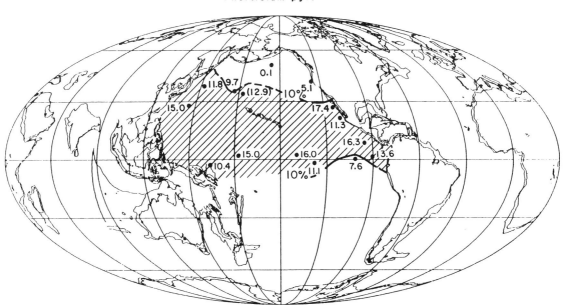

Figure 16. Percentages (from total number of Radiolaria) of *Phorticium pylonium* at each site, 8 Ma North Pacific. Areas of values greater than 10 percent are shaded.

Didymocyrtis spp.

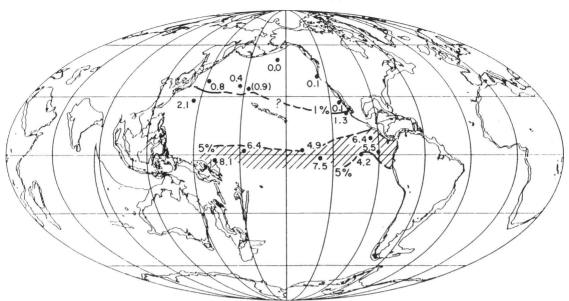

Figure 17. Percentages (from total number of Radiolaria) of *Didymocyrtis* spp., 8 Ma North Pacific. Areas of values greater than 5 percent are shaded.

Disolenia & Otosphaera spp.

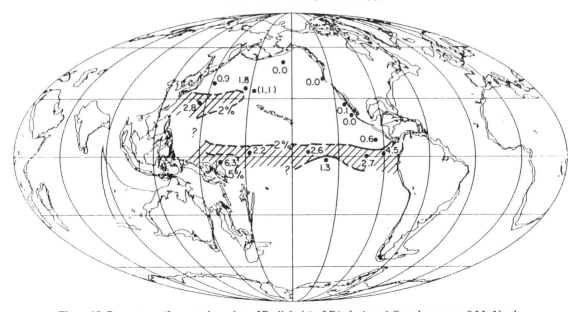

Figure 18. Percentages (from total number of Radiolaria) of *Disolenia* and *Otosphaera* spp., 8 Ma North Pacific. Areas of greater than 2 percent are shaded.

Heliodiscus asteriscus

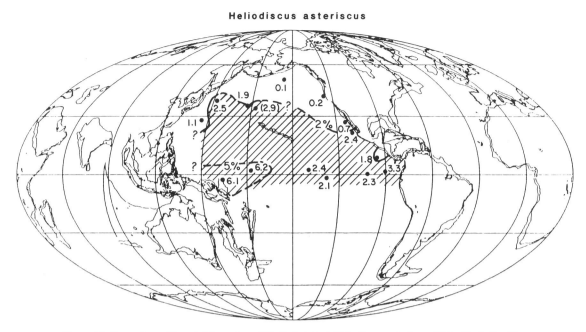

Figure 19. Percentages (from total number of Radiolaria) of *Heliodiscus asteriscus,* 8 Ma North Pacific. Areas of greater than 5 percent are shaded.

Larcospira quadrangula

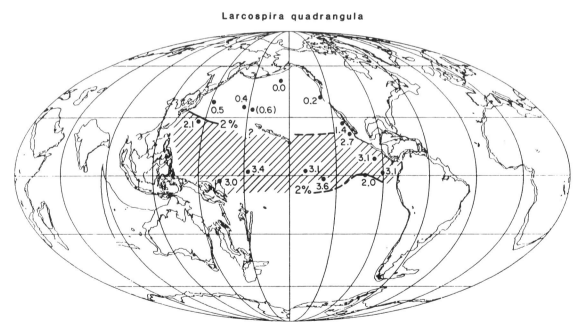

Figure 20. Percentages (from total number of Radiolaria) of *Larcospira quadrangula,* 8 Ma North Pacific. Areas of greater than 2 percent are shaded.

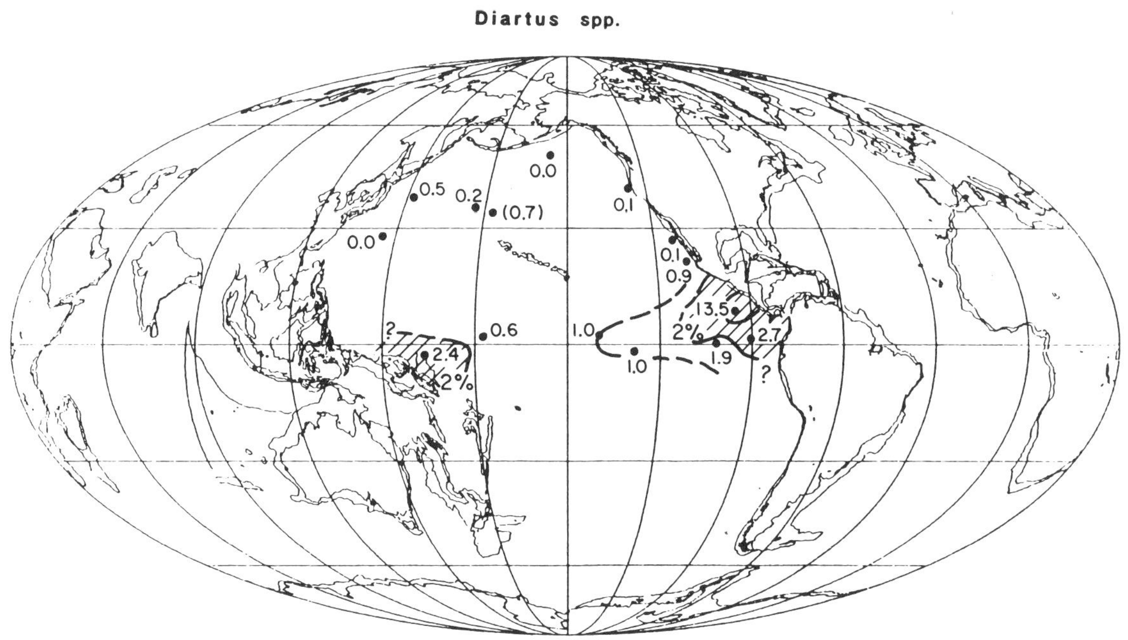

Diartus spp.

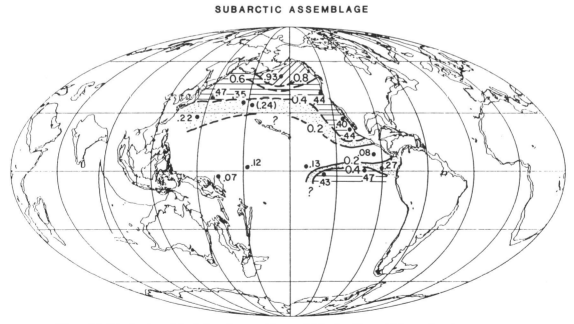

Figure 21. Percentages (from total number of Radiolaria) of *Diartus* spp. (*D. hughesi* and *D. petterssoni*), 8 Ma North Pacific. Areas of greater than 2 percent are shaded.

SUBARCTIC ASSEMBLAGE

Figure 22. Distribution of the Subarctic assemblage at 8 Ma. See Figure 15 for explanation.

Stichocorys delmontensis

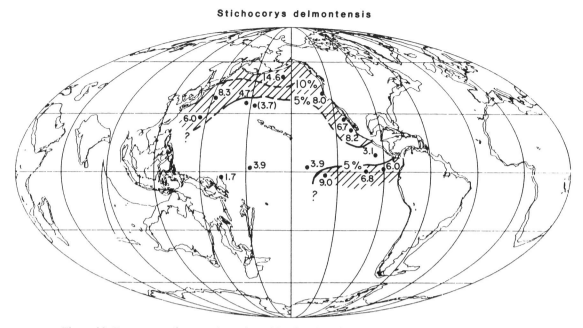

Figure 23. Percentages (from total number of Radiolaria) of *Stichocorys delmontensis*, 8 Ma North Pacific. Areas of greater than 5 percent are shaded.

Stichocorys peregrina

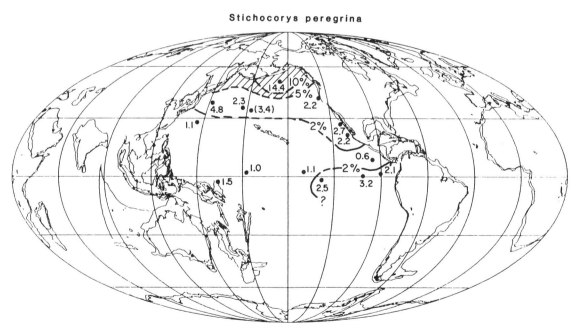

Figure 24. Percentages (from total number of Radiolaria) of *Stichocorys peregrina*, 8 Ma North Pacific. Areas of greater than 5 percent are shaded.

Cornutella profunda

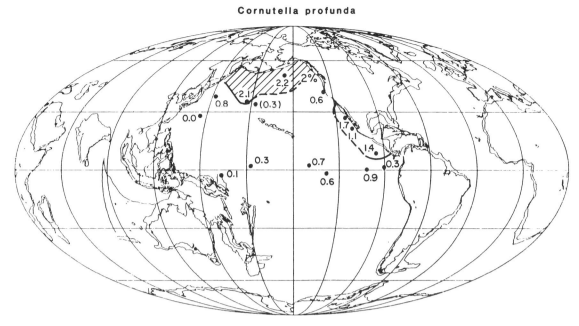

Figure 25. Percentages (from total number of Radiolaria) of *Cornutella profunda,* 8 Ma North Pacific. Areas of greater than 2 percent are shaded.

peregrina became extinct before Modern times; thus, there are distinct changes in the Subarctic assemblage between late Miocene and the Recent.

Factor 3: Eastern Transitional/California Current assemblage (15% of the variance; Figure 26). A California Current and North Pacific transitional zone is well-defined by several radiolarian groups, all present in recent deep-sea sediments and none changed greatly in their test morphology since 8 Ma. Many of these species were noted in Part 1 as members of the Modern Subpolar and N.W. Tropical assemblages. The species composing this assemblage include: *Stylochlamidium asteriscus* (Figure 27), *S. venustum, Actinomma* spp. (Figure 28), *Spongurus* (?) sp. (Figure 29), *Spongopyle osculosa* (Figure 30), *Larcopyle butschlii, Spongotrochus glacialis, Theocalyptra bicornis* (Figure 31), *Theocalyptra davisiana.* These species are abundant in the transition zone where the warm North Pacific and cold Subarctic Currents mix as they flow eastward. In the California Current region they also indicate the limits of southward advection of subarctic and/or transitional waters (as compared with the Modern ocean) between 25° and 45°N latitude.

Figure 4: Western Transitional assemblage (6% of the variance; Figure 32). The transitional region of factor 3 is shared by an assemblage composed of species that prefer the western half of the transition zone. An admixture of warm and cold-loving species make up the assemblage, consistent with the oceanographic character of this region where today the cold Oyashio and warm Kuroshio Currents meet and begin to mix as they turn and flow eastward. Several of these species are actually more important components in the first three assemblages but together they delineate this unique region of convergence and mixing. The species that compose the assemblage are: *Phorticium pylonium*

(Figure 16), *Stichocorys peregrina* (Figure 24), *Actinomma* spp. (Figure 28), *Heliodiscus asteriscus* (Figure 19), and *Spongopyle osculosa* (Figure 30).

Factor 5: Dissolution and Oxygen Minimum assemblages (2% of the variance; Figure 33). The fifth assemblage has 2 aspects indicated on the figure by regions of negative and positive factor values. The region in the southwestern tropical Pacific with the highest positive factor value is dominated by *Disolenia* spp. (Figure 18) and *Heliodiscus asteriscus* (Figure 19). This assemblage is probably a dissolution-residual assemblage since the most abundant group, *Disolenia* spp., has been found to be extremely dissolution-resistant (Johnson 1974).

An area in the northeastern tropical Pacific is delineated by the highest negative factor values. That the sign is negative probably indicates better preserved radiolarian assemblages than in the southwestern tropical Pacific. Three species dominate the factor in samples from this region: *Phorticium pylonium* (Figure 16), *Diartus hughesi* (Figure 21), and *Cornutella profunda* (Figure 25). *P. pylonium* is a cosmopolitan species and is a more important component of factors 1 and 4. The remaining species, however, have distribution patterns that are closely aligned with that of the subsurface oxygen minimum characteristic of this region today (Figure 34). Evidence for the existence of a late Miocene oxygen minimum in this region has been presented by Summerhayes (1981).

***Discussion.** Paleoecology.* This paper so far has only considered the horizontal distribution of species in sediments and their apparent (inferred) relationships to water masses or currents in the overlying water. Studies on the ecology of modern Radiolaria in the last decade have greatly improved our knowledge of how different species are distributed within the

EASTERN TRANSITIONAL/CALIFORNIA CURRENT ASSEMBLAGE

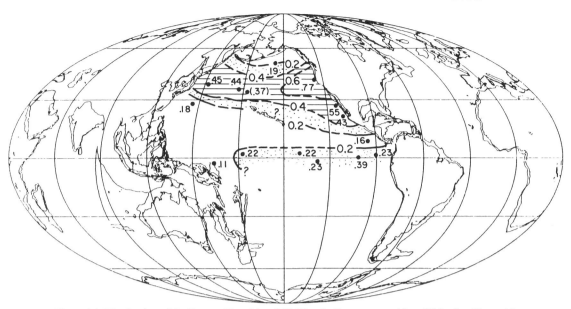

Figure 26. Distribution of the Eastern Transitional/California Current assemblage 8 Ma. See Figure 15 for explanation.

Stylochlamidium asteriscus

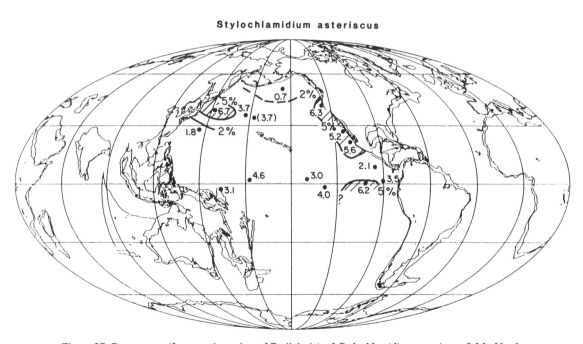

Figure 27. Percentages (from total number of Radiolaria) of *Stylochlamidium asteriscus,* 8 Ma North Pacific. Areas of greater than 5 percent are shaded.

Actinomma spp.

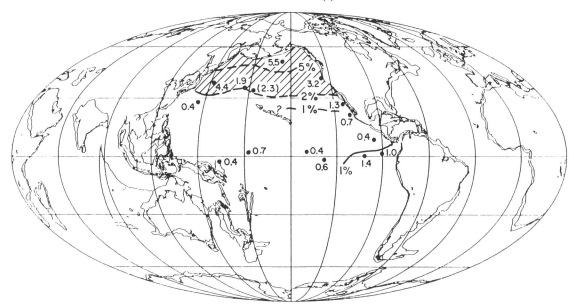

Figure 28. Percentages (from total number of Radiolaria) of *Actinomma* spp., 8 Ma North Pacific. Areas of greater than 2 percent are shaded.

Spongurus spp.

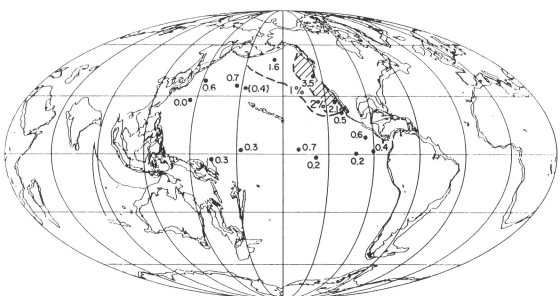

Figure 29. Percentages (from total number of Radiolaria) of *Spongurus* (?) sp., 8 Ma North Pacific. Areas of greater than 2 percent are shaded.

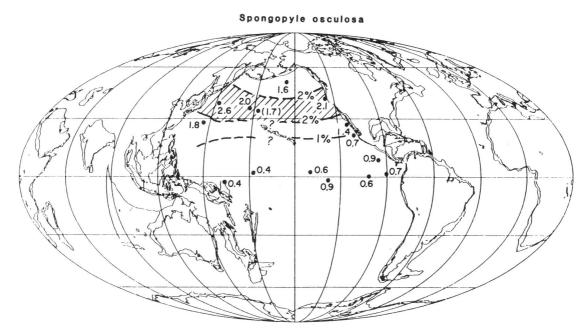

Figure 30. Percentages (from total number of Radiolaria) of *Spongopyle osculosa,* 8 Ma North Pacific. Areas of greater than 2 percent are shaded.

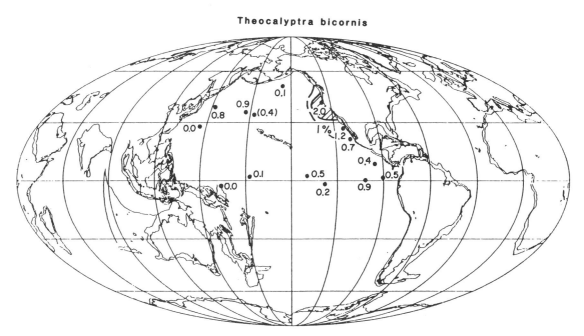

Figure 31. Percentages (from total number of Radiolaria) of *Theocalyptra bicornis,* 8 Ma North Pacific. Areas of greater than 2 percent are shaded.

WESTERN TRANSITIONAL ASSEMBLAGE

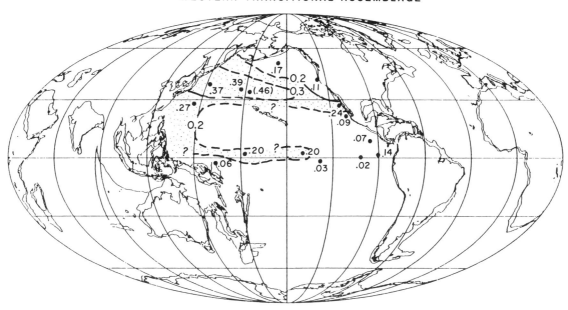

Figure 32. Distribution of the Western Transitional assemblage at 8 Ma. See Figure 15 for explanation.

DISSOLUTION–OXYGEN MINIMUM ASSEMBLAGE

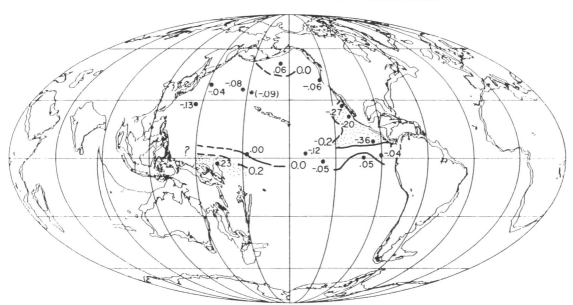

Figure 33. Dissolution-Oxygen Minimum assemblage at 8 Ma. See Figure 15 for explanation. (Note: In this factor highest negative factor values are shaded similarly to highest positive values.)

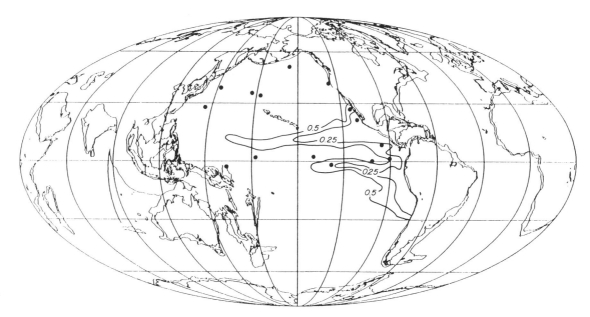

Figure 34. Oxygen concentration in milliliters per liter at about 400 m water depth. Dots indicate sample locations used in survey (adapted from Reid 1965).

water column (Casey 1971a, b; Casey et al. 1979a; Kling 1979; McMillen and Casey 1978; Petrushevskaya 1971a, b; Renz 1976; Spaw 1979) and what these depth distributions might indicate about the ecological tolerances or preferences of individual radiolarian species (Casey et al. 1979 a, b; Spaw 1979).

Using the modern species present in late Miocene sediments as a guide, inferences can be made about the assemblages described by factor analysis (see Table 4). Only the first three assemblages actually grouped species as to their apparent depth preferences. The Western Transitional and Dissolution-Oxygen Minimum assemblages are composed of species of mixed depth preferences. Factor 1, the Tropical assemblage is made up of shallowest-water dwellers. Two of the species groups, *Didymocyrtis* spp. and *Diartus* spp., (*D. hughesi* and *D. petterssoni*), are closely-related ancestral groups of *Didymocyrtis tetrathalamus* (*Ommatartus tetrathalamus* in Casey et al. 1979a, b) which has been found in Modern plankton samples to possess zooxanthellae (Casey et al. 1979b). This characteristic allows these zooplankton to obtain nutrients and survive in regions where oligotrophic conditions prevail and nutrient availability is low. It also requires them to live in the photic zone, usually in warmer surface waters in the tropics where daylight hours are maximal. The *Diartus* species group is an offshoot of the main lineage of the Didymocyrtids. The abundance of *Diartus* species in the region of the subsurface oxygen minimum in association with *Cornutella profunda,* a known deep water dweller in the Modern ocean (see factor 5) suggests that *Diartus* species may have been deep-water dwellers as well. Since the group was an evolutionary "dead-end," unlike its successful relative, the Didymocyrtid group, it is possible that they may not have had zooxanthellae and that possession of zooxanthellae may have been the key to the success of the Didymocyrtid group.

TABLE 4. DOMINANT SPECIES IN THE FIRST THREE FACTORS AND THEIR ECOLOGIC PREFERENCES*

	Ecologic preference
Tropical Assemblage	
Phorticium pylonium	Unknown
Didymocyrtis spp.	Warm, shallow water (0-200 m); possess zooxanthellae
Disolenia spp.	Warm, shallow water (0-200 m)
Heliodiscus asteriscus	Shallow water (100-200 m)
Larcospira quadrangula	Warm, shallowest water (<100 m)
Diartus spp.	(May be similar to Didymocyrtids?)
Subarctic Assemblage	
Stichocorys delmontensis	(Not extant)
Stichocorys peregrina	(Not extant)
Cornutella profunda	Subarctic, intermediate-deep water (500 m to >2000 m)
Eastern Transitional/ California Current Assemblage	
Stylochlamidium asteriscus	
Spongotrochus venustum	
Actinomma spp.	
Spongurus (?) sp.	
Spongopyle osculosa	Intermediate and deep water (200-2000 m)
Larcopyle butschlii	
Spongotrochus glacialis	Upwelling indicator; intermediate and deep water (200-2000 m)
Theocalyptra bicornis	Intermediate and deep water (200-2000 m)
Theocalyptra davisiana	(In the Quaternary this species is important in the Northwest Subarctic Pacific.)

*According to Casey, et al. (1979a, b); Spaw (1979); and Kling (1979).

The Subarctic assemblage is dominated by two extinct species, *Stichocorys delmontensis* and *S. peregrina* and one extant species, *Cornutella profunda* which is associated with Subarctic Intermediate Water in the California Current region and Antarctic Intermediate Water and/or North Atlantic Deep Water in the open ocean Gulf of Mexico (Casey et al. 1979a). Figure 22 indicates the presence of this assemblage in the eastern equatorial Pacific. In the Modern ocean, oxygenated Antarctic Intermediate Water is known to spread north toward the Equator in the southeast Pacific to at least 10°S latitude and possibly as far north as the equatorial divergence (Sverdrup et al. 1942; Johnson 1972).

The Eastern Transitional/California Current assemblage has a large number of species associated with it, all of which are extant today. The general makeup of the species can largely be characterized as cold water, deeper dwelling species (Table 4). The association with true upwelling is supported by the presence of *Spongotrochus glacialis* which is considered by Casey et al. (1979a), Spaw (1979), and Leavesley et al. (1978) to be an upwelling indicator. *Theocalyptra bicornis, Spongopyle osculosa*, and *Spongotrochus glacialis* have been associated with Antarctic Intermediate Water (or North Atlantic Deep Water) in the Gulf of Mexico and Sargasso Sea (Spaw 1979; Casey et al. 1979a) which is probably also upwelling in the equatorial Pacific. This assemblage is composed of species associated with cold intermediate water masses that are upwelling at oceanic divergences and along the eastern boundary regions of the Pacific.

Paleoceanography. The late Miocene data (for 44 species) indicate that there is latitudinal zonality in the biogeographic patterns with warm and cold end-members similar to the Modern ocean. A cold, subarctic water mass and both warm western and cold eastern boundary currents are indicated in the first four factors (Figures 15, 22, 26, and 32). A zone of poor silica preservation is also represented in the same general region in the Southern Hemisphere as in recent surface sediments, suggesting similar gyral configuration and deep water transport in both the late Miocene and Modern ocean. The regions of poor preservation in both hemispheres occur at the centers and inner margins of the subtropical gyres. The choice of sites for this study and the areal coverage by coretops in the Modern analysis (Moore and Lombari 1981) were dependent on the presence or absence of siliceous sediment. Therefore, it is not a coincidence that the late Miocene choices for sites outline the North Pacific gyre. There was little silica preserved in the middle and low-latitude South Pacific Ocean in the late Miocene and no sites were found that were useful to this study.

At 8 Ma, the Subarctic assemblage was separated from the tropics by a well-developed transitional assemblage (both factors 3 and 4), which was more important in the California Current region than today (Figure 35, a, c). This suggests a diminished role for subarctic water (factor values are equal to or less than 0.4 south of 40°N latitude) in the California Current 8 Ma ago, also indicated by distribution patterns discussed in Part 1. In contrast, along the North American coast the transitional assemblage was more important over a broader zone than today. In fact, the range

of the late Miocene "transition zone" encompasses the extremes in position of this zone in the Modern (interglacial) and 18 Ma (glacial) Pacific (Figure 35). Figure 35 indicates that away from the coast (along 150° W) the Modern and late Miocene assemblage distributions are generally very similar in their latitudinal relationships, but that the Subarctic assemblage at 8 Ma decreases in importance southward more rapidly and the transitional assemblage again occupies a broader zone than in Modern times. They are both quite different from the Quaternary ice-age ocean where the Subarctic assemblage extends its distribution further to the south, displacing the northern limit of the Tropical assemblage southward as well. The Tropical assemblage occupies a very large area at 8 Ma, occurring as far north as 40°N latitude in the western North Pacific and 35°N in the east (Figure 15). This distribution indicates a warm late Miocene North Pacific suggested by sea surface temperature estimates in sites used in this study (sites 310, 173, 158, 77B, 289; Moore and Lombari 1981). Barron and Keller (1983) also presented evidence from diatom and foraminiferal assemblage regions in the California Current that indicates that the 8 Ma interval was warm.

The late Miocene Eastern Transitional/California Current assemblage extends southward along the California coast to about 15°N latitude (Figure 26). This biogeographic pattern suggests that the frontal region between colder California Current water and warmer Pacific Equatorial Water existed somewhat further to the south than its present position at about 23°N. This is consistent with the inferred increase in volume transport in the California Current.

The existence of the Subarctic assemblage along the equator at 8 Ma (Figure 22) could indicate upwelling of cooler and deeper waters along the equatorial divergence. Ekman divergence occurs along the equator as a result of a dynamic balance between the Coriolis force, an apparent force that causes current transport to be deflected to the right in the Northern Hemisphere and to the left in the Southern Hemisphere, and the wind stress caused by the southeast tradewinds (Knauss 1978). Divergence and upwelling occur in association with wind-driven westward flow along the equator. The rates and sources of water for the divergence and upwelling, however, can be variable. The Equatorial Undercurrent is a major source of the upwelled water along the Modern equatorial divergence (Wyrtki 1966, 1967) and has been associated with a specific radiolarian assemblage (Romine and Moore 1981). In the Paleogene and early Neogene, Australia migrated northward, blocking Pacific-Indian Ocean communication through the Indonesian region. Prior to this event, the Equatorial Undercurrent probably did not exist. Van Andel et al. (1975) suggest the development of this current by approximately 12 Ma to explain the rapid carbonate deposition that occurred along the Equator at a time when the Pacific equatorial CCD was shallowing. At 8 Ma bulk accumulation rates and calcium carbonate content of the sediments were relatively high at sites 83 and 77B (van Andel et al 1975). Since most of the sediments deposited were biogenic, this implies high productivity and/or increased preservation and probably increased upwelling at the

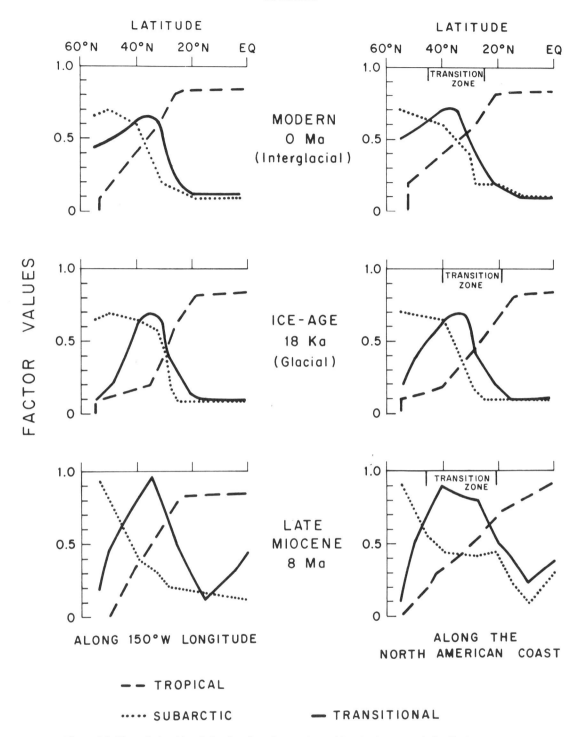

Figure 35. The relationship of the first four factors (assemblages) along two latitudinal transects: a transect along 150°W longitude; and a transect in the California Current region along the western North American coast. Factors 3 and 4 are both transitional and have been summed in the late Miocene plots. The factor values are plotted on the z-axis for comparison of the relative proportions of each assemblages. (a) Modern Pacific Ocean assemblages (data from Moore 1978); (b) Quaternary ice-age assemblages (18,000 years B.P.; data from Moore 1978); (c) Late Miocene assemblages (8 Ma, this paper).

divergence. However, the late Miocene subarctic species that seem to indicate the existence of colder upwelled water at the equator (Figures 22, 23, and 24) are associated with intermediate to deep water masses, suggesting that the source for the upwelled water at the divergence may have been dominantly Subarctic Intermediate Water and not the Equatorial Undercurrent. This seems unlikely when one considers the dynamics of Modern equatorial upwelling that indicate the source depth of upwelled waters to be less than 200 m (Wyrtki, 1981). It is probable that the subarctic species were either advected into the equatorial region by eastern boundary circulation or were *not* associated with upwelling if brought to the equatorial region in intermediate water masses.

Warmer temperatures along the California coast at 8 Ma (Figure 14) indicate that cold subarctic water was not contributing significantly to the California Current. The subarctic gyre probably supplied much less water to the California Current, and what little it did contribute was possibly somewhat warmer, unlike the Modern situation. Instead, cool transitional waters whose character was a mixture of subarctic and subtropical waters were dominant in the California Current region. Figure 35 (a and c) also demonstrates that the source of waters along the coast at 8 Ma was dominantly transitional with less subarctic contribution than occurs there today.

Evidence from the record of calcium carbonate concentration indicates a warm North Pacific at 8 Ma. This time-slice was stratigraphically determined by its coincidence with an interval of maximum calcium carbonate content in the downcore records of various DSDP sites. Studies of Late Quaternary carbonate records from the Pacific Ocean have shown that high carbonate percentages correspond to glacial intervals (Hays et al. 1969; Berger 1970; Broecker 1971) while those from the Atlantic show the reverse, high carbonate percentages during interglacials (Olausson 1965, 1971; Broecker 1971; Berger 1973; Gardner 1975). Recent studies on carbonate stratigraphy in the late Miocene and Pliocene indicate that prior to Northern Hemisphere glaciation (approximately 2.5–3.0 Ma) Pacific equatorial carbonate records show the Atlantic-type relationship of high carbonate percentages during interglacial intervals (Dunn and Moore 1981; Dunn 1982). Therefore, coincidence of the late Miocene time-slice with high carbonate percentages would lead one to predict that the time-slice was a warm interval, a prediction confirmed by the distribution of radiolarian assemblages.

SUMMARY AND CONCLUSIONS

The study presented here has examined the radiolarian biogeography of the North Pacific Ocean at 8 Ma and made inferences about the paleoceanography of the late Miocene Pacific. Using 27 species categories that ranged from 8 Ma through the Present, a comparison of the 8 Ma versus Modern time-slice indicated a general similarity, but with some differences in the eastern tropical and subarctic regions. Equatorial flow from the Atlantic Ocean entered the eastern tropical Pacific through the Central American Seaway resulting in greater transport of warm water to the temperate and subarctic Pacific and generally warmer temperatures throughout the North Pacific Ocean. To further evaluate the biogeographic patterns, a multivariate factor analysis of species representing the majority of the most abundant late Miocene radiolaria was completed, providing a more complete picture of the biogeography of the existing population. Several conclusions concerning Miocene biogeographic and oceanographic patterns can be reached by comparing these data with those of the Modern ocean and their relationship to Modern oceanography.

(1) Factor analysis of 44 late Miocene radiolarian species groups resulted in assemblages that are distinct paleoecologically and show latitudinal zonality. The analysis appeared to group the species not only into biogeographic provinces but also into general depth-related categories. Deep (Subarctic assemblage), intermediate (Transitional assemblages) and shallow (Tropical assemblage) water provinces were distinguished.

(2) Extant species marking the cold subpolar and transitional region appear to have survived without great morphologic change since the late Miocene. These species may have developed a tolerance for changes in depth habitat to cope with changes in temperature. They can survive at lower latitudes by living in deeper water masses or by "diving" with the water mass they prefer to live in. The sedimentary record seems to support this idea since these species are always found in greatest relative abundance in sediments beneath regions of upwelling at lower latitudes and subpolar water in high latitudes.

(3) Changes in the biogeography and abundances of several of the extant species groups that dominated assemblages determined by factor analysis became apparent in a comparison of the late Miocene and the Modern Pacific. The absence of a late Miocene Eastern Tropical assemblage, the splitting up of a late Miocene trans-equatorial assemblage into the Modern S.W. and N.W. Tropical assemblages and the movement northward along the California coast of the Eastern Subpolar assemblage strongly suggest major changes in water mass structure in the equatorial and, in particular, eastern equatorial Pacific. The middle Pliocene appearance of the Isthmus of Panama as a barrier to Atlantic-Pacific surface water communication is the most likely cause of such water mass changes.

(4) The abundance of subarctic and transitional species at site 472 (21°N latitude) at 8 Ma suggests a more southern position for the front between the cold California Current and warm Pacific Equatorial water (which today exists at approximately 23°N). Flow of Atlantic water into the equatorial Pacific appears to have disrupted the flow of the Equatorial Countercurrent so the warm water it normally carries back to the eastern tropical Pacific was instead diverted northward into the late Miocene equivalent to the Kuroshio Current. This resulted in increased transport of warmer water by the subtropical gyre, warmer sea surface temperatures in the 8 Ma North Pacific than in the Modern, and expansion of the subtropics. The combined effect of these changes resulted in greater penetration of the California

Current to the south of its Modern limits and a latitudinally broader zone of influence.

In summary, the flow of water from the Atlantic into the equatorial Pacific probably influenced the flow pattern of the Equatorial Countercurrent causing some of the warm water that normally returns to the eastern tropical Pacific in the Countercurrent to be diverted into the northward-flowing western limb of the subtropical gyre (Figure 13). The amount of warmer water in the North Pacific gyral system increased, probably inducing expansion of the transitional region and warmer sea-surface temperatures at higher latitudes (Ingle 1973; Moore and Lombari 1981; Barron and Keller 1983). As flow from the Atlantic diminished through the latest Miocene and early Pliocene, the Pacific subarctic regions became progressively cooler and probably expanded their seasonal influence as times of increased input of warm tropical water into the subpolar regions became less frequent. The gradual cessation of Atlantic-Pacific surface water communication accelerated the global cooling trend initiated by paleogeographic and paleoclimatic changes in the Southern Hemisphere in the latest Paleogene and early Neogene. The termination of this flow in the middle Pliocene (Keigwin 1978) set the scene for a Quaternary world whose climate was determined by periodic changes in insolation that resulted in cyclic fluctuations in continental ice volume.

ACKNOWLEDGMENTS

I thank T. C. Moore, Jr., N. G. Pisias, R. E. Casey, and J. P. Kennett for thorough critical reviews and suggestions that have substantially improved this manuscript. I am also grateful to G. Lombari for her help in regard to taxonomy and determination of radiolarian counting categories. J. Barron and D. Dunn generously provided preprints and unpublished data that allowed the determination of a precise stratigraphy.

Thanks go to S. Silvia, B. Watkins, M. McHugh, and L. DeFusco for quality illustration and photographic services, to K. Hazard for typing the manuscript, and to N. Meader for editing the final draft. This research was supported by NSF grant OCE79-14594, Cenozoic Paleoceanography (CENOP), awarded to the University of Rhode Island.

```
             APPENDIX - MIOCENE FORMS
           TAXONOMIC LIST AND REFERENCES
```

A listing of the species categories used in this study is presented based on a recently compiled compendium of Radiolaria from the Miocene (Nigrini and Lombari, 1983). Illustrations of the species listed can be found in Nigrini and Moore (1979) and Nigrini and Lombari (1984).

```
Phylum Protozoa
Class Actinopoda
Subclass Radiolaria Muller, 1858
Order Polycystina Ehrenberg, 1838 emend. Riedel, 1967

Suborder Spumellaria Ehrenberg, 1875
Disolenia spp.
      Disoleniaguadrata (Ehrenberg), 1872
      Disolenia zanguebarica (Ehrenberg), 1872
      Otosphaera auriculata Haeckel, 1872
      Otosphaera polymorpha Haeckel, 1887
      Siphonosphaera polysiphonia Haeckel, 1887
      Solenosphaera omnitubus omnitubus Riedel and
         Sanfilippo, 1971
      Solenosphaera omnitubus procera Sanfilippo and Riedel,
         1974

Actinomma spp.
      Actinomma antarcticum (Haeckel), 1887
      Actinomma arcadophorum Haeckel, 1887
      Actinomma medianum Nigrini, 1967

Cenosphaera cristata Haeckel, 1887

Hexacontium spp.
      Hexacontium enthacanthum Jorgenson, 1899
      Hexacontium laevigatum Haeckel, 1887

Stylatractus universus Hays, 1970

Spongurus (?) sp. Petrushevskaya, 1967

Spongurus sp. B Nigrini and Lombari, 1984

Styptosphaera spumacea Haeckel, 1887
```

APPENDIX - MIOCENE FORMS
TAXONOMIC LIST AND REFERENCES
(continued)

Heliodiscus asteriscus Haeckel, 1887

Diartus spp.
    Diartus petterssoni Sanfilippo and Riedel, 1980
    Diartus hughesi Sanfilippo and Riedel, 1980

Didymocyrtis spp.
    Didymocyrtis laticonus Sanfilippo and Riedel, 1980
    Didymocyrtis antepenultimus Sanfilippo and Riedel, 1980
    Didymocyrtis penultimus Sanfilippo and Riedel, 1980
    Didymocyrtis tetrathalamus Sanfilippo and Riedel, 1980

Hymeniastrum euclidis Haeckel, 1887

Stylodictya validispina Jorgensen, 1905

Circodiscus microporus (Stohr) group Petrushevskaya and
    Kozlova, 1972

Stylochlamidium asteriscus Haeckel, 1887

Spongopyle osculosa Dreyer, 1889

Spongotrochus glacialis Popofsky group Petrushevskaya, 1975

Spongotrochus venustum (Bailey), Nigrini and Moore, 1979

Phorticium polycladum Tan and Tchang, 1976

Phorticium pylonium (Haeckel) Cleve in Benson, 1966

Tetrapyle octacantha Muller, 1858

Larcopyle butschlii Dreyer, 1889

Larcospira quadrangula Haeckel, 1887
    Larcospira moschkovskii Kruglikova, 1978

Lithelius minor Jorgensen, 1899

Lithelius sp. Nigrini and Lombari, 1984

Pylospira octopyle Haeckel, 1887

Acrosphaera spp.
    Acrosphaera lappacea Bjorklund and Goll, 1979
    Acrosphaera flammabunda
    Acrosphaera spinosa

Suborder Nassellaria Ehrenberg, 1875

Zygocircus spp.
    Zygocircus productus capulosus (Popofsky) Goll, 1980
    Zygocircus productus tricarinatus Goll, 1980

Antarctissa spp.
    Antarctissa deflandrei (Petrushevskaya) 1975
    Antarctissa longa (Popofsky), Petrushevskaya, 1967
    Antarctissa sterelkovi Petrushevskaya, 1967

Ceratocyrtis histricosa (Jorgensen), Petrushevskaya, 1971

Giraffospyris angulata (Haeckel), Goll, 1969
    Giraffospuris angulata Goll, 1969

Lophospyris pentagona pentagona (Ehrenberg), Goll, 1977

Phormospyris stabilis stabilis (Goll), Goll, 1977

APPENDIX - MIOCENE FORMS
TAXONOMIC LIST AND REFERENCES
(continued)

Dendrospyris bursa Sanfilippo and Riedel, Sanfilippo, et al,
        1973

Cornutella profunda Ehrenberg, Nigrini, 1967

Eucyrtidium hexagonatum Haeckel, 1887

Eucyrtidium acuminatum (Ehrenberg), 1844

Stichocorys delmontensis (Campbell and Clark), Sanfilippo and
        Riedel, 1970

Stichocorys peregrina (Riedel), Sanfilippo and Riedel, 1970

Theocalyptra bicornis (Popofsky) Riedel, 1958

Theocalyptra bicornis var. - see Nigrini and Lombari, 1984

Theocalyptra davisiana davisiana (Ehrenberg), Riedel, 1958

Theocorysredondoensis (Campbell and Clark), Kling, 1973

Anthocyrtidium ophirense (Ehrenberg), Nigrini, 1967
        Anthocyrtidium ehrenbergi ehrenbergi (Stohr) Riedel, et
        al., 1974

Calocycletta caepa Moore, 1972

Lamprocyrtis maritalis Haeckel group, Nigrini, 1967

Pterocorys cf. zancleus (Muller) Nigrini and Moore, 1979

Botryostrobus aquilonaris (Bailey), Nigrini, 1977

Phormostichoartus corbula (Harting), Nigrini, 1977
        Phormostichoartus doliolum (Riedel and Sanfilippo),
        Nigrini, 1977
        Phormostichoartus marylandicus (Martin), Nigrini, 1977

Siphostichartus corona (Haeckel), Nigrini, 1977

REFERENCES CITED

Barron, J. A., 1980, Lower Miocene to Quaternary diatom biostratigraphy of DSDP Leg 57, off northeastern Japan, in Langseth, M., Okada, H., and others, Initial Reports of the Deep Sea Drilling Project, Volume 56 and 57, Pt. 2: Washington, D.C., U.S. Government Printing Office, p. 641–686.

—— 1981, Late Cenozoic diatom biostratigraphy and paleoceanography of the middle-latitude eastern North Pacific, Deep Sea Drilling Project Leg 63, in Yeats, R. S., Haq, B.U., et al., Initial Reports of the Deep Sea Drilling Project, Volume 63: Washington, D.C., U.S. Government Printing Office, p. 507–537.

Barron, J. A., and Keller, G., 1983, Paleotemperature oscillations in the middle and late Miocene of the northeastern Pacific: Micropaleontology, v. 29, no. 2, p. 150–181.

Bé, A.W.H., and Tolderlund, D. S., 1971, Distribution and ecology of living planktonic foraminifera in surface waters of the Atlantic and Indian Oceans, in Funnell, B. M., and Riedel, W. R., eds., The Micropaleontology of Oceans: London, Cambridge University Press, p. 105–149.

Berger, W. H., 1970, Selective solution and the lysocline: Marine Geology, v. 8, p. 111–138.

—— 1973, Deep-sea carbonates: Pleistocene dissolution cycles: Journal of Foraminiferal Research, v. 3, p. 187–195.

Broecker, W. S., 1971, Calcite accumulation rates and glacial to interglacial changes in oceanic mixing, in Turekian, K. K., ed., The Late Cenozoic Glacial Ages: New Haven, Conn., Yale University Press, p. 239–265.

Casey, R. E., 1971a, Distribution of polycystine Radiolaria in the oceans in relation to physical and chemical conditions, in Funnell, B. M., and Riedel, W. R., eds., The Micropaleontology of Oceans: London, Cambridge University Press, p. 151–160.

—— 1971b, Radiolaria as indicators of past and present water masses, in Funnell, B. M., and Riedel, W. R., eds., The Micropaleontology of Oceans: London, Cambridge University Press, p. 331–342.

—— 1982, Lamprocyrtis and Stichocorys lineages: Biogeographical and ecological perspectives relating to the tempo and mode of Polycystine radiolarian evolution, in Proceedings, Third North American Paleontological Convention, Volume 1, p. 77–82.

Casey, R. E., Spaw, J. M., Kunze, F., Reynolds, R., Duis, T., McMillen, K., Pratt, D., and Anderson, V., 1979a, Radiolaria ecology and the development of the radiolarian component in Holocene sediments, Gulf of Mexico and adjacent seas with potential paleontological applications: Transactions of the Gulf Coast Association Geological Society, v. 29, p. 228–237.

Casey, R. E., Gust, L., Leavesley, A., Williams, D., Reynolds, R., Duis, T., and Spaw, J. M., 1979b. Ecological niches of radiolarians planktonic foraminiferans and pteropods inferred from studies on living forms in the Gulf of

Mexico and adjacent waters: Transactions of the Gulf Coast Association Geological Society, v. 29, p. 216–223.

Casey, R. E., Spaw, J. M., and Kunze, F. R., 1982, Polycystine radiolarian distributions and enhancements related to oceanographic conditions in a hypothetical ocean: Transactions of the Gulf Coast Association Geological Society, v. 32, p. 319–322.

CLIMAP, 1976, The surface of the ice-age earth: Science, v. 191, p. 1131–1136.

—— 1982, Seasonal reconstructions of the earth's surface at the last glacial maximum, *in* McIntyre, A., (Leader), Geological Society of America Map and Chart Series.

Dunn, D. A., 1982, Change from "Atlantic-type" to "Pacific-type" carbonate stratigraphy in the middle Pliocene equatorial Pacific Ocean: Marine Geology, v. 0, p. 41–60.

Dunn, D. A., and Moore, T. C., Jr., 1981, Late Miocene (Magnetic Epoch 5–Epoch 9) calcium carbonate stratigraphy of the equatorial Pacific Ocean: Geological Society of America Bulletin, v. 92, p. 408–451.

Embley, R. W., and Johnson, D. A., 1980, Acoustic stratigraphy and biostratigraphy of Neogene carbonate horizons in the north equatorial Pacific: Journal of Geophysical Research, v. 85, p. 5423–5437.

Firing, E., and Lukas, R., 1983, El Niño at the Equator and 159°W: Tropical Ocean-Atmosphere Newsletter, v. 21, p. 9–11.

Gardner, J. V., 1975, Late Pleistocene carbonate dissolution cycles in the eastern equatorial Atlantic, *in* Sliter, W. V., Bé, A.W.H., and Berger, W. H., eds., Dissolution of deep-sea carbonates: Cushman Foundation for Foraminiferal Reserach, Special Publication, v. 13, p. 129–141.

Haq, B. U., Worsley, T. R., Burckle, L. H., Douglas, R. G., Keigwin, L. D., Jr., Opdyke, N. D., Savin, S. M., Sommer, M. A., II, Vincent, E., and Woodruff, F., 1980, Late Miocene marne carbon-isotopic shift and synchroneity of some phystoplanktonic biostratigraphic events: Geology, v. 8, p. 427–431.

Hays, J. D., Saito, T., Opdyke, N. D., and Burckle, L. H., 1969, Plio-Pleistocene sediments of the equatorial Pacific: their paleomagnetic, biostratigraphic and climatic record: Geological Society of America Bulletin, v. 80, p. 1481–1514.

Imbrie, J., and Kipp, N. G., 1971, A new micropaleontological method for quantitative paleoclimatology; application to a late Pleistocene Caribbean core, *in,* Turekian, K. K., ed., Late Cenozoic Glacial Ages: New Haven, Conn., Yale University Press, p. 71–181.

Ingle, J. C., Jr., 1973, Neogene Foraminifera from the northeastern Pacific Ocean, Lcg 18, Dccp Sca Drilling Project, *in* Kulm, L. D., von Huene, R., and others, Initial Reports of the Deep Sea Drilling Project, Volume 18: Washington, D.C., U.S. Government Printing Office, p. 517–567.

Janacek, T. R., and Rea, D. K., 1983, Eolian deposition in the northeast Pacific Ocean: Cenozoic history of atmospheric circulation: Geological Society of America Bulletin, v. 94, p. 730–738.

Johnson, R. E., 1972, Antarctic Intermediate Water in the South Pacific Ocean, *in* Oceanography of the South Pacific, compiled by R. Fraser, Wellington, N. Z., UNESCO, p. 55–69.

Johnson, T. C., 1974, The dissolution of siliceous microfossils in surface sediments of the eastern tropical Pacific: Deep Sea Research, v. 21, p. 851–864.

Keigwin, L. D., Jr., 1978, Pliocene closing of the Panama Isthmus based on biostratigraphic evidence from nearby Pacific Ocean and Caribbean cores: Geology, v. 6, p. 630–634.

Keller, G., Barron, J. A., and Burckle, L. H., 1982, North Pacific late Miocene correlations using microfossils, stable isotopes, percent $CaCO_3$, and magnetostratigraphy: Marine Micropaleontology, v. 7, p. 327–357.

Kipp, N. G., 1976, New transfer function for estimating past sea surface conditions from sea-bed distribution of planktonic foraminiferal assemblages in the North Atlantic, *in,* Cline, R. M., and Hays, J. D., eds., Investigations of Late Quaternary Paleoceanography and Paleoclimatology: Geological Society of America Memoir, v. 145, p. 3–41.

Kling, S. A., 1979, Vertical distribution of polycystine radiolarians in the central North Pacific: Marine Micropaleontology, v. 4, p. 295–318.

Klovan, J. E., and Imbrie, J., 1971, An algorithm and Fortran IV program for large-scale Q-mode factor analysis: Journal of the International Association of Mathematical Geologists, v. 3, p. 61–67.

Knauss, J., 1978, Introduction to Physical Oceanography: Englewood Cliffs, New Jersey, Prentice-Hall, Inc., 338 p.

Leavesley, A., Bauer, M., McMillen, K., and Casey, R., 1978, Living shelled microzooplankton (radiolarians, foraminiferans and pteropods) as indicators of oceanographic processes in water over the outer continental shelf of south Texas: Transactions of the Gulf Coast Association Geological Society, v. 28, p. 229–238.

Leinen, M., and Heath, G. R., 1981, Sedimentary indicators of atmospheric activity in the Northern Hemisphere during the Cenozoic: Palaeogeography, Palaeoclimatology, and Palaeoecology, v. 36, p. 1–21.

Lombari, G., and Boden, G., 1982, Paleobiogeography and diversity of radiolaria: Recent versus Miocene: Geological Society of America Abstracts with Programs, v. 14, p. 548.

Lozano, J. A., and Hays, J. D., 1976, Relationship of radiolarian assemblages to sediment types and physical oceanography in the Atlantic and Western Indian Ocean sectors of the Antarctic Ocean, *in* Cline, R. M., and Hays, J. D., eds., Investigation of Late Quaternary Paleoceanography and Paleoclimatology: Geological Society of America Memoir, v. 145, p. 303–336.

Luyendyk, B. P., Forsyth, D., and Phillips, J. D., 1972, Experimental approach to the paleocirculation of the oceanic surface waters: Geological Society of America Bulletin, v. 83, p. 2649–2664.

McGowan, J. A., 1971, Oceanic biogeography of the Pacific, *in,* Funnell, B. M., and Riedel, W. R., eds., The Micropaleontology of Oceans: London, Cambridge University Press, p. 3–74.

McMillen, K., and Casey, R. E., 1978, Distribution of living polycystine radiolarians in the Gulf of Mexico and Caribbean Sea, and comparison with the sedimentary record: Marine Micropaleontology, v. 3, p. 121–145.

Molina-Cruz, A., 1977, Late Quaternary oceanic circulation along the Pacific Coast of South America [Ph.D. thesis]: Corvallis, Oregon State University, 246 p.

Moore, T. C., Jr., 1973, Method of randomly distributing grains for microscopic examination: Journal of Sedimentary Petrology, v. 43, p. 904.

—— 1978, The distribution of radiolarian assemblages in the Modern and ice-age Pacific: Marine Micropaleontology, v. 3, p. 229–266.

Moore, T. C., Jr., and Lombari, G., 1981, Sea-surface temperature changes in the North Pacific during the late Miocene: Marine Micropaleontology, v. 6, p. 581–597.

Nigrini, C., 1970, Radiolarian assemblages in the North Pacific and their application to a study of a Quaternary sediment core, *in* Hays, J. D., ed., Geological Investigations of the North Pacific: Geological Society of America Memoir, v. 126, p. 139–183.

Nigrini, C., and Lombari, G., 1984, A Guide to Miocene Radiolaria: Cushman Foundation for Foraminifera Research, Special Publication No. 22.

Nigrini, C., and Moore, T. C., Jr., 1979, A Guide to Modern Radiolaria: Cushman Foundation for Foraminiferal Research, Special Publication No. 16.

Olausson, E., 1965, Evidence of climatic changes in North Atlantic deep-sea cores, with remarks on isotopic paleotemperature analysis: Progress in Oceanography, v. 3, p. 221–252.

—— 1971, Quaternary correlations and the geochemistry of oozes, *in,* Funnell, B. M., and Riedel, W. R., eds., The Micropaleontology of Oceans: Massachusetts, Cambridge University Press, p. 375–398.

Pak, H., and Zaneveld, J. R., 1973, The Cromwell Current on the east side of the Galapagos Islands: Journal of Geophysical Research, v. 78, p. 7485–7859.

Petrushevskaya, M. G., 1971a, Spumellarian and Nassellarian Radiolaria in the plankton and bottom sediments of the central Pacific, *in,* Funnell, B. M., and Riedel, W. R., eds., The Micropaleontology of Oceans: London, Cambridge University Press, p. 309–317.

—— 1971b, Radiolaria in the plankton and Recent sediments from the Indian Ocean and Antarctic, *in,* Funnell, B. M., and Riedel, W. R., eds., The Micropaleontology of Oceans: London, Cambridge University Press, p. 319–329.

Rea, D. K., and Janacek, T. R., 1982, Late Cenozoic changes in atmospheric circulation deduced from North Pacific eolian sediments: Marine Geology,

v. 49, p. 149–167.

Reid, J. L., 1965, Intermediate waters of the Pacific Ocean, *in* Johns Hopkins Oceanographic Studies, v. 2, Baltimore, The John Hopkins University Press, 85 p.

Renz, G. E., 1976, The distribution and ecology of Radiolaria in the central Pacific plankton and surface sediments: Bulletin of the Scripps Institution of Oceanography, v. 22, p. 1–267.

Robertson, J. H., 1975, Glacial to interglacial oceanographic changes in the northwest Pacific, including a continuous record of the last 400,000 years [Ph.D. thesis]: New York, Columbia University, 355 p.

Romine, K., 1982, Late Quaternary history of atmospheric and oceanic circulation in the eastern equatorial Pacific: Marine Micropaleontology, v. 7, p. 163–187.

Romine, K., and Moore, T. C., Jr., 1981, Radiolarian assemblage distributions and paleoceanography of the eastern equatorial Pacific Ocean during the last 127,000 years: Palaeogeography, Palaeoclimatology, and Palaeoecology, v. 35, p. 281–314.

Sachs, H. M., 1973, North Pacific radiolarian assemblages and relationship to oceanographic parameters: Quaternary Research, v. 3, p. 73–88.

Saito, T., 1976, Geologic significance of coiling direction in the planktonic foraminifera *Pulleniatina:* Geology, v. 4, p. 305–309.

Sancetta, C., 1979, Oceanography of the North Pacific during the last 18,000 years: evidence from fossil diatoms: Marine Micropaleontology, v. 4, p. 103–123.

Sclater, J., Meinke, L., Bennett, A., and Murphy, C., 1985, Geological Society of America Memoir 163 p. 000.

Shaw, A. B., 1964, Time in Stratigraphy: New York, McGraw-Hill, 365 p.

Spaw, J. M., 1979, Vertical distribution, ecology and preservation of Recent polycystine Radiolaria of the North Atlantic Ocean (southern Sargasso Sea region) [Ph.D. thesis]: Houston, Rice University, 231 p.

Summerhayes, C. P., 1981, Oceanographic controls on organic matter in the Miocene Monterey Formation, offshore California, *in* The Monterey Formation and Related Siliceous Rocks of California: Society of the Economic Paleontologiss and Mineralogists, p. 213–220.

Sverdrup, H. U., Johnson, M. W., and Fleming, R. H., 1942, The Oceans: their physics, chemistry and general biology: WHERE, New Jersey, Prentice-Hall, Inc., 1087 p.

Theyer, F., Mato, C. Y., and Hammond, S. R., 1978, Paleomagnetic and geochronologic calibration of latest Oligocene to Pliocene radiolarian events, equatorial Pacific: Marine Micropaleontology, v. 3, p. 377–395.

van Andel, T. H., Heath, G. R., and Moore, R. C., Jr., 1975, Cenozoic history and paleoceanography of the central equatorial Pacific Ocean: Geological Society of America Memoir, v. 143, 134 p.

Vincent, E., 1981, Neogene carbonate stratigraphy of Hess Rise (central North Pacific) and paleoceanographic implications, *in* Thiede, J., Vallier, T. L., and others, Initial Reports of the Deep Sea Drilling Project, Volume 62: Washington, D.C., U.S. Government Printing Office, p. 571–606.

Westberg, M. J., and Riedel, W. R., 1978, Stratigraphy and evolution of tropical Cenozoic radiolarians: Micropaleontology, v. 23, p. 61–96.

Wyrtki, K., 1964, Upwelling in the Costa Rica Dome: Fisheries Bulletin, v. 63, p. 355–372.

——1965, Surface currents of the eastern tropical Pacific Ocean: International-American Tropical Tuna Commission Bulletin, v. 9, p. 207–304.

——1966, Oceanography of the eastern equatorial Pacific Ocean: Oceanographic Marine Biological Annual Review, v. 4, p. 33–68.

——1967, Circulation and water masses in the eastern equatorial Pacific Ocean: International Journal of Oceanology and Limnology, v. 1, p. 117–147.

——1981, An estimate of equatorial upwelling in the Pacific: Journal of Physical Oceanography, v. 11, p. 1205–1214.

MANUSCRIPT ACCEPTED BY THE SOCIETY DECEMBER 17, 1984

Geological Society of America
Memoir 163
1985

Evolution of Pacific circulation in the Miocene: Radiolarian evidence from DSDP Site 289

Karen Romine*
Gail Lombari*
University of Rhode Island
Graduate School of Oceanography
Narragansett, Rhode Island 02882-1197

ABSTRACT

A time-series of abundance data on Radiolaria in the Miocene at Site 289, western equatorial Pacific, shows variations that appear to be related to the evolution of oceanic circulation patterns and climate. In the early Miocene, unusually large abundances (up to 70 percent of the population) of two highly silicified radiolarian species, *Stichocorys wolffii* and *Calocycletta robusta* group, occurred during a period of relatively high silica accumulation rates, with highest abundance at the maximum in accumulation rate at about 17 Ma. This species dominance is difficult to explain, but may have been due to the development of a specific ecological niche that these species were able to dominate. Their rapid proliferation and decline suggests that the niche was a temporary development that probably occurred as certain threshold conditions were satisfied during progressive changes in circulation patterns in the equatorial Pacific. As both climate and circulation continued to evolve, the niche changed character or perhaps disappeared, resulting in the rapid decline and extinction of *S. wolffii* and the *C. robusta* group.

The extinction of *Stichocorys wolffii* and the evolution of *Stichocorys peregrina* in Site 289 occurred in conjunction with the rapid development of an assemblage whose species characterize the western portion of the transitional water mass of the late Miocene North Pacific. The timing of these faunal developments is contemporaneous with a large increase in accumulation of continental ice on East Antarctica during the middle Miocene (15 to 13 Ma). A major increase in the transitional assemblage, a decrease in abundances of *Stichocorys* spp., a decrease in silica accumulation rates in both the central and western equatorial Pacific, and an increase in mass accumulation rates of eolian dust in the central North Pacific are coincident with the closure of the Indo-Pacific passage to significant westward flow from the Pacific (about 12 to 10 Ma). The faunal changes indicate the development or intensification of a North Pacific transitional water mass at this time, a result of the intensification of the subtropical gyre due to the diversion of westward equatorial flow towards the poles and probably influenced by the intensification of the westerly winds.

An assemblage of species resistant to dissolution indicates three major and one minor interval of increased silica dissolution in the middle and late Miocene, 15–13 Ma, 12–11 Ma, 9–8 Ma (minor), and 7–6 Ma. The major intervals correlate with significant paleoceanographic events: a significant increase in $\delta^{18}O$ interpreted as an increase in accumulation of ice on East Antarctica (15–13 Ma); closure of the Indo-Pacific passage (12–11 Ma); and the Messinian and/or $\delta^{13}C$ shift (7–6 Ma). Two of these events

*Present addresses: Romine, Exxon Production Research Company, P.O. Box 2189, Houston, Texas 77001. Lombari, 20 Knight St., Coventry, Rhode Island 02816.

occurred during a time of increased hiatus abundance in the western Pacific (11 Ma) and during a time of low silica accumulation (11 Ma and 7–6 Ma).

INTRODUCTION

During the Cenozoic, paleogeographic changes brought about by global plate tectonics gradually modified the shape and size of the world's oceans. Such changes resulted in the development of high-latitude circum-global circulation in the Southern Hemisphere by latest Oligocene time and in the fragmentation and final destruction of an ancient circum-global, low-latitude seaway, the Tethys, by the middle Pliocene (Berggren and Hollister 1977); Keigwin 1978; Kennett 1977, 1978, 1980). Global climate was certainly affected by these changes in paleogeography.

It has been suggested that the Oligocene development of the Circum-Antarctic Current (as Australia and South America moved away from Antarctica) progressively isolated the Antarctic continent from subtropical influence, resulting in cooling polar temperatures, an increase in Antarctic sea-ice, cooler bottom-water temperatures, steeper pole-to-equator thermal gradients and, ultimately, the development of a major continental ice sheet on East Antarctica by the middle Miocene (Kennett 1977, 1978, 1980). At about the same time, Indo-Pacific equatorial circulation was restricted as the northward movement of Australia brought it into contact with southeast Asia in the Indonesian region. Global cooling continued into the late Miocene, culminating in the late Pliocene with the formation of Northern Hemisphere ice. This event was partially a result of closure of the Central American Seaway, which terminated inter-oceanic circulation between the equatorial Atlantic and Pacific Oceans. This scenario assumes an essentially ice-free world prior to the middle Miocene and a "glacial" world with a permanent Southern Hemisphere ice cap by late Miocene. Variations in the extent of Southern Hemisphere continental ice influenced late Miocene climate, whereas the glacial-interglacial climatic cycles of the late Pliocene and Quaternary appear to have been dominated by fluctuations in the volume of Northern Hemisphere ice.

It is important to point out that the viewpoint of an "ice-free" world prior to 15 Ma is largely based on a particular interpretation of the Tertiary record of oxygen isotopes (Shackleton and Kennett 1975). Matthews and Poore (1980) challenged this interpretation of Tertiary $\delta^{18}O$ and suggested that it is more likely that the major ice sheets had already developed by the Eocene and that the middle Miocene oxygen isotopic shift was largely the result of temperature change. Recently, isotopic and geologic data (Denton et al. 1984; Keigwin and Keller 1984) indicate that significant accumulations of continental ice may have existed at times during the Oligocene. This interpretation of Tertiary $\delta^{18}O$ must be kept in mind in any study that attempts to reconstruct Miocene climate and/or oceanography.

Many events critical to the development of our modern oceans took place during the Miocene. For this reason the Miocene was chosen as the focus of a multi-faceted study concerning the evolution of the oceans by the CENOP (Cenozoic Paleoceanography) project. The objective of the following investigation is to examine the evolution of Pacific circulation in the Miocene using a time-series of radiolarian census data from the western equatorial Pacific.

The paleoceanography of the Miocene equatorial Pacific has not been investigated in as much detail as in the southwest Pacific, especially considering studies of the latter region which focus on development of circum-Antarctic circulation and Antarctic glaciation (Kennett et al. 1975; Hayes et al. 1975; Kemp et al. 1975; Shackleton and Kennett 1975; Kennett 1977, 1978, 1980). The equatorial Pacific was part of the last remnant of circum-equatorial Tethys circulation, and the isolation of the region by the creation of geographic barriers in the west (Indonesian arc system) and the east (Panama Isthmus) must have changed the circulation patterns radically. The evolution of modern equatorial Pacific circulation and its role in the development of both global climate and modern Pacific circulation are the major focus of this investigation.

Radiolarian census data were used as the basic tool for this study. The distributions of Radiolaria in the plankton and in Recent deep-sea sediments have been found to follow the distribution of modern water masses and currents (Casey 1971a, b; Petrushevskaya 1971). This relationship has been used in the Quaternary to monitor changes in water mass and current boundaries in association with glacial-interglacial fluctuations in climate (Sachs 1973; Robertson 1975; Molina-Cruz 1977; Moore 1978; Romine and Moore 1981). Moore and Lombari (1981) and Romine (1985) have been successful in determining latest Miocene variations in sea-surface temperatures and in reconstructing the paleoceanography of a time-slice at 8 Ma using late Miocene radiolarian assemblages from the North Pacific Ocean. In this paper, the variations in distribution and abundance of radiolarian species in a complete Miocene section from the western equatorial Pacific (Deep Sea Drilling Project Site 289, Ontong-Java Plateau, 2206 m water depth; Figure 1) have been examined with these questions in mind:

(1) In general, do temporal changes in radiolarian species abundances coincide with global changes in oceanic circulation and climate and if so, is there a relationship?

(2) More specifically, what do the faunal changes indicate about the evolution of Pacific circulation through the Miocene?

METHODS

Deep Sea Drilling Site 289 (Figure 1) was chosen for this particular study because it contains an unbroken record of Miocene sedimentation, making it suitable for detailed time-series analysis. The site was sampled at varying intervals but samples averaged one for every 2.25 m (see Appendix on microfiche in pocket inside back cover for sample depths and data).

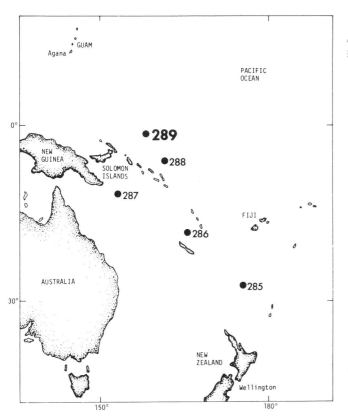

Figure 1. Location of DSDP Site 289, western equatorial Pacific (reproduced from Andrews et al. 1975).

TABLE 1. MIOCENE SPECIES USED IN TIME-SERIES ANALYSIS, SITE 289*

Species	Code	
S2	Didymocyrtis spp. - D. violina, D. mammifer, D. laticonus, D. antepenultimus, D. penultimus	
S4	Actinomma spp. (late Miocene); Echinomma spp. (early Miocene)	
S6	Heliodiscus asteriscus	
S9	Hexastylus spp., Hexastylus thaletis	
S12	Lithelius nautiloides	
S13	Lithelius minor, L. spiralis	
S14	Phorticium pylonium	
S16	Acrosphaera spp.	
S19	Dictyocoryne malagaense, Hymeniastrum euclidis, H. koellikeri	
S20	Spongopyle osculosa	
S21	Spongotrochus glacialis	
S23	Stylochlamidium asteriscus	
S24	Stylodictya aculeata	
S26	Phorticium polycladum	
S27	Stylosphaera spp.	
S30	Disolenia spp., Otosphaera spp., Solenosphaera spp.	
S35	Spongurus sp. B	
S36	Spongurus (?) sp.	
N2	Calocycletta robusta group - C. virginis, C. robusta, C. caepa	
N15	Dendrospyris pododendros	
N18	Lophospyris pentagona	
N20	Giraffospyris circumflexa, G. angulata	
N21	Ceratocyrtis histricosa	
N29	Phormostichoartus marylandicus, P. doliolum, P. corbula	
N36	Stichocorys delmontensis, S. wolffii, S. peregrina	
N38	Theocorys redondoensis	
N44	Zygocircus spp.	
N52B	Eucyrtidium hexagonatum	
N64	Theocalyptra davisiana, T. davisiana cornutoides	
N65	Theocalyptra bicornis, T. bicornis var.	
N71	Siphostichoartus corona	

*Explanation of species groupings and synonomies may be found in Nigrini and Lombari (1983), A Guide to Miocene Radiolaria, that was prepared as a standard for taxonomic consistency to be used by members of the CENOP project.
    Key: S - Spumellaria; N - Nassellaria.

Sample Preparation

Samples were prepared for slide mounts by acid treatment to remove calcium carbonate and further treatment with hydrogen peroxide to disaggregate the residue and oxidize any organic matter present. Random slide-mounts of the size fraction larger than 64 microns were then prepared according to a technique developed by Moore (1973) with modifications by Molina-Cruz (1977).

Data Treatment

Census data on approximately 100 radiolarian species categories was generated for 238 samples in Site 289 with counts of approximately 500 to 600 individuals per slide mount (Appendix). Abundance of individual species varied widely. A number of species that were rare (<0.5%) in all samples were eliminated from further data analysis. For the time-series analysis of the entire Miocene at Site 289, 31 species/species groups ranging from lower to upper Miocene that represented the most abundant species were selected from the original 100 categories (Table 1). Several of the 31 species categories necessarily consisted of groupings of species within lineages that were evolving during the Miocene interval. The census data from individual species within such evolutionary lineages were lumped in the factor analysis.

These data were analyzed using CABFAC, a factor analysis program, (Klovan and Imbrie 1971; Imbrie and Kipp 1971) that resolves multivariate data into a set of mathematically-grouped species assemblages or factors. This technique has been used in time-slice studies because of its ability to resolve a large amount of multivariate, faunal census data into a few assemblages whose geographic distributions are oceanographically meaningful. The advantage of using factor analysis on time-series data is that temporal changes in relative abundances are often related to variations in environmental parameters that respond to fluctuations in climate. By grouping a large amount of multivariate data into smaller subsets, (in this case, species assemblages) that account for most of the variation in the data, it is possible to examine relationships of individual subsets to oceanographic changes and other subsets of the data. Further, examination of the variations in abundance of the individual species within a subset has been utilized in an attempt to define changes in timing and rate of evolution that may also be related to oceanography or climatic change.

OCEANOGRAPHY

The modern equatorial Pacific is an oceanographically complex region (Figure 2). Cold eastern boundary currents flow toward the equator and meet warm Pacific Equatorial Water

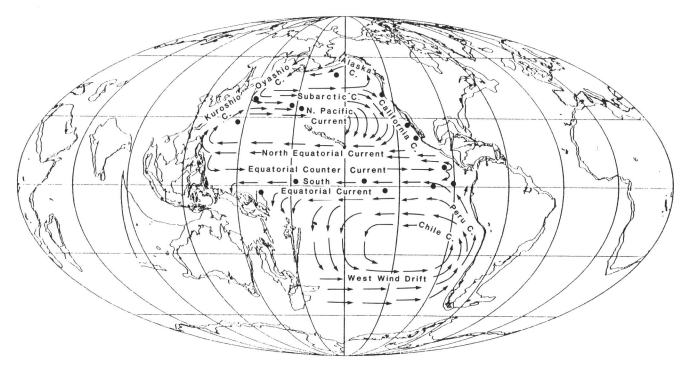

Figure 2. Patterns of surface circulation in the Pacific Ocean (from Romine 1985).

before they turn westward joining the North and South Equatorial Currents. The Equatorial Countercurrent, a surface current that flows eastward between the North and South Equatorial Currents (between 4° and 10°N), brings warm water back into the eastern Pacific. Geostrophic and mass balance requires some return-flow from the western equatorial Pacific where water from the westward-flowing equatorial currents piles up: the Equatorial Undercurrent, a subsurface current, flows eastward along the equator at depths between 150 and 250 m in the western Pacific and rises to less than 50 m depth in the eastern Pacific where it becomes involved in wind-induced divergence and upwelling (Pak and Zaneveld 1974; Wyrtki 1966).

A major control on both past and present circulation patterns is the geographic configuration of the continents forming the boundaries of the ocean basins. Paleogeographic configurations of the Miocene suggest that equatorial Pacific circulation patterns were quite different. In earliest Miocene, westward tropical flow along the path of the ancient Tethys Seaway continued from the Atlantic, through the Pacific, and into the Indian Ocean (Figure 3). This westward flow separated the North and South Pacific gyres and presumably contributed little to the volume transport of these gyres until the Indo-Pacific passage closed as the Australian continental mass moved further northward (Edwards 1975). The closure of the passage to significant flow probably began to affect circulation in mid-early Miocene with the gradual diversion of warm, tropical surface water into the western boundary currents of both the Northern and Southern Hemispheres. Once a permanent barrier formed in the middle Miocene (Edwards 1975; Kennett 1977, 1980) return-flow in the Equatorial Countercur-

rent strengthened and the Equatorial Undercurrent evolved as equatorial Pacific circulation assumed essentially its modern form.

Climatic cooling increased in the middle Miocene (14–15 Ma), coincident with an increase in accumulation of ice on East Antarctica. However, surface-water exchange between the equatorial Atlantic and the Pacific continued until the late Pliocene (Keigwin 1978). Romine (1985) presents evidence suggesting that the gradual closing of this passage during the late Miocene amplified the effect of global cooling on the North Pacific.

STRATIGRAPHY

The completeness of the Miocene record at Site 289 has made it the subject of intensive stratigraphic investigation. Planktonic foraminiferal biostratigraphy in this site has been examined in detail by Srinivasan and Kennett (1981a, b), who applied the tropical N-zonation scheme of Blow (1969), with modifications of their own to some of the zonal definitions. A new time-scale has recently been generated that ties Blow's tropical foraminiferal zonation scheme to the paleomagnetic reversal record (Berggren et al. 1984). Ages for the foraminiferal N-zone boundaries from this scale have been applied to the planktonic foraminiferal zonation scheme of 289 in the time-series study presented here (Figure 4).

Radiolarian stratigraphy in Site 289 is also shown on Figure 4. Zone boundary definitions are from Riedel and Sanfilippo (1978). The zonation of Site 289 has been taken from Westberg and Riedel (1978) and Holdsworth (1975). There is disagreement

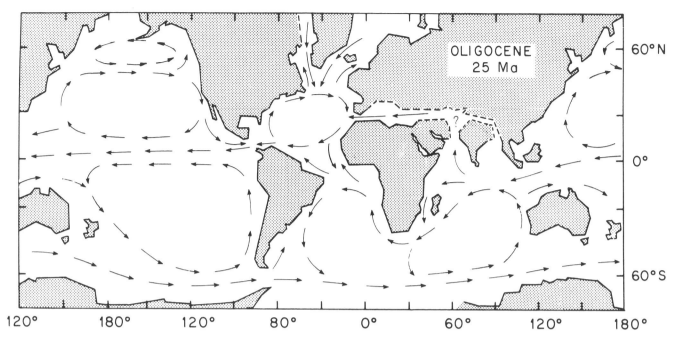

Figure 3. Hypothetical surface circulation in latest Oligocene time (25 Ma; Haq 1981).

in the early Miocene between the foraminiferal and radiolarian zonation schemes. The base of the *Stichocorys delmontensis* zone (defined by the morphotypic top of the range of *Theocyrtis annosa*) is approximately 20 m below the Oligocene/Miocene boundary in Site 289, but has been dated at 22.2 Ma by Theyer et al. (1978). The Oligocene/Miocene boundary falls in the middle of the *Lychnocanoma elongata* zone (Riedel and Sanfilippo 1978) and both this zone and the *Cyrtocapsella tetrapera* Zone (that lies between the *L. elongata* and *S. delmontensis* zones) occur in the latest Oligocene at Site 289. It is because of this apparent inconsistency that the ages applied to the foraminiferal zonation scheme have been utilized.

Stable Isotopes

A study by Woodruff et al. (1981) produced a detailed record of fluctuations in the stable isotope ratios of oxygen ($^{18/16}$O) and carbon ($^{13/12}$C) measured in tests of benthic foraminifera preserved in the Miocene sediments at Site 289 (Figure 5). Isotopic measurements on planktonic foraminifera from a detailed study of the early to middle Miocene (14.5–20.0 Ma; Shackleton 1982) have also been included in the figure in order to compare the isotopic variation in surface and bottom waters.

When precipitating calcite during formation of their tests, the foraminifera incorporate oxygen in an isotopic ratio dependent on the temperature and isotopic ratio of seawater. Changes in the oxygen isotopic ratio (δ^{18}O) of seawater are controlled by evaporation and by precipitation. Evaporation and subsequent precipitation on high latitude continents in the form of ice and snow fractionates the isotopes, preferentially removing the lighter

isotope of oxygen from the ocean and precipitating them on high-latitude continental ice sheets. As large continental ice sheets grow there is an appreciable change in the isotopic composition of the oceanic reservoir. It is generally believed that the major control on the δ^{18}O ratio of seawater in the Paleogene and Early Miocene was temperature (Woodruff et al. 1981; Savin 1977; Shackleton and Kennett 1975). However, recent studies (Keigwin and Keller 1984; Denton et al. 1984) suggest that significant ice accumulation may have occurred at times prior to the middle Miocene. The large increase in δ^{18}O in the middle Miocene has been attributed to the formation of a permanent ice sheet on East Antarctica (Shackleton and Kennett 1975). New data suggest that Antarctica may have had an ice sheet at times during the Oligocene. Nevertheless, the middle Miocene (16–14 Ma) increase in δ^{18}O (Woodruff et al. 1981) almost certainly represents the beginning of permanent major ice sheets on Antarctica with associated global climatic cooling. Thus, Miocene δ^{18}O fluctuations probably record both variation in the volume of continental ice as well as a paleotemperature signal making the oxygen isotope record at 289 useful as both a record of paleoclimate and a stratigraphy of climatic events.

The mechanisms that control the variation of the carbon isotopic composition (δ^{13}C) of the deep ocean are not well understood. The δ^{13}C of benthic foraminiferal calcite is thought to reflect the average δ^{13}C of the oceans. Bottom-water masses acquire a δ^{13}C signature at their origin in high latitude surface waters. As the water moves through the ocean basins, it is modified by the addition of carbon derived from the oxidation of organic matter and the dissolution of calcium carbonate. This carbon is depleted in δ^{13}C through biotic discrimination against

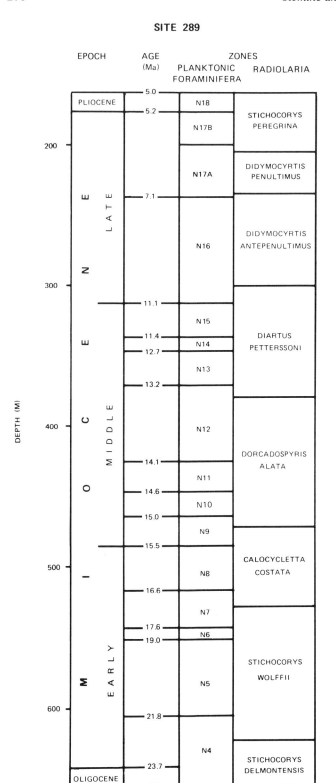

Figure 4. Stratigraphy of Site 289. Foraminiferal zonation is from Srinivasan and Kennett (1981a, b). Ages of N-zone boundaries are from Berggren et al. (1984). Radiolarian zonation scheme is from Riedel and Sanfilippo (1978). Radiolarian stratigraphy in Site 289 from Westberg and Riedel (1978) and Holdsworth (1975).

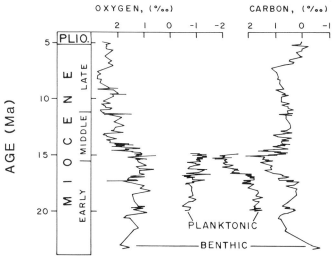

Figure 5. Stable isotopes of oxygen and carbon at Site 289. Benthic data are from Savin et al. (1981). Planktonic data are from Shackleton (1982).

the heavy isotope in surface waters during photosynthesis and test formation (Kroopnick et al. 1972; Kroopnick 1980; Bender and Graham 1981). Time-series records of $\delta^{13}C$ variations show similarities within and between ocean basins that suggest a global mechanism controlling the $\delta^{13}C$ composition of the deep ocean. Changes in the available sources (marine or terrestrial) of organic matter, in fluxes of organic matter to the deep sea, in the residence time of deep water masses, and in upwelling rates, can all affect the $\delta^{13}C$ measured in the deep ocean. Controls on these variations are generally tectonic and/or climatic, making the $\delta^{13}C$ record of interest to deep-sea stratigraphers and paleoceanographers (Loutit et al. 1983a, b). It is our intention here to utilize the $\delta^{13}C$ record as a paleoceanographic tool to provide information on variations in the age and sources of bottom waters in the western equatorial Pacific.

RESULTS

Factor analysis of 31 species categories, whose ranges span the Miocene, resulted in 6 factors that account for 96 percent of the variability in the data (Table 2). Table 3 lists the varimax factor scores for each species category within each factor. Highest "scores" indicate the dominant species within each assemblage.

Factor 1: Stichocorys *Assemblage (Figure 6)*

The first factor accounts for 46 percent of the variance and is dominated solely by a lineage composed of 3 species: *Stichocorys delmontensis, S. wolffii,* and *S. peregrina* (taxonomic definitions and identification of species based on Riedel and Sanfilippo 1978). Downcore plots of abundance of each species demonstrate the changing patterns of dominance as evolution occurs within

TABLE 2. VARIMAX FACTOR MATRIX, SITE 289

Depth (m)	Communality	Factors					
		1	2	3	4	5	6
156.8	0.9000	0.0290	0.6181	0.6853	0.1264	0.1774	-0.0094
162.4	0.8938	0.0977	0.4451	0.7922	0.1901	0.1168	0.0936
167.1	0.9452	0.5447	0.5112	0.5754	0.1242	0.1791	0.0921
173.9	0.8794	-0.0107	0.3716	0.5444	0.0934	0.6580	0.0558
186.9	0.9059	0.0028	0.3437	0.6478	0.1051	0.5960	0.0444
192.1	0.9140	-0.0173	0.2299	0.2612	0.0134	0.8901	0.0131
204.5	0.8732	0.0240	0.1743	0.2259	0.0210	0.8884	0.0389
212.5	0.9882	0.2718	0.3145	0.4042	0.0708	0.8044	0.0034
217.8	0.9488	0.2508	0.6472	0.2789	0.0404	0.6207	-0.0487
223.0	0.9396	0.1688	0.6079	0.3783	0.0390	0.6234	-0.0908
227.6	0.9789	0.1843	0.5396	0.2527	-0.0143	0.7611	-0.1021
232.9	0.9456	0.3181	0.5740	0.4537	0.0381	0.5458	0.0986
239.3	0.9461	0.5826	0.3021	0.4231	0.0433	0.5754	-0.0591
244.6	0.9476	0.7200	0.2621	0.5560	0.0359	0.2171	-0.0537
249.6	0.9183	0.3680	0.2455	0.7647	0.1344	0.3386	-0.0712
254.5	0.9060	0.6078	0.2347	0.6741	0.0733	0.1354	-0.0582
285.4	0.8870	0.3945	0.4314	0.6679	0.0792	0.2960	-0.0725
264.5	0.9039	0.3881	0.3428	0.6682	0.1530	0.4030	-0.0588
269.4	0.8870	0.1828	0.5027	0.6875	0.0337	0.3278	-0.1401
279.4	0.9124	0.1598	0.3764	0.8378	0.0841	0.1790	-0.0645
289.5	0.8160	0.2491	0.2961	0.7822	0.0675	0.1771	-0.1358
294.1	0.8685	0.2684	0.2543	0.8164	0.0705	0.2050	-0.1352
300.0	0.9130	0.2390	0.3304	0.8565	0.0643	0.0723	-0.0612
306.6	0.8883	0.1086	0.3090	0.8623	0.0692	0.1541	-0.0948
308.9	0.9408	0.3430	0.4708	0.5832	0.1455	0.4880	0.0447
318.3	0.9183	0.2013	0.5025	0.6647	0.0658	0.4211	-0.0418
327.8	0.8874	0.3348	0.3714	0.6715	0.1380	0.3982	-0.0944
337.3	0.9634	-0.0163	0.9143	0.3203	-0.0087	0.1251	-0.0939
338.9	0.9057	0.5751	0.6612	0.3520	0.0428	0.0641	-0.0884
343.8	0.9859	0.7596	0.3662	0.4785	0.1202	0.1501	0.0942
348.3	0.9417	0.5953	0.4991	0.4786	0.1498	0.2893	0.0558
351.9	0.9260	0.5991	0.3002	0.6830	0.1003	0.0198	0.0048
356.4	0.9774	0.8462	0.3317	0.3472	0.1482	0.0622	0.0703
365.9	0.9436	0.6562	0.3422	0.6033	0.1275	0.0942	0.0822
373.9	0.9790	0.6883	0.5912	0.3512	0.1398	0.0623	0.0945
378.4	0.9739	0.8220	0.3287	0.4091	0.0834	0.0649	0.1075
381.8	0.9798	0.7633	0.3408	0.5115	0.1276	0.0398	0.0386
394.3	0.9646	0.7948	0.2628	0.4793	0.1349	-0.0014	0.1260
403.1	0.9810	0.7477	0.5569	0.2571	0.0233	0.0635	-0.0041
404.6	0.9855	0.7755	0.4955	0.3271	0.0707	0.1625	0.0146
406.1	0.9923	0.7615	0.6230	0.1473	-0.0103	0.0428	0.0260
407.6	0.9800	0.6866	0.6799	0.2022	-0.0369	0.0597	0.0233
409.6	0.9682	0.5346	0.7702	0.2941	0.0381	0.0259	-0.0154
411.1	0.9598	0.6352	0.6666	0.3166	0.0626	0.0868	-0.0150
412.6	0.9550	0.3547	0.8617	0.2180	0.1664	0.0551	-0.0915
413.4	0.9903	0.4409	0.8800	0.0906	0.0505	0.0935	-0.0436
413.9	0.9756	0.7502	0.6009	0.2134	0.0359	0.0590	-0.0374
414.4	0.9787	0.2783	0.9279	0.1540	0.0515	0.0345	-0.1128
414.9	0.9707	0.6272	0.6983	0.2558	0.0873	0.0390	-0.1229
415.4	0.9794	0.6346	0.7298	0.1788	0.0274	0.0217	-0.1043
415.9	0.9900	0.4745	0.8578	0.1550	-0.0212	0.0373	-0.0567
416.4	0.9685	0.4585	0.8420	0.1709	0.0227	0.0756	-0.1177
416.9	0.9756	0.4620	0.8496	0.1921	0.0271	0.0034	-0.0518
417.4	0.9811	0.3044	0.9383	0.0631	-0.0361	0.0413	-0.0320
418.2	0.9918	0.5869	0.7970	0.0766	0.0053	0.0465	-0.0648
418.5	0.9879	0.5382	0.8209	0.1401	0.0346	0.0578	0.0141
418.8	0.9929	0.3412	0.9307	0.0647	-0.0434	0.0302	-0.0572
419.0	0.9859	0.3914	0.8967	0.1342	-0.0242	0.0840	-0.0539
419.3	0.9921	0.2123	0.9648	-0.0076	-0.0155	0.0769	-0.1001
419.5	0.9818	0.3743	0.9064	0.0785	0.0417	0.1084	-0.0209
419.7	0.9702	0.4158	0.8241	0.3190	0.0564	0.1017	0.0532
420.0	0.9736	0.2432	0.9315	0.1946	-0.0001	0.0916	-0.0214
420.2	0.9856	0.4256	0.8894	0.0567	-0.0195	0.0988	-0.0126
420.5	0.9818	0.5345	0.8127	0.1536	0.0176	0.1014	-0.0395
420.7	0.9871	0.4583	0.8776	0.0230	-0.0279	0.0581	-0.0463
421.0	0.9912	0.5918	0.7868	0.1285	-0.0015	0.0677	-0.0282
421.2	0.9860	0.4124	0.9017	0.0102	-0.0038	0.0496	-0.0144
421.5	0.9926	0.4311	0.8932	0.0506	-0.0013	0.0618	-0.0492
421.8	0.9937	0.3596	0.9273	0.0383	-0.0256	0.0347	-0.0329
422.1	0.9911	0.6607	0.7327	0.1218	0.0061	0.0270	0.0463
422.7	0.9798	0.8225	0.4878	0.2269	0.0205	0.0777	0.0858
423.0	0.9877	0.7578	0.6018	0.1627	0.0948	0.0859	0.0909
423.3	0.9834	0.7671	0.5600	0.2725	0.0474	0.0694	0.0062
423.5	0.9858	0.7760	0.5782	0.1689	0.1050	0.0565	0.0809
423.7	0.9873	0.6317	0.7430	0.1796	0.0023	0.0626	-0.0073
424.0	0.9955	0.6385	0.7475	0.1425	0.0381	0.0615	0.0594

TABLE 2. VARIMAX FACTOR MATRIX, SITE 289 (continued)

Depth (m)	Communality	Factors					
		1	2	3	4	5	6
424.2	0.9940	0.7772	0.5998	0.1633	0.0050	0.0526	0.0271
424.5	0.9864	0.6898	0.6525	0.2726	0.0594	0.0490	0.0682
424.8	0.9910	0.5192	0.8431	0.0274	0.0005	0.0974	-0.0196
425.0	0.9928	0.7116	0.6761	0.1322	0.0122	0.0944	0.0512
425.2	0.9790	0.6294	0.7519	0.0705	0.0492	0.0935	0.0364
425.4	0.9852	0.3292	0.9269	-0.0220	-0.0033	0.1310	0.0020
425.7	0.9930	0.1452	0.9775	-0.0826	-0.0372	0.0687	-0.0589
426.0	0.9875	0.8075	0.5134	0.2491	0.0437	0.0342	0.0826
426.3	0.9866	0.8646	0.3955	0.2688	0.0529	0.0359	0.0790
426.4	0.9413	0.5991	0.6630	0.3462	0.0861	0.0741	0.0996
426.7	0.9875	0.7516	0.5903	0.2640	0.0270	0.0433	0.0443
427.0	0.9857	0.7836	0.5551	0.2303	0.0311	0.0573	0.0795
428.3	0.9911	0.5204	0.8353	0.1410	-0.0147	0.0489	0.0096
428.5	0.9815	0.6403	0.7076	0.2404	0.0350	0.0831	0.0700
428.8	0.9758	0.4773	0.8620	-0.0315	-0.0035	0.0582	0.0246
429.0	0.9881	0.4931	0.8565	0.0466	0.0504	0.0675	0.0450
429.2	0.9801	0.5710	0.7893	0.1606	0.0010	0.0410	0.0603
429.5	0.9927	0.5300	0.8208	0.1754	0.0050	0.0830	0.0227
429.8	0.9921	0.4581	0.8775	0.0405	-0.0282	0.0986	0.0089
430.0	0.9448	0.3717	0.8503	0.1173	0.0935	0.2114	0.1284
430.3	0.9708	0.6215	0.6654	0.2664	0.1822	0.1637	0.1042
430.4	0.9877	0.3087	0.9168	0.1225	0.1126	0.1430	0.0611
437.6	0.9731	0.5505	0.7872	0.1187	0.1384	0.1138	0.0648
437.8	0.9281	0.2078	0.6002	0.6560	0.2733	0.0634	0.1247
438.0	0.9903	0.4101	0.8420	0.2749	0.1380	0.1242	0.0561
438.2	0.6651	0.5878	0.2332	0.4125	0.2482	0.0435	0.1776
438.4	0.9001	0.1547	0.8758	0.2220	0.1852	0.1320	0.0903
438.7	0.9739	0.4097	0.8401	0.1883	0.1905	0.1504	0.0767
438.9	0.9705	0.1486	0.9597	0.1165	0.0855	0.0791	0.0140
439.3	0.9344	0.2284	0.8260	0.4062	0.1859	0.0119	0.0155
439.5	0.9760	0.3241	0.7645	0.4933	0.1836	0.0799	0.0548
439.7	0.9649	0.2081	0.8730	0.3405	0.1676	0.1242	-0.0086
440.0	0.9077	0.1087	0.7335	0.5557	0.1636	0.1413	-0.0475
440.2	0.9604	-0.0620	0.9398	0.1835	0.1290	0.1514	0.0077
440.5	0.9521	-0.0523	0.7477	0.5922	0.1808	0.0395	0.0733
440.8	0.9614	0.2200	0.7736	0.5104	0.2217	0.0658	0.0231
441.0	0.9885	0.1087	0.9536	0.1196	0.0240	0.2263	-0.0338
441.2	0.9462	0.0965	0.9380	0.0868	0.0885	0.1888	-0.0778
441.5	0.9314	0.0300	0.8207	0.4733	0.1066	0.0874	-0.1180
441.7	0.9755	0.0113	0.9477	0.2022	0.0975	0.1545	-0.0537
442.0	0.9781	0.0919	0.8860	0.3929	0.1401	0.1026	-0.0027
442.3	0.9756	0.0821	0.7622	0.6045	0.1449	0.0104	-0.0367
442.5	0.9564	-0.0312	0.8010	0.5239	0.1450	0.1235	-0.0228
442.7	0.9646	-0.0659	0.8087	0.5215	0.1630	0.0705	0.0525
443.0	0.9749	-0.1479	0.9408	0.2302	0.1125	0.0473	0.0056
443.2	0.9553	-0.0959	0.8556	0.4445	0.1226	0.0382	-0.0037
443.5	0.9786	-0.1184	0.9696	0.1373	0.0295	0.0261	-0.0636
443.8	0.9827	-0.1413	0.9660	0.1459	0.0246	0.0765	-0.0424
444.0	0.9799	-0.1328	0.8762	0.4310	0.0848	0.0060	-0.0406
444.2	0.8891	-0.0893	0.8655	0.1595	0.3194	0.0260	-0.0621
444.5	0.9832	0.1211	0.8940	0.3725	0.1706	0.0278	0.0241
444.8	0.9474	-0.0346	0.6315	0.7293	0.1220	-0.0241	0.0070
445.0	0.9227	-0.0544	0.7642	0.5624	0.1272	-0.0545	0.0180
445.3	0.9240	-0.0606	0.6676	0.6633	0.1769	-0.0571	0.0136
445.5	0.9446	0.0272	0.5852	0.7445	0.2076	-0.0304	0.0548
445.7	0.8852	-0.0686	0.5850	0.6906	0.2378	-0.0273	0.0628
446.0	0.9889	-0.0736	0.9034	0.3849	0.1220	0.0280	-0.0594
447.0	0.9202	0.6090	0.6968	0.2348	0.0830	0.0406	-0.0067
447.5	0.9958	0.9467	0.2975	0.0771	-0.0074	0.0546	0.0448
448.0	0.9958	0.8696	0.4820	0.0468	-0.0196	0.0652	-0.0218
448.5	0.9954	0.7013	0.7040	0.0346	-0.0071	0.0768	-0.0277
449.0	0.9962	0.8678	0.4843	0.0589	0.0008	0.0531	-0.0468
449.5	0.9847	0.8256	0.5343	0.0680	0.0344	0.0534	-0.0944
459.3	0.9198	0.8283	0.4274	0.2039	0.0935	0.0271	-0.0070
460.8	0.9639	0.8940	0.3075	0.1964	0.1761	0.0185	-0.0097
462.3	0.9963	0.9728	0.1955	0.0941	0.0135	0.0431	0.0290
470.3	0.9778	0.9176	0.3254	0.1356	0.0900	0.0328	-0.0489
479.9	0.9970	0.9738	0.1848	0.1011	0.0483	0.0425	-0.0119
489.3	0.9935	0.9710	0.1645	0.1420	0.0510	0.0284	-0.0216
498.7	0.9981	0.9755	0.1732	0.0994	0.0468	0.0553	-0.0373
508.3	0.9968	0.9803	0.1512	0.0865	-0.0060	0.0612	0.0413
513.3	0.9987	0.9582	0.2726	0.0520	0.0125	0.0471	-0.0347
513.8	0.9967	0.9680	0.2101	0.1110	0.0339	0.0425	-0.0138
514.8	0.9989	0.9681	0.2301	0.0795	0.0177	0.0434	-0.0147
515.3	0.9776	0.9033	0.3580	0.0906	0.0938	0.0424	-0.1215
516.3	0.9975	0.9800	0.1695	0.0778	0.0245	0.0400	-0.0078

TABLE 2. VARIMAX FACTOR MATRIX, SITE 289 (continued)

Depth (m)	Communality	Factors					
		1	2	3	4	5	6
516.8	0.9983	0.9816	0.1586	0.0869	0.0112	0.0430	0.0017
517.8	0.9956	0.9618	0.2151	0.1341	0.0611	0.0462	-0.0232
518.3	0.9950	0.9657	0.2030	0.0903	0.0575	0.0400	-0.0899
519.3	0.9977	0.9789	0.1713	0.0884	0.0054	0.0462	0.0043
519.8	0.9980	0.9828	0.1494	0.0852	0.0083	0.0459	0.0175
520.8	0.9985	0.9699	0.2241	0.0695	0.0170	0.0457	-0.0181
521.3	0.9992	0.9835	0.1592	0.0656	0.0086	0.0472	0.0016
552.8	0.9982	0.9840	0.1450	0.0826	0.0054	0.0452	0.0123
523.3	0.9960	0.9675	0.2041	0.0606	0.0571	0.0464	-0.0961
524.3	0.9989	0.9820	0.1553	0.0869	0.0265	0.0461	-0.0125
524.8	0.9978	0.9775	0.1750	0.0715	0.0535	0.0475	-0.0382
525.8	0.9993	0.9697	0.2197	0.0832	0.0310	0.0501	-0.0195
526.3	0.9985	0.9809	0.1704	0.0693	0.0174	0.0455	-0.0135
527.3	0.9979	0.9841	0.1466	0.0717	0.0017	0.0436	0.0291
527.8	0.9977	0.9715	0.1928	0.1147	0.0394	0.0448	0.0032
532.3	0.9983	0.9839	0.1455	0.0803	-0.0035	0.0427	0.0276
532.8	0.9975	0.9688	0.2157	0.0948	-0.0329	0.0463	-0.0137
533.8	0.9979	0.9831	0.1522	0.0769	0.0043	0.0440	0.0215
534.3	0.9952	0.9705	0.1422	0.0643	0.0827	0.0383	-0.1436
535.3	0.9980	0.9838	0.1445	0.0772	-0.0090	0.0415	0.0370
535.8	0.9977	0.9824	0.1468	0.0882	-0.0023	0.0449	0.0360
536.8	0.9985	0.9826	0.1503	0.0888	0.0028	0.0440	0.0217
537.3	0.9981	0.9804	0.1502	0.0987	-0.0042	0.0452	0.0505
538.3	0.9967	0.9804	0.1540	0.0867	-0.0054	0.0433	0.0500
538.8	0.9976	0.9823	0.1490	0.0821	-0.0095	0.0431	0.0436
539.8	0.9982	0.9829	0.1468	0.0879	0.0000	0.0440	0.0297
540.3	0.9976	0.9799	0.1616	0.0927	0.0022	0.0431	0.0278
541.8	0.9984	0.9824	0.1513	0.0592	0.0383	0.0410	-0.0613
542.3	0.9992	0.9849	0.1448	0.0772	0.0158	0.0434	-0.0067
543.3	0.9897	0.9624	0.1333	0.0447	0.1006	0.0327	-0.1805
543.8	0.9885	0.9571	0.1612	0.1044	0.1077	0.0313	-0.1519
544.8	0.9893	0.9616	0.1424	0.0512	0.0937	0.0362	-0.1779
545.2	0.9649	0.9075	0.1358	0.0464	0.1555	0.0282	-0.3095
546.3	0.9541	0.8499	0.1705	0.0414	0.2254	0.0457	-0.3849
546.7	0.9949	0.9732	0.1472	0.0690	0.0692	0.0382	-0.1231
547.8	0.9936	0.9683	0.1455	0.0479	0.0839	0.0356	-0.1555
548.3	0.9981	0.9809	0.1486	0.0619	0.0540	0.0395	-0.0745
549.3	0.9965	0.9757	0.1494	0.0589	0.0686	0.0372	-0.1129
549.8	0.9985	0.9709	0.2048	0.0766	0.0630	0.0519	-0.0368
551.5	0.9744	0.9057	0.1508	0.0655	0.2208	0.0399	-0.2769
551.8	0.9857	0.9369	0.1476	0.0627	0.1611	0.0373	-0.2342
552.8	0.9927	0.9677	0.1439	0.0627	0.1151	0.0384	-0.1305
553.5	0.9799	0.9164	0.1402	0.0707	0.1976	0.0386	-0.2737
561.5	0.9819	0.9601	0.1676	0.1174	0.1124	0.0556	0.0509
562.3	0.9991	0.9817	0.1521	0.0880	0.0490	0.0454	0.0037
562.8	0.9979	0.9838	0.1459	0.0818	0.0096	0.0424	0.0147
563.8	0.9974	0.9840	0.1454	0.0755	0.0203	0.0412	0.0149
564.3	0.9983	0.9828	0.1488	0.0917	0.0200	0.0377	-0.0006
565.3	0.9980	0.9792	0.1522	0.0895	0.0735	0.0426	-0.0284
565.8	0.9970	0.9825	0.1482	0.0873	0.0063	0.0407	0.0200
566.8	0.9967	0.9834	0.1435	0.0734	-0.0027	0.0463	0.0374
567.3	0.9973	0.9824	0.1452	0.0782	-0.0111	0.0440	0.0535
568.3	0.9980	0.9834	0.1454	0.0812	-0.0042	0.0437	0.0366
568.8	0.9976	0.9836	0.1458	0.0776	0.0072	0.0428	0.0295
570.3	0.9032	0.6418	0.0935	0.0200	0.3993	0.0417	-0.5665
570.8	0.9738	0.8534	0.1387	0.1076	0.2464	0.0111	-0.3922
571.8	0.9871	0.8976	0.1515	0.1286	0.2155	0.0188	-0.3085
572.2	0.9714	0.7996	0.1441	0.1519	0.3110	-0.0028	-0.4376
573.3	0.9567	0.6458	0.1435	0.2093	0.4409	0.0080	-0.5298
573.8	0.9594	0.6151	0.1193	0.1637	0.4492	0.0152	-0.5813
577.7	0.9930	0.9672	0.1638	0.1285	0.1061	0.0533	-0.0055
581.2	0.9351	0.8502	0.1808	0.2068	0.3146	0.0737	-0.1799
590.8	0.9685	0.8644	0.2243	0.2970	0.2753	0.0609	0.0569
601.8	0.9465	0.8533	0.2268	0.3088	0.2480	0.0725	0.0694
611.3	0.9683	0.8664	0.2201	0.2689	0.2753	0.0939	0.1111
619.3	0.9649	0.8724	0.2393	0.1853	0.2996	0.0533	0.1399
628.8	0.9187	0.5899	0.2007	0.4063	0.5052	0.0112	-0.3317
634.8	0.8977	0.4418	0.2108	0.3261	0.7397	0.0349	-0.0572
638.3	0.8512	0.7865	0.2148	0.1675	0.3845	0.0713	0.0743
641.3	0.9625	0.9517	0.1652	0.0812	0.1453	0.0329	0.0259
644.3	0.9793	0.9559	0.1722	0.1254	0.1173	0.0431	0.0666
647.7	0.8205	0.5388	0.2331	0.3844	0.5723	0.0028	-0.0211
650.7	0.9347	0.8426	0.2266	0.4002	0.1074	0.0134	0.0376
652.2	0.8603	0.6640	0.1583	0.1585	0.5748	0.0503	-0.1903
657.2	0.8905	0.4963	0.1684	0.2710	0.6412	0.0061	-0.3622
660.3	0.8746	0.7118	0.1683	0.0300	0.5794	0.0472	0.0296

TABLE 2. VARIMAX FACTOR MATRIX, SITE 289 (continued)

Depth (m)	Communality	Factors					
		1	2	3	4	5	6
663.3	0.8591	0.6225	0.2050	0.2275	0.6088	0.0079	-0.0837
665.3	0.7705	0.1326	0.1332	0.0043	0.8430	0.1191	0.1011
666.8	0.8049	0.0244	0.1926	0.1661	0.8139	0.0991	0.2595
669.8	0.6254	0.0258	0.1001	-0.0017	0.7838	-0.0116	0.0162
676.3	0.6444	0.0615	0.1178	0.0565	0.7631	0.1210	0.1631
679.3	0.7871	0.2616	0.1430	0.1835	0.7781	0.0494	-0.2381
682.3	0.9244	0.1150	0.1099	0.1776	0.8030	-0.0060	-0.4719
685.8	0.8671	0.3359	0.1897	0.2624	0.7631	0.0102	-0.2588
689.2	0.8044	0.0724	0.1687	0.3267	0.8100	0.0689	-0.0564

The <u>communality</u> of a sample is a quantitive measure of the proportion of variance in the species percentage data (for that sample) that has been accounted for by the mathematical data treatment. Ideally, the value is equal to 1.0 if all the information contained in the sample is described by the factor model. The <u>factors</u> are essentially mathematically grouped assemblages of radiolarian <u>species</u>. The highest factor value for a sample indicates the dominant assemblage. <u>Varimax</u> refers to the type of vector rotation used in the analysis to achieve the best fit of the sample vectors to an n-vector model. See Klovan and Imbrie (1971) or Imbrie and Kipp (1971) for a detailed discussion of factor analysis.

TABLE 3. VARIMAX FACTOR SCORES

Species*	1	2	3	4	5	6
S2	-0.0090	0.0120	0.2186	-0.0826	-0.0518	-0.1211
S4	-0.0001	0.0002	0.0144	0.0064	-0.0021	-0.0128
S6	-0.0477	0.1105	0.3765	-0.0546	-0.1070	-0.1295
S9	-0.0050	0.0061	0.0071	0.0516	0.0091	0.0075
S12	-0.0050	0.0150	0.0213	0.0364	-0.0153	0.0119
S13	-0.0046	0.0021	0.0083	0.0270	0.0189	0.0050
S14	-0.0381	0.0901	0.5155	0.0751	0.1204	-0.1855
S16	-0.0446	-0.0090	0.0491	-0.0279	0.9363	-0.0157
S19	-0.0112	0.0624	0.2581	0.0419	-0.0545	-0.0514
S20	-0.0055	0.0065	0.0608	0.0257	-0.0125	0.0549
S21	-0.0036	0.0720	0.0489	0.0600	-0.0100	0.0463
S23	-0.0667	0.1917	0.5735	0.1014	-0.0645	0.2541
S24	-0.0020	0.0065	0.0091	0.0157	0.0208	0.0137
S26	-0.0054	0.0706	0.1775	0.0474	-0.0837	0.0115
S27	-0.0006	0.0007	0.0093	0.0019	0.0106	-0.0023
S30	-0.1205	0.9425	-0.2405	-0.1197	0.0320	-0.0996
S35	-0.0124	0.0229	0.0867	0.0160	-0.0475	-0.0269
S36	-0.0016	-0.0032	0.0126	-0.0021	0.0226	0.0019
N2	0.0904	-0.0124	-0.0830	0.3597	-0.0281	-0.7842
N15	-0.0012	-0.0005	-0.0511	0.4919	-0.0440	-0.1698
N18	-0.0036	0.0581	-0.0357	0.3162	0.0091	0.1866
N20	-0.0122	0.0113	-0.0344	0.3599	0.1464	0.0985
N21	-0.0010	0.0420	-0.0672	0.3024	0.0309	0.1993
N29	0.0039	0.0678	-0.0383	0.1028	0.0018	0.1117
N36	0.9829	0.1407	0.0643	-0.0234	0.0464	0.0657
N38	-0.0053	0.0869	-0.0568	0.2146	-0.0141	0.2395
N44	-0.0092	-0.0352	0.0983	-0.0567	0.2101	-0.1035
N52B	-0.0129	0.0273	-0.0829	0.3955	0.0246	0.1753
N64	-0.0005	0.0003	-0.0011	0.0094	0.0003	-0.0017
N65	0.0002	0.0016	0.0018	0.0101	-0.0012	0.0112
N71	0.0003	0.0053	-0.0283	0.1170	0.0233	0.0802

*Species designations listed in Table 1.
The factor scores indicate the relative importance of each species category to each factor (assemblage). The highest 'score' indicates the dominating species category in the assemblage.

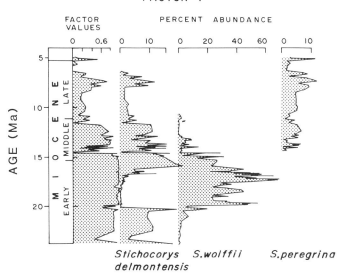

FACTOR 1

Figure 6. Factor 1: *Stichocorys* lineage, Site 289; percent abundance scales are equivalent for *S. delmontensis* and *S. peregrina. S. wolffii* is on one-half scale relative to the other species because of its extremely high abundances. Factor values are in increments of 0.2. Percentages are from the total radiolarian population of the sample.

the lineage. The importance of the *Stichocorys* lineage in the western equatorial Pacific is obvious both from the factor plot and the percentages (Figure 6). The lineage becomes important in the earliest Miocene, soon after *S. delmontensis* evolved in the late Oligocene, and declines by 12 Ma during the early evolution of *S. peregrina* in the late Miocene. *S. delmontensis* appears to be the "parent" stock for the evolution of *S. wolffii* and *S. peregrina*. Its abundance is variable and reaches maximum values of greater than 10 percent of the total fauna (includes both identified and unidentified species categories) in the earliest Miocene (20–24

Ma) and again in the middle Miocene between 12 and 16 Ma. The low values between 16 and 20 Ma are obviously due to the rapid evolution and tremendous proliferation of *S. wolffii* that maintains abundances of 25 percent and ranges upwards to as much as 70 percent of the fauna (at about 17–18 Ma) in this interval. It is very unusual for open-ocean tropical radiolarian species to exceed abundances of 50 percent. For example, in the modern ocean, the most abundant tropical species, *Tetrapyle octacantha,* achieves maximum abundances no greater than 20–25 percent of the total fauna in the eastern tropical Pacific (Gail

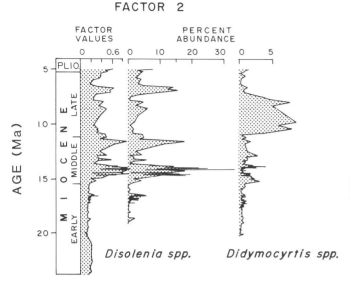

FACTOR 2

Figure 7. Factor 2: Dissolution assemblage, Site 289; percent abundance scale of *Disolenia* spp. is one-half that of *Didymocyrtis* spp. Factor values are in increments of 0.2. Percentages are from the total radiolarian population of the sample.

Lombari unpublished data). *S. wolffii* abundance declines sharply between 17 and 14.5 Ma when *S. peregrina* makes its first appearance. The abundance of *S. peregrina* in this region never reaches the high values for the genus seen in the early and early-middle Miocene, attaining a maximum of only 10–12 percent at about 7.0–7.5 Ma.

Factor 2: Dissolution Assemblage (Figure 7)

This assemblage accounts for 30.4 percent of the variance and is dominated by the *Disolenia* species group. This group is formed of several related species that are members of the family Collosphaeridae. In laboratory experiments collosphaerids have been found to be resistant to dissolution (Johnson 1974); and, in the Initial Report (DSDP Leg 30) of Site 289, Holdsworth and Harker (1975) discussed the possibility that unusually high abundances of collosphaerids in some parts of the sedimentary record at this site may indicate times of increased dissolution of silica. The downcore record of a more solution-susceptible species group, *Didymocyrtis* spp. (Moore 1969; Johnson 1974; note that Moore and Johnson use older generic designations: *Panarium, Cannartus, Ommatartus*) generally shows high abundances when collosphaerid abundances are lowest, lending support to the hypothesis that high abundances of collosphaerids at this site are associated with times of increased silica dissolution. The *Disolenia* spp. group first appeared in the latest early Miocene (approximately 17 Ma), hence, if abundance of these species do signify the amount of silica dissolution, use of this factor to illustrate this effect is restricted to the middle and late Miocene. The record of *Disolenia* spp. abundance (Figure 7) indicates that two significant dissolution events (with factor values > 0.8) occurred in the

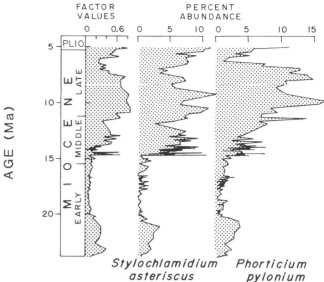

FACTOR 3

Figure 8. Factor 3: Western Tropical/Transitional assemblage, Site 289; percent abundance scales are equivalent and reflect percent of total radiolarian population. Factor values are in increments of 0.2.

middle Miocene at 14–15 Ma and 11–12 Ma. In the late Miocene, a minor dissolution event occurred between 8 and 9 Ma, and a more significant one at 6 to 7 Ma. These events (with the exception of the minor one at 8–9 Ma) coincide with dissolution intervals determined by Holdsworth and Harker (1975) using percent abundances of both spyrids and all collosphaerid groups.

Factor 3: Western Tropical/Transitional Assemblage (Figure 8)

The assemblage of species in this factor accounts for 10 percent of the variance and is composed of a diverse group of very long-ranging Spumellarians that occur from earliest Miocene to the Recent. The two dominant species, *Stylochlamidium asteriscus* and *Phorticium pylonium,* were important in a Western Transitional assemblage in the late Miocene (Figure 9; Romine 1985). This assemblage is associated with the western tropics where westward-flowing equatorial circulation begins to diverge and turn north and south into western boundary circulation, carrying water of tropical character into higher latitudes. The assemblage as not important in Site 289 until the middle Miocene, just after 15 Ma when factor values began to increase. They reached maximum values between 7–11 Ma. A brief decrease occurred between 6 and 7 Ma followed by a rapid increase extending to the Miocene/Pliocene boundary.

Factors 4, 5, and 6

The last 3 factors make up the remaining 9 percent of the variance and have separated out a group of Nassellarian species that occurred in significant numbers only in the latest Oligocene

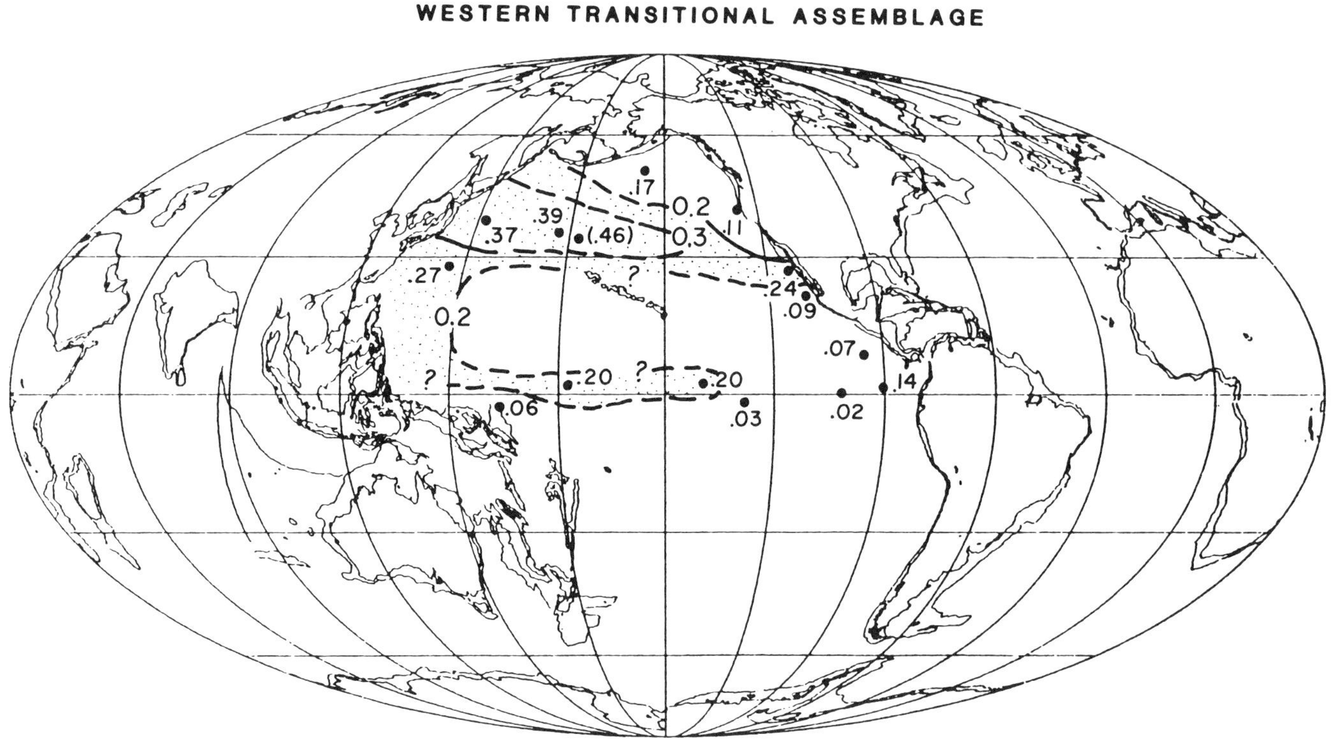

Romine and Lombari

WESTERN TRANSITIONAL ASSEMBLAGE

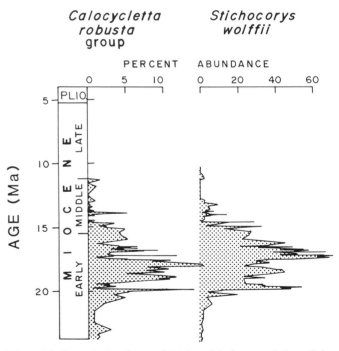

Figure 9. Distribution of the Western Transitional assemblage at 8 Ma (late Miocene), (Romine 1985). Highest factor values occur in the North Pacific Transitional watermass.

at the Oligocene/Miocene boundary (Factor 4; Table 2); a single species group that was important only during a short interval in the latest Miocene (Factor 5; Table 2); and a species category whose major proliferation occurred at the same time as *Stichocorys wolffii* (Factor 6; Figure 10). The latter species group, the *Calocycletta robusta* group, will be discussed along with the evolution of *Stichocorys* species. Factors 4 and 5 will not be discussed as they are primarily of stratigraphic interest and do not appear to bear a clear relationship to oceanographic changes in the equatorial region.

DISCUSSION

Early Miocene Faunal History, 20 to 15 Ma

Until about 12 Ma, the Miocene at Site 289 was dominated almost solely by the *Stichocorys* lineage (Figure 6). In fact, between 15 and 20 Ma, as much as 25–70 percent of the entire radiolarian population (including unidentified species) consisted of one species, *Stichocorys wolffii*. This species evolved from *S. delmontensis* in the early Miocene (20–21 Ma) and rapidly became the most abundant radiolarian at Site 289. This early Miocene proliferation has been recorded in other western tropical DSDP sites (specifically, Site 55, 65, and 448; Lombari 1985), in Sites 62, 63, and 64 of DSDP Leg 7 (Riedel and Sanfilippo 1971), and in Core 91 of the Swedish Deep-Sea Expedition, 1947–1948 (Riedel 1957). The acme event in which the abun-

Figure 10. Percent abundance of total radiolarian population, *Calocycletta robusta* group and *S. wolffii*, Site 289. Scale for *C. robusta* group is four times that of *S. wolffii* due to the high abundances of the latter species.

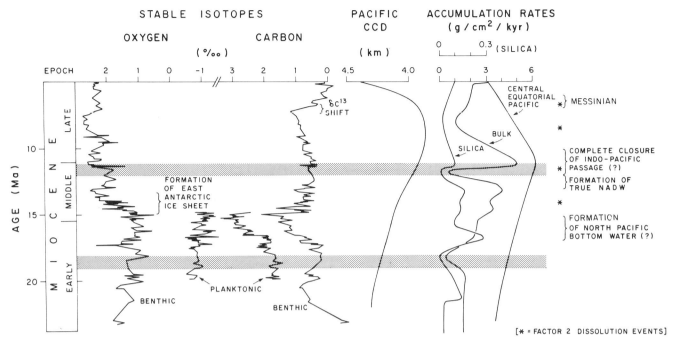

Figure 11. Elements of Miocene oceanographic history: oxygen and carbon isotopes from Site 289 (see Figure 5); Pacific CCD (calcium carbonate compensation depth; van Andel et al. 1975); bulk and silica accumulation rates in Site 289; central equatorial Pacific opal accumulation rates from Leinen (1979). Peaks in western Pacific hiatus abundance are indicated by stippled regions. Asterisks indicate the times of increased dissolution of silica suggested by the abundance variations of *Disolenia* spp. (Factor 2, Figure 7). The timing of events of paleoceanographic significance is indicated on the figure.

dance of *S. wolffii* is suddenly large relative to *S. delmontensis* does not appear to be restricted to the western tropical Pacific, however, but also has been recorded in Central Pacific sites from DSDP Leg 8 (Moore 1971), in Site 206 (Southwest Pacific; Westberg and Riedel 1978), and in the Caribbean (Trinidad; Riedel and Sanfilippo 1971). Only one other species occurred in significant numbers in the western Pacific at this time, the *Calocycletta robusta* group (Figure 10; factor 6), which exhibits an almost identical pattern of evolution and abundance at Site 289 and, together with *S. wolffii*, makes up nearly the whole of the radiolarian population at 17 Ma. The *Calocycletta robusta* group, however, is most dominant in the eastern equatorial Pacific (CENOP unpublished data) so these two equatorial species together dominate the entire equatorial region for a period of about 5 Ma. This is a unique phenomenon in the tropics where diversities are usually high and individual species abundances consequently low relative to those at high latitudes.

What could have caused this unusual condition of dominance? Silica dissolution and subsequent concentration of these two species in sediments does not appear to have been a factor, since the pattern of abundance of a solution-resistant group, the spyrids, indicates little significant silica dissolution during this interval in Site 289 (Holdsworth and Harker 1975). This evidence is substantiated by laboratory experiments that indicated that both *S. wolffii* and the *C. robusta* group are relatively solution-susceptible species when ranked with 40 other Cenozoic species (Moore 1969). Winnowing or tidal current concentration

of the species does not appear to be a viable mechanism to explain these data since the sediments of Site 289 show no evidence of reworking or significant current activity during this period (Andrews et al. 1975). Post- and syn-depositional processes, then, do not seem to explain this unusual occurrence of species dominance.

Although they appear to have been relatively dissolution-susceptible, *S. wolffii* and the *C. robusta* group are heavily silicified species; *S. wolffii* exhibits pore-filling and ridge development while the *C. robusta* group possesses thickened test walls and a heavy apical horn (see plates in Riedel and Sanfilippo 1978, for examples). This silicification may indicate that these species are deep-living, perhaps within nutrient-rich (i.e. silica-rich) waters, similar to heavily silicified phyto- and zooplankton of the modern oceans found in cold, nutrient-rich waters of high latitudes.

Early Miocene Paleoceanography, 20 to 15 Ma (Figure 11)

If the supply of silica/nutrient-rich water in the western tropical Pacific is connected with the proliferation of some radiolarian species, then what was the source of this water? In order to determine the plausibility of an oceanographic explanation, information on the paleoceanography and paleoclimates of the early Miocene has been collected and examined (Figure 11). During the early Miocene, surface waters (planktonic $\delta^{18}O$) were warming and bottom waters (benthic $\delta^{18}O$) were at their warmest for the entire Miocene. Warm oceans and, consequently, a relatively low latitudinal thermal gradient suggests that both oceanic and atmospheric circulation were relatively sluggish. However, central equatorial productivity was relatively high in the early Miocene (Figure 11, opal accumulation curve of Leinen 1979). A peak in silica accumulation at about 17 Ma in Site 289 (calculated from sediment properties data in Van der Lingen and Packham 1975) during this interval suggests at least a local (perhaps western tropical) increase in productivity. This timing also coincides with peak abundances of *S. wolffii,* implying a relationship to the supply of silica. Since silica is most concentrated in old, intermediate, and deep waters, for example, the Modern North Pacific Ocean, a source for more silica-rich deep water must be looked for. One possibility has been discussed by Blanc and Duplessy (1982). Closure of the eastern Tethys initiated the formation of the Mediterranean basin in the early Miocene (about 18–19 Ma; Berggren and Van Couvering 1974). Creation of the mid-latitude Mediterranean Sea during the warm early Miocene may have led to formation of dense, saline water that began to flow out into the Atlantic much as it does today, contributing dense salty water to the deep water circulation. Some of the water would have been entrained in Atlantic equatorial circulation at intermediate depths and transported into the Pacific equatorial region through the Central American Seaway (Straits of Panama; see Figure 3). Warmer early Miocene temperatures accompanied by high rates of evaporation in higher latitudes of the North Pacific, and the diversion of deep, more saline equatorial intermediate waters to the North Pacific as Indo-Pacific circulation gradually became restricted, may have combined to provide water dense enough to sink as deep or bottom water in the northwest Pacific (cf. Blanc and Duplessy 1982). The observed increase in $\delta^{13}C$ at 17 Ma (Figure 11) in both bottom and surface waters in the Pacific Ocean can be explained by this hypothesis, since a North Pacific bottom-water source would contribute water enriched in $\delta^{13}C$ (from equilibration with atmospheric CO_2 before sinking).

Another, simpler possibility exists. Riedel and Sanfilippo (1973) point out the cessation in the deposition of silica in the Caribbean that occurred at the time of the peak in western Pacific hiatus abundance (18–19 Ma; Moore et al. 1978), suggesting a new source or a rerouting of chemically old deep waters that were rich in dissolved silica and nutrients. Perhaps the timely development of a "proto-Mediterranean" basin provided a source of deep waters that "aged" enroute to the Pacific equatorial region, indirectly resulting in relatively high rates of opal productivity and accumulation in the central and western equatorial Pacific.

A source of old, deep water enriched in silica probably existed in early Miocene. The problem remains, however, to explain how two radiolarian species were able to become so overwhelmingly abundant throughout the equatorial and western tropical Pacific. It is possible that some special ecologic niche developed in early Miocene as certain threshold conditions were met during the progression of changes in circulation patterns, paleogeography and climate. The rapid proliferation, domination

and decline of *S. wolffii* and *C. costata* group suggests that conditions within the niche were initially favorable to the 2 species, but later became unfavorable, or the niche disappeared. However, available data are not adequate to determine what that niche was or how it developed.

Middle to Late Miocene Faunal History, 15 to 5 Ma

S. delmontensis evolved in latest Oligocene, and appears to have been rather evenly distributed across the Pacific equatorial region in early Miocene (Lombari 1985). This species ranged throughout the Miocene, being least abundant at Site 289 when *S. wolffii* was dominant (Figure 6). The progressive restructuring of western equatorial flow patterns as the Indo-Pacific passage became more restrictive to equatorial flow may have produced conditions suitable for the appearance and proliferation of *S. wolffii.* The extinction of *S. wolffii* (13–15 Ma) occurred during the interval of the sharp increase in $\delta^{18}O$ attributed to major accumulation of ice on East Antarctica (Shackleton and Kennett 1975; Figure 11). The first evolutionary appearance of the late Miocene representative of this lineage, *S. peregrina,* occurred between 14 and 15 Ma in Site 289 (Figure 6). Its first appearance in North Pacific DSDP Sites has been recorded at much younger ages (younger than 8 Ma; Theyer 1972; Theyer et al. 1978; Weaver et al 1981). In the North Pacific, *S. peregrina* has been found in association with transitional water masses. The transition region in the North Pacific forms a hydrographic boundary between the subpolar and subtropical water masses. Intermediate waters forming along this boundary sink below the warmer, subtropical waters and flow towards the equator; Casey (1982) suggests that species such as *S. peregrina* "dive" with these intermediate waters and live at greater depths at lower latitudes of the Pacific. The data at 289 indicate that *S. peregrina* evolved in the middle Miocene in the western equatorial Pacific and must have migrated into the North Pacific transition region. The persistence of *S. delmontensis* through the Miocene suggests that it may have had a wide range of ecologic tolerance and, therefore, could migrate in response to environmental changes in its habitat. Thus, the decrease in *S. delmontensis* percent abundance at about 11.5 Ma in Site 289 and the reported abundances in the North Pacific at 8 Ma (Romine 1985) may also support migration of the lineage into the North Pacific. This migration may have been facilitated by a restructuring of circulation in the western equatorial Pacific during the middle Miocene.

The third factor (Figure 8), representing a western tropical/transitional assemblage, increased in importance at 289 in two stages: first at 14.5 and then at 11 Ma. The first increase in this assemblage may be partly related to an intensification in eastern boundary circulation and the subsequently increased advection of transitional assemblages from east to west in the equatorial region. Increased advection would have occurred as pole-to-equator thermal gradients increased, brought about by the rapid decrease in polar temperatures coincident with the formation of the East Antarctic ice sheet. The second increase is most likely

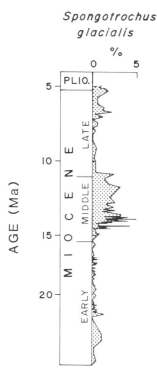

Spongotrochus glacialis %

Figure 12. Percent abundance of *Spongotrochus glacialis,* Site 289. Percentages are from the total radiolarian population of the sample.

due to the final blocking of Indo-Pacific equatorial circulation that was completed by 12 Ma (Audley-Charles, et al. 1972; Edwards 1975). Westward-flowing, warm equatorial surface currents were diverted into the northward-flowing western boundary current (ancestral Kuroshio) causing an overall intensification of gyral circulation in the North Pacific. An assemblage preferring the warm, western transitional region (i.e. the western limb of the North Pacific subtropical gyre and the western part of the transitional water mass) thus developed in this setting (Figure 9). The diversion of equatorial water in the North Pacific provided an easy avenue of transport allowing species to migrate northward (e.g. *S. peregrina*) into the transitional water mass.

Immediately following formation of the East Antarctic ice sheet (14–15 Ma), abundances of individual species within the western tropical (transitional) radiolarian population increased. An initial intensification of circulation brought cooler, eastern boundary current/transitional elements like *S. glacialis,* which is an indicator of upwelling in the Modern ocean (Casey et al., 1979a, b), across the equatorial expanse to the western equatorial region where they were reasonably abundant between 15 and 11 Ma (Figure 12). Warmer, western tropical/transitional elements (Figure 8, 9) increased slowly during this initially cool period and then increased rapidly at 10–12 Ma when equatorial Indo-Pacific communication totally ceased and development of intensified western boundary circulation was complete (factor 3; Figures 8 and 9). The evolution of essentially modern equatorial circulation was completed at this time by the further development of west to east counterflow in order to compensate for the pile-up of

westward-flowing equatorial circulation (Equatorial Undercurrent). A peak in opal accumulation and productivity at about 11 Ma (van Andel et al. 1975; Leinen 1979) may be related to the development of the Equatorial Undercurrent which, in the modern ocean, is involved in eastern equatorial divergence and upwelling. The decrease in opal accumulation after this time may have been linked to the development of a North Pacific transitional water mass and high-latitude silica sinks.

A decrease in the western transitional assemblage occurs at 6–7 Ma, approximately coincident with the timing of the δ^{13}C shift and the postulated timing of closure of the Central American Seaway to deep-water flow between the Atlantic and Pacific equatorial regions. It was also a time of increased silica dissolution, indicated by factor 2. Low productivity and increased syn- and post-depositional modification may be responsible for the decreased abundances within the assemblage.

High abundances of the collosphaerid group, *Disolenia* spp., which dominates factor 2 (Figure 7) have been associated with times of increased dissolution of silica (Holdsworth and Harker 1975). After their first appearance, just prior to 15 Ma, peaks in the assemblage occur at 14–15 Ma, 11–12 Ma, 8–9 Ma, and 6–7 Ma. In general, these peaks coincide with times of depressed bulk and silica accumulation rates. These correlations suggest a decreased silica supply to the equatorial region and low productivity. Low burial rates of siliceous microfossils might also have led to increased dissolution through longer exposure to deep water undersaturated in silica.

Middle to Late Miocene Paleoceanography, 15 to 5 Ma

Bulk and silica accumulation rates in Site 289 increased between 15 and 13 Ma, with a major low occurring at approximately 11.5 to 12.5 Ma. This low coincided with a peak in hiatus abundance (Moore et al. 1978) at about 11 to 12 Ma. Accumulation rates peaked at about 11 Ma and decreased immediately afterwards; however, no peak in hiatus abundance younger than 11 Ma was associated with the late Miocene low in sediment accumulation suggesting that this decrease in accumulation rates was probably the result of a reduction in preservation of calcium carbonate due to the shallow Pacific CCD (Figure 11). The record of central equatorial Pacific opal accumulation (Leinen 1979; van Andel et al. 1975) also peaked at about 11 Ma and decreased thereafter, similar tot he record of western equatorial bulk accumulation. The post-11 Ma decline may indicate a decrease in productivity and/or in the supply of silica.

The decrease in accumulation rates at the equator after 11 Ma accompanies a decrease in abundances of *Stichocorys* species (Figures 6, 8 and 11). These data may be evidence of intensification of circulation in the North Pacific. How this intensification may have been accomplished is suggested in the following argument.

The middle Miocene formation of the East Antarctic ice sheet was accompanied by increased production of cold Antarctic Bottom Water (AABW) which induced a vertical (depth) and horizontal (latitudinal) stratification of water masses as thermal gradients steepened. Atmospheric and oceanic circulation were intensified due to the steepening of these thermal gradients. It is possible that prior to the Middle Miocene Antarctic ice sheet formation and the termination of Indo-Pacific communication, silica productivity was relatively high in the equatorial regions because of equatorial divergence associated with an essentially uninterrupted circum-tropical flow that tapped nutrient-rich intermediate waters. As circum-tropical flow began to fragment, nutrients in intermediate and deeper waters were funnelled into higher latitudes. This began in early Miocene in the Southern Hemisphere, conveniently timed with early Miocene development of the Antarctic Convergence (Kennett 1977; 1980; 1982). "Trapping" of silica (and other nutrients) may have begun at this time in the Antarctic region as the siliceous biogenic province developed and expanded, especially after formation of a major ice sheet on East Antarctica with the associated climatic cooling and increased upwelling. Eventual concentration of silica in the Antarctic region would occur through internal cycling within the polar water mass, so that once nutrients were collected, they remained within the system unless transported out by Antarctic Bottom Water (Brewster 1980). An increase in the intensity of the mid-latitude westerlies (Janacek and Rea 1983) and blocking of the Indo-Pacific surface circulation by the late middle Miocene (11–12 Ma) completed the sequence of events that resulted in intensification of North Pacific gyral circulation and, as high latitude cooling continued, the creation and development of a true North Pacific transitional water mass. Development of this transitional water mass partitioned off the Northern Hemisphere subpolar region that became a sink for silica as productivity increased and provided a habitat for radiolarian species such as the *Stichocorys* group to expand and migrate into the North Pacific. The strong dominance of the Southern Ocean as a silica sink by the latest Miocene (Brewster 1980) may have resulted from the much earlier development of the Antarctic Convergence, and therefore, an earlier establishment of a system for concentrating and recycling nutrients.

SUMMARY AND CONCLUSIONS

Paleogeographic changes in the distribution of continental masses have had profound influence on the evolution of oceanic circulation patterns and climate. The gradual fragmentation of the once circum-global, tropical current system, the Tethys Seaway, and the creation of a circumpolar current in the Southern Hemisphere, the Circum-Antarctic Current, were major factors in the evolution of Miocene circulation and climate. The deep-sea sediments from Site 289 in the western equatorial Pacific (once a part of the Tethys Seaway) contain a record of this evolution.

(1) The radiolarian fauna in the western tropical Pacific was dominated by two species, *Stichocorys wolffii* and *Calocycletta robusta* group, which made up from 25 to 75 percent of the population during the early Miocene. The highly silicified nature of the tests of these two species suggests that they may have lived

at depth within nutrient-rich intermediate and deep waters in the equatorial region. The acme of their abundance (17 Ma) coincides with a high in silica accumulation rates in Site 289 which may indicate increased productivity at this time. The unusual dominance of these two species may be related to the development of a particular ecological niche. The creation and destruction of this niche was probably controlled by gradual changes in circulation patterns in the equatorial region that were influenced by changes in paleogeography and climate. The end of this species dominance is coincident with middle Miocene formation of the East Antarctic ice sheet, 15–13 Ma.

(2) A major increase in a western transitional assemblage, a decrease in abundances of the *Stichocorys* species group, a decrease in silica accumulation rates in both the central and western equatorial Pacific, and an increase in mass accumulation rates of eolian dust in the central North Pacific (Janacek and Rea 1983) are linked to the late-middle Miocene closure of the Indo-Pacific passage (12–10 Ma). These changes signify the development of a North Pacific transitional water mass due to an increased transport in the subtropical gyre and an intensification of the westerly winds.

(3) Three major intervals of silica dissolution are indicated by the increased abundances of a dissolution-resistant assemblage at Site 289. These intervals coincide with significant paleoceanographic events: formation of the East Antarctic ice sheet (15–13 Ma); closure of the Indo-Pacific passage (12–10 Ma); the Messinian and $\delta^{13}C$ shift (7–6 Ma). Lows in rates of silica accumulation are associated with the two younger events; a peak in hiatus abundance in the western Pacific at about 11 Ma may be related to the late-middle Miocene dissolution event at 12–10 Ma.

ACKNOWLEDGMENTS

Many thanks to T. C. Moore, Jr., J. P. Kennett, L. A. Mayer, K.-Y. Wei, S. Wise and W. R. Riedel for critical reviews and helpful suggestions. We appreciate the help of N. Meader in preparing the final draft of the manuscript. This research was supported by an NSF grant OCE79-14594 (CENOP) awarded to the University of Rhode Island for study of Cenozoic paleoceanography.

APPENDIX

Percentages for 31 species, Site 289
    (See microfiche in pocket inside back cover)

REFERENCES CITED

Andrews, J. E., et al., 1975, Initial Reports of the Deep Sea Drilling Project, Volume 30: Washington, D.C., U.S. Government Printing Office, 753 p.

Audley-Charles, M. G., Carter, D. J., and Milson, J. S., 1972, Tectonic development of eastern Indonesia in relation to Gondwanaland dispersal: Nature Physical Science, v. 239, p. 35–39.

Bender, M. L., and Graham, D. W., 1981, On late Miocene abyssal hydrography: Marine Micropaleontology, v. 6, p. 451–464.

Berggren, W. A., and Hollister, C. D., 1977, Plate tectonics and paleocirculation—Commotion in the ocean: Tectonophysics, v. 38, p. 11–48.

Berggren, W. A., and van Couvering, J. A., 1974, The late Neogene biostratigraphy, geochronology and paleoclimatology of the last 15 m.y.: Palaeogeography, Palaeoclimatology, and Palaeoecology, v. 16, p. 1–216.

Berggren, W. A., Kent, D. V., and Van Couvering, J. A., 1984, Neogene geochronology and chronostratigraphy: Geological Society of London Bulletin, *in* Snelling, N. J., ed., Geochronology and the Geologic Time Scale, Geological Society, London, Special Paper, in press.

Blanc, P.-L., and Duplessy, J.-C., 1982, The deep-water circulation during the Neogene and the impact of the Messinian salinity crisis: Deep-Sea Research, v. 29, p. 1391–1414.

Blow, W. H., 1969, Late middle Eocene to Recent planktonic foraminiferal biostratigraphy, *in* Bronniman, P., and Renz, H. H., eds., First International Conference on Planktonic Microfossils: Geneva, 1967, p. 199–421.

Brewster, N. A., 1980, Cenozoic biogenic silica sedimentation in the Antarctic Ocean, based on two Deep Sea Drilling Project Sites: Geological Society of America Bulletin, v. 91, p. 337–347.

Casey, R. E., 1971a, Distribution of polycrystine Radiolaria in the oceans in relation to physical and chemical conditions, *in* Funnell, B. M., and Riedel, W. R., eds., The Micropaleontology of Oceans: Cambridge University Press, London, p. 151–160.

—— 1971b, Radiolaria as indicators of past and present water masses, *in* Funnell, B. M., and Riedel, W. R., eds., The Micropaleontology of Oceans: Cambridge University Press, London, p. 331–342.

Casey, R. E., 1982, *Lamprocyrtis* and *Stichocorys* lineages: Biogeographical and ecological perspectives relating to the tempo and mode of Polycystine radiolarian evolution: Third North American Paleontological Convention Proceedings, v. 1, p. 77–82.

Casey, R. E., Gust, L., Leavesley, A., Williams, D., Reynolds, R., Duis, T., and Spaw, J. M., 1979b, Ecological niches of radiolarians, planktonic foraminiferans and pteropods inferred from studies on living forms in the Gulf of Mexico and adjacent waters: Transactions of the Gulf Coast Association Geological Society: v. 29, p. 216–223.

Casey, R. E., Spaw, J. M., Kunze, F., Reynolds, R., Duis, T., McMillen, K., Pratt, D., and Anderson, V., 1979a, Radiolarian ecology and the development of the radiolarian component in Holocene sediments, Gulf of Mexico and adjacent seas with potential paleontological applications: Transactions of the Gulf Coast Association Geological Society, v. 29, p. 228–237.

Cita, M. B., 1976, Early Pliocene paleoenvironment after the Messinian Salinity Crisis: VI. African Micropaleontology Colloquium, Tunis, 1974.

Denton, G. H., Prentice, M. L., Kellogg, D. E., and Kellogg, T. B., 1984, Late Tertiary history of the Antarctic ice sheet: Evidence from the Dry Valleys: Geology, v. 12, p. 263–267.

Edwards, A. R., 1975, Southwest Pacific Cenozoic paleogeography and an integrated Neogene paleocirculation model, *in* Andrews, J. E., Packham, G., et al., eds., Initial Reports of the Deep Sea Drilling Project, Volume 30: Washington, D.C., U.S. Government Printing Office, p. 667–684.

Haq, B. U., 1981, Paleogene paleoceanography—Early Cenozoic oceans revisited: Oceanologica Acta, Special Publication, Proceedings of the 26th International Geological Congress, Paris, 1980, p. 71–82.

Hayes, D. E., Frakes, L. A., and others, 1975, Initial Reports of the Deep Sea Drilling Project, Volume 28: Washington, D.C., U.S. Government Printing Office, 1017 p.

Holdsworth, B. K., 1975, Cenozoic radiolarian biostratigraphy, Leg 30: Tropical and Equatorial Pacific, *in* Andrews, J. E., Packham, G. et al., eds., Initial Reports of the Deep Sea Drilling Project, Volume 30: Washington, D.C., U.S. Government Printing Office, p. 499–538.

Holdsworth, B. K., and Harker, B. M., 1975, Possible indicators of degree of radiolaria dissolution in calcareous sediments of the Ontong-Java Palteau, *in* Andrews, J. E., Packham, G., et al., eds., Initial Reports of the Deep Sea Drilling Project, Volume 30: Washington, D.C., U.S. Government Printing Office, p. 489–498.

Imbrie, J., and Kipp, N. G., 1971, A new micropaleontological method for quantitative paleoclimatology; application to a late Pleistocene Caribbean core, *in* Turekian, K. K., eds., Late Cenozoic Glacial Ages: New Haven, Conn., Yale University Press, p. 71–181.

Janacek, T. R., and Rea, D. K., 1983, Eolian deposition in the northeast Pacific Ocean: Cenozoic history of atmospheric circulation: Geological Society of America Bulletin, v. 94, no. 6, p. 730–738.

Johnson, T. C., 1974, The dissolution of siliceous microfossils in surface sediments of the eastern tropical Pacific: Deep-Sea Research, v. 21, p. 851–864.

Keigwin, L. D., 1978, Pliocene closing of the Panama Isthmus based on biostratigraphic evidence from nearby Pacific Ocean and Caribbean cores: Geology, v. 6, p. 630–634.

Keigwin, L. D., and Keller, G., 1984, Middle Oligocene cooling from equatorial Pacific DSDP Site 77B: Geology, v. 12, p. 16–19.

Kemp, E. M., Frakes, L. A., and Hayes, D. E., 1975, Paleoclimatic significance of diachronous biogenic facies, Leg 28, Deep Sea Drilling Project: *in* Hayes, D. E., Frakes, L. A., et al., eds., Initial Reports of the Deep Sea Drilling Project, Volume 28: Washington, D.C., U.S. Government Printing Office, p. 909–917.

Kennett, J. P., 1977, Cenozoic evolution of Antarctic glaciation, the Circum-Antarctic Ocean, and their impact on global paleoceanography: Journal of Geophysical Research, v. 82, p. 3843–3859.

—— 1978, The development of planktonic biogeography in the Southern Ocean during the Cenozoic: Marine Micropaleontology, v. 3, p. 301–345.

—— 1980, Paleoceanographic and biogeographic evolution of the Southern Ocean during the Cenozoic, and Cenozoic microfossil datums. Palaeogeography, Palaeoclimatology, and Palaeoecology, v. 31, p. 123–152.

Kennett, J. P., et al. 1975, Initial Reports of the Deep Sea Drilling Project, Volume 29: Washington, D.C., U.S. Government Printing Office, 1197 p.

Klovan, J. E., and Imbrie, J., 1971, An algorithm and Fortran IV program for large-scale Q-mode factor analysis: Journal of the International Association for Mathematics and Geology, v. 3, p. 61–67.

Kroopnick, P., 1980, The distribution of $\delta^{13}C$ in the Atlantic: Earth and Planetary Science Letters, v. 49, p. 469–484.

Kroopnick, P., Weiss, R. F., and Craig, H., 1972, Total CO_2, $\delta^{13}C$, and dissolved oxygen—$\delta^{18}O$ at Geosecs II in the North Atlantic: Earth and Planetary Science Letters, v. 16, p. 103–110.

Leinen, M., 1979, Biogenic silica accumulation in the central equatorial Pacific and its implications for Cenozoic paleoceanography. Geological Society of America Bulletin, v. 90, p. 1310–1376.

Lombari, G., 1985, Biogeographic Trends in Neogene Radiolaria from the Northern and Central Pacific, *in* Kennett, J. P., ed., The Miocene ocean: Paleoceanography and biogeography: Geological Society of America.

Loutit, T. S., Pisias, N. G., and Kennett, J. P., 1983, Pacific Miocene carbon isotope stratigraphy using benthic foraminifera: Earth and Planetary Science Letters, v. 66, p. 48–62.

Loutit, T. S., Kennett, J. P., and Savin, S. M., 1983, Miocene equatorial and southwest Pacific paleoceanography from stable isotope evidence: Marine Micropaleontology, v. 8, no. 3, p. 21–233.

Matthews, R. K., and Poore, R. Z., 1980, Tertiary $\delta^{18}O$ record and glacioeustatic sea-level fluctuations: Geology, v. 8, p. 501–504.

Molina-Cruz, A., 1977, Late Quaternary oceanic circulation along the Pacific Coast of South America [Ph.D. dissertation]: Oregon State University, Corvallis, Oregon, 246 p.

Moore, T. C., Jr., 1969, Radiolaria: change in skeletal weight and resistance to solution: Geological Society of America Bulletin, v. 80, p. 2103–2108.

—— 1971, Radiolaria, *in* Tracey, J. I., Jr., and others, Initial Reports of the Deep Sea Drilling Project, Volume 8: Washington, D.C., U.S. Government Printing Project, Volume 8.

—— 1973, Method of randomly distributing grains for microscopic examination: Journal of Sedimentology, v. 43, p. 904.

—— 1978, The distribution of radiolarian assemblages in the modern and ice-age Pacific: Marine Micropaleontology, v. 3, p. 229–266.

Moore, T. C., Jr., and Lombari, G., 1981, Sea-surface temperature changes in the North Pacific during the late Miocene: Marine Micropaleontology, v. 6, p. 581–597.

Moore, T. C., Jr., van Andel, T. H., Sancetta, C., and Pisias, N. G., 1978, Cenozoic hiatuses in pelagic sediments: Micropaleontology, v. 24, no. 2, p. 113–138.

Nigrini, C. A., and Lombari, G., 1984, A Miocene Guide to Radiolaria: Cushman Foundation for Foraminiferal Research, Special Publication No. 22.

Pak, H., and Zaneveld, J. R., 1974, Equatorial front in the eastern Pacific Ocean: Journal of Physical Oceanography, v. 4, p. 570–578.

Petrushevskaya, M. G., 1971, Spumellarian and Nassellarian Radiolaria in the plankton and bottom sediments of the central Pacific, *in* Funnell, B. M., and Riedel, W. R., eds., The Micropaleontology of Oceans: London, Cambridge University Press, p. 309–317.

Riedel, W. R., 1957, Radiolaria: a preliminary stratigraphy: Reports of the Swedish Deep-Sea Expedition, v. 6, no. 3, 59 p.

Riedel, W. R., and Sanfilippo, A., 1971, Cenozoic Radiolaria from the western tropical Pacific, Leg 7, *in* Winterer, E. L., Riedel, W. R., and others, eds., Initial Reports of the Deep Sea Drilling Project, Volume 7: Washington, D.C., U.S. Government Printing Office, p. 1529–1672.

——1973, Cenozoic Radiolaria from the Caribbean, DSDP Leg 15, *in* Edgar, N. T., Saunders, J. B., and others, Initial Reports of the Deep Sea Drilling Project, Volume 15: Washington, D.C., U.S. Government Printing Office, p. 705–751.

——1978, Stratigraphy and evolution of tropical Cenozoic radiolarians: Micropaleontology, v. 23, no. 1, p. 61–96.

Robertson, J. H., 1975, Glacial to interglacial oceanographic changes in the northwest Pacific, including a continuous record of the last 400,000 years [Ph.D. Dissertation]: Columbia University, New York, 355 p.

Romine, K., 1985, Radiolarian biogeography and paleoceanography of the North Pacific at 8 Ma: Geological Society of America Memoir, this volume.

Romine, K., and Moore, T. C., Jr., 1981, Radiolarian assemblage distributions and paleoceanography of the eastern equatorial Pacific Ocean during the last 127,000 years: Palaeogeography, Palaeoclimatology, and Palaeoecology, v. 35, p. 281–314.

Sachs, H. M., 1973, North Pacific radiolarian assemblages and relationship to oceanographic parameters: Quaternary Research, v. 3, p. 73–88.

Savin, S. M., 1977, The history of the earth's surface temperature during the past 100 million years: Annual Reviews of Earth and Planetary Science Letters, v. 5, p. 319–355.

Savin, S. M., Douglas, R. G., Keller, G., Killingley, J. S., Shaughnessy, L., Sommer, M. A., Vincent, E., and Woodruff, F., 1981, Miocene benthic foraminiferal isotope records: A synthesis: Marine Micropaleontology, v. 6, p. 423–450.

Shackleton, N. J., 1982, The deep-sea sediment record of climate variability.

Progress in Oceanography, v. 11, p. 199–218.

Shackleton, N. J., and Kennett, J. P., 1975, Paleotemperature history of the Cenozoic and the initiation of Antarctic glaciation: oxygen- and carbon-isotope analyses in DSDP Sites 277, 279, and 281, *in* Kennett, J. P., Houtz, R. E., and others, eds., Initial Reports of the Deep Sea Drilling Project, Volume 29: Washington, D.C., U.S. Government Printing Office, p. 743–755.

Srinivasan, M. S., and Kennett, J. P., 1981a, A review of Neogene planktonic foraminiferal biostratigraphy: Applications in the equatorial and South Pacific: SEPM Special Publication No. 32, p. 395–432.

——1981b, Neogene planktonic foraminiferal biostratigraphy and evolution: Equatorial to Subantarctic, South Pacific: Marine Micropaleontology, v. 6, p. 499–533.

Theyer, F., 1972, Late Neogene paleomagnetic and planktonic zonation, southeast Indian Ocean-Tasman Basin [Ph.D. Dissertation]: University of Southern California, Los Angeles, 198 p.

Theyer, F., Mato, C. Y., and Hammond, S. R., 1978, Paleomagnetic and geochronologic calibration of latest Oligocene to Pliocene radiolarian events, equatorial Pacific: Marine Micropaleontology, v. 3, p. 377–395.

van Andel, T. H., Heath, G. R., and Moore, T. C., Jr., 1975, Cenozoic history and paleoceanography of the central equatorial Pacific Ocean: Geological Society of America Memoir No. 143, 134 p.

Van der Lingen, G. J., and Packham, G. H., 1975, Relationships between diagenesis and physical properties of biogenic sediments of the Ontong-Java Plateau (Sites 288 and 289, Deep Sea Drilling Project), *in* Andrews, J. E., Packham, G., et al., eds., Initial Reports of the Deep Sea Drilling Project, Volume 30: Washington, D.C., U.S. Government Printing Office, p. 443–481.

Weaver, F. M., Casey, R. E., and Perez, A. M., 1981, Stratigraphic and paleoceanographic significance of early Pliocene to middle Miocene radiolarian assemblages from northern to Baja California, *in* Garrison, R. E., Douglas, R. G., eds., The Monterey Formation and Related Siliceous Rocks of California: SEPM Special Publication, p. 71–86.

Westberg, M. J., and Riedel, W. R., 1978, Accuracy of radiolarian correlations in the Pacific Miocene. Micropaleontology, v. 24, no. 1, p. 1–23.

Woodruff, F., Savin, S. M., and Douglas, R. G., 1981, Miocene stable isotope record: a detailed deep pacific Ocean study and its paleoclimatic implications: Science, v. 212, p. 665–668.

Wyrtki, K., 1966, Oceanography of the eastern equatorial Pacific Ocean: Oceanographic Marine Biology Annual Review, v. 4, p. 33–68.

MANUSCRIPT ACCEPTED BY THE SOCIETY DECEMBER 17, 1984

Geological Society of America
Memoir 163
1985

Biogeographic trends in Neogene radiolaria from the Northern and Central Pacific

Gail Lombari*
Graduate School of Oceanography
University of Rhode Island
Narragansett, Rhode Island 02882-1197

ABSTRACT

Radiolarian assemblages have been quantitatively examined for two equatorial Pacific time-slices (22 and 16 Ma) and three time-series (DSDP Sites 71, 173 and 289) spanning the early Miocene from 22–15 Ma. A subset of 43 radiolarian species that range throughout the 22–15 Ma interval was used to compare the time-slices. Census data tabulated for the time-series were treated with a Q-mode factor analysis that identifies five faunal provinces.

The early Miocene was a time of very high evolutionary radiation, especially among the suborder Nassellaria. However, by the end of the middle Miocene 72 percent of the Nassellarian species present in the greater than $63 \mu m$ fraction of the sediment at 15 Ma had become extinct. Therefore, the middle Miocene was a time of Nassellarian extinctions. In contrast, the Spumellaria were relatively unchanging throughout the early to middle Miocene, and only 16 percent had become extinct by the end of the middle Miocene. These Spumellaria are inferred to have been mainly cooler water/deeper dwellers and were ubiquitous. Their position as generalists in the fauna probably allowed them to be less affected by changing ocean conditions.

By 16 Ma, diversity of the radiolarian assemblage was lower than at any other time during the Neogene in the tropical Pacific. At this time a dramatic change was observed in the western tropical assemblage. One species, *Stichocorys wolffii,* reached maxima of over 50 percent of the population and other species that had previously been abundant disappeared. These characteristics seemed to have resulted from extinctions of Nassellaria and biogeographic changes within the assemblage.

The radiolarian faunal groups responded to and reflect broadscale paleoceanographic changes. As Neogene Indo-Pacific circulation became restricted, some radiolarian species, such as the spyroids, moved eastward and some, such as the stichocorids, moved into the temperate/subarctic.

INTRODUCTION

A major goal of the CENOP project has been to reconstruct the paleoceanographic conditions of the Miocene oceans, using available deep-sea cored materials. In this study, the radiolarian assemblages in Central and North Pacific sediments (Figure 1) have been tabulated for three early Miocene time-series (DSDP Sites 71, 173, and 289 only) and for two time-slices (DSDP Sites 55, 65, 71, 289, 448, and 495 at 22 Ma and DSDP Sites 55, 63.1, 65, 71, 173, 289, 448, and 495 at 16 Ma). The earliest Miocene was relatively warm and tropical Tethys circulation was unobstructed between the Pacific Ocean and both the Indian and Atlantic Oceans. A time-slice was chosen at 22 Ma to investigate the fauna during a time of relatively stable climates. Another time-slice was chosen at the end of the early Miocene prior to the beginning of a trend of climatic cooling connected with major accumulation of ice on East Antarctica (Shackleton and Kennett

---

*Present address: 20 Knight St., Coventry, Rhode Island 02816.

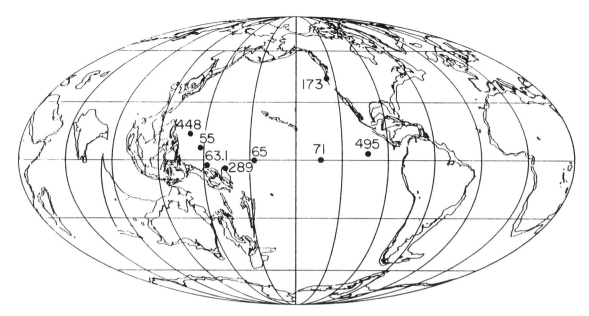

Figure 1. Location of DSDP Sites.

1975; Kennett 1982). Data on relative proportions of the two suborders of polycystine radiolaria (Lombari and Boden 1982) have been included and show changes in the radiolarian assemblages due to variations in oceanic circulation and climate.

Studies of the polycystine radiolarian suborders, Spumellaria and Nassellaria, have shown that water masses can be characterized by their species assemblages (Sachs 1973; Molina-Cruz 1977; Dow 1978; Maurrasse 1978; Moore 1978; Casey et al. 1979; Moore et al. 1980). Biogeographic investigations of Radiolaria in surface sediments show spatial relationships that in the modern oceans match water-mass circulation patterns (Johnson and Nigrini 1980 and 1982). Their distribution in the past provides information on paleoenvironmental conditions and water-mass circulation.

The objective of this investigation is to study the radiolarian census data from each time-slice as well as the time-series to evaluate biogeographic patterns. Thus, they may assist in understanding the paleoceanographic history during the early Miocene. In comparing these early Miocene trends to other available data, general changes in the radiolarian assemblages throughout the Neogene can be determined.

METHODS

Sample Preparation

Samples were treated with concentrated hydrogen peroxide and a 50 percent solution of hydrochloric acid to aid dissaggregation and rid the sample of any carbonate material. The samples were then sieved over a 63μm screen, followed by treatment for 10 to 15 seconds in an ultrasonic bath with a solution of concentrated sodium hydroxide. This last procedure was omitted if the Radiolaria were free of amorphous silica and clay. The remaining coarse fraction was mounted following the random settling technique of Moore (1973). Radiolarian counts were made on 400–600 known specimens for each sample.

Stratigraphy and Sample Selection

Intervals to be counted for each of the two time-slices (Table 1) were selected according to the stratigraphy described in Barron et al. (this volume). The 22 Ma time-slice is within the foraminiferal biostratigraphic zone N4B and the 16 Ma time-slice is within the foraminiferal biostratigraphic zone N8 (Blow 1969).

Six sites, all tropical, were found to contain sufficient radiolarian abundances for counting purposes in the N4B time-slice. For the N8 time-slice, eight sites were chosen. As with the earliest Miocene interval, this provided an equatorial suite of samples but also included one subtropical mid-latitude site (DSDP Site 173). Samples for three early Miocene time-series were collected from DSDP Sites 71, 173, and 289. The oldest sediment recovered from Site 173 is approximately 18–17 Ma and samples were taken from this level through the 15 Ma interval.

Taxonomy and Species Selection

The taxonomic framework used in this study was established by Nigrini and Lombari (1984). Of the 150 species groups described in that text, 96 were utilized for counting purposes. Only categories that contributed at least one percent to the total population in one or more sites were used. Although census data were tabulated for all 96 groups, species that did not range throughout the entire 22–16 Ma interval were eliminated from the factor analysis. To compare the time-slices, it was essential that each group be present in both the 22 and 16 Ma time-slice. In this way, the changes seen in temporal and spatial distribution

TABLE 1. SAMPLES USED FOR THIS STUDY

DSDP Site	Location* (Lat., Long.)	Interval Sampled (Core-Sec.: Depth-)	No. of Samples	Time-Slice
55	5.15N, 151.30E	13-3:55cm--13-1:45cm	4	22Ma
65	1.22S, 174.73W	10-3:117cm--11-6:46cm	10	22Ma
71	0.18S, 132.16W	30-6:91cm--31-6:93cm	5	22Ma
289	5.53S, 166.69E	65-3:42cm--69-5:36cm	10	22Ma
448	12.71N, 144.25E	6-1:82cm--6-2:144cm	5	22MA
495**	3.89N, 105.20W	37-3:82cm--37-3:144cm	2	22Ma
55	6.46N, 148.91E	10-2:95cm--10-6:134cm	5	16Ma
63A	2.25S, 153.82E	10-5:51cm--11-6:50cm	5	16Ma
65	0.38N, 177.03W	7cc--8-4:25cm	5	16Ma
71	1.02N, 134.35W	20-4:81cm--11-6:81cm	10	16Ma
173	41.48N, 127.44W	28-1:105cm--28-3:124cm	5	16Ma
289	4.02S, 164.39E	52-4:34cm--55-3:90cm	10	16Ma
448	13.89N, 141.70E	3-2:44cm--3-2:98cm	2	16Ma
495	3.89N, 105.20W	25-4:102cm--26-6:56cm	9	16Ma
71	1.02N, 134.35W	20-4:81cm--31-4:91cm	35	Time-Series
173	41.48N, 127.44W	18-1:105cm--31cc	24	Time-Series
289	4.02S, 164.39E	51-4:35cm--73-2:36cm	35	Time-Series

*Paleolatitude
**Paleolatitude unavilable at 22Ma

of the selected, long-ranging species are more likely to result from oceanographic and ecologic changes with time, than from evolutionary changes in total radiolarian assemblage. So that a maximum number of species groups could be used for this study, an attempt was made to combine species within the same genus having short ranges. Since morphotypes of the same genus may be found in different niches, a preliminary factor analysis was performed to determine if any combinations were possible. *Calocycletta robusta* and *C. virginis* were similarly described by the factor analysis. Since this would imply that these species behaved the same in a statistical sense, they could safely be combined. Other groups tested and combined are *Didymocyrtis mammifera* with *D. violina*, *Spongurus* (?) sp. A with *Spongurus* (?) sp. B, *Dendrospyris damaecornis* with *D. pododendros* and *Theocalyptra bicornis* with *T. davisiana* var. *cornutoides*. A total of 43 species groups were used in this study (Table 2). Table 3 lists the species groups that have a maximum relative abundance of greater than 5 percent in the samples studied. The creation of assemblage subsets by Q-mode factor analysis is dependent upon the relative abundances of the individual species in each sample. The most abundant species will dominate the factor analysis. Therefore, the species listed in Table 3 control the assemblages in a statistical sense. However, valuable biogeographic information may also be derived from the less common forms.

RESULTS

Early Miocene Time-series

The Radiolaria from three early Miocene time-series for DSDP Sites 71, 173, and 289 were examined in order to con-

TABLE 2. SPECIES USED FOR TIME-SERIES ANALYSIS

1. Acrosphaera spp.
2. Actinomma spp.
3. Didymocyrtis mammifera (Haeckel)
   Didymocyrtis violina (Haeckel)
4. Disolenia spp.
   Otosphaera spp.
5. Heliodiscus asteriscus Haeckel
6. Hexastylus spp.
7. Hymeniastrum spp.
8. Lithelius minor Jorgensen
9. Lithelius nautiloides Popofsky
10. Phorticium polycladum Tan and Tchang
11. Phorticium pylonium Haeckel
12. Spongopyle osculosa Dreyer
13. Spongotrochus glacialis Popofsky
14. Spongurus (?) sp.A
    Spongurus (?) sp.B
15. Stylatractus spp.
16. Stylochlamydium asteriscus Haeckel
17. Stylodictya aculeata Jorgensen
18. Stylosphaera spp.
19. Calocycletta robusta Moore
    Calocycletta virginis (Haeckel)
20. Carpocanarium sp.
21. Ceratocyrtis histricosa (Jorgensen)
22. Ceratocyrtis stigi (Bjorklund)
23. Cyrtocapsella cornuta (Haeckel)
24. Cyrtocapsella tetrapera (Haeckel)
25. Dendrospyris damaecornis (Haeckel)
    Dendrospyris pododendros (Carnevale)
26. Eucyrtidium cienkowskii Haeckel
27. Eucyrtidium diaphanes Sanfilippo and Riedel
28. Giraffospyris circumflexa Goll
29. Liriospyris geniculosa Goll
30. Liriospyris mutuaria Goll
31. Lophophaena pentagona pentagona (Ehrenberg) emend. Goll
32. Phormostichoartus marylandicus (Martin)
33. Rhodospyris (?) spp. De 1 Goll grp.
34. Siphostichartus corona (Haeckel)
35. Stichocorys delmontensis (Campbell and Clark)
36. Stichocorys diploconus (Haeckel)
37. Stichocorys wolffii Haeckel
38. Theocalyptra bicornis (Popofsky)
    Theocalyptra davisiana var. cornutoides Kling
39. Theocorys redondoensis (Campbell and Clark)
40. Tholospyris anthophora (Haeckel)
41. Tholospyris kantiana (Haeckel)
42. Tholospyris mammillaris (Haeckel)
43. Zygocircus spp.

TABLE 3. SPECIES WITH MAXIMUM ABUNDANCE GREATER THAN
FIVE PERCENT OF THE POPULATION

Species Group	Site of Maximum Abundance	Time of Maximum Abundance	Maximum Relative Abundance
Stichocorys wolffii	289	16Ma	50%
Calocycletta robusta/virginis	495	16	26
Stichocorys delmontensis	289	16	21
Liriospyris mutuaria	55	22	16
Disolenia/Otosphaera spp.	495	16	13
Cyrtocapsella tetrapera	495	22	11
Stylatractus spp.	173	16	11
Eucyrtidium diaphanes	289	22	10
Tholospyris kantiana	289	22	10
Cyrtocapsella cornuta	65	22	9
Stylochlamydium asteriscus	173	16	8
Theocalyptra bicornis/ davisiana var. cornutoides	173	16	8
Tholospyris mammillaris	289,55	22	7
Didymocyrtis mammifera/violina	495	16	7
Tholospyris anthophora	65	22	7
Stylosphaera spp.	173	16	7
Phorticium pylonium	71	16	6
Theocorys redondoensis	71	22	6
Eucyrtidium cienkowskii	71	16	5

struct a complete 22–16 Ma sequence. Q-mode factor analysis (CABFAC; Imbrie and Kipp 1971) was used to derive early Miocene radiolarian "assemblages." Q-mode techniques are most often used in processing paleontological data and the resulting factors group species that tend to vary in abundance together. Five factors or assemblage groups account for 86 percent of the variance in the census data (Table 4). This analysis also established an assemblage matrix that could then be used to describe additional samples (time-slice samples) in terms of the defined assemblages (Klovan and Imbrie 1971). The factor values for each site are shown in Figure 2.

Factor 1: Stichocorys Wolffii Assemblage. This factor (19.6% of faunal variation) is almost entirely described by *S.*

wolffii, a form that dominates the late Early Miocene of the tropical western Pacific (Figure 3). *S. wolffii* has long been used as a stratigraphic indicator to correlate Early Miocene tropical Pacific sediments (Westberg and Riedel 1978). However, recent work (Theyer et al. 1978; Nigrini personal communication; Romine and Lombari 1985) has shown that *S. wolffii* appears some 1 to 1.5 my earlier in western tropical Pacific sediments, where it apparently evolved. This species reaches maximum abundances of over 50 percent of the population in the late Early Miocene. No other species in this study approaches these frequencies (Table 3). Central and eastern equatorial sediments possess only moderate relative percentages of *S. wolffii.* Holdsworth (1975) suggested that *S. wolffii* might be a dimorph of *S. delmontensis* and that its abundance is determined by local environmental factors. I have not observed any obvious geographic morphological variations of *S. wolffii.* The only other species which contributes significantly to Factor 1 is the combined group of *Calocycletta robusta* and *C. virginis.* This group has the second highest maximum relative abundance of the species used herein and these maxima are also reached in the latest early Miocene.

Factor 2: Temperate Assemblage. This assemblage (19.9% of faunal variation) is primarily described by Radiolaria of the suborder Spumellaria. The species found in this factor are commonly considered to be associated with subtropical to temperate water masses in the North Pacific. *Stylatractus* spp., one of the more important groups in this assemblage, was shown to be most significant in the subtropics (Romine and Moore 1981). Although cosmopolitan in distribution, *Stylochlamydium asteriscus* lives in intermediate waters (McMillen and Casey 1978) and may be associated with the equatorial undercurrent (Molina-Cruz

TABLE 4. SPECIES DISTRIBUTION FROM Q-MODE ANALYSIS OF TIME SERIES SAMPLES

		1	2	3	4	5
				Factors (species no. – factor value)		
PRIMARY SPECIES		37 – 6.3	15 – 3.2	39 – 3.0	35 – 5.9	24 – 3.7*
		19 – 1.4	16 – 2.7	43 – 3.0*	19 – 1.2	41 – 2.8
			28 – 2.3	19 – 2.2*	11 – 1.2	42 – 2.7
			24 – 2.0	25 – 1.7	3 – 1.0	23 – 1.8
			12 – 1.8*	7 – 1.4		27 – 1.5
			18 – 1.7	16 – 1.4		29 – 1.4*
			9 – 1.6	10 – 1.3		40 – 1.3*
			2 – 1.4	40 – 1.1		
			8 – 1.0	28 – 1.1		
				29 – 1.0		
SECONDARY SPECIES			21 – 0.7	12 – 0.8	4 – 0.8	31 – 0.6
				13 – 0.8	26 – 0.7	
				14 – 0.8		
				22 – 0.8		
				30 – 0.8		
Variance accounted for		19.6%	19.9	17.2	19.2	10.1

*Denotes species highest factor value.
(For species name refer to species number on Table 2.)

FACTOR SCORES

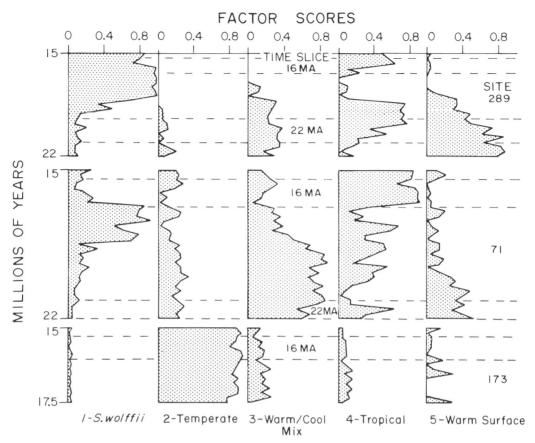

Figure 2. Factor scores from Q-mode analysis of time-series data.

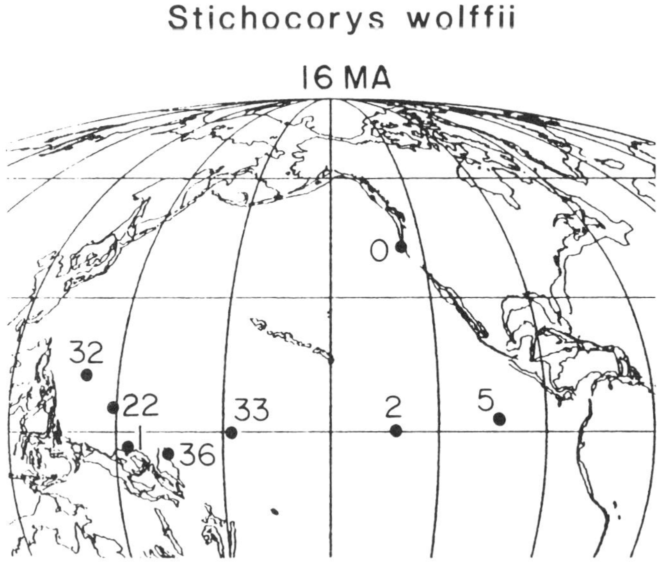

Stichocorys wolffii

16 MA

Figure 3. *Stichocorys wolffii* average percentage of assemblage (16 Ma).

1977). *Theocalyptra bicornis* and *T. davisiana* var. *cornutoides* are members of the suborder Nassellaria and are most abundant in subpolar waters and regions of coastal upwelling (Lombari and Boden 1985). Another species significantly contributing to this factor is *Spongopyle osculosa* which has been shown to live in deep water (McMillen and Casey 1978), reaching maxima in the transition zone of the Central Pacific (Casey 1971).

Factor 3: Tropical-Subtropical Assemblage. This assemblage (17.2 percent of faunal variation) is composed of Nassellarian species such as *Zygocircus* spp., *Calocycletta robusta* and *C. virginis, Dendrospyris damaecornis* and *D. pododendros, Giraffospyris circumflexa,* and a Spumellarian, *Hymeniastrum euclidis.* These species are known to be subtropical to tropical in distribution. Some of the secondary species describing this factor, *Ceratocyrtis stigi, Spongurus* (?) sp. A and B and *Spongotrochus glacialis,* are cooler water species. This mixture of warm and cool water dwellers is perhaps representative of a tropical fauna influenced by cooler upwelled waters.

Factor 4: Stichocorys delmontensis Assemblage. *Stichocorys delmontensis* is the primary component of this assemblage (19.2 percent of faunal variation) along with *Phorticium pylonium, Didymocyrtis mammifera* and *D. violina* and *Calocycletta robusta* and *C. virginia.* These species appear to be associated with tropical water masses. *S. delmontensis* reaches maxima of over 20 percent in the equatorial western Pacific of the Early Miocene and has a relatively high abundance of all equatorial regions (6–12 percent) of the total population) as seen in Figure 4a, b. This species became more restricted to the subarctic by the late Miocene (Figure 4c). *S. delmontensis* also occurred in high frequencies in the Subantarctic during the Late Miocene (Petrushevskaya 1975).

Factor 5: Cyrtocapsella-Spyroid Assemblage. This factor (10.1 percent of faunal variation) is described by a group of nassellarian species that range throughout the entire span of the time-series. The first seven species listed for this factor on Table 4 all evolved in the earliest Miocene and reached their maximum abundances within the 22–16 Ma interval. *Liriospyris geniculosa, L. mutuaria, Tholospyris anthophora, T. kantiana* and *T. mammillaris* belong to the same family, *Trissocyclidae* (commonly called spyroids). Modern representatives of this family appear to live symbiotically with zooxanthellae (Casey, personal communication). Since this ecological relationship would require that this assemblage live in the photic zone, it is assumed that this fauna is indicative of surface waters.

Time-Slices

Using a "transfer-function" technique developed by Klovin and Imbrie (1971) the radiolarian assemblage present in each time-slice sample was mathematically compared to the assemblage composition of each factor from the time-series analysis. Factor values were assigned to the time-slice samples based on the compositional similarity of their assemblages to those of the time-series. Thus, the following results compare each time-slice to the entire early Miocene record (see Table 5 for time-slice factor values).

22 Ma Time-Slice. Sites 448, 55, 289, and 65 are most closely associated with the Spyroid assemblage (Factor 5) described by the time-series analysis (Figure 5a). This is evidence that the spyroid assemblage favors the western equatorial Pacific at 22 Ma. Warm surface waters during the early Miocene may be associated with the success of this group. The combined relative abundance of the spyroids reaches 30 percent during the 22 Ma time-slice (Figure 6a).

Site 71 has highest factor scores in association with the Tropical-Subtropical Assemblage (Factor 3) (Figure 5b). As previously stated, this assemblage is a mixture of warm and cool water species. At 22 Ma, Site 71 is equatorial and it is evident from the fauna that cooler upwelled waters were present.

Site 495 has maximum values associated with the *S. delmontensis* assemblage (Factor 4) (Figure 5c). Sites 65 and 289 also show rather strong affinity to this assemblage that suggests that the assemblages represented by both Factors 4 and 5 from the time-series are definitely tropical assemblages. The exception of Site 71 is easily explained by equatorial divergence that should not affect the other equatorial sites.

Factors 1 and 2 were not representative of any 22 Ma time-slice material. This is not surprising since *S. wolffii* (Factor

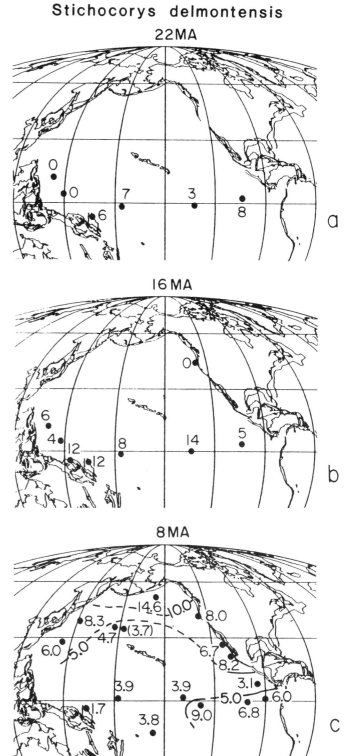

Figure 4. *Stichocorys delmontensis* average percentage of assemblages (a. 22 Ma, b. 16 Ma, c. 8 Ma). The data for the 8 Ma time-slice are from Romine (1985).

22 MA
FACTOR 5

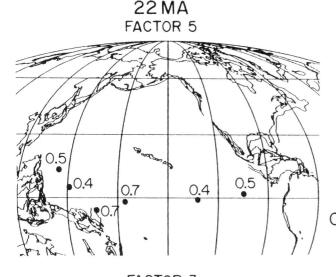

FACTOR 3

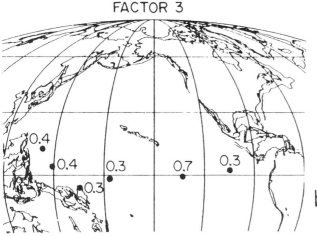

FACTOR 4

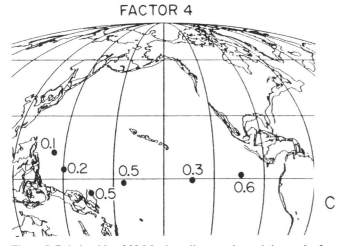

Figure 5. Relationship of 22 Ma time-slice samples and time-series factors (a. Factor 5, b. Factor 3, c. Factor 4).

1) had not yet become dominant and there was no temperate site in this data set to reflect Factor 2.

16 Ma Time-Slice. Time-slice samples from Site 173 were associated very strongly with the temperate assemblage (Factor 2) (Figure 7a). This is not surprising since the temperate assemblage is defined in the time series by samples from Site 173 (Figure 2).

The *S. wolffii* assemblage (Figure 7b) controls the western tropical sites at 16 Ma. The dominance of *S. wolffii* in this region overshadows all other influences. Site 63A, also a western equatorial site, is not in agreement with neighboring sediments. This disagreement may represent a taxonomic problem or unrepresentative sampling of the time-slice interval. Riedel and Sanfilippo (1971) found *S. wolffii* to be common (greater than 10 percent) in most samples from the *Calocyletta costata* zone which spans 15.5 to 17.0 Ma, suggesting that this site actually does contain an abundance of *S. wolffii* that is consistent with those nearby.

The Tropical-Subtropical Assemblages (Figure 7c) characterizes Site 63A as well as Sites 71 and 495. At 16 Ma, Site 71 lost some of the cooler water species that were present at 22 Ma making the 16 Ma assemblage at Site 71 very similar to that of the Tropical-Subtropical Assemblage from the time-series analysis.

DISCUSSION

From an examination of the census data collected for this study and data supplied by Lombari and Boden (1982), long-term biogeographic trends could be detected. Figures 8a and b (from Lombari and Boden (1982) illustrate the pattern of increasing spumellarian dominance with time. This trend is especially apparent in the tropical Pacific. In the relatively stable equatorial Pacific of the early Miocene, Nassellaria were important in describing the assemblages. The *S. wolffii* Assemblage and the Spyroid Assemblage (Factors 1 and 5) are entirely nassellarian dependent. The Tropical-Subtropical Assemblage and the *S. delmontensis* Assemblage (Factors 3 and 4) are dominated by Nassellaria. Only the Temperate Assemblage (Factor 2) is dominated by Spumellaria. The dominant species of Factor 2 are extant including the two Nassellaria important to this factor, *T. bicornis* (with *T. davisiana* var. *cornutoides*) and *Ceratocyrtis histricosa*. Of the 43 species groups used in this analysis, 18 belong to the Spumellaria and 25 to the Nassellaria. By the late Miocene one of the spumellarian groups, *D. mammifera* (with *D. violina*), became extinct, while 18 of the Nassellaria were extinct. In the modern ocean all of the remaining 17 spumellarian groups are extant (see Nigrini and Moore 1979 for illustrations of modern Radiolaria), but only four of the nassellarian species survive, including the two species from Factor 2. This translates to a 94 percent survival rate for the Spumellaria and a 16 percent survival rate for the Nassellaria. The spumellarian species groups used for this study are mainly cooler-water forms and have remained stable throughout the Neogene. It appears that cool-water dwelling radiolarians had the advantage of more stable niches or were

Spyroids

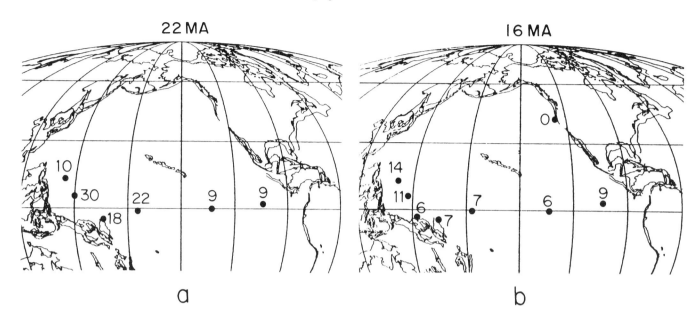

Figure 6. Combined Spyroids average percentage of assemblages (a. 22 Ma, b. 16 Ma).

more easily able to adapt to the changing environmental conditions during the Neogene. It is obvious from the dramatic difference in the survival rates of the two suborders of polycystine Radiolaria that Nassellaria were not as successful in adapting to changes in their environment.

After 16 Ma, Nassellaria did not contribute to the radiolarian assemblages to the degree that they did in the early Miocene. Plankton provinces in the Pacific show great changes between 18 and 12 Ma (Sancetta 1978) and Radiolaria were not exceptions. The changing Neogene oceans caused extinctions, migrations, and, in general, reorganization of the radiolarian assemblages. Figure 9 (from Lombari and Boden 1982) illustrates that at 16 Ma the tropical Pacific has a minimum number of species when compared to the entire late Neogene record. This is not an artifact of dissolution processes, since relatively dissolution susceptible species, such as *S. wolffii* (Moore 1969) were found in great abundance. This is most likely due to loss of nassellarian species. Since 16 Ma, diversity increased, probably due to an increase in Spumellaria.

The western tropical Pacific appears to have favored the development of the spyroids (describing Factor 5) where they reached a combined maxima of up to 30 percent of the population. This group of Nassellaria are also known to make up a large proportion of the early Miocene fauna in the equatorial Indian Ocean (Srinivasan and Lombari 1983). The spyroids are one of the most diverse families of Radiolaria, with rapid evolution taking place during the Miocene. One genus, *Tholospyris*, has at least eleven taxa co-existing during the Late Miocene (Goll 1969, 1972). By late early Miocene the spyroids contributed 13 percent to the total population in the western tropics (Figure 6b). This sharp decrease is not due to extinctions because the spyroids remained relatively unchanged at Site 495 in the eastern Pacific.

Other species groups, such as *Eucyrtidium cienkowskii* (Figure 10a, b) also decreased in abundance in the western tropical Pacific. As seen from Factor 1, by 16 Ma *S. wolffii* was the dominant species in the western tropics. One possibility is that as *S. wolffii* migrated in greater abundance to this region other species that had been successful at 22 Ma moved elsewhere.

TABLE 5. Q-MODE FACTOR LOADINGS OF TIME-SLICE SAMPLES
ON THE TIME-SERIES MATRIX

		Factors (Site: Age: Factor Score)			
1	2	3	4	5
55: 16Ma: 0.9	173: 16Ma: 0.9	71: 22Ma: 0.7	63A: 16Ma: 0.7	65: 22Ma: 0.7
65: 16Ma: 0.9		55: 22Ma: 0.4	71: 16Ma: 0.9	239: 22Ma: 0.7
289: 16Ma: 0.9		448: 22Ma: 0.4	495: 22Ma: 0.6	448: 22Ma: 0.5
448: 16Ma: 0.9		63A: 16Ma: 0.4	495: 16Ma: 0.5	495: 22Ma: 0.5
495: 16Ma: 0.4			289: 22Ma: 0.5	55: 22Ma: 0.4
			65: 22Ma: 0.5	

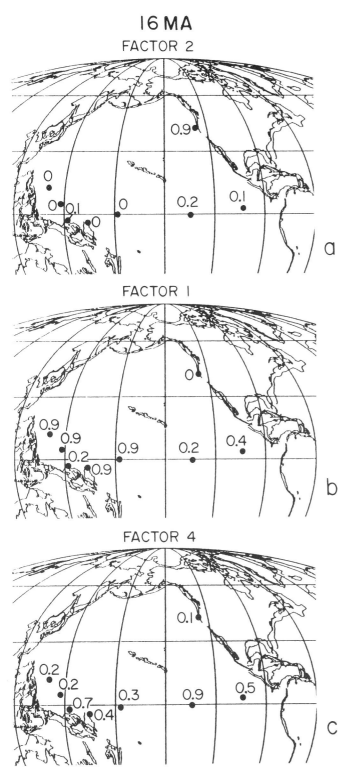

16 MA

FACTOR 2

a

FACTOR I

b

FACTOR 4

c

Figure 7. Relationship of 16 Ma time-slice samples and time-series factors, (a. Factor 2, b. Factor 1, c. Factor 4).

Another consideration is that if *S. wolffii* was indeed a deeper-dweller (as evidenced by its low frequencies in shallow water Site 63A), it could have been exploiting an influx of nutrients to intermediate waters (van Andel et al. 1975). Meanwhile, surface circulation became more restricted through the Indonesian Seaway and species favoring these waters were forced to adapt, migrate, or face extinction. By the late Miocene, *S. delmontensis* appears to have migrated to the subpolar oceans and *S. wolffii*, which had been so successful during the early Miocene, became extinct.

Eastern tropical Pacific radiolarian assemblages also experienced some changes. This region (Site 71) was influenced by cool westward flowing upwelled water at 22 Ma, as evidenced by the mixed assemblage of Factor 3. The Peru-Chile current had developed by this time (Sancetta 1978) although this was probably still weak because of low latitudinal thermal gradients. This current was most likely the source of the cooler water dwellers found at Site 71. Site 495 in the far eastern Pacific was in the path of warm westerly flowing Atlantic waters that probably diverted the upwelled water toward Site 71. By 16 Ma, the eastern upwelling system was more restricted. Although upwelling increased by the middle Miocene (van Andel et al. 1975), the cooler water Radiolaria found at Site 71 during the 22 Ma time-slice did not contribute to the assemblage found there at 16 Ma, probably because of paleoceanographic changes related to the more restrictive Indo-Pacific circulation. This closure eventually caused the formation of the eastward flowing Equatorial Undercurrent (Cromwell Current). A general absence of cooler-water forms was taken up by higher frequencies of tropical forms such as *Calocycletta virginis.* The spyroids appear to have followed an eastern migratory route, since this group is distributed in the eastern tropics in the modern ocean (Lombari and Boden 1985).

One of the most striking observations made in examining the census data collected for this study is the small amount of change that occurred in the cool water Spumellaria during the Miocene. These forms underwent the least change through time of any radiolarian group. This is partly due to the fact that a well developed temperate assemblage was formed by the early Miocene. A pattern of spumellarian dominance began in the latest early Miocene as nassellarians became less successful. This trend continued through the Neogene to the Modern ocean.

During the earliest Miocene, there arose many nassellarian species and one of these, *S. wolffii*, reached maxima unparalleled by any other Neogene Radiolaria. However, the middle Miocene was a time of nassellarian extinctions, creating a Neogene low in radiolarian diversity at 16 Ma. What happened in the Miocene to permanently alter the nassellarian component of the radiolarian assemblages? The disruption of Indo-Pacific circulation is the most likely explanation for changes in the western tropics. Warm water surface dwellers such as the spyroids migrated eastward, and deeper dwellers such as *S. delmontensis* moved into the subpolar regions. Other species, such as *S. wolffii*, became extinct.

In the eastern tropics, cool water forms of Radiolaria found at 22 Ma had disappeared by 16 Ma. Upwelled waters were more

G. Lombari

DSDP SITES

NO. OF SPUMELLARIAN SPECIES ÷ NO. OF NASSELLARIAN SPECIES

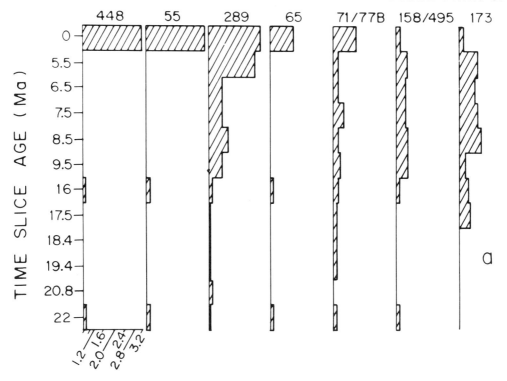

% OF SPUMELLARIAN INDIVIDUALS ÷ % NASSELLARIAN INDIVIDUALS

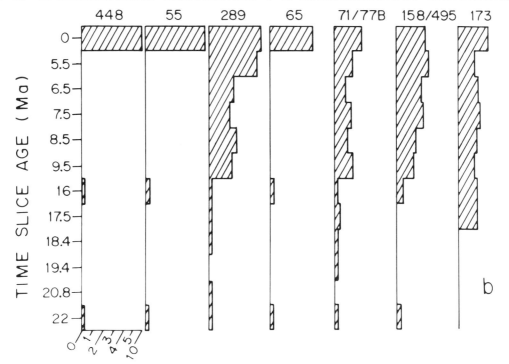

Figure 8. a. Ratio of Spumellarian species: Nassellarian species. b. Ratio of Spumellarian individuals: Nassellarian individuals.

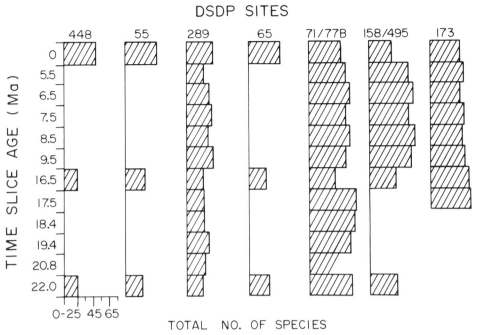

Figure 9. Total number of species through the Neogene.

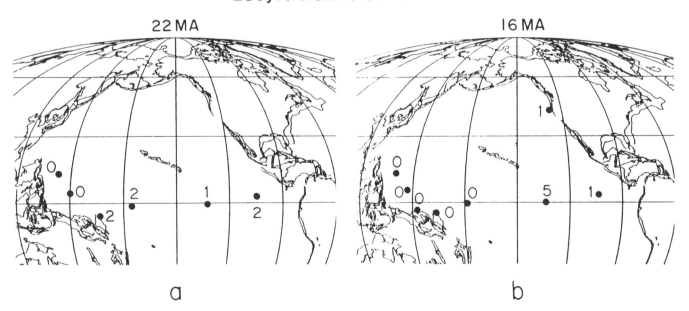

Figure 10. *Eucyrtidium cienkowskii* average percentage of assemblages (a. 22 Ma, b. 16 Ma).

restricted at 16 Ma. Tropical Radiolaria migrating out of the western Pacific found suitable conditions for survival in the eastern Pacific at this time.

As seen in Figures 8 and 9, major changes occurred in radiolarian faunal composition between 16 and 9.5 Ma. Further investigation is needed to explore the causes of these changes and to determine how individual radiolarian species responded.

CONCLUSIONS

1. Radiolarian assemblages have been examined in two equatorial Pacific time-slices (22 and 16 Ma) and three early Miocene time-series (DSDP Sites 71, 173, and 289). Faunal trends were detected by treating census data collected from a subset of 43 radiolarian species with a Q-mode factor analysis producing five faunal provinces.

2. The early Miocene was a time of nassellarian evolutionary radiation, while the middle Miocene was a time of nassellarian extinction. In contrast, Spumellaria were relatively stable throughout the Neogene. A decrease in radiolarian diversity at 16 Ma resulted from a loss of nassellarian species.

3. High diversity among Nassellaria is exhibited at 22 Ma in the western tropical Pacific. By 16 Ma, one species, *S. wolffii*, dominated the assemblage and other, once successful, species rapidly decreased. Changes in surface-water circulation in the western tropical Pacific restructured the radiolarian assemblages between 22 and 16 Ma.

4. Site 71 in the Central Pacific was affected by equatorial divergence at 22 Ma, but not at 16 Ma. The eastern equatorial upwelling system was more restricted at 16 Ma.

5. Many radiolarian species followed an eastern migratory route as western Pacific circulation changed and central Pacific waters warmed.

ACKNOWLEDGMENTS

Sincere thanks go to T. Moore for the opportunity to complete this study and to J. Kennett for his many helpful suggestions and advice. Appreciation is also given to C. Nigrini and J. P. Caulet for their useful reviews of this manuscript.

REFERENCES CITED

Barron, J. A. et al., this vol., A multiple microfossil biochronology for the Miocene.

Blow, W. H., 1969, Late middle Eocene to Recent planktonic foraminiferal biostratigraphy, *in* Bronniman, P. and Renz, H., eds., First International Conference on Planktonic Microfossils, Geneva, 1967, p 199–421.

Casey, R. E., 1971, Radiolaria as indicators of past and present water masses, *in* Funnell, B. M. and Riedel, W. R., eds., The Micropaleontology of Oceans: London, Cambridge University Press, p. 331–342.

Casey, R. E. et al., 1979, Radiolarian ecology and development of the radiolarian component in Holocene sediments, Gulf of Mexico and adjacent seas with potential paleontological applications: Transactions of the Gulf Coast Association Geological Society, v. 29, p. 228–237.

Dow, R. L., 1978, Radiolarian distribution and the Late Pleistocene history of the southeastern Indian Ocean: Marine Micropaleontology, v. 3, p. 203–227.

Goll, R. M., 1969, Classification and phylogeny of Cenozoic Trissocyclidae (Radiolaria) in the Pacific and Caribbean basins, part II: Journal of Paleontology, v. 43, p. 322–339.

——1972, Systematics of eight *Tholospyris* taxa (Trissocyclidae; Radiolaria); Micropaleontology, v. 18, p. 338–349.

Holdsworth, B. K., 1975, Cenozoic radiolarian biostratigraphy: Leg 30: Tropical and equatorial Pacific, *in* Andrews, J. E. et al., Initial reports of the Deep Sea Drilling Project, v. 30: Washington, D.C., U.S. Government Printing Office, p. 449–537.

Imbrie, J. and Kipp, N., 1971, A new micropaleontological method for quantitative paleoclimatology: Application to a Late Pleistocene Caribbean core in Turekian, K., ed., Late Cenozoic Glacial Ages: New Haven and London, Yale University Press, p. 71–182.

Johnson, D. A., and Nigrini, C. A., 1980, Radiolarian biogeography in surface sediments of the western Indian Ocean: Marine Micropaleontology, v. 5, p. 111–152.

——1982, Radiolarian biogeography in surface sediments of the eastern Indian Ocean: Marine Micropaleontology, v. 7, p. 237–281.

Kennett, J. P., 1982, Marine Geology: Prentice Hall, Englewood Cliffs, New Jersey, 813 p.

Klovan, J. E., and Imbrie, J., 1971, An algorithm and Fortran IV program for large-scale Q-mode factor analysis: Journal of the International Association of Mathematical Geologists, v. 3, p. 61–67.

Lombari, G., and Boden, G., 1982, Radiolarian paleobiogeography and diversity: Recent vs. Miocene: Geological Society of America Abstracts with Program, v. 14, p. 548.

——1985, Modern radiolarian global distributions: Cushman Foundation for Foraminiferal Research, Special Publication No. 16A.

Maurrasse, F., 1978, Cenozoic radiolarian paleobiogeography: Implications concerning plate tectonics and climatic cycles: Paleogeography: Paleoclimatology, and Paleoecology, v. 26, p. 253–289.

McMillen, K., and Casey, R. E., 1978, Distribution of living polycystine Radiolaria in the Gulf of Mexico and Caribbean Sea, and comparison with the sedimentary record: Marine Micropaleontology, v. 3, p. 121–145.

Molina-Cruz, A., 1977, Late Quaternary oceanic circulation along the Pacific coast of South America [Ph.D. thesis]: Corvallis, Oregon, Oregon State University, 246 p.

Moore, T. C., Jr., 1969, Radiolaria: Change in skeletal weight and resistance to solution: Geological Society of America Bulletin, v. 80, p. 2103–2108.

——1973, Method of randomly distributing grains for microscopic examination: Journal of Sedimentary Petrology, v. 43, p. 904–908.

——1978, The distribution of radiolarian assemblages in the modern and ice-age Pacific: Marine Micropaleontology, v. 3, p. 229–266.

Moore, T. C., Jr., et al., 1980, The reconstruction of sea-surface temperatures in the Pacific Ocean of 18,000 B.P.: Marine Micropaleontology, v. 5, p. 215–247.

Nigrini, C. A., and Lombari, G., 1984, A Miocene Guide to Radiolaria: Cushman Foundation for Foraminiferal Research, Special Publication No. 22.

Nigrini, C. A., and Moore, T. C., Jr., 1979, A Modern Guide to Radiolaria: Cushman Foundation for Foraminiferal Research, Special Publication No. 16.

Petrushevskaya, M., 1975, Cenozoic radiolarians of the Antarctic, Leg 29, DSDP, in Kennett, J. P. et al., Initial reports of the Deep Sea Drilling Project, Vol. 29: Washington, D.C., U.S. Government Printing Office, p. 541–675.

Riedel, W. R. and San Filippo, 1971, Cenozoic Radiolaria from the western tropical Pacific, Leg 7, *in* Winterer, E. L. et al., eds., Initial Reports of the Deep Sea Drilling Project, Vol. 7, Washington, D.C., U.S. Government Printing Office, p. 1529–1672.

Romine, K., 1985, Radiolarian biogeography and paleoceanography of the North Pacific at 8 Ma, *in* Kennett, J. P., ed., The Miocene ocean: Paleoceanography and biogeography: Geological Society of America Memoir 163.

Romine, K., and Lombari, G., 1985, Evolution of Pacific circulation in the Miocene: Radiolarian evidence from Site 289, *in* Kennett, J. P., ed., The Miocene ocean: Paleoceanography and biogeography: Geological Society of America Memoir 163.

Romine, K., and Moore, T. C., Jr., 1981, Radiolarian assemblage distributions and paleoceanography of the eastern equatorial Pacific Ocean during the last 127,000 years: Paleogeography, Paleoclimatology, and Paleoecology, v. 35, p. 281–314.

Sachs, H. M., 1973, North Pacific radiolarian assemblages and their relationship to oceanographic parameters: Quaternary Research, v. 3, p. 73–88.

Sancetta, C., 1978, Neogene Pacific microfossils and paleoceanography: Marine Micropaleontology, v. 3, p. 347–376.

Shackleton, N. J., and Kennett, J. F., 1975, Paleotemperature history of the Cenozoic and the initiation of Antarctic glaciation: Oxygen and carbon isotope analyses in DSDP Sites 277, 279, and 281, *in* Kennett, J. P. et al.

eds., Initial Reports of the Deep Sea Drilling Project: Washington, D.C., U.S. Government Printing Office, v. 29, p. 743–755.

Srinivasan, M. S., and Lombari, G., 1983, Early Miocene planktonic foraminiferal and radiolarian zonation, Colebrook Island, Andaman Sea: Journal Geological Society of India, v. 3, p. 1–18.

Theyer, F. et al., 1978, Paleomagnetic and geochronologic calibration of latest Oligocene to Pliocene radiolarian events, equatorial Pacific: Marine Micropaleontology, v. 3, p. 377 395.

van Andel, Tj. H. et al., 1975, Cenozoic history and paleoceanography of the central equatorial Pacific Ocean: Geological Society of America Memoir 143, 134 p.

Westberg, J. and Riedel, W. R., 1978, Accuracy of radiolarian correlations in the Pacific Miocene: Micropaleontology, v. 24, p. 1–23.

MANUSCRIPT ACCEPTED BY THE SOCIETY DECEMBER 17, 1984

Geological Society of America
Memoir 163
1985

The post-Eocene sediment record of DSDP Site 366:
Implications for African climate and plate tectonic drift

*Ruediger Stein**
Geologisch-Palaeontologisches Institut
Universitaet Kiel
Olshausenstr. 40, 2300 Kiel
F.R. Germany

ABSTRACT

The post-Eocene sediment record of Site 366, eastern equatorial Atlantic, provides important information about the evolution of paleoenvironments at the Sierra Leone Rise, which changed in response to the continuous northward drift of the African plate and the history of African climate. Changes in the accumulation of biogenic opal and the composition of the terrigenous sediment fraction clearly reflect the motion of Site 366 from the Southern to the Northern Hemisphere during Oligocene/Miocene times.

Distinct maxima of eolian sediment supply and coarser grain-sizes suggest development of an intensified meridional wind circulation and a more arid to semi-arid climate in the South Saharan and Sahelian regions towards the end of the Miocene between 6 and 5 Ma and during the last 2.5 Ma.

INTRODUCTION

This paper is part of a major study of Neogene sediments along the northeast Atlantic continental margin (Stein, 1984a). Based on a detailed sediment record of DSDP-Site 366 (Fig. 1), this study focuses on the Oligocene to Miocene evolution of the paleoenvironment at Site 366 as this site drifted across the equator (Sclater et al. 1977). This region is marked by contrasting depositional regimes including high and low oceanic productivity as well as a substantial dust input from the northeasterly and/or southeasterly trade winds (Figure 1A). Accordingly, the gradual northward drift of the African plate should be reflected by distinct changes of paleoenvironments and sedimentary regimes at Site 366.

Earlier studies of sediments on the Sierra Leone Rise (Dean et al. 1978; 1981; Diester-Haass 1978a; Fuetterer 1978; Shipboard Scientific Party 1978) show that deposition of eolian dust and calcareous biogenic input have largely controlled the sediment accumulation at Site 366 (Figure 1A) during Cenozoic time. The Neogene sequence of Site 366 can be compared with "Meteor" core 13519 (Fig. 1A) that has been studied in detail for its stable-isotope stratigraphy, fluctuations of terrigenous sedi-

ment supply, and calcium carbonate dissolution during the last 750,000 years (von Grafenstein 1982; Sarnthein et al. 1984).

Investigations of the terrigenous, especially eolian, sediment fraction can provide a useful record of paleoclimate and paleoatmospheric circulation. Clay minerals can be used as proxy indicators of different source areas and climatic regimes (e.g., Biscaye 1965; Griffin et al. 1968; Leinen and Heath 1981; Sarnthein et al. 1982). For example, regions with high relief and/or active tectonism resulting in enhanced erosion, as well as areas with hot-dry or cold climates with dominant physical weathering favor the formation of illite and chlorite (Chamley 1979; Robert 1982; Stein 1984a). On the other hand, kaolinite is preferentially associated with intensive chemical weathering of low latitude terrains such as the lateritic zones in equatorial Africa (Biscaye 1965; Griffin et al. 1968; Chester et al. 1972). Mass accumulation rates of terrigenous matter can provide further information about the climates of the source area (Leinen and Heath 1981; Rea and Janecek 1982; Sarnthein et al. 1982). Estimations of wind intensity can be deduced from grain sizes of the eolian sediment (Parkin 1974; Leinen and Heath, 1981; Sarnthein et al. 1981; Rea and Janecek 1982; Lever and McCave 1983).

Indicators of high productivity stored in deep-sea sediments may be higher accumulation rates of organic matter, biogenic

*Present address: Institute of Petroleum and Organic Geochemistry, KFA Jülich, P.O. Box 1913, 5170 Jülich, F.R. Germany.

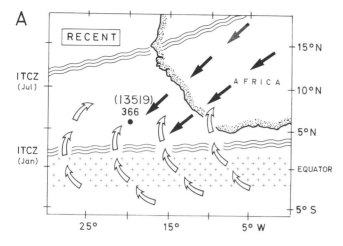

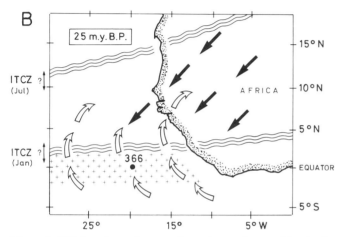

Figure 1. Wind regimes, dust supply, and oceanic upwelling in the equatorial east Atlantic. Black arrows indicate the dust supply by northeasterly trade winds during northern winter when the Inter-Tropical Convergence Zone (ITCZ) reaches its southernmost position. Open arrows indicate dust supply by the southeasterly trade winds during northern summer when the ITCZ reaches its northernmost position (southeasterly trades continuing north as southwesterly monsoonal winds after crossing the equator). Mean position of the recent ITCZ is according to Flohn (personal communication, 1983). *Crosses mark equatorial upwelling area.* A possible shift of the ITCZ would not affect the position of the center of the equatorial upwelling.
A. Scheme of the modern situation with the present position of DSDP-Site 366 (5°40.7′N/19°51.1′W, water depth: 2853 m; Shipboard Scientific Party 1978). The "Meteor" core 13519 was raised from the same position (Sarnthein et al. 1984).
B. Scheme of the Oligocene/Miocene situation. Paleolatitude of the African continent and Site 366 according to Sclater et al. (1977).

opal, and fish debris as well as low $\delta^{13}C$-values (Diester-Haass 1978b; Sarnthein et al. 1984; Mueller et al. 1983; Stein 1985).

THE MODERN ANALOGUE

Today, Site 366 lies north of the actual region influenced by equatorial upwelling that extends to about 2°N (Dietrich et al. 1975). It is currently situated in an area of low productivity. This

regime is indicated by a negligible input of biogenic opal accumulation (Figure 2D; Diester-Haass 1978a, b; von Grafenstein 1982; Sarnthein et al. 1984) and $\delta^{13}C$-values of benthic foraminifera typical of the North Atlantic Deep Water (NADW) in the subtropical low productivity zone (Kroopnick, 1980; Sarnthein et al. 1984). The terrigenous sediment supply consists mainly of eolian dust because the most proximal part of the Sierra Leone Rise lies 800 km offshore and is separated from the African continental margin by the Kane Gap (4600 m water depth), which prevents turbidity currents from reaching the rise. The eolian sediment flux is controlled by both the northeasterly and the southeasterly trade winds. During the northern winter, when the Inter-Tropical Convergence Zone (ITCZ) reaches its southernmost position, Site 366 lies in the belt of the northeasterly trade winds that bring the bulk of dust from the South Saharan and Sahelian zones (Figure 1A). Alternatively, during the northern summer when the ITCZ assumes a more northerly position (Figure 1A), additional eolian sediment is supplied in minor quantities by the southeasterly trade winds from the Namib and Kalahari regions. Northeasterly eolian sediment supply is dominant because the primary dust loading of the northeasterly trade winds is more than an order of magnitude higher than that of the southeasterly trade winds (Chester et al. 1972). Furthermore, the distance between the Kalahari region and the Sierra Leone Rise (about 6000 km) is much greater.

Fortunately, the different composition of northeasterly and southeasterly trade wind dust has enabled us to distinguish the dominant eolian transport system in the eolo-marine sediments (Chester et al. 1972; Aston et al. 1973; Parkin and Padgham 1975; Whalley and Smith 1981; McTainsh and Walker, 1982; Robert 1982; Sarnthein et al. 1982; Wilke et al. 1983). The northeasterly trade wind dust at the Sierra Leone Rise has its origin in the South Sahara and Sahel zones and is thus characterized by reddish stained quartz, a high kaolinite content, and relatively coarse grain sizes due to the shorter distance of transport. In contrast, dust supplied by the southeasterly trade winds is marked by abundant illite and much finer grain sizes.

MATERIAL AND METHODS

The 220 samples investigated were taken discontinuously, concentrating on the Oligocene and late Miocene to middle Quaternary time intervals. Samples were washed through a 63 μm sieve. The coarse sediment fraction was dried at 40°C. Benthic foraminiferal species *Uvigerina peregrina, Cibicides wuellerstorfi,* and *Pyrgo murrhina* were picked for oxygen- and carbon-isotope measurements from the >250 μm size fraction. Preparation of foraminiferal tests and isotope analysis were according to Ganssen (1983). The isotope data supplement the isotopic curves of Blanc (1981) and Savin et al. (1981) and are discussed in detail in Stein (1984a). The content of biogenic opal of the coarse fraction was determined semi-quantitatively under a microscope using the following classifications: abundant (25% to 50%), common (5% to 25%), rare (1% to 5%), and traces (<1%). Calcium

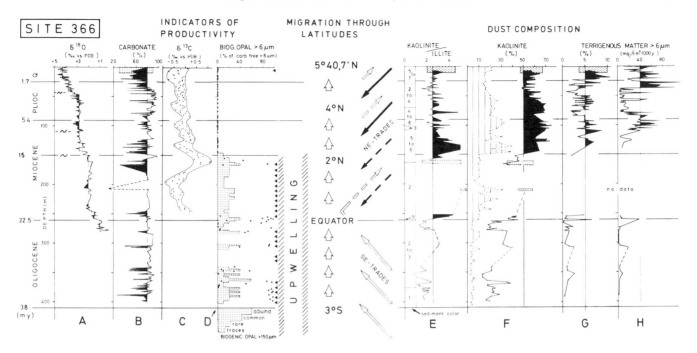

Figure 2. Long-term evolution of the biogenic and terrigenous sediment record of Site 366 in response to plate tectonic drift. Paleoposition of Site 366 according to Sclater et al. (1977). The influence of the northeasterly (southeasterly) trade winds is indicated by solid (open) arrows. Hatched area marks equatorial upwelling. A. δ^{18}O-values, adjusted to *Uvigerina-peregrina*-values (Blanc 1981; Savin et al. 1981, supplemented by own data; Oligocene values from Vergnaud-Grazzini and Rabussier-Lointier 1980). Hiatuses are indicated by "~". B. Calcium-carbonate content. Dotted box marks range of carbonate fluctuations of "Meteor" core 13519 (Sarnthein et al., 1984). C. δ^{13}C-values, adjusted to *U. peregrina*-values (Blanc 1981; Savin et al. 1981, supplemented by own data). D. Content of biogenic opal in the total sediment fraction >150 μm (dotted area) and in the carbonate-free sediment fraction 6 to 150 μm (large single points). Black triangles mark intervals with amounts of freshwater diatoms (Schrader 1978). E. Ratio of kaolinite versus illite. Numbers at the left border indicate the dominant color of the carbonate-free clay fraction. Number code (according to rock-color chart, GSA 1975): 1: yellowish gray (5Y8/1), 2: yellowish gray (5Y7/2), 3: light olive gray (5Y6/1), 4: pale yellowish brown (10YR6/2), 5: light brown (5YR6/4), 6: moderate brown (5YR4/4), 7: dark yellowish brown (10YR4/2). F. Content of kaolinite in the total clay minerals. Dotted boxes in E and F mark range of short-term fluctuations in early Miocene intervals (Lange unpublished) and in the Quaternary section ("Meteor" core 13519; von Grafenstein 1982). Dashed curve indicates data from Robert (1982). G. Terrigenous sediment fraction >6 μm in percentage of the total carbonate-free sediment fraction. Dotted box indicates range of Quaternary fluctuations ("Meteor" core 13519; von Grafenstein 1982). H. Mass accumulations rates of the terrigenous sediment fraction >6 μm.

carbonate was determined by infrared absorption of CO_2 released by phosphoric acid treatment.

The sediment fraction less than 63 μm was treated with hydrogen peroxide and acetic acid in order to remove the organic matter and carbonate, respectively. The grain sizes of the carbonate-free sediment fraction of samples from middle Miocene to Quaternary time intervals, (i.e., from sediments with a negligible content of biogenic opal. See Table 2, Figure 2D) were measured using a SediGraph (Stein 1984b). After grain-size analysis, the carbonate-free sediment fraction was separated into silt and clay fractions (>6 μm, 2 to 6 μm, <2 μm) with the Atterberg method (Mueller 1967). Silt fractions were studied by smear-slide counts under a microscope to determine their content of biogenic opal (method according to Koopmann 1981). The mass accumulation rates (MAR) of the terrigenous matter (as-

sumed as carbonate-free sediment fraction minus biogenic opal) were calculated according to Thiede et al. (1982), using the porosity (Po) and wet bulk density (ρ_{wet}) data from Trabant (1978):

$$MAR \ (g \cdot cm^{-2} \ 10^{-3} \ y^{-1}) = SR \times \rho_{dry}$$

$$= SR \times \left(\rho_{wet} - 1.026 \times \frac{Po}{100} \right)$$

where SR = sedimentation rates (cm $\cdot$ 10^{-3} y^{-1}), ρ_{dry} = dry bulk density (g $\cdot$ cm^{-3}), and 1.026 = density of interstitial water (g $\cdot$ cm^{-3}). The mass accumulation rates of the terrigenous matter (MART) can then be calculated by apportioning the MAR according to the concentration data of the terrigenous matter:

$$MART = MAR \times \frac{\% \ terrigenous \ matter}{100}$$

TABLE 1. SEDIMENTATION RATES OF SITE 366*

Depth (m)	Age (my. BP.)	SR (cm/1000 y.)
0.00	0.000	
		1.78
0.24	0.0135	
		1.00
0.38	0.027	
		1.57
0.85	0.057	
		1.08
0.98	0.069	
		1.48
1.84	0.127	
		1.08
2.77	0.213	
		1.41
3.70	0.279	
		1.24
4.16	0.316	
		1.48
5.06	0.377	
		1.41
5.58	0.414	
		1.94
6.22	0.447	
		0.89
6.85	0.518	
		2.19
7.44	0.545	
		1.56
7.58	0.554	
		1.34
8.25	0.604	
		1.16
8.84	0.655	
		1.60
9.16	0.675	
		1.44
9.68	0.711	
		1.14
9.92	0.732	
		1.29
10.50	0.780	
		Hiatus
10.50	0.950	
		1.82
20.50	1.500	
		1.22
25.00	1.870	
		3.48
36.50	2.200	
		1.06
45.00	3.000	
		2.22
55.00	3.450	
		1.71
75.00	4.620	
		1.67
83.00	5.100	
		3.00
92.00	5.400	
		2.57
110.00	6.100	
		Hiatus
110.00	10.750	
		1.22
132.50	12.600	
		0.90
137.00	13.100	
		1.00
141.00	13.500	
		0.94
148.50	14.300	
		Hiatus
148.50	17.000	
		2.04
230.00	21.000	
		2.33
265.00	22.500	
		1.00
300.00	26.000	
		1.44
357.50	30.000	
		1.40
385.50	32.000	
		0.50
415.00	38.000	

*Used for calculation of the mass accumulation rates. The data of the uppermost 10.5 m sub-bottom depth are from "Meteor" core 13519.

TABLE 2. CLAY MINERALS AND AMOUNTS OF BIOGENIC OPAL >6 μM*

Sample	Depth	K	I	M	K/I	Opal	Zeol.
2 1 088	6.89	57	21	22	2.74	0.5	-
2 3 092	9.93	58	17	25	3.53		
2 5 085	12.86	65	22	14	2.98		
3 1 84	16.35	55	23	23	2.44	0.2	-
3 2 77	17.78	52	19	29	2.77		
3 2 144	18.45	60	21	19	2.79	0.7	-
3 3 80	19.31	58	26	16	2.24	-	-
3 3 145	19.96	57	20	23	2.93	0.3	-
3 4 78	20.79	55	18	27	2.97	0.3	-
3 4 143	21.44	60	23	17	2.64		
3 5 90	22.41	57	20	23	2.78		
3 6 26	23.27	59	25	17	2.36		
4 1 23	25.24	55	19	26	2.87		
4 2 70	27.21	55	22	23	2.45		
4 3 29	28.30	53	23	24	2.27	0.1	-
4 4 72	30.23	61	24	15	2.50	-	-
4 5 43	31.44	53	18	29	2.96		
4 6 28	32.77	52	23	25	2.25	0.2	-
5 1 16	34.67	52	19	29	2.70	-	-
5 1 090	35.41	60	31	9	1.93	0.2	-
5 2 037	36.68	49	16	35	3.10	-	-
5 3 029	37.80	49	33	19	1.48		
5 4 110	40.11	56	24	20	2.32		
5 5 026	40.77	51	24	25	2.12	0.2	-
5 6 043	42.44	60	27	14	2.20		
5 6 091	42.92	66	25	9	2.62		
6 1 070	44.71	59	21	20	2.76		
6 1 138	45.39	57	22	21	2.55	0.5	-
6 2 055	46.06	57	23	20	2.50		
6 3 054	47.55	60	18	22	3.33	0.2	-
6 4 091	49.42	52	26	22	2.02		
6 5 108	51.09	56	22	22	2.48		
6 6 053	52.04	56	22	22	2.54		
7 1 090	54.41	66	19	15	3.53	-	-
7 2 027	55.28	59	26	15	2.28		
7 3 120	57.71	63	25	12	2.56		
7 5 111	60.62	66	22	12	3.03	0.2	-
7 6 113	62.14	57	24	19	2.41		
8 1 016	63.17	70	23	7	3.06		
8 3 130	67.31	68	22	10	2.09	0.2	-
8 5 076	69.77	68	26	6	2.65		
9 1 111	73.62	70	30	0	2.31	0.2	-
9 4 025	77.26	68	20	12	3.31		
9 5 017	78.68	67	14	20	4.85		
9 6 086	80.78	58	20	22	2.86		
10 1 15	82.16	61	23	16	2.66		
10 1 101	83.02	59	27	14	2.17	-	-
10 1 136	83.37	61	27	12	2.23	0.4	-
11 1 118	92.69	67	20	12	3.32	0.6	-
11 2 40	93.40	72	22	7	3.34		
11 2 88	93.89	71	22	8	3.28	-	-
11 3 15	94.66	63	26	11	2.41	0.8	-
11 3 72	95.23	62	27	12	2.31	0.2	-
11 4 15	96.16	63	33	4	1.91	0.5	-
11 4 145	98.96	70	22	8	3.24	0.4	-
11 6 23	99.24	51	25	24	2.03	-	-
11 6 136	100.37	75	18	7	4.17	0.4	-
12 1 138	102.39	62	29	9	2.17	0.2	-
12 2 43	102.94	66	27	8	2.49	0.5	-
12 3 101	105.02	73	20	7	3.58	0.4	-
12 4 50	106.01	66	26	8	2.50	0.4	-
12 4 128	106.79	70	23	7	3.06	0.4	-
12 6 122	109.73	69	25	7	2.80	0.7	-
14 1 37	120.38	61	19	21	3.20	0.5	-
14 1 84	120.85	61	25	15	2.46		
14 2 80	122.31	55	17	28	3.28		
14 3 50	123.51	71	18	11	3.99	0.6	-
14 2 125	124.26	70	20	10	3.44	0.2	-
14 4 51	125.02	69	17	14	4.01	0.3	-
14 4 132	125.83	69	18	13	3.93	1.4	-
14 5 72	126.73	74	17	9	4.40	-	-
14 6 45	127.96	61	20	20	3.04	0.1	-
14 6 131	128.82	63	18	20	3.57		
15 4 35	134.36	57	11	32	5.02	1.8	-
16 3 90	142.91	65	13	22	4.98	0.3	-
17 3 25	151.76	43	15	43	2.91		
17 5 090	155.41	46	13	41	3.45	13.3	-
18 1 77	158.78	45	21	34	2.13	56.0	-
18 4 80	163.31	38	24	37	1.57		<
18 6 12	165.63	32	14	54	2.37	89.8	-
18 6 89	166.40	49	16	35	3.02	9.2	-

TABLE 2. CLAY MINERALS AND AMOUNTS OF
BIOGENIC OPAL >6 µM (continued)

Sample			Depth	K	I	M	K/I	Opal	Zeol.
23	4	63	210.64	45	14	41	3.22		
28	1	62	253.63	43	10	47	4.28	6.1	-
28	2	24	254.75	39	11	50	3.74	89.0	-
28	2	98	255.49	44	11	45	4.06	80.0	-
28	4	30	257.81	36	12	52	3.03	95.0	-<
28	4	111	258.62	36	13	52	2.79	98.0	-
28	5	71	259.72	36	14	49	2.60	91.0	-
28	6	43	260.94	45	19	36	2.33	16.9	-
29	1	95	263.46	21	13	66	1.58	60.0	36.0<
29	2	55	264.56	30	15	55	2.00	87.0	-<
29	3	64	266.15	23	12	65	1.81	76.8	20.0<
29	5	64	269.15	18	14	68	1.31	0.8	85.0<
30	1	86	272.87	20	14	67	1.43	0.9	83.0<
30	2	56	274.07	19	12	70	1.61	95.0	-<
30	3	30	275.31	26	11	64	2.44	5.0	85.0<
30	4	55	277.06	17	11	72	1.62	4.0	83.0<
30	5	58	278.59	31	12	58	2.65	55.0	28.0<
30	6	58	280.09	35	13	53	2.70	70.0	3.0<
33	1	135	301.86	31	10	59	3.08	94.0	-
33	4	72	305.73	35	12	54	2.97	85.0	-
37	2	55	340.56	21	14	66	1.46	30.0	70.0<
37	6	103	347.04	21	11	68	1.94	30.0	50.0<
38	4	28	352.79	13	16	71	0.78	0.4	81.0<
39	1	124	358.75	13	12	75	1.08		<
39	4	116	363.17	17	9	75	1.94	87.0	6.5<
5	1	14	366.15	35	15	50	2.36	96.0	-<
5	2	15	367.66	39	17	44	2.29	50.0	40.0<
5	4	10	370.61	19	18	63	1.10	3.0	90.0<
5	4	116	371.67	19	14	68	1.35	1.5	95.0<
5	6	51	374.02	22	17	61	1.30	1.0	94.0<
6	2	29	377.30	24	10	66	2.31		
6	2	140	378.41	19	13	68	1.48	96.0	1.5
6	4	63	380.64	22	10	68	2.25	95.0	3.0
6	5	37	381.88	22	14	64	1.57	97.0	1.0
6	6	61	383.63	21	10	69	2.14	96.0	1.0
7	2	12	386.63	28	14	58	1.93	90.0	-
7	4	89	390.35	20	12	68	1.65	98.0	-
8	2	50	396.51	16	9	75	1.75	95.0	-
8	3	46	397.97	20	12	69	1.70	98.0	-

*Samples from sub-bottom depths between 6.89 m and 363.17 m are from Hole 366A, samples from sub-bottom depths between 366.15 m and 397.97 m from Hole 366.
In all samples the amount of chlorite is negligible. Amounts of biogenic opal >6 µm and zeolites >6 µm are listed as percentages of the carbonate-free silt fraction > 6 µm. Left arrows (<) mark samples with significant amounts of clinoptilolite, measured by X-ray diffraction.
Key: K = Kaolinite; I = Illite; M = Montmorillonite; K/I = Kaolinite/illite-ratio.

Clay minerals were analysed from the clay fraction by X-ray diffraction following Biscaye (1965) and Lange (1982). The deviations of absolute kaolinite values between our own data and the clay mineral data of Robert (1982) (Figure 2F) were caused by different techniques; however, both curves show a general parallelism. On the other hand, these differences must be considered, if the MAR of the clay minerals are discussed.

STRATIGRAPHY AND SEDIMENTATION RATES

Stratigraphic ages are based on planktonic foraminiferal and coccolith datum levels (Bukry 1978; Cepek et al. 1978; Krasheninnikov and Pflaumann 1978a, b; Pflaumann and Krasheninnikov 1978; Samtleben 1978). The ages used for the determination of sedimentation rates are according to Berggren and van Couvering (1974), Berggren et al. (1980), Haq et al. (1980), Keller (1981), Thunell (1981), and Backman and Shackleton (1983).

According to Krasheninnikov and Pflaumann (1978a), the core interval between 415 and 265m sub-bottom depth is of Oligocene age. The Oligocene/Miocene boundary occurs in 265m sub-bottom depth, the first appearance datum (FAD) of *G. kugleri* and *D. deflandrei*. The transition between the early and middle Miocene is not documented in the sediment sequence of Site 366. The absence of the *P. glomerosa* zone (N8) and the *O. saturalis/G. peripheroronda* zone (N9) (Krasheninnikov and Pflaumann 1978b) suggest a hiatus in 148.5m sub-bottom depth, lasting from about 17 to 14.3 Ma (Table 1). According to the nannofossil stratigraphy used for the middle Miocene time interval, a further distinct hiatus occurs in about 110 m sub-bottom depth between the *D. quinqueramus* zone and the *D. hamatus* zone (Samtleben personal communication). This hiatus lasted from 10.75 to 6.1 Ma (Table 1). The Miocene/Pliocene boundary is recorded in 92 m sub-bottom depth (FAD *G. dehiscens*), the Pliocene/Pleistocene boundary in about 25 m sub-bottom depth (FAD *G. truncatulinoides,* LAD [last appearance datum] *D. brouweri*). The detailed stratigraphy derived from "Meteor" core 13519 suggests a third hiatus in 10.5 m sub-bottom depth, which lasted from 0.95 to 0.78 Ma (Sarnthein et al. 1984). The sedimentation rates (Table 1), which vary between 0.5 cm/1000y (early Oligocene) and 3.5 cm/1000y (late Pliocene), were applied for calculation of the mass accumulation rates. Details and differences of age assignment are discussed in Stein (1984a).

RESULTS

The results of the oxygen and carbon isotope analyses from Site 366 are presented and discussed in detail by Vergnaud-Grazzini and Rabussier-Lointier (1980), Savin et al. (1981), Blanc and Duplessy (1982), and Stein (1984a). All stable isotope data presented in Figures 2A and C are adjusted to $\delta^{18}O$ and $\delta^{13}C$ of *Uvigerina peregrina* according to Shackleton and Opdyke (1973) and Shackleton and Cita (1979). Distinct shifts of 0.7–1.0‰ to heavier $\delta^{18}O$-values are observed near the Oligocene/Miocene boundary, in the middle Miocene (near 14 Ma), in the late Pliocene (near 3 Ma), and in the Quaternary (near 1 Ma) (Figure 2A). An interval with significantly enriched $\delta^{18}O$ also occurred in the latest Miocene between 6 and 5 Ma (Figure 2A). The middle and late Miocene increase of $\delta^{18}O$ has been interpreted as related to the build-up of major Antarctic ice sheets (Shackleton and Kennett 1975; Kennett 1977; Woodruff et al. 1981), whereas the late Pliocene increase of $\delta^{18}O$ was probably caused by the formation of major northern hemisphere ice sheets (Shackleton and Opdyke 1977; Shackleton et al. 1984). The $\delta^{13}C$-values show a drastic increase of about 1.5 percent in the late early Miocene near 17.5 Ma, followed by a drastic decrease of about 1 percent in the middle Miocene near 14 Ma (Figure 2C). The latter coincides with the middle Miocene increase of $\delta^{18}O$.

The contents of calcium carbonate vary between 30 and 90 percent (Figure 2B). The Oligocene samples show $CaCO_3$ fluc-

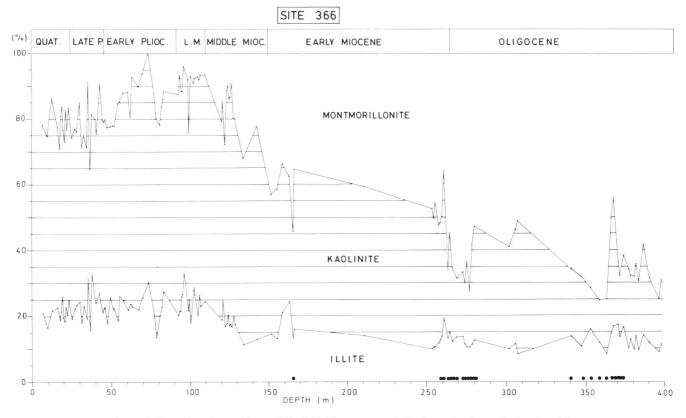

Figure 3. Clay mineral assemblages of Site 366. The amounts of chlorite are in all samples less than 2%.
Black dots mark samples with significant amounts of clinoptilolite.

tuations similar to those of Pleistocene time (Figure 2B; von Grafenstein 1982). Low carbonate values are mostly recorded during the late middle Miocene and during the late Pliocene, whereas higher $CaCO_3$ values are dominant during the early Pliocene.

The contents of biogenic opal, consisting primarily of marine diatoms, radiolarians, and sponge spicules, are high in core intervals between 400 m and 148 m sub-bottom depth (i.e. during Oligocene to early middle Miocene times) and are characterized by high-amplitude variations (Table 2, Figure 2D). The coarse fraction, >150 μm, of the total sediment fraction consists up to 20 percent biogenic opal and the carbonate-free silt fraction up to 100 percent biogenic opal. In the upper 145 m sub-bottom depth (i.e. during the last 14 m.y.) biogenic opal accumulation is almost negligible (Table 2, Figure 2D). Freshwater diatoms and phytoliths are recorded for the late early Eocene to early Miocene and the Pleistocene (Figure 2D; Schrader 1978; Fenner 1981).

Dominant clay minerals at Site 366 are montmorillonite, kaolinite, and illite (Table 2, Figure 3). Chlorite only occurs in trace amounts. The clay mineral assemblages of Oligocene samples are dominated by montmorillonite (Figure 3, 50–75%). Kaolinite (12–35%) and illite (8–18%) are of secondary importance. The kaolinite/illite (K/I)-ratios vary between 1 and 3, and the mean value lies near 1.9 (Figure 2E, Table 3). Significant amounts of clinoptilolite are recorded in core intervals of 375 to

340 m and 280 m to 258 m sub-bottom depth (Table 2, Figure 3). The first distinct increase of kaolinite (up to 45%) occurs near the Oligocene/Miocene boundary, while montmorillonite decreases. The amount of illite varies at the same level as before (Figure 3). The K/I-ratios increase up to more than 4. A drastic change in the clay mineral assemblages occurs at 150 m sub-bottom depth (near the early/middle Miocene boundary) (Figure 3). During the last 15 m.y., kaolinite became dominant (50–75%), montmorillonite decreased to amounts of 5 to 35%, and illite varied between 15 and 33%. The K/I-ratios fluctuated between 2 and 5, within the range of the middle Pleistocene to Holocene values recorded in "Meteor"-core 13519 (Table 3, Figure 2E).

The mass accumulation rates of total clay minerals show distinct maxima of up to 1.5 g cm^{-2} 10^{-3} y^{-1} between 22 and 18, 6 and 5, and 2.5 and 2 Ma (Figure 4). The former maximum is caused by increased supplies of kaolinite and montmorillonite, whereas the two younger maxima are mainly related to an increased supply of kaolinite. Generally, the mass accumulation rates of the terrigenous silt fraction (MART) were low during the Oligocene (less than 0.010 g cm^{-2} 10^{-3} y^{-1}) as well as during the middle Miocene and the early Pliocene (less than 0.020 g cm^{-2} 10^{-3} y^{-1}) (Figure 2H, Table 3). Maximum MART of up to 0.100 g cm^{-2} 10^{-3} y^{-1} occurred between 6 and 5 Ma and during the last 2.5 m.y. (Figure 2H, Table 3).

TABLE 3. MEAN VALUES OF CLAY MINERALS, GRAIN SIZES OF THE TERRIGENOUS MATTER,
AND MASS ACCUMULATION RATES OF THE TERRIGENOUS SILT FRACTION (MART) >6 μM.*

Site 366 Time intervals	Clay minerals (%)				Grain sizes (%)					MART ($g/cm^2 \times 10^3 y$) >6 μm
	M	I	K	K/I	<2	2-6	6-20	>20 m	>6 μm	>6 μm
Surface (M-13519)	13	20	67	3.4	77	17	5.5	0.5	6	0.020
Quaternary/Late Pliocene (2.5-0 m.y. BP.)	23	21	56	2.7	85	8	6.3	0.7	7	0.045
Early/Late Pliocene (5-2.5 m.y. BP.)	16	23	61	2.7	88	6.7	5.1	0.2	5.3	0.010
Latest Miocene (6-5 m.y. BP.)	10	25	65	2.6	87	7	5.9	0.1	6	0.040
Middle Miocene (14-10.7 m.y. BP.)	18	18	64	3.6	89	6	4.9	0.1	5	0.012
Early Miocene (22-18 m.y. BP.)	43	12	45	3.7	-	-	-	-	2	0.005
Late Oligocene (32-23 m.y. BP.)	62	13	25	1.9	-	.-	-	-	1.2	0.005

*Detailed grain-size analyses of the carbonate-free sediment fraction were not useful from Oligocene and early Miocene samples because of the high content of biogenic opal. From those samples, only the >6 μm values are determined. The listed values of the terrigenous matter >6 μm are percentages of the carbonate-free sediment fraction >6 μm minus percentages of biogenic opal >6 μm.

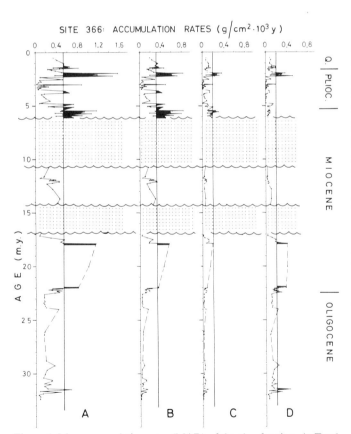

Figure 4. Mass accumulation rates (MAR) of the clay fraction. A: Total clay fraction, B: Kaolinite, C: Illite, D: Montmorillonite. The MAR of the clay minerals are based on the data obtained by evaluation technique according to Biscaye (1965). If other techniques are used, the MAR can change significantly. Dotted intervals indicate hiatuses.

The grain sizes of terrigenous matter are calculated as %>6 μm (carbonate-free sediment fraction >6 μm minus biogenic opal >6 μm (Figure 2G). Additionally, results of more detailed grain-size measurements of carbonate-free sediment fractions from samples with a negligible content of biogenic opal are listed in Table 3. The Oligocene is marked by very fine-grained terrigenous sediments, only less than 3% is coarser than 6 μm. Maximum grain-sizes of terrigenous matter occur between 140 and 90 m (middle Miocene), 70 m and 30 m (early Pliocene), and in the upper 40 m sub-bottom depth (late Pliocene to Quaternary).

DISCUSSION

Both the biogenic and the terrigenous fractions of the Oligocene and Miocene sediments of Site 366 are strongly influenced by the paleopositions of the depositional area (Figure 2).

Indicators of Paleoproductivity

At about 25 Ma, Site 366 lay at the equator (Sclater et al. 1977) in the center of the equatorial upwelling and high productivity region (Figure 1B). This is reflected by the very high content of biogenic opal (Figure 2D). High paleoproductivity at Site 366 is assumed for the middle and late Eocene, the Oligocene, and the early Miocene (Figure 2; Fenner 1981). Major decreases of biogenic opal observed in several Oligocene samples (Figure 2D) were probably caused by secondary dissolution rather than a decrease of the primary opal flux. This was deduced from the clearly negative correlation between the content of biogenic opal and the amount of zeolites (Table 2; Stein 1984a). The silica required for the formation of zeolites, mainly clinoptilolite (Figure 3), was supplied by dissolution of biogenic silica, as also suggested from investigations of diatoms (Fenner 1981).

Near 15 Ma, the position of Site 366 moved away from the area influenced by equatorial upwelling. This is indicated by the decrease of biogenic opal to almost zero (Table 2, Figure 2D; see also Diester-Haass, 1978a; Johnson, 1978; Schrader, 1978). Dissolution of opal, as observed in the Oligocene, is unlikely because of the absence of zeolites (Table 2). Dilution by siliciclastic matter is also almost negligible; the increased accumulation rates of terrigenous matter (Figure 2H) were not recorded until 4 m.y. later. The change in sediment composition (Figure 2D) is exaggerated by a hiatus lasting about 2.7 m.y., from 17 to 14.3 Ma (Table 1). The hiatus may have been caused by a strongly enhanced deep-water circulation as a consequence of the initiation of the Norwegian Overflow Water into the North Atlantic, which presumably intensified the production of North Atlantic Deep Water (NADW) (Shor and Poore 1979; Keller and Barron 1983). Keller and Barron (1983) proposed that the enhanced NADW advection may have acted as a barrier restricting the northward flow of the silica-rich Antarctic Bottom Water (AABW). Thus it caused a general reduction of biogenic silica in the North Atlantic. This would explain the reduced flux of biogenic opal, but not the disappearance of biogenic opal at Site 366, which was related to the drift out of the equatorial divergence zone.

The $\delta^{13}C$ record, which shows a distinct increase of about $1.5^o/_{oo}$ near 17.5 Ma and a distinct decrease of about $1^o/_{oo}$ near 14 Ma, did not coincide with the evolution of the biogenic opal record (Figure 2C and D). Instead, the $\delta^{13}C$ shifts were synchronous to similar $\delta^{13}C$ shifts observed in other ocean regions (Shackleton and Kennett 1975; Keigwin 1979; Savin et al. 1981; Woodruff et al. 1981). Therefore, these conditions suggest a change in the isotopic composition of the bulk available carbon in the oceans (Shackleton and Kennett 1975; Blanc and Duplessy 1982) rather than a change in local paleoproductivity.

Eolian Sediment Supply

The changes of the terrigenous sediment input also substantiate the drift of Site 366 from the Southern to the Northern Hemisphere.

Prior to 25 Ma, Site 366 lay in the Southern Hemisphere, probably outside of the northeasterly trade wind belt (Figure 1B). The ITCZ hindered the supply of kaolinite-rich dust from the South Sahara and Sahelian zone throughout the year. During this time, the terrigenous sediment supply was restricted to southeasterly trade wind dust. This terrigenous matter was characterized by extremely low content of kaolinite (Table 3, Figure 2E and F), fine grain sizes (Table 3, Figure 2G), and low mass accumulation rates of silt fraction (Table 3, Figure 2H). The clay minerals were similar to those of both recent southeasterly trade wind dust samples (Aston et al. 1973) and DSDP-Sites (Robert 1982) from the South Atlantic. Source areas of the Oligocene dust were

probably the Namib and the Kalahari deserts. Due to the great distance of about 6000 km between the source area and depositional site, coarser grain sizes of the dust would have been removed before reaching the Sierra Leone Rise. The composition of the Oligocene terrigenous matter of Site 366 should only yield information about climatic conditions in South Africa and not North Africa. For example, the relatively high content of fresh-water diatoms in these Oligocene sediments of Site 366 (Figure 2D; Fenner 1981) may indicate a semi-arid climate with a strong seasonal wet-dry cycle in the Kalahari area.

In the lower early Miocene (i.e. between 22 and 18 Ma), the first distinct increase of kaolinite (Figure 2E, 2F and 4, Table 3; Robert 1982) may have been the result of a first influence of the northeasterly trade winds at the Sierra Leone Rise. This change coincided with a first occurrence of light-brown colored sediments at Site 366 (Figure 2E), a color typical of the dust derived from the South Sahara and Sahelian area. However, the intensity of the northeasterly trade winds was still low during the early Miocene compared to the following time intervals, as deduced from the fine-grained terrigenous matter (Table 3). The fresh-water diatoms observed in early Miocene samples (Schrader 1978) may now have been derived from North African regions. According to a paleoposition close to the equator, between 0° and 1°N, the dust supply at Site 366 should actually have been restricted to northeasterly trade wind dust if a position of the ITCZ similar to that of today is assumed (Figure 1B). The still observed northeasterly trade wind dust may suggest a more southerly position of the ITCZ, a less distinct asymmetry of the atmospheric circulation compared to the present situation. This shift of the ITCZ to the south is plausible if temperature gradient between pole and equator was reduced due to the lack of major polar ice sheets (Flohn 1981; Woodruff et al. 1981).

A distinct change in the composition of the terrigenous matter, which parallels the decrease of biogenic opal, is observed at about 15 Ma (Figure 2E to H). Above the hiatus, the terrigenous sediment of Site 366 was more similar to that of today; that is, kaolinite was the dominant clay mineral (Figure 2E and F, Table 3; cf. also Lever and McCave, 1983). Pale yellowish brown to moderate brown sediment color prevailed (Figure 2E). Furthermore, the grain size of the terrigenous matter was much coarser compared to the Oligocene samples (Figure 2G, Table 3). These values are similar to present-day northeasterly trade wind dust (Chester et al. 1972; Parkin and Padgham, 1975; Wilke et al. 1983). These results suggest a paleoposition of Site 366 in the belt of the northeasterly trade winds that may have had similar speeds as today.

The time intervals between 14.4 and 10.7 Ma (150 m to 110 m sub-bottom depth) and between 5 and 2.5 Ma (83 m to 40 m sub-bottom depth) were characterized by a reduced supply of the terrigenous matter (Figure 2H, Table 3). This may indicate reduced availability of eolian sediment in the source area due to more humid climatic conditions.

Distinct maxima of eolian mass accumulation rates and a coarsening of grain size are observed in the latest Miocene, between 6 and 5 Ma, and in the late Pliocene and Quaternary, in the last 2.5 m.y. (Figure 2G, H, and 4, Table 3). The coarsening of eolian grain size is interpreted in terms of enhanced meridional atmospheric circulation. For both time intervals, paleo-wind speeds of the northeasterly trade winds calculated from the grain sizes of the terrigenous sediments of DSDP-Site 397 (NE-Atlantic, 27°N), already show values similar to those of today, that is 5 to 15 m/s (Stein, 1984a; 1985). Enhanced atmospheric circulation is also deduced from North Pacific eolian sediments during the last 3 to 2 m.y. (Rea and Harrsch 1981; Rea and Janecek 1982). This intensified atmospheric circulation was probably caused by an increased temperature gradient between the north pole and equator due to extended Northern Hemisphere Glaciation (NHG), indicated by the contemporaneous increase of $\delta^{18}O$ at those times (Figure 2A). These results may imply that already prior to the late Pliocene NHG (e.g., Shackleton et al. 1984), in the late Miocene, ice sheets may have existed in high northern latitudes (see Mercer and Sutter 1982; Mudie and Helgason 1983). At the same times, a more arid to semi-arid climate in the South Sahara and Sahelian area may have caused the increased dust supply. However, the terrigenous sediment flux rates at Site 366 still remained low compared to other deep-sea sediment cores off NW-Africa (Koopmann 1981; Stein 1984a). They are too low to explain the large carbonate fluctuations (Figure 2B) which were mainly influenced by dissolution (Dean et al. 1978; 1981; Sarnthein et al. 1984).

CONCLUSIONS

At DSDP-Site 366, the long-term evolution of paleoenvironments during the last 40 m.y. was clearly controlled by the northward drift of the African plate. Prior to 25 Ma, the terrigenous input at the paleolatitude of Site 366 was restricted to eolian sediment supply from South Africa by southeasterly trade winds, as shown by dominant illite in the clay fraction and extremely fine-grained terrigenous matter. Near the Oligocene/Miocene boundary, Site 366 drifted across the equator into the belt of the northeasterly trade winds, as inferred from the increased content of kaolinite and coarser grain sizes of the terrigenous sediment fraction. Until the middle Miocene, Site 366 lay beneath the zone of equatorial upwelling, as reflected by the high content of biogenic silica. This signal terminated at about 15 m.y. ago when the site drifted beyond 2°N. The change of sedimentation appears drastically shortened by a hiatus of about 2.7 m.y. which lasted from 17 to 14.3 Ma.

Enhanced meridional atmospheric circulation and a more arid to semi-arid climate in the South Sahara and Sahelian Zone towards the end of the Miocene, between 6 and 5 Ma, and in the last 2.5 m.y. are inferred from distinct maxima of mass accumulation rates and coarsening of terrigenous sediment fraction.

ACKNOWLEDGMENTS

I would like to thank Drs. M. Sarnthein, G. Wefer, and K. Winn for critical remarks and Drs. G. A. Jones and D. K. Rea for reviewing the manuscript. I am indebted to Dr. H. Lange for providing unpublished clay-mineral data. For technical assistance I would like to thank Ms. M. Baumann, Mr. F. Sirocko, and Mr. M. Stransky. This study was part of my doctoral thesis which was supervised by Prof. Dr. M. Sarnthein and supported by the "Deutsche Forschungsgemeinschaft."

REFERENCES CITED

Aston, S. R., Chester, R., Johnson, L. R., and Padgham, R. C., 1973, Eolian dust from the lower atmosphere of the eastern Atlantic and Indian Oceans, China Sea and Sea of Japan: Marine Geology, v. 14, p. 15–28.

Backman, J. and Shackleton, N. J., 1983, Quantitative biochronology of Pliocene and Early Pleistocene calcareous nannofossils from the Atlantic, Indian and Pacific Ocean: Marine Micropaleontology, v. 8, p. 141–170.

Berggren, W. A., Burckle, L. H., Cita, M. B., Cooke, H.B.S., Funnel, B. M., Gartner, S., Hays, J. D., Kennett, J. P., Opdyke, N. D., Pastouret, L., Shackleton, N. J., and Takayanag, Y., 1980, Towards a Quaternary Time Scale: Quaternary Research, v. 13, p. 277–302.

Berggren, W. A. and van Couvering, J. A., 1974, The Late Neogene: Biostratigraphy and geochronology and paleoclimatology of the last 15 million years in marine and continental sequences: Palaeogeography, Palaeoclimatology, and Palaeoecology, v. 16, p. 1–216.

Biscaye, P. E., 1965, Mineralogy and sedimentation of recent deep-sea clay in the Atlantic Ocean and adjacent seas and oceans: Geological Society of America Bulletin, v. 76, p. 803–832.

Blanc, P., 1981, Paleoclimatologie isotopique et histoire de l'eau profonde Atlantique depuis 15 millions d'annees [Thesis]: University Paris XI–Orsay, 190 p.

Blanc, P., and Duplessy, J.-C., 1982, The deep water circulation during the Neogene and the impact of the Messinian salinity crisis: Deep Sea Research, in press.

Bukry, D., 1978, Cenozoic coccolith and silicoflagellate stratigraphy offshore Northwest Africa, DSDP Leg XLI; *in* Lancelot, Y., Seibold, E., et al., Initial Reports of the Deep Sea Drilling Project, v. 41: Washington, D.C., U.S. Government Printing Office, p. 689–708.

Cepek, P., Johnson, D., Krasheninnikov, V. A., and Pflaumann, U., 1978, Synthesis of the Leg 41 biostratigraphy and paleontology, Deep Sea Drilling Project; *in* Lancelot, Y., Seibold, E., et al., Initial Reports of the Deep Sea Drilling Project, v. 41: Washington, D.C., U.S. Government Printing Office, p. 1181–1198.

Chamley, H., 1979, North Atlantic Clay Sedimentation and Paleoenvironment since the Late Jurassic, *in* Talwani, E., Hay, W. W., and Ryan, W.B.F., eds., Deep Drilling Results in the Atlantic Ocean: Continental Margins and Paleoenvironment, Maurice Ewing Series 3, American Geophysical Union Publication, p. 342–361.

Chester, R., Elderfield, H., Griffin, J. J., Johnson, L. R., and Padgham, R. C., 1972, Eolian dust along Atlantic eastern margins: Marine Geology, v. 13, p. 91–106.

Dean, W. E., Gardner, J. V., and Cepek, P., 1981, Tertiary carbonate-dissolution cycles on the Sierra Leone rise, eastern equatorial Atlantic Ocean: Marine Geology, v. 39, p. 81–101.

Dean, W. E., Gardner, J. V., Jansa, L. F., Cepek, P., and Seibold, E., 1978, Cyclic sedimentation along the continental margin of NW-Africa, *in* Lancelot, Y., Seibold, E., et al., Initial Reports of the Deep Sea Drilling Project, v. 41: Washington, D.C., U.S. Government Printing Office, p. 965–986.

Diester-Haass, L., 1978a, Influence of carbonate dissolution, climate, sea-level changes, and volcanism on Neogene sediments of Northwest Africa (Leg 41), in Lancelot, Y., Seibold, E. et al., Initial Reports of the Deep Sea Drilling Project, v. 41: Washington, D.C., U.S. Government Printing Office, p. 1033–1047.

—— 1978b, Sediments as an indicator of upwelling; in Boje, R., and Tomczak, M., eds., Upwelling ecosystems: Berlin, Springer Verlag, p. 261–281.

Dietrich, G., Kalle, K., Krauss, W., and Siedler, G., 1975, Allgemeine Meereskunde: Berlin, Verlag Gebrueder Borntraeger, 593 p.

Fenner, J., 1981, Diatoms in the Eocene and Oligocene sediments off NW-Africa, their stratigraphic and paleogeographic occurrences [Thesis]: University of Kiel, 230 p.

Flohn, H., 1981, A hemispheric circulation asymmetry during Late Tertiary: Geologische Rundschau, v. 70, p. 725–736.

Fuetterer, D., 1978, Late Neogene silt at the Sierra Leone Rise (Leg 41 Site 366): Terrigenous and biogenous components: in Lancelot, Y., Seibold, E., et al., Initial Reports of the Deep Sea Drilling Project, v. 41, Washington, D.C., U.S. Government Printing Office, p. 1049–1059.

Ganssen, G., 1983, Dokumentation von Kuestenauftrieb anhand stabiler Isotope in rezenten Foraminiferen vor Nordwest-Afrika: "Meteor" Forschungsergebnisse, C, v. 37, p. 1–46.

Griffin, J. J., Windom, H., and Goldberg, E. D., 1968, The distribution of clay minerals in the World Ocean: Deep-Sea Research, v. 15, p. 433–459.

Haq, B. U., Worsley, T. R., Burckle, L. H., Douglas, R. G., Keigwin, L. U., Jr., Opdyke, N. D., Savin, S. M., Sommer, M. A., II, Vincent, E., and Woodruff, F. 1980, Late Miocene marine carbon-isotope shift and synchroneity of some phytoplanktonic biostratigraphic events: Geology, v. 8, p. 427–431.

Johnson, D. A., 1978, Cenozoic radiolaria from the eastern tropical Atlantic DSDP Leg 41: in Lancelot, Y., Seibold, E., et al., Initial Reports of the Deep Sea Drilling Project, v. 41, Washington, D.C., U.S. Government Printing Office, p. 763–789.

Keigwin, L. D., Jr., 1979, Late Cenozoic stable isotope stratigraphy and paleoceanography of DSDP sites from the east equatorial and north central Pacific Ocean: Earth and Planetary Science Letters, v. 45, p. 361–382.

Keller, G., 1981, Miocene biochronology and paleoceanography of the North Pacific: Marine Micropaleontology, v. 6, p. 535–551.

Keller, G., and Barron, J. A., 1983, Paleoceanographic implications of Miocene deep-sea hiatuses: Geological Society of America Bulletin, v. 94, p. 590–613.

Kennett, J. P., 1977, Cenozoic Evolution of Antarctic Glaciation, the Circum-Antarctic Ocean, and Their Impact on Global Paleoceanography: Journal of Geophysical Research, v. 82, p. 3843–3860.

Koopmann, B., 1981, Sedimentation von Saharastaub im subtropischen Nordatlantik waehrend der letzten 25,000 Jahre: Meteor Forschungsergebnisse, C, v. 35, p. 23–59.

Krasheninnikov, V. A., and Pflaumann, U., 1978a, Zonal stratigraphy and planktonic foraminifers of Paleogene deposits of the Atlantic Ocean to the west off Africa (Deep Sea Drilling Project, Leg 41); in Lancelot, Y., Seibold, E., et al., Initial Reports of the Deep Sea Drilling Project, v. 41: Washington, D.C., U.S. Government Printing Office, p. 581–612.

—— 1978b, Zonal stratigraphy and Neogene deposits of the eastern part of the Atlantic Ocean by means of planktonic foraminifers, Leg 41, Deep Sea Drilling Project, in Lancelot, Y., Seibold, E., et al., Initial Reports of the Deep Sea Drilling Project, v. 41: Washington, D.C., U.S. Government Printing Office, p. 613–658.

Kroopnick, P., 1980, The distribution of ^{13}C in the Atlantic Ocean: Earth and Planetary Science Letters, v. 49, p. 469–484.

Lange, H., 1982, Distribution of chlorite and kaolinite in eastern Atlantic sediments off North Africa: Sedimentology, Vol. 29, p. 427–432.

Leinen, M., and Heath, G. R., 1981, Sedimentary indicators of atmospheric activity in the Northern Hemisphere during the Cenozoic: Palaeogeography, Palaeoclimatology, Palaeoecology, v. 36, p. 1–21.

Lever, A., and McCave, I. N., 1983, Eolian components in Cretaceous and Tertiary North Atlantic sediments: Journal of Sedimentary Petrology, v. 53, p. 811–832.

McTainsh, G. H., and Walker, P. H., 1982, Nature and distribution of Harmattan dust: Zeitschrift Geomorphologie, v. 26, p. 417–435.

Mercer, J. H., and Sutter, J. F., 1982, Late Miocene–earliest Pliocene glaciation in southern Argentina: implications for global ice-sheet history: Palaeogeography, Palaeoclimatology, Palaeoecology, v. 38, p. 185–206.

Mudie, P. J., and Helgason, J., 1983, Palynological evidence for Miocene climatic cooling in eastern Iceland about 9.8 Myr ago: Nature, v. 303, p. 689–692.

Mueller, G., 1967, Methods in Sedimentary Petrology, in W. v. Engelhardt, W. v. Fuechtbauer, H., and Mueller, G., Sedimentary Petrology, Part I: Stuttgart, Schweizerbart'sche Verlagsbuchhandlung, 283 p.

Mueller, P. J., Erlenkeuser, H., and von Grafenstein, R., 1983, Glacial-interglacial cycles in oceanic productivity inferred from organic carbon contents in eastern north Atlantic sediment cores; in Thiede, J. and Suess, E., eds., Coastal Upwelling: Its Sediment Record, Part B; New York, Plenum Press, p. 365–398.

Parkin, D. W., 1974, Trade winds during the glacial cycles: Proceedings of the Royal Society of London, Series A, v. 337, p. 73–100.

Parkin, D. W., and Padgham, R. C., 1975, Further studies on trade winds during glacial cycles: Proceedings of the Royal Society of London, A, v. 346, p. 245–260.

Pflaumann, U., and Krasheninnikov, V. A., 1978, Quaternary stratigraphy and planktonic foraminifers of the eastern Atlantic, Deep Sea Drilling Project, Leg 41; in Lancelot, Y., Seibold, E., et al., Initial Reports of the Deep Sea Drilling Project, v. 41, Washington, D.C., U.S. Government Printing Office, p. 883–911.

Rea, D. K., and Harrsch, E. C., 1981, Mass-accumulation rates of the non-authigenic inorganic crystalline (eolian) components of deep-sea sediments from Hess Rise, Deep-Sea Drilling Project Sites 464, 465, and 466; in Thiede, J., Vallier, T. L., et al., Initial Reports of the Deep Sea Drilling Project, v. 62, Washington, D.C., U.S. Government Printing Office, p. 661–668.

Rea, D. K., and Janecek, T. R., 1982, Late Cenozoic changes in atmospheric circulation deduced from North Pacific eolian sediments: Marine Geology, v. 49, p. 149–167.

Robert, C., 1982, Modalite de la sedimentation argileuse en relation avec l'histoire geologique de l'Atlantique sud, [Unpublished thesis]: Marseille, 141 p.

Samtleben, C., 1978, Pliocene-Pleistocene coccolith assemblages from the Sierra Leone Rise—Site 366, Leg 41; in Lancelot, Y., Seibold, E., et al., Initial Reports of the Deep Sea Drilling Project, v. 41, Washington, D.C., U.S. Government Printing Office, p. 913–921.

Sarnthein, M., Erlenkeuser, H., v. Grafenstein, R., and Schroeder, C., 1984, Stable isotope stratigraphy for the last 750,000 years: "Meteor" core 13519 from the eastern equatorial Atlantic; 'Meteor' Forschungsergebnisse, C, v. 38, p. 9–24.

Sarnthein, M., Tetzlaff, G., Koopmann, B., Wolter, K., and Pflaumann, U., 1981, Glacial and Interglacial wind regimes over the eastern subtropical Atlantic and NW-Africa: Nature, v. 293, p. 193–196.

Sarnthein, M., Thiede, J., Pflaumann, U., Erlenkeuser, H., Fuetterer, D., Koopmann, B., Lange, H., and Seibold, E., 1982, Atmospheric and oceanic circulation patterns off Northwest Africa during the past 25 million years; in v. Rad, U., Hinz, K., Sarnthein, M. and Seibold, E., eds., Geology of the Northwest African Continental Margin: Berlin, Springer Verlag, p. 545–604.

Savin, S. M., Douglas, R. G., Keller, G., Killingley, J. S., Shaughnessy, L., Sommer, M. A., Vincent, E., and Woodruff, F., 1981, Miocene benthic foraminiferal isotope records: a synthesis: Marine Micropaleontology, v. 6, p. 423–450.

Schrader, H. -J., 1978, Diatoms in DSDP Leg 41 Sites; in Lancelot, Y., Seibold, E., et al., Initial Reports of the Deep Sea Drilling Project, v. 41, Washington, D.C., U.S. Government Printing Office, p. 791–812.

Sclater, J. G., Hellinger, S., and Tapscott, C., 1977, The paleobathymetry of the Atlantic Ocean from the Jurassic to the Recent: Journal of Geology, v. 85, p. 509–552.

Shackleton, N. J., Backman, J., Zimmerman, H., Kent, D. V., Hall, M. A., Roberts, D. G., Schnitker, D., Baldauf, J. G., Desprairies, A., Homrighausen,

R., Huddlestun, P., Keene, J. B., Kaltenback, A. J., Krumsiek, K.A.O., Morton, A. C., Murray, J. W., and Westberg-Smith, J., 1984, Oxygen isotope calibration of the onset of ice-rafting and history of glaciation in the North Atlantic region: Nature, v. 307, p. 620–623.

Shackleton, N. J., and Cita, M. B., 1979, Oxygen and Carbon Isotope Stratigraphy of Benthic Foraminifers at Site 397: Detailed History of Climatic Change during the Late Neogene; *in* v. Rad, U., Ryan, W.B.F., et al., Initial Reports of the Deep Sea Drilling Project, v. 47A, Washington, D.C., U.S. Government Printing Office, p. 433–445.

Shackleton, N. J., and Kennett, J. P., 1975, Paleotemperature history of the Cenozoic and the initiation of Antarctic glaciation: oxygen and carbon isotope analyses in DSDP Sites 277, 279, and 281; *in* Kennett, J. P., Houtz, R. E., et al., Initial Reports of the Deep Sea Drilling Project, v. 29, Washington, D.C., U.S. Government Printing Office, p. 743–755.

Shackleton, N. J., and Opdyke, N. D., 1973, Oxygen isotope and paleomagnetic stratigraphy of equatorial Pacific core V28-238: oxygen isotope temperatures and ice volumes on a 10^5 year and 10^6 year scale: Quaternary Research, v. 3, p. 39–55.

——1977, Oxygen isotope and paleomagnetic evidence for early northern hemisphere glaciation: Nature, v. 270, p. 216–219.

Shipboard Scientific Party, 1978, Site 366, Sierra Leone Rise; *in* Lancelot, Y., Seibold, E., et al., Initial Reports of the Deep Sea Drilling Project, v. 41, Washington, D.C., U.S. Government Printing Office, p. 21–162.

Shor, A. N., and Poore, R. Z., 1979, Bottom currents and ice rafting in the North Atlantic: Interpretation of Neogene depositional environments of Leg 49 cores; *in* Luyendyk, B. P., Cann, J. R., et al., Initial Reports of the Deep Sea Drilling Project, v. 49, Washington, D.C., U.S. Government Printing Office, p. 859–872.

Stein, R., 1984a, Zur neogenen Klimaentwicklung in Nordwest-Afrika und Palaeo-Ozeanographie im Nordost-Atlantik: Ergebnisse von DSDP-Sites 141, 366, 397, und 544B [doctoral thesis]: Berichte-Reports, Geologisch-Palaeontologisches Institut, Universitaet Kiel, Nr. 4, 210 S.

——1984b, Rapid grain-size analyses of silt and clay fractions by SediGraph 5000D: Comparison with Coulter Counter and Atterberg methods: Journal of Sedimentary Petrology, in press.

——1985, Late Neogene Changes of Paleoclimate and Paleoproductivity off Northwest Africa (DSDP-Site 397): Palaeogeography, Palaeoclimatology, and Palaeoecology, v. 49, p. 47–59.

Stein, R., and Sarnthein, M., 1984, Late Neogene events of atmospheric and oceanic circulation offshore Northwest Africa: High-resolution record from deep-sea sediments: Palaeoecology of Africa, v. 16, p. 9–36.

Thiede, J., Suess, E., and Mueller, P., 1982, Late Quaternary Fluxes of Major Sediment Components to the Sea Floor at the Northwest African Continental Slope; *in* von Rad, U., Hinz, K., et al., eds., Geology of the Northwest African Continental Margin: Berlin, Springer Verlag, p. 605–631.

Thunell, R. C., 1981, Late Miocene–Early Pliocene Planktonic Foraminiferal Biostratigraphy and Paleoceanography of Low-Latitude Marine Sequences: Marine Micropaleontology, v. 6, p. 71–90.

Trabant, P. K., 1978, Synthesis of physical properties data from DSDP Leg 41, *in* Lancelot, Y., Seibold, E., et al., Initial Reports of the Deep Sea Drilling Project, v. 41, Washington, D.C., U.S. Government Printing Office, p. 1199–1213.

Vergnaud-Grazzini, C., and Rabussier-Lointier, D., 1980, Paleotemperatures et paleocourants tertiaires en Atlantique equatorial: compositions isotopiques de l'oxygene et du carbone des foraminiferes au site 366 du D.S.D.P.: Revue de Géologie Dynamique et de Géographie Physique, Vol. 22, p. 63–74.

von Grafenstein, R., 1982, 750,000 Jahre Klimageschichte und Palaeo-Ozeanographie an einem Tiefseekern vom Sierra Leone Ruecken (Aequatorialer Ostatlantik) [Unpublished Diplomarbeit]: University of Kiel, 68 p.

Whalley, W. B., and Smith, B. J., 1981, Mineral content of harmattan dust from northern Nigeria examined by scanning electron microscopy: Journal of Arid Environments, v. 4, p. 21–29.

Wilke, B. M., Duke, B. J., and Jimon, W.L.O., 1983, Investigations on Harmattan dust in Northern Nigeria, in preparation.

Woodruff, F., Savin, S. M., and Douglas, R. G., 1981, Miocene Stable Isotope Record: A Detailed Deep Pacific Ocean Study and Its Paleoclimate Implications: Science, v. 212, p. 665–667.

Manuscript Accepted by the Society December 17, 1984

Printed in U.S.A.

Geological Society of America
Memoir 163
1985

Miocene paleoceanography of the South Atlantic Ocean at 22, 16, and 8 Ma

David A. Hodell
James P. Kennett
Graduate School of Oceanography
University of Rhode Island
Narragansett, Rhode Island 02882-1197

ABSTRACT

Planktonic foraminiferal assemblages and stable isotopic ratios were analyzed along a longitudinal transect across the South Atlantic Ocean for three Miocene time slices: 22, 16, and 8 Ma. The sites are distributed so that they intersect three distinct paleoceanographic regimes in the South Atlantic basin: western boundary current, central gyre region, and eastern boundary current. From the faunal and isotopic data, we infer some first order paleocirculation features of the South Atlantic anticyclonic gyre system during the Miocene. Circulation patterns were most similar at 22 and 16 Ma, but markedly changed between 16 and 8 Ma, following the development of large permanent ice-sheets on Antarctica.

During the 22 and 16 Ma time slices: 1. Site 362 exhibits the most depleted ("warmest") ^{18}O values and highest frequencies of tropical-warm subtropical species reflecting the lower latitude position of this site; 2. No evidence exists for coastal upwelling in Sites 360 or 362 along the eastern boundary; and 3. Sites 357 and 516A along the western boundary contain high frequencies of warm-water species and relatively depleted ^{18}O values indicating southward transport of warm water by the Brazil Current that flows along the South American margin.

By 8 Ma, marked changes occurred in faunal, isotopic, and sedimentologic patterns, especially along the eastern boundary of the South Atlantic. Sites 360 and 362 along the eastern boundary record: 1. an increase in the frequency of the polar-form *Neogloboquadrina pachyderma;* 2. a decrease in the oxygen isotopic gradient between these sites; 3. a depletion in average surface water ^{13}C values; and 4. an increase in sedimentation rates. These changes are interpreted as recording the initiation of the Benguela upwelling system between 16 and 8 Ma.

During all Miocene time-slices, the central gyre sites (14, 15, 16, and 18) exhibit the most enriched ^{18}O values, and contain a greater frequency of cool, subantarctic species. The anomalously cool central gyre region may reflect a northward expression of the Subtropical Convergence, or may alternatively reflect increased dissolution due to greater water depths of these sites.

Between 22 and 16 Ma, the difference in planktonic ^{18}O values indicate that surface water temperatures warmed by an average of 2°C in the mid-latitude South Atlantic. Between 16 and 8 Ma, all sites (except 360) show an enrichment in average surface water ^{18}O reflecting an increase in global ice volume and/or cooling of surface waters. If an ice volume effect of $0.5^0/_{00}$ is assumed between 16 and 8 Ma, then surface water temperatures in the mid-latitude of the South Atlantic remained nearly constant between these time-slices. Global CENOP isotopic data suggest that the response of sea

surface temperatures to increased glaciation of Antarctica was a warming of the tropics,
little or no temperature change in the mid-latitudes, and a cooling at high-latitudes. The
net effect was a marked intensification of the planet's latitudinal temperature gradient.

INTRODUCTION

Planktonic foraminiferal assemblage and stable isotopic data were collected on samples from the mid-latitude (19° to 35°S) South Atlantic as part of a global mapping program of the Miocene oceans undertaken by the CENOP (Cenozoic Paleoceanography) project. Three Miocene time-slices were investigated: 22 Ma (Zone N4B, earliest Miocene), 16 Ma (Zone N8, late early Miocene), and 8 Ma (Zone N16/17, late Miocene). This South Atlantic study augments the global CENOP data base by providing mid-latitude coverage along a longitudinal transect across the South Atlantic Ocean. The sites are distributed so that they intersect three different oceanographic regimes in the South Atlantic: western boundary current (Sites 357 and 516A); central gyre region (Sites 14, 15, 16, and 18), and eastern boundary current (Sites 360, 362, and 526A). Changes in planktonic foraminiferal assemblages and stable isotopic ratios between sites and time slices provide a history of the South Atlantic subtropical gyre system through the Miocene. The South Atlantic data have been also interpreted in a global context by comparison with other CENOP sites.

SITE LOCATIONS AND STRATIGRAPHY

The Deep Sea Drilling Project (DSDP) sites used in this study form an east-west transect across the South Atlantic extending from the Rio Grande Rise in the western basin, to the Mid Atlantic Ridge in the central basin, to the Walvis Ridge in the eastern basin (Figure 1). The modern position, water depth, and time-slices contained in each site are listed in Table 1. These sites were backtracked by Sclater et al. (1985), and the maximum latitudinal displacement of any site during the Miocene was less than 3°.

Several microfossil groups and geochemical indices were used to define the three CENOP time-slices (Barron et al. 1985). Each time-slice spans approximately one million years: earliest Miocene (21.2 to 20.1 Ma), late early Miocene (16.3 to 15.2 Ma), and late Miocene (8.9 to 8.2 Ma). The placement of time-slices in the South Atlantic is based upon planktonic foraminiferal biostratigraphy from the *Initial Reports of the Deep Sea Drilling Project* (Maxwell et al. 1970; Boersma 1977; Jenkins 1978; Berggren et al. 1983). The placement of time-slices has been also compared with calcareous nannofossil biostratigraphy of Haq

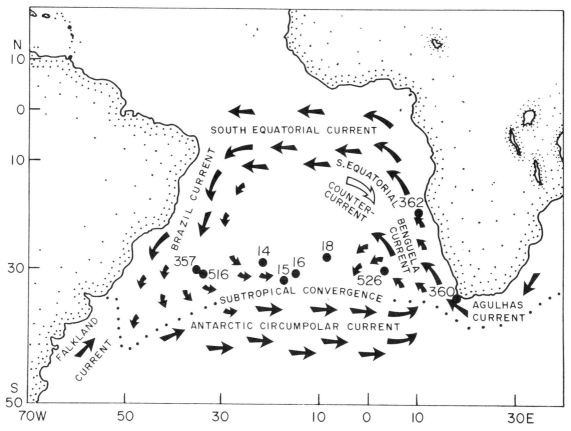

Figure 1. Location map and surface water circulation patterns of the South Atlantic Ocean showing modern position of DSDP sites used in this study.

TABLE 1. SITE POSITIONS, WATER DEPTH, AND TIME-SLICES
CONTAINED IN EACH SITE

Site	Latitude	Longitude	Depth (m)	22 Ma	16 Ma	8 Ma
Rio Grande Rise:						
357	30°00.25'S	35°33.59'W	2086	+		
516	30°16.59'S	35°17.11'W	1313		+	+
Mid-Atlantic Ridge:						
14	28°19.89'S	20°56.46'W	4343	+		
15	30°53.38'S	17°58.99'W	3927		+	
16	30°20.15'S	15°42.79'W	3527			+
18	27°58.72'S	08°00.70'W	4018	+		
Walvis Ridge:						
360	35°50.75'S	18°05.79'E	2967	+	+	+
362	19°45.45'S	10°31.95'E	1326	+	+	+
526	30°07.36'S	03°08.28'E	1054	+	+	+

TABLE 2. PLACEMENT OF TIME-SLICES CHOSEN IN EACH SITE USED IN THIS STUDY*

Site	Planktonic foraminifera	Calcareous nannoplankton
	22 Ma Time-Slice Placement	
14	2-1, 145-150 cm to 2-4, 143-148 cm	Below 1-6, 117-119 cm
18	4-2, 145-150 cm to 5-5, 145-150 cm	Below 5-2, 9- 11 cm
357	12-1, 145-150 cm to 13-6, 145-150 cm	10-1, 103-105 cm to 12-1, 100-102 cm
360	26-1, 100-107 cm to 26-2, 145-150 cm	Below 26-2, 141-145 cm
362	39-3, 145-150 cm to 40-6, 145-150 cm	40-1, 21- 24 cm to 41-1, 12- 15 cm
526A	27-1, 50- 54 cm to 39-3, 50- 54 cm	....
	16 Ma Time-Slice Placement	
15	6-6, 46- 50 cm to 7-4, 48- 57 cm	7-1, 27- 29 cm to 7-6, 123-125 cm
360	22-2, 74- 79 cm to 22-6, 146-150 cm	At 22-6, 140-142 cm
362	36 cc to 37-2, 145-159 cm	At 36-1, 97-100 cm
516	21-1, 51- 55 cm to 22-2, 50- 54 cm	....
526A	21-1, 50- 54 cm to 21-4, 11- 16 cm	....
	8 Ma Time-Slice Placement	
16	9-1, 58- 65 cm to 10-5, 50- 55 cm	....
360	8-2, 50- 55 cm to 11-6, 50- 55 cm	11-6, 140-143 cm to 12-1, 50- 53 cm
362	24-1, 52- 56 cm to 27-6, 52- 57 cm	23-6, 121-124 cm to 24-1, 10- 13 cm
516	13-1, 70- 72 cm to 14-3, 70- 72 cm	....
526A	9-1, 50- 54 cm to 11-3, 55- 59 cm	....

*Note: Time-slice positions are given in core-section, and interval within each of
the sites. Planktonic foraminiferal placement of the time-slices (this study) is
compared with time-slice placement based upon calcareous nannofossils of Haq (1980).

(1980) (Table 2). The general agreement between foraminiferal and nannofossil biostratigraphy suggests that the time-slices are sufficiently well defined for paleoceanographic interpretation.

The selection of time-slices was based upon a revised Miocene foraminiferal zonation for the South Atlantic (Berggren et al. 1983). The 22 Ma time slice occurs in planktonic foraminiferal Zone N4B (Blow 1969) and calcareous nannofossil Zone NN2 (Martini 1971). In the South Atlantic, this time-slice is defined by the concurrent ranges of *Globoquadrina dehiscens* and *Globorotalia kugleri* (Berggren et al. 1983). The 16 Ma time slice corresponds to foraminiferal Zone N8 (Blow 1969; Srinivasan and Kennett 1981a, b) and to the upper part of calcareous nannofossil Zone NN4 to lower NN5 (Martini 1971). In the South Atlantic, the 16 Ma time-slice is constrained as occurring between the First

Appearance Datum (FAD) of *Praeorbulina glomerosa* and the FAD of *Orbulina suturalis*. The 8 Ma time-slice falls within lowermost planktonic foraminiferal Zone N17 (Blow 1969) and uppermost calcareous nannofossil Zone NN10 (Martini 1971). In the South Atlantic, the 8 Ma time-slice is defined as occurring between the FAD of *Dentoglobigerina altispira* and the FAD of *Globorotalia conomiozea/mediterranea* (Berggren et al. 1983). This time-slice also encompasses the partial ranges of *Globorotalia miozea* and *Globorotalia conoidea*.

The stratigraphic placement of the South Atlantic time-slices are less certain than that of the Pacific where a resolution approaching 100,000 years was obtained (Keller 1981b; Keller et al. 1982; Barron et al. 1985). Since the time-slices were selected to examine intervals of relative climatic stability, the averaging of

South Atlantic faunal and isotopic data over a slightly broader time interval should have little effect on the results.

METHODS

Census data of planktonic foraminiferal assemblages in the >150 μm size fraction were generated on samples from each site for each of the time-slices. Samples were oven dried at 50°C, weighed, disaggregated in hot Calgon solution, washed over a 63 μm Tyler screen, dried, and reweighed. Samples were microsplit to obtain an average of 300 specimens per sample, and specimens were identified and counted according to the taxonomy of Kennett and Srinivasan (1983). The raw data are given in Appendix 1.

Isotopic analyses were performed on at least two planktonic foraminiferal species from numerous samples within each time-slice. *Globigerina praebulloides* and *Globoquadrina dehiscens* were analyzed at 22 Ma, *Globigerinoides* spp. and *Gq. dehiscens* at 16 Ma, and *Globigerinoides* spp. and *Gr. conoidea* at 8 Ma.

Specimens for isotopic analysis were ultrasonically cleaned in methanol, loaded into stainless steel carrying boats, and roasted *in vacuo* at 400°C for 1 hour to vaporize organic matter. The samples were loaded into a reaction vessel and reacted to completion in purified phosphoric acid at 50°C. The resulting CO_2 gas was distilled four times to remove water and analyzed using an on-line VG Micromass 602D mass spectrometer. All isotopic results are expressed as per mil difference from PDB and are given in Appendix 2. Based upon replicate analyses of a powdered sample (B-1), analytic precision (1σ of the mean) was ±0.11 for ^{18}O and ±0.09 for ^{13}C.

HYDROGRAPHY

The surface water circulation of the South Atlantic is a large subtropical anticyclonic gyre composed of the South Equatorial Current, Brazil Current, Antarctic Circumpolar Current and Benguela Current (Figure 1). In the north, the South Equatorial Current flows to the west encompassing the latitudinal area between 10°S and 3°N. Along the western boundary, the Brazil Current flows along the South American coast to about 40°S where it encounters the northward-flowing Falkland Current (Reid et al. 1976).

The southern limb of the gyre is driven by the eastward-flowing Antarctic Circumpolar Current. A sharp transition zone, the Subtropical Convergence, marks the boundary between relatively warm subtropical waters to the north of 40°S and cooler subantarctic waters to the south. The Subtropical Convergence is farther south in the western and eastern South Atlantic, and bends equatorward in the central gyre region. The position of the Subtropical Convergence varies seasonally over several degrees of latitude.

The surface circulation along the eastern boundary consists of the Agulhas Current, the Benguela Current, and the South Equatorial Countercurrent. The Agulhas Current transports warm tropical-subtropical waters in a series of major eddies from the Indian Ocean around the southern tip of Africa into the South Atlantic. The northward-flowing Benguela Current and southward-flowing South Equatorial Countercurrent meet between 10° and 20°S to produce a complex confluence of surface currents and gyres. The Benguela Current flows northward along the coast of west Africa until it reaches 23°S where the main branch of the current swings to the northwest. Seasonal wind stress drives surface waters offshore and draws cold, nutrient-enriched waters from depths of 150 to 400 m to the surface. This upwelling and high productivity forms a broad band extending from the tip of South Africa to 23°S where upwelling markedly decreases as the Benguela Current shifts westward to join the South Equatorial Current.

RESULTS

Numerous time series records of oxygen isotopes have documented the major changes in Miocene paleoclimates (Shackleton and Kennett 1975; Boersma and Shackleton 1977; Savin et al. 1981, Woodruff et al. 1981). The CENOP time-slices at 22, 16, and 8 Ma were selected to contrast three differing states of the Miocene ocean. The 22 Ma time-slice was chosen to examine an oligotaxic ocean immediately prior to a major Neogene evolutionary radiation of planktonic foraminifera (Keller 1981a, b; Kennett and Srinivasan, 1983). Isotopic records in the early Miocene show a distinct depletion trend between 22 and 16 Ma, which culminates to a warm climatic optimum at 16 Ma at the end of the early Miocene. From 16 to 13 Ma, a 1‰ enrichment in ^{18}O values has been interpreted to represent the growth of a large, permanent ice-sheet on Antarctica (Shackleton and Kennett 1975; Savin et al. 1975; Woodruff et al. 1981), although an alternative explanation is that the enrichment resulted mostly from a cooling at high latitudes (Matthews and Poore 1980). Oxygen isotopic values in the late Miocene are enriched relative to pre-Middle Miocene values, but late Miocene oxygen isotopic records exhibit considerable variability suggesting that temperature and/or ice volume changes were occurring during this time (Savin et al. 1981). The 8 Ma time slice was chosen to characterize the late Miocene ocean following the climatic changes associated with the middle Miocene oxygen isotopic event.

For each of the time-slices, the average of the planktonic ^{18}O values for each species is plotted on each of the time-slice maps in Figures 2, 4, and 6. The mean, standard deviation, and 95% confidence limits of the oxygen isotopic data for each time-slice is shown in Tables 3-5. The average percent frequency of selected species is plotted and contoured in Figures 3, 5, and 7. Species of *Globigerinoides* have been grouped for mapping purposes. Most of the faunal data represent an average of only two samples per time-slice, and the variability between samples is sometimes large. The faunal data are used only to interpret general trends, and provide a qualitative check on the oxygen isotopic paleotemperature interpretations.

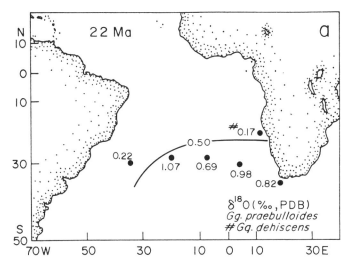

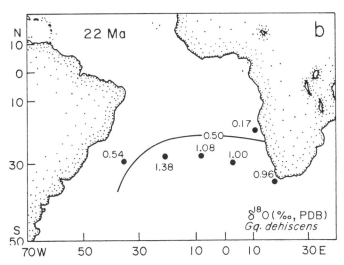

Figure 2. (a) Average ^{18}O of *Globigerina praebulloides* at 22 Ma. Average ^{18}O of *Globoquadrina dehiscens* is plotted at Site 362 since *Gg. praebulloides* was not analyzed at that site. (b) Average ^{18}O of *Gq. dehiscens* at 22 Ma. Note relatively depleted ^{18}O value at central gyre Site 14.

22 Ma (earliest Miocene)

Oxygen isotopic values of *Gg. praebulloides* and *Gq. dehiscens* at 22 Ma reveal a distinct pattern for the South Atlantic (Table 3; Fig. 2a, b). Site 362, the lowest latitude site, exhibits the most depleted ("warmest") ^{18}O values, while Site 357 in the western basin has the second most depleted values. Site 14 in the central gyre region has the most enriched ("coolest") mean oxygen isotopic value of all sites for both species analyzed.

Contours of average frequencies of *Gg. glutinata* show a similar pattern as the oxygen isotopic data (Fig. 3a). The highest frequencies of *Gg. glutinata* are found in sites with relatively enriched ^{18}O values. Low frequencies of this species occur in lower-latitude Site 362 and western boundary Site 357. These

sites also exhibit the most depleted ^{18}O values. Sites 14 and 360 contain lower frequencies of *Gg. woodi* (Fig. 3b), and central gyre Sites 14 and 18 have lower frequencies of *Gr. continuosa/nana* (Fig. 3c). *Gg. praebulloides* is an important species in all sites and ranges up to 36% of the total assemblage (Fig. 3d), but shows no obvious geographic trends. *Catapsydrax dissimilis* and *Cs. unicavus* are also present at all sites in moderate to low frequencies (Figure 3e).

16 Ma (late early Miocene)

The oxygen isotopic pattern at 16 Ma is similar to that at 22 Ma (Figure 4a, b; Table 4), although all sites record an average depletion of 0.5$^{0}/_{00}$ between 22 and 16 Ma. The isotopic results at 16 Ma show similar patterns for both species analyzed, including shallow-dwelling *Globigerinoides* spp. and deeper-dwelling *Gq. dehiscens*. The most depleted isotopic values occur at lower-latitude Site 362, while the most enriched values are found at Site 15 in the central gyre (Figure 4a, b; Table 4). Site 516A in the western basin remains significantly more depleted in ^{18}O than any other site at similar latitude.

The biogeographic distribution of planktonic foraminiferal species at 16 Ma supports the isotopic results. Sites exhibiting the most depleted ^{18}O values have higher frequencies of warmer-water forms, whereas isotopically enriched sites contain cooler water assemblages. For example, a sharp demarcation exists between high frequencies of the warm-water form *Globigerinoides* found in Sites 362 and 516A compared with low frequencies found in other sites (Figure 5a). In contrast, *Gr. miozea,* a temperate-subantarctic species, has highest frequencies in central gyre Site 15 and high-latitude Site 360 (Figure 5b). Central gyre Site 15 also contains a greater number of deep-dwelling *Gq. dehiscens* and *Gq. venezuelana,* and lower frequencies of the warmer-water form *Gg. woodi* (Figure 5c, d, e).

8 Ma (late Miocene)

At 8 Ma, eastern basin Sites 360 and 362 show anomalously depleted ^{18}O values for *Globigerinoides* (Figure 6a, Table 5). The ^{18}O data for *Gr. conoidea* indicate that Site 362 has the most depleted ("warmest") ^{18}O values (Figure 6b, Table 5), but the magnitude of the ^{18}O gradient between Site 362 and the remaining sites to the south is diminished at 8 Ma. compared with earlier Miocene time slices. Western basin Site 516A is no longer isotopically more depleted ("warmer") than other sites at similar latitude. Central gyre Site 16 remains the most isotopically enriched site for both species analyzed. Between 16 and 8 Ma, all sites except 360 show an enrichment in average surface water ^{18}O values (Table 6).

A distinct change in planktonic carbon isotopic patterns occurred between 16 and 8 Ma. A significant depletion occurred in average δ^{13}C values in eastern basin Sites 360 and 362. No change in average carbon isotopic values occurred in the remaining sites. In Sites 360 and 362, *Globigerinoides* became depleted

22 Ma

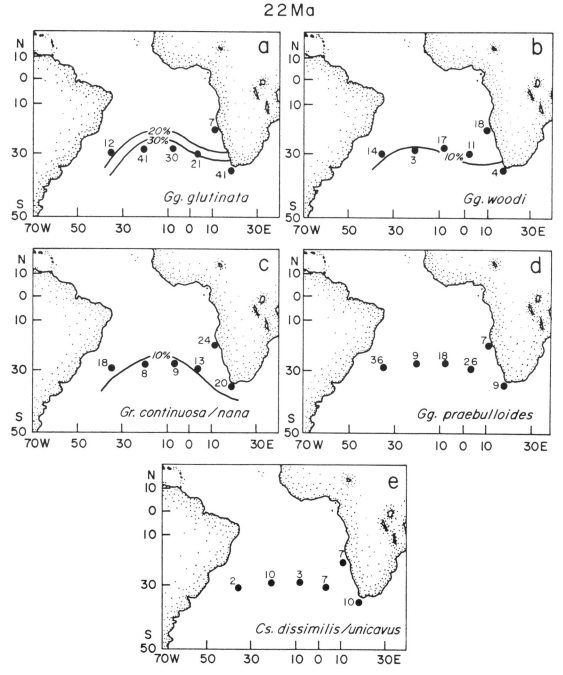

Figure 3. Species percentage data at 22 Ma. for: (a) *Globigerinita glutinata*—lower frequencies occur in Sites 362 and 357; (b) *Globigerina woodi*—lower frequencies occur in central gyre Site 14 and high-latitude Site 360; (c) *Globorotalia continuosa/nana*—lower frequencies occur in central gyre Sites 14 and 18; (d) *Gg. praebulloides* is an important species in all sites but shows no obvious geographic trends; (e) *Catapsydrax dissimilis/unicavus*—moderate to low frequencies occur in all sites.

TABLE 3. SUMMARY OF STATISTICS ON OXYGEN ISOTOPIC DATA FOR
GLOBIGERINA PRAEBULLOIDES AND GLOBOQUADRINA DEHISCENS AT 22 MA*

Level	N	Mean	Standard Deviation	Individual 95% Confidence Intervals for Mean Based on Pooled Standard Deviation

$\delta^{18}O$ (O/oo, PDB) **Globigerina praebulloides**

```
                                           ---------+---------+---------+-------
Site 14     4    1.3800    0.1160                               (---*--)
Site 18     6    1.0833    0.1959                           (--*--)
Site 357   11    0.5409    0.2094                 (-*-)
Site 360    4    0.9575    0.2133                       (--*---)
Site 362   10    0.1710    0.1396             (-*--)
Site 526    2    0.9950    0.0778                     (----*----)
                                           ---------+---------+---------+-------
Pooled Standard Deviation = 0.1786            0.50      1.00      1.50
```

$\delta^{18}O$ (O/oo, PDB) **Globoquadrina dehiscens**

```
                                          ----------+---------+---------+------
Site 14     4    1.0650    0.0645                              (----*----)
Site 18     4    0.6925    0.2482                        (----*----)
Site 357    4    0.2200    0.2709              (---*----)
Site 360    3    0.8233    0.1721                     (-----*----)
Site 526    5    0.9780    0.0901                       (---*----)
                                          ----------+---------+---------+------
Pooled Standard Deviation = 0.1843            0.40      0.80      1.20
```

*Note: 95% confidence intervals are plotted about the mean oxygen isotopic value at each site based upon the pooled standard deviation.

TABLE 4. SUMMARY OF STATISTICS ON OXYGEN ISOTOPIC DATA FOR
GLOBIGERINOIDES SPP. AND GLOBOQUADRINA DEHISCENS AT 17 MA*

Level	N	Mean	Standard Deviation	Individual 95% Confidence Intervals for Mean Based on Pooled Standard Deviation

$\delta^{18}O$ (O/oo, PDB) **Globigerinoides**

```
                                       ----+---------+---------+---------+--
Site 15     3    0.467     0.235                                  (---*--)
Site 360    7   -0.014     0.152                          (-*-)
Site 362    5   -1.254     0.166       (--*--)
Site 516    4   -0.203     0.071                     (--*--)
Site 526    4    0.277     0.213                             (--*--)
                                       ----+---------+---------+---------+--
Pooled Standard Deviation = 0.168      -1.20    -0.60      0.00      0.60
```

$\delta^{18}O$ (O/oo, PDB) **Globoquadrina dehiscens**

```
                                        ---+---------+---------+---------+---
Site 15     2    0.9250    0.0354                         (-----*-----)
Site 360    9    0.5789    0.3545                    (--*-)
Site 362    5   -0.5100    0.1089        (--*---)
Site 516    5    0.0580    0.1274              (---*---)
Site 526    7    0.7443    0.2048                   (--*---)
                                        ---+---------+---------+---------+---
Pooled Standard Deviation = 0.2441      -0.60      0.00      0.60      1.20
```

*Note: 95% confidence intervals are plotted about the mean oxygen isotopic value at each site based upon the pooled standard deviation.

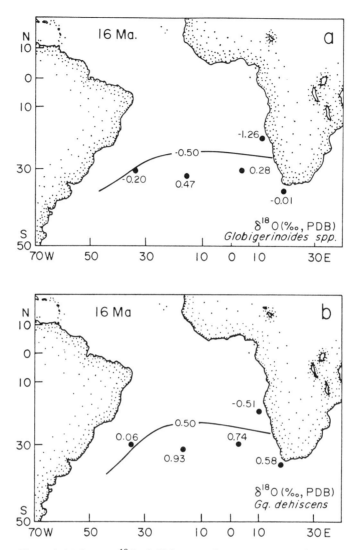

Figure 4. (a) Average [18]O of *Globigerinoides* spp. at 16 Ma. Sites 362 and 516A have the most depleted mean [18]O values, while central gyre Site 15 has the most enriched mean [18]O value. (b) Average [18]O of *Gq. dehiscens* at 16 Ma. Note similar isotopic pattern as 5a.

by 1.07⁰/₀₀ and the deeper-dwelling species became depleted by 1.30⁰/₀₀ (Table 7).

The most distinctive aspect of the biogeographic distribution of planktonic foraminifera at 8 Ma is the anomalous assemblage in eastern boundary Sites 360 and 362 (Figure 7). Planktonic foraminiferal assemblages from these sites were modified by dissolution as evidenced by intense fragmentation of foraminiferal tests. Abundances of dissolution-susceptible forms (e.g. *Globigerinoides*) are markedly reduced (Figure 7a), while dissolution-resistant forms (e.g. *N. pachyderma*) dominate the assemblage (Figure 7b).

Central gyre Site 16 contains high frequencies of *Globigerina druryi* and *Gg. nepenthes* that are relatively dissolution-resistant (Figure 7c). Western boundary Site 516A has the highest frequencies of *Globigerinoides* and *Gg. woodi* (Figure 7a,

d), and eastern boundary Sites 360, 362, and 526A contain greater frequencies of the subpolar-form *Gg. praebulloides/bulloides* (Figure 7e).

DISCUSSION

Some common patterns emerge from the oxygen isotopic data at 22, 16, and 8 Ma. Site 362 always has the most depleted ("warmest") [18]O values reflecting its lower latitude position. A sharp faunal and oxygen isotopic gradient existed between Site 362 (20°S) and the remaining sites to the south (~30°S) during the early Miocene. This gradient marked the transition between the warm subtropics to the north and the temperate region to the south. The sharp gradient between Sites 360 and 362 also indicates that the Benguela Current was not effectively transporting

16 Ma

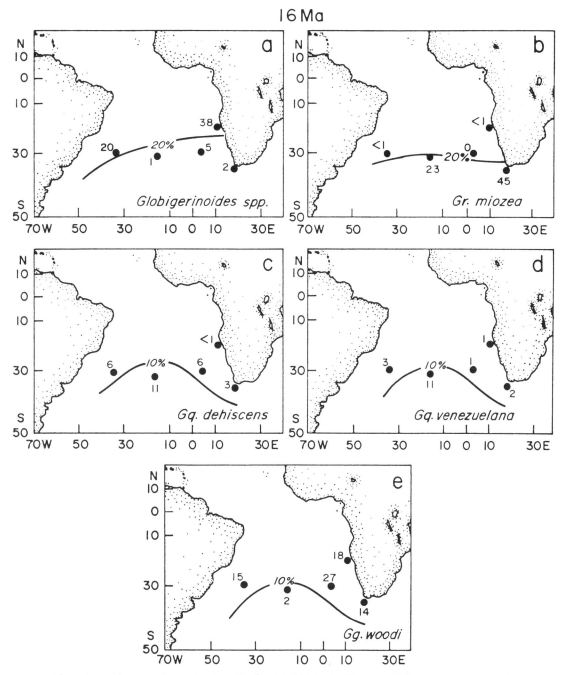

Figure 5. Species percentage data at 16 Ma. for: (a) *Globigerinoides* spp.—higher frequencies occur in low-latitude Site 362 and Site 516A to the west; (b) *Globorotalia miozea*—higher frequencies occur in high-latitude Site 360 and central gyre Site 15; (d) *Globoquadrina venezuelana*—highest frequencies occur in Site 15; (e) *Gg. woodi*—lowest frequencies occur in Site 15.

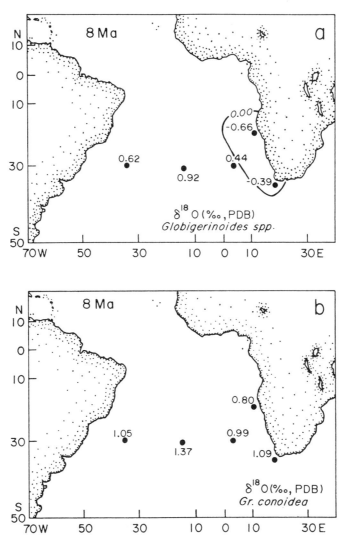

Figure 6. (a) Average [18]O of *Globigerinoides* at 8 Ma. [18]O values are anomalously depleted in Site 360. (b) Average [18]O of *Globorotalia conoidea* at 8 Ma. Note decreased contrast in [18]O values between Site 362 and remaining sites to the south as compared to the early Miocene time-slices.

cold subantarctic waters along the eastern boundary to 20°N during the early Miocene. A similar faunal and isotopic gradient was noted for all three Miocene time-slices in the southwest Pacific between Sites 206 (32°S) and 208 (26°S) (Kennett et al. 1985; Savin et al. 1985).

An oxygen isotopic gradient is also apparent between the central gyre sites and remaining sites for each of the time-slices. The central gyre site always has slightly more enriched mean oxygen isotopic values than other sites regardless of the species analyzed. These enriched [18]O values may reflect cooler temperatures due to the proximity of the Subtropical Convergence that bends to the north in the central gyre region of the modern South Atlantic (Figure 1). Since the central gyre sites (14, 15, 16, and 18) are also the deepest (4343, 3927, 3527, and 4018 m respectively), an alternative explanation could be that the anomalously

enriched [18]O values result from increased dissolution at depth. Enhanced dissolution could selectively remove delicate forms that lived closer to the sea surface and had more depleted [18]O values. Although we cannot conclusively reject the latter interpretation, the consistently enriched [18]O values obtained on both resistant and susceptible species suggest that the enriched [18]O values of the central gyre reflects cooler sea surface temperatures rather than any bias imparted by dissolution.

During the early Miocene (22 and 16 Ma), average planktonic [18]O values in western boundary Sites 357 and 516A were significantly more depleted ("warmer") than values at other sites located at the same latitude in the central or eastern basins. The planktonic foraminiferal assemblages in the western sites also contain higher percentages of warm-water forms such as *Globigerinoides*. The relative warmth along the western boundary during

8 Ma

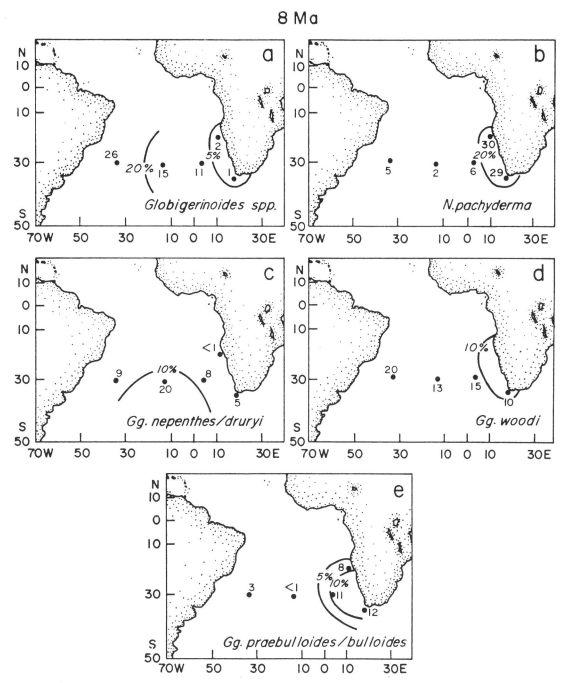

Figure 7. Species percentage data at 8 Ma. for: (a) *Globigerinoides* spp.—low frequencies occur in sites along the eastern basin of the South Atlantic. (b) *Neogloboquadrina pachyderma*—highest frequencies occur in Sites 360 and 362 along the west coast of Africa; (c) *Globigerina druryi/nepenthes*—high frequencies occur in central gyre Site 16; (d) *Gg. woodi*—lowest frequencies occur in Sites 360 and 362; (e) *Gg. praebulloides/bulloides*—higher frequencies occur to the east.

328 Hodell and Kennett

TABLE 5. SUMMARY OF STATISTICS ON OXYGEN ISOTOPIC DATA FOR
GLOBIGERINOIDES SPP. AND GLOBOROTALIA CONOIDEA AT 8 MA*

Level	N	Mean	Standard Deviation	Individual 95% Confidence Intervals for Mean Based on Pooled Standard Deviation

$\delta^{18}O$ (O/oo, PDB) Globigerinoides spp.

```
                                          ------+---------+---------+---------+
Site 16    5   0.9160    0.1150                                      (---*---)
Site 360   8  -0.3425    0.2583                (--*---)
Site 362   5  -0.6560    0.4628           (---*---)
Site 516   5   0.6160    0.2681                             (---*---)
Site 526   6   0.4433    0.1384                            (--*---)
                                          ------+---------+---------+---------+
Pooled Standard Deviation = 0.2708        -0.60      0.00      0.60      1.20
```

$\delta^{18}O$ (O/oo, PDB) Globorotalia conoidea

```
                                          ----------+---------+---------+------
Site 16    5   1.3700    0.1027                              (-------*------)
Site 360   9   1.0944    0.3964                      (----*-----)
Site 362   6   0.8017    0.1357           (------*-----)
Site 516   5   1.0480    0.1121               (------*------)
Site 526   7   0.9914    0.0847               (-----*-----)
                                          ----------+---------+---------+------
Pooled Standard Deviation = 0.2345           0.90      1.20      1.50
```

*Note: 95% confidence intervals are plotted about the mean oxygen isotopic value at each site based upon the pooled standard deviation.

TABLE 6. THE DIFFERENCE IN OXYGEN ISOTOPIC VALUES ($\Delta^{18}O$)
BETWEEN EACH OF THE 3 CENOP TIME-SLICES*

Site	Gg. praebuloides 22 Ma $\overline{x}(\delta^{18}O)$	Globigerinoides spp. 16 Ma $\overline{x}(\delta^{18}O)$	22-16 Ma $\Delta^{18}O$	Gq. dehiscens 22 Ma $\overline{x}(\delta^{18}O)$	Gq. dehiscens 16 Ma $\overline{x}(\delta^{18}O)$	22-16 Ma $\Delta^{18}O$
14, 15, 16	1.07	0.47	0.60	1.38	0.93	0.45
360	0.82	-0.01	0.83	0.96	0.58	0.38
362	---	---	---	0.17	-0.51	0.68
357, 515A	0.22	-1.26	0.42	0.54	0.06	0.48
526A	0.98	0.28	0.70	1.00	0.74	0.26
			$\overline{x} = \overline{0.64}$			$\overline{x} = \overline{0.45}$

Site	Globigerinoides spp. 16 Ma $\overline{x}(\delta^{18}O)$	Globigerinoides spp. 8 Ma $\overline{x}(\delta^{18}O)$	16-8 Ma $\Delta^{18}O$	Gq. dehiscens 16 Ma $\overline{x}(\delta^{18}O)$	Gr. conoidea 8 Ma $\overline{x}(\delta^{18}O)$	16-8 Ma $\Delta^{18}O$
14, 15, 16	0.47	0.92	-0.45	0.93	1.38	-0.44
360	-0.01	-0.34	+0.33*	0.58	1.09	-0.51
362	-1.26	-0.66	-0.60	-0.51	0.80	-1.31
357, 516A	-0.20	0.62	-0.82	0.06	1.05	-0.99
526A	0.30	0.44	-0.14	0.74	0.99	-0.25
			$\overline{x} = \overline{-0.51}$			$\overline{x} = \overline{-0.70}$

*Note the average change in $\delta^{18}O$ between 22 and 16 Ma was a depletion of about 0.5 O/oo. Between 16 and 8 Ma, $\delta^{18}O$ values increased by about 0.6 O/oo.

TABLE 7. THE DIFFERENCE IN CARBON ISOTOPIC VALUES ($\Delta^{13}C$)
BETWEEN EACH OF THE 3 CENOP TIME-SLICES*

Site	Gg. praebuloides 22 Ma $\bar{x}(\delta^{13}C)$	Globigerinoides spp. 16 Ma $\bar{x}(\delta^{13}C)$	22-16 Ma $\Delta^{13}C$	Gq. dehiscens 22 Ma $\bar{x}(\delta^{13}C)$	Gq. dehiscens 16 Ma $\bar{x}(\delta^{13}C)$	22-16 Ma $\Delta^{13}C$
14, 15, 16	1.66	2.37	-0.71	1.79	2.15	-0.36
360	1.98	2.44	-0.46	1.79	2.32	-0.53
362	---	2.69	---	1.18	2.15	-0.97
357, 516A	0.97	1.80	-0.83	1.28	1.55	-0.27
526A	1.50	1.99	-0.49	1.68	1.38	+0.30

Site	Globigerinoides spp. 16 Ma $\bar{x}(\delta^{13}C)$	Globigerinoides spp. 8 Ma $\bar{x}(\delta^{13}C)$	16-8 Ma $\Delta^{13}C$	Gq. dehiscens 16 Ma $\bar{x}(\delta^{13}C)$	Gr. conoidea 8 Ma $\bar{x}(\delta^{13}C)$	16-8 Ma $\Delta^{13}C$
14, 15, 16	2.37	2.02	0.35	2.15	1.90	0.25
360	2.44	1.36	1.08*	2.32	0.98	1.34*
362	2.69	1.63	1.06*	2.15	0.88	1.27*
357, 516A	1.80	1.77	0.03	1.55	1.55	0.00
526A	1.97	1.95	0.02	1.38	1.33	0.05

*Note the distinct depletion in carbon isotopic values in Sites 360 and 362 between 16 and 8 Ma.

the early Miocene was probably caused by transport of tropical-warm subtropical waters within the Brazil Current, which in the Modern ocean extends as far south as 40°S. At 8 Ma, oxygen isotopic values at Site 516A were no longer significantly more depleted than other sites, although *Globigerinoides* remained about twice as abundant in the west as in the east. With our sample coverage, it was not possible to determine whether the strength of the Brazil Current actually diminished between 16 and 8 Ma or whether the axis of the current shifted away from Site 516A. Other CENOP data suggest that gyre circulation markedly intensified during the late Miocene (Kennett et al. 1985; Barrera et al 1985), and it is possible that the Brazil Current also increased in velocity, but became narrower and shifted closer to South America between 16 and 8 Ma.

Between 22 and 16 Ma, planktonic ^{18}O values become more depleted by an average of 0.5‰ in all sites (Table 6). Assuming this oxygen isotopic change represented only a temperature effect, surface waters in the mid-latitude South Atlantic would have warmed by an average of 2°C between 22 and 16 Ma. This estimate is in good agreement with global CENOP oxygen isotopic results that also suggest surface waters globally warmed by an average of 2°C between these time slices. In the tropics, sea surface temperatures may have warmed by as much as 4°C between 22 and 16 Ma (Savin et al. 1985).

At 8 Ma, eastern boundary Sites 360 and 362 show an increase in sedimentation rates, an increase in foraminiferal fragmentation, high abundances of the polar-form *N. pachyderma*, a reduction in the oxygen isotopic gradient between Sites 360 and 362, and a marked depletion in average surface water ^{13}C. These changes are interpreted as resulting from an increase in

northward transport of the Benguela Current and increased intensity of the Benguela upwelling system. The late Miocene increase in sedimentation rates at Sites 360 and 362 reflect increased surface water productivity induced by upwelling. Increased fragmentation and dissolution during the late Miocene may be due to a shoaling of the lysocline beneath the Benguela upwelling system as organic matter rain-rate increased. Dissolution would have resulted from in situ oxidation of organic matter and formation of corrosive, metabolic CO_2 within the sediments. The high percentage of *N. pachyderma* and a diminished oxygen isotopic gradient between Sites 360 and 362 may be due to increased transport of cool subantarctic waters to lower latitudes by the Benguela Current. In the Modern ocean, high frequencies of *N. pachyderma* continue to be found in surface sediments off the southwest coast of Africa reflecting northward transport of cool water within the Benguela Current (Bé and Tolderlund 1977). The depletion in average surface water ^{13}C between 16 and 18 Ma is consistent with increased upwelling intensity since deep waters are more depleted in ^{13}C than surface waters. The use of planktonic foraminiferal ^{13}C as an upwelling indicator is not definative, however, in the Modern ocean (Prell and Curry 1981).

Our interpretation of increased upwelling off the southwest African coast between 16 and 8 Ma is supported by Seisser (1980) and more recently by the results of DSDP Leg 75 (Dean et al. 1984), which indicate an age of early late Miocene for initiation of the Benguela upwelling system. Thunell and Belyea (1982) also provide evidence for intensified eastern boundary currents and upwelling along northwest Africa during the late Miocene based upon planktonic foraminiferal biogeographic patterns.

330 Hodell and Kennett

At 8 Ma, average oxygen isotopic values for *Globigerinoides* are anomalously depleted in Sites 360 and 362. Although upwelling of cold deep water is expected to produce an enrichment in surface water ^{18}O, upwelling is a seasonal phenomenon and may not occur during summer months when *Globigerinoides* proliferate as in the Panama Basin (Thunell et al. 1983). Oxygen isotopic values for deeper-dwelling *Gr. conoidea* indicate a decrease in the oxygen isotopic gradient between Sites 360 and 362. The relatively enriched oxygen isotopic values at Site 362 are consistent with upwelling and advection of cooler waters at this low-latitude site. The anomalously depleted oxygen isotopic values for *Globigerinoides* may reflect southward transport of warm water from the Indian Ocean by the Agulhas Current.

Between 16 and 8 Ma, all sites show an enrichment in surface water [18]O except *Globigerinoides* in Site 360. Excluding data from Site 360, an average [18]O enrichment of 0.51%o is recorded for *Globigerinoides,* whereas an average enrichment of 0.66%o is recorded by *Gq. dehiscens/Gq. conoidea* between 16 and 8 Ma (Table 6). This enrichment probably reflects a combined global ice volume increase as well as cooling of surface waters. If an ice volume of 0.5%o is assumed between 16 and 8 Ma as suggested by Savin et al. (1985), then sea surface temperatures in the mid-latitude South Atlantic remained nearly constant between 16 and 8 Ma. During this time, surface waters warmed by as much as 4°C in the equatorial Pacific and Indian Oceans, while surface waters cooled at high latitudes (Savin et al. 1985). The growth of ice-sheets on Antarctica between middle and late Miocene time strengthened latitudinal temperature gradients whereby the tropics warmed, the mid-latitudes maintained near constant temperature, and the high latitudes cooled.

CONCLUSIONS

1. Planktonic foraminiferal and stable isotopic patterns were used to infer changes in the South Atlantic anticyclonic gyre system for 3 time-slices in the Miocene: 22, 16, and 8 Ma. Circulation patterns were similar at 22 and 16 Ma, but major changes occurred in the subtropical gyre system between 16 and 8 Ma.

2. At 22 and 16 Ma, a sharp faunal and isotopic gradient existed between Site 362 (20°S) and the remaining sites to the south (~30°S). This gradient marked the transition between the warm subtropics to the north and temperate waters to the south. No evidence exists for coastal upwelling along the eastern margin of the South Atlantic at 22 or 16 Ma.

3. At 22 and 16 Ma, western boundary Sites 357 and 516A had more depleted ("warmer") [18]O values and contained higher

frequencies of warm-water planktonic foraminifera. The warm surface waters along the western boundary suggest that the Brazil Current was effective at transporting tropical-warm subtropical waters as far south as the Rio Grande Rise during the early Miocene.

4. Between 22 and 16 Ma, surface water temperatures in the mid-latitude South Atlantic warmed by an average of 2°C.

5. The most pronounced change in the South Atlantic subtropical gyre system occurred between 16 and 8 Ma following climatic deterioration associated with the middle Miocene oxygen isotopic event. Increased velocities of the Benguela Current transported cold subantarctic waters to lower latitudes along the eastern boundary of the South Atlantic. This is evidenced by an assemblage dominated by the polar-form *N. pachyderma* in Sites 360 and 362 as well as a decrease in the oxygen isotopic gradient between these sites. The intensity of the Benguela upwelling system also increased by 8 Ma as indicated by increased sedimentation rates, increased foraminiferal fragmentation, and a depletion of average surface water ^{13}C at Sites 360 and 362.

6. Between 16 and 8 Ma, all sites except 360 show an enrichment in average surface water [18]O reflecting an increase in global ice volume and decrease in surface water temperatures. Global CENOP isotopic data suggest that the high latitudes cooled between 16 and 8 Ma, while the tropics warmed by as much as 4°C (Savin et al 1985). If an ice volume effect of 0.5%o is assumed between 16 and 8 Ma, then surface water temperatures remained nearly constant in the mid-latitude South Atlantic between these time-slices.

7. In all three Miocene time-slices, the central gyre sites exhibited the most enriched oxygen isotopic values and contained a greater frequency of cold-water forms. Although this may reflect dissolution at greater water depths, it is more probable that the enriched [18]O values of the central gyre region reflect cooler temperatures due to the proximity of the Subtropical Convergence.

ACKNOWLEDGMENTS

The authors thank J. Ingle, W. Prell, K. Romine, and R. Thunell for critically reviewing the manuscript. We also thank C. Cipolla, S. Cynar, and K. Elmstrom for preparation of samples for isotopic analyses, and N. Meader for typing the manuscript. This research was supported by NSF Grant OCE79-14594 (CENOP) and OCE82-14937 to J. P. Kennett. Samples were provided through the Deep Sea Drilling Project with funding from the National Science Foundation.

APPENDIX 1A. RAW SPECIES COUNTS ON SAMPLES BELONGING TO THE 22 MA TIME-SLICE.
Generic abbreviations are according to Kennett and Srinivasn (1983).

Core Interval	14-2-2 69-74 cm	14-2-3 145-150 cm	18-4-4 68-73 cm	18-5-2 145-150 cm	357-12-3 73-78 cm	357-13-2 73-78 cm	360-26-2 45-50 cm	360-26-2 93-97 cm	362-39-6 72-77	362-40-4 145-150 cm	526A-27-2 50-54 cm	526A-29-1 50-54 cm
Gg. angustiumbilicata	3	0	2	8	3	6	0	2	3	3	9	3
Gg. praebulloides	24	23	45	46	80	117	41	11	24	28	195	41
Gg. brazieri	0	1	2	3	2	2	1	0	1	3	2	3
Gg. connecta	6	4	8	3	16	10	1	3	3	0	7	2
Gg. woodi	15	1	47	37	70	11	10	14	52	76	16	46
Gg. euapertura	0	1	0	0	0	3	4	0	0	0	5	1
Gs. primordius	0	7	1	13	1	5	0	0	0	0	1	0
Gs. triloba	0	3	0	0	0	0	0	0	0	1	1	1
Gs. parawoodi	0	0	0	0	0	0	0	0	0	1	2	0
Gs. subquadratus	0	0	0	0	1	0	2	0	0	0	1	0
Gr. kugleri	46	8	2	2	5	0	0	1	0	1	0	2
Gr. nana	6	12	8	12	31	40	21	22	39	154	49	10
Gr. semivera	0	0	0	0	0	0	0	0	0	0	1	0
Gr. mayeri	1	0	0	0	7	0	0	0	2	1	0	0
Gq. venezuelana	30	24	6	5	1	1	14	11	0	5	11	8
Gq. binaiensis	3	1	0	0	1	1	0	0	0	0	0	2
Gq. praedehiscens	3	4	20	15	0	0	14	6	0	0	22	15
D. altispira altispira	1	2	0	0	4	0	0	0	0	0	1	1
N. continuosa	4	12	12	12	13	13	32	37	56	89	52	7
Gd. suteri	1	3	3	3	8	6	5	1	24	6	26	0
Ga. uvula	1	2	2	0	0	0	1	0	0	1	0	1
Ga. glutinata	142	88	86	67	41	25	114	121	12	47	91	60
C. dissimilis	5	12	5	3	1	2	1	5	5	10	1	1
C. unicavus	11	23	7	1	4	2	10	42	11	26	20	21
Total	302	231	266	227	288	244	271	271	234	451	524	224

APPENDIX 1B. RAW SPECIES COUNTS ON SAMPLES BELONGING TO THE 16 MA TIME-SLICE

Core Interval	15-7-1 65-70 cm	15-7-2 55-59 cm	360-22-2 142-147 cm	360-22-3 145-150 cm	362-37-1 145-150 cm	362-37-2 145-150 cm	516-22-1 51-55 cm	516-21-3 86-90 cm	526A-21-2 50-54 cm	526A-21-3 130-134 cm
Cg. chiplensis	0	0	0	0	0	2	0	0	0	0
Cs. parvulus	0	2	0	0	1	1	0	12	4	12
Gg. angustiumbilicata	0	2	0	0	0	0	0	0	0	0
Gg. quinqueloba	0	0	1	0	0	0	0	0	0	3
Gg. praebulloides	2	12	15	4	13	9	33	105	46	104
Gg. falconensis	0	1	11	10	12	9	40	75	10	44
Gg. brazieri	0	4	7	0	16	23	10	26	30	25
Gg. woodi	3	15	75	20	89	84	52	174	170	153
Gg. connecta	7	42	21	14	35	21	6	77	100	49
Gg. druryi	7	7	9	4	1	47	7	46	29	29
Gs. obliquus	0	4	2	5	73	94	13	83	3	1
Gs. triloba	1	0	8	4	40	0	20	65	19	0
Gs. sicanus	0	0	0	0	15	3	2	22	1	0
Gs. immaturus	0	0	0	0	4	34	0	2	1	1
Gs. quadrilobatus	7	2	2	2	23	0	8	25	6	3
Gs. sacculifer	0	0	0	0	0	0	0	2	0	0
Gs. subquadratus	0	0	0	0	25	0	1	13	0	0
Gs. diminutus	0	0	0	0	0	0	0	3	0	0
Gs. mitra	0	1	0	0	7	0	1	9	0	0
Praeorbulina	3	4	5	5	0	2	1	0	0	0
Gr. peripheroronda	0	4	2	2	18	3	0	4	4	1
Gr. birnageae	0	2	0	4	0	6	0	0	14	0
Gr. miozea	15	144	245	70	1	1	1	8	0	0
Gr. praescitula	0	0	34	18	0	22	10	29	2	0
Gr. zealandica	0	0	1	0	5	6	0	7	8	3
Gr. incognita	0	0	0	4	1	30	4	19	15	58
Gr. semivera	0	1	0	0	1	3	0	13	13	1
Gr. siakensis	0	0	0	0	1	0	0	1	0	0
Gr. bella	0	0	0	0	0	1	0	1	2	0
Gr. mayeri	0	0	0	4	13	28	5	3	9	8
Gr. acrostoma	0	0	0	0	0	5	4	0	0	13
Gr. venezuelana	28	67	13	4	6	3	1	29	9	6
Gq. dehiscens	21	70	17	10	3	6	7	64	38	24
Gq. baroemoenensis	0	0	3	0	0	0	0	14	7	1
D. altispira globosa	0	0	0	0	0	0	0	0	0	0
D. altispira altispira	0	0	0	0	0	1	0	1	0	0
N. continuosa	0	0	0	0	2	0	0	0	0	0
Ss. disjuncta	6	0	0	1	8	4	0	1	1	0
Gd. suteri	0	4	2	3	0	7	2	45	6	12
Gd. variabilis	0	0	0	0	0	1	0	0	0	0
Gd. hexagona	0	0	0	0	0	1	0	2	0	0
Cl. bermudezi	0	0	0	0	0	1	0	0	0	0
Ga. uvula	0	0	0	1	0	0	0	0	0	1
Ga. glutinata	13	72	58	79	44	64	31	123	64	113
Gt. insueta	0	0	0	0	0	0	0	0	0	0
Ge. obesa	0	0	0	0	0	1	0	7	0	0
Total	113	474	531	261	457	533	259	1114	608	665

APPENDIX 1C. RAW SPECIES COUNTS ON SAMPLES BELONGING TO THE 8 MA TIME-SLICE

Core / Interval	16-9-6 51-56 cm	16-10-6 50-55 cm	360-9-2 50-55 cm	360-9-5 43-48 cm	362-24-1 52-56 cm	362-24-3 52-56 cm	362-25-6 51-55 cm	516A-13-3 20-22 cm	516A-14-1 20-22 cm	526A-10-3 50-54 cm	526A-11-3 55-59 cm
Gg. bulloides	0	3	39	0	0	0	35	17	7	68	23
Gg. decoraptera	1	4	0	1	0	0	7	11	10	3	9
Gg. falconensis	1	5	5	6	4	1	2	24	19	47	11
Gg. nepenthes	119	99	1	0	0	2	0	57	26	21	10
Gg. druryi	79	84	25	10	0	0	13	71	41	44	18
Gg. praebulloides	1	8	16	33	27	25	0	8	12	24	7
Gg. quinqueloba	0	0	0	1	0	0	0	0	0	0	0
Gg. woodi	59	51	96	5	22	12	85	111	148	129	80
Gg. apertura	0	3	4	1	1	3	12	21	19	17	3
Gg. bollii	0	0	0	0	0	0	0	2	10	5	1
Gs. conglobatus	20	4	3	0	1	0	0	29	27	16	3
Gs. extremus	0	0	0	0	0	0	0	0	1	1	0
Gs. immaturus	1	0	0	0	0	0	1	0	2	2	0
Gs. obliquus	1	5	2	0	6	0	1	32	44	10	6
Gs. sacculifer	5	4	0	1	1	0	0	5	2	10	4
Gs. quadrilobatus	10	10	0	1	4	0	1	9	47	11	5
Gs. seiglii	0	0	0	0	0	0	0	16	0	0	0
Gs. trilobus	29	8	1	0	2	0	0	4	52	30	4
Gs. ruber	0	1	0	0	0	0	0	0	2	4	5
Gr. cibaoensis	1	1	7	1	0	1	6	34	50	15	3
Gr. conoidea	51	67	21	17	4	19	79	1	0	109	33
Gr. conomiozea	0	1	12	0	0	0	0	77	10	10	1
Gr. lenguaensis	0	0	0	0	0	0	0	25	14	0	1
Gr. menardii	20	20	4	4	1	14	1	0	28	41	11
Gr. merotumida	0	0	1	0	0	0	0	0	0	0	0
Gr. plesiotumida	0	0	0	0	0	0	0	2	0	1	0
Gr. scitula	0	0	7	5	5	0	4	0	1	42	16
Gq. altispira altispira	0	0	0	0	0	0	1	0	2	0	0
Gq. altispira globosa	0	0	1	1	1	0	0	4	0	0	0
Gq. dehiscens	0	0	0	0	0	0	0	0	0	1	37
Gq. venezuelana	0	0	0	1	0	0	0	0	8	0	2
N. acostaensis	0	0	11	14	19	27	22	0	1	6	0
N. continuosa	0	0	79	33	92	51	3	0	13	0	2
N. humerosa	0	1	0	1	0	0	59	12	0	0	0
N. pachyderma (sinistral)	6	14	138	84	144	72	27	12	27	17	10
N. pachyderma (dextral)	3	7	5	1	5	2	0	0	12	30	9
Gd. varibailis	0	0	0	2	1	0	0	1	0	0	0
Ge. aequilateralis	0	0	4	0	1	0	0	0	3	3	2
Ge. obesa	0	0	24	1	1	1	0	1	11	3	2
Ga. glutinata	3	33	8	11	25	11	28	29	39	48	51
Ga. uvula	0	0	0	0	2	1	0	3	0	0	0
O. suturalis	0	0	0	0	4	1	0	0	0	0	0
O. universa	6	18	17	8	19	15	37	10	15	29	18
Ss. kochi	0	5	0	0	0	0	0	1	14	32	11
Ss. seminulina	16	27	3	0	1	1	2	14	0	0	0
Gd. suteri	0	0	0	0	0	20	0	0	2	2	2
C. nitida	1	0	1	0	0	0	0	0	2	2	1
Total	413	480	535	242	394	278	425	643	717	823	387

APPENDIX 2A. OXYGEN AND CARBON ISOTOPIC DATA RELATIVE TO
PDB ON SAMPLES BELONGING TO THE 22 MA TIME-SLICE

Site	Core/Section	Interval	$\delta^{18}O$ PDB	$\delta^{13}C$ PDB	$\delta^{18}O$ PDB	$\delta^{13}C$ PDB	$\delta^{18}O$ PDB	$\delta^{13}C$ PDB
			Globoquadrina dehiscens		Globigerina praebulloides			
14	2-1	145-150	1.35	1.90	1.01	1.64		
	2-2	69-74	1.53	1.86	1.02	1.69		
	2-3	75-80	1.25	1.72	1.08	1.63		
	2-4	143-148	1.39	1.69	1.15	1.66		
		n	4	4	4	4		
		x̄	1.38	1.79	1.07	1.66		
		s	0.12	0.10	0.06	0.03		
18	4-2	145-150	1.02	1.33	0.35	1.29		
	4-3	124-129	1.29	1.90	0.71	1.73		
	4-4	142-148	1.35	2.00				
	4-5	145-148	1.04	1.88	0.77	1.93		
	5-2	145-150	0.95	1.84				
	5-3	145-150			0.94	1.82		
	5-5	145-150	0.85	1.82				
		n	6	6	4	4		
		x̄	1.08	1.80	0.69	1.69		
		s	0.02	0.24	0.25	0.28		
			Globorotalia kugleri		Globigerina praebulloides			
357	12-1	145-150	0.65	1.04			0.56	0.96
	12-2		0.45	0.91	-0.41	1.38		
	12-3		0.49	1.13	-0.72	1.50	-0.04	1.13
	12-5		0.26	0.89			0.31	0.75
	12-6		0.59	1.41	-0.55	1.82		
	13-1		0.56	1.32	-0.70	1.81		
	13-2		0.44	1.33	-0.50	2.00		
	13-3		0.46	0.85				
	13-4		0.27	1.67				
	13-5		0.88	1.97	0.13	2.09	0.05	1.04
	13-6		0.90	1.57	-0.02	2.02		
		n	11	11	7	7	4	4
		x̄	0.54	1.28	-0.40	1.80	0.22	0.97
		s	0.21	0.36	0.33	0.27		
			Catapsydrax					
362	39-3	145-150	0.34	1.13	0.61	1.07		
	39-4		0.24	1.17	0.72	1.18		
	39-5		0.13	0.95				
	39-6		0.31	1.08	0.94	0.90		
	40-1		0.21	1.13				
	40-2		0.05	1.27	0.99	1.22		
	40-3		-0.07	1.46	0.63	1.34		
	40-4		0.05	1.05	0.63	1.26		
	40-5		0.34	1.35	0.78	1.08		
	40-6		0.11	1.21	0.67	0.97		
		n	10	10	8	8		
		x̄	0.17	1.18	0.75	1.13		
		s	0.14	0.15	0.15	0.15		
			Catapsydrax		Globigerina praebulloides		Globoquadrina dehiscens	
526A	27-1	50-54	1.56	1.80	1.11	1.81		
	27-2		1.70	2.01			1.05	1.87
	27-3		1.56	1.82				
	28-2		1.11	1.64	0.93	1.38	0.94	1.48
	28-3		1.34	1.83	0.97	1.68		
	29-1		1.34	1.71	1.01	1.33		
	29-2		1.48	1.53	0.87	1.29		
	29-3		0.97	1.40				
		n	8	8	5	5	2	2
		x̄	1.38	1.72	0.98	1.50	1.00	1.68
		s	0.25	0.19	0.09	0.23		
			Globoquadrina dehiscens		Globigerina praebulloides			
360	26-1	100-107	1.18	1.94	1.02	2.23		
		142-150	1.10	1.74				
	26-2	45-50	0.77	1.75	0.70	1.80		
		93-97	0.78	1.73	0.75	1.91		
		n	4	4	3	3		
		x̄	0.96	1.79	0.82	1.98		
		s	0.21	0.10	0.17	0.22		

APPENDIX 2B. OXYGEN AND CARBON ISOTOPIC DATA RELATIVE TO
PDB ON SAMPLES BELONGING TO THE 16 MA TIME-SLICE

Site	Core/ Section	Interval	$\delta^{18}O$ PDB	$\delta^{13}C$ PDB	$\delta^{18}O$ PDB	$\delta^{13}C$ PDB
			Globoquadrina dehiscens		*Globigerinoides* spp.	
15	6-6	46-50			0.70	1.86
	7-1	65-70	0.90	2.28	0.23	2.73
	7-2	55-59	0.95	2.01		
	7-4	48-57			0.47	2.52
		n	2	2	3	3
		x̄	0.93	2.15	0.47	2.37
		s			0.24	0.45
360	22-2	74-79	0.42	2.60		
		142-147	0.83	2.39	0.16	2.15
	22-3	71-76	1.40	2.32	-0.21	2.33
		145-150	0.42	2.03	-0.12	2.27
	22-4	21-26	0.54	2.21	0.19	2.34
		145-150	0.44	2.47	0.05	2.30
	22-5	79-83	0.61	2.41	-0.05	2.54
	22-6	70-74	0.35	2.31		
		146-150	0.20	2.18	-0.12	3.17
		n	9	9	7	7
		x̄	0.58	2.32	-0.01	2.44
		s	0.35	0.17	0.15	0.34
362	36	cc	-0.39	2.18	-1.31	2.87
	37-1	72-77	-0.47	1.99	-1.08	2.46
		145-150	-0.63	2.29	-1.10	2.78
	37-2	70-75	-0.44	2.02	-1.30	2.56
		145-150	-0.62	2.28	-1.48	2.76
		n	5	5	5	5
		x̄	-0.51	2.15	-1.26	2.69
		s	0.11	0.14	0.17	0.17
516	21-1	51-55	0.14	1.47		
	21-2	50-54	0.00	1.59	-0.14	2.02
	21-3	52-56	0.06	1.78	-0.30	1.93
	22-1	51-55	0.21	1.40		
		126-130	-0.12	1.49	-0.16	1.51
	22-2	50-54			-0.21	1.72
		n	5	5	4	4
		x̄	0.06	1.55	-0.20	1.80
		s	0.13	0.15	0.07	0.23
526A	21-1	50-54	0.68	1.49	0.27	2.07
		130-134	0.62	1.24	0.58	2.01
	21-2	50-54	0.95	1.48	0.15	2.02
		130-134	0.39	1.50		
	21-3	50-54	0.74	1.24	0.11	1.85
		130-134	0.97	1.38		
	21-4	11-16	0.86	1.34		
		n	7	7	4	4
		x̄	0.74	1.38	0.28	1.99
		s	0.20	0.11	0.21	0.10

APPENDIX 2C. OXYGEN AND CARBON ISOTOPIC DATA RELATIVE TO
PDB ON SAMPLES BELONGING TO THE 8 MA TIME-SLICE

Site	Core/ Section	Interval	$\delta^{18}O$ PDB	$\delta^{13}C$ PDB	$\delta^{18}O$ PDB	$\delta^{13}C$ PDB
			Globorotalia conoidea		Globigerinoides spp.	
16	9-1	58-65	1.45	1.73	0.85	2.05
	9-4	51-57	1.35	1.93	0.92	2.14
	9-5				1.04	1.98
	10-1	50-55	1.44	1.97	0.76	1.84
	10-3	48-53	1.41	1.93	1.01	2.09
	10-5	50-55	1.20	1.94		
		n	5	5	5	5
		x̄	1.37	1.90	0.92	2.02
		s	0.10	0.10	0.12	0.12
360	8-2	50-55	1.55	0.84		
	9-1		0.88	0.54	-0.29	1.08
	9-3	38-43	1.35	1.19	-0.42	0.78
	9-5		0.34	0.53	-0.93	1.73
	10-1	45-50	1.51	1.24	-0.20	1.40
	10-3		0.68	0.50	-0.36	1.39
	11-2	50-56	1.19	1.34	-0.27	1.74
	11-4	50-55	1.14	1.23	-0.16	1.39
	11-6	50-55	1.21	1.43	-0.11	1.40
		n	9	9	8	8
		x̄	1.09	0.98	-0.34	1.36
		s	0.40	0.38	0.26	0.32
362	24-1	52-56	0.82	-0.02	-1.31	1.58
	24-3	52-56	0.94	0.99		
	25-1	50-54	0.70	1.05		
	26-3	50-59	0.78	1.15	-0.84	1.57
	26-6	50-54			-0.07	1.56
	27-1	52-56	0.61	0.72	-0.63	1.53
	27-6	52-57	0.96	1.37	-0.43	1.93
		n	6	6	5	5
		x̄	0.80	0.88	-0.66	1.63
		s	0.14	0.49	0.46	0.17
516	13-1	70-72			0.78	2.01
	13-2	70-72	1.21	1.40	0.51	1.86
	13-3	60-62	0.94	1.56	0.27	1.50
	14-1	70-72	1.04	1.42	0.55	1.56
	14-2	70-72	1.10	1.88		
	14-3	70-72	0.95	1.51	0.97	1.90
		n	5	5	5	5
		x̄	1.05	1.55	0.62	1.77
		s	0.11	0.19	0.27	0.22
526A	9-1	50-54	1.05	1.23		
	9-2	50-54	0.91	1.34	0.48	1.53
	9-3	51-55	0.89	1.29	0.71	2.29
	10-1	50-54	1.08	1.33	0.36	2.00
	10-2	50-54	1.09	1.60		
	10-3	50-54	1.00	1.26	0.37	1.84
	11-1	50-45			0.38	2.19
	11-3	55-59	0.92	1.25	0.36	1.84
		n	7	7	6	6
		x̄	0.99	1.33	0.44	1.95
		s	0.08	0.13	0.14	0.27

REFERENCES CITED

Barrera, E., Keller, G., and Savin, S. M., 1985, Evolution of the Miocene Ocean in the eastern north Pacific as inferred from oxygen and carbon isotopic ratios of Foraminifera, *in* Kennett, J. P., ed., The Miocene ocean: Paleoceanography and biogeography: Geological Society of America Memoir 163 (this volume).

Barron, J. A., Keller, G., and Dunn, D. A., 1985, A multiple microfossil biochronology for the Miocene, *in* Kennett, J. P., ed., The Miocene ocean: Paleoceanography and biogeography: Geological Society of America Memoir 163 (this volume).

Bé, A.W.H., and Tolderlund, D. S., 1977, Distribution and ecology of living planktonic foraminifera in surface waters of the Atlantic and Indian Oceans, *in* Funnell, B. M., and Reidel, W. R., eds., The Micropaleontology of the Oceans: London, Cambridge University Press, p. 105–149.

Berggren, W. A., Aubry, M.-P., and Hamilton, N., 1983, Neogene magnetobiostratigraphy of DSDP Site 516A (Rio Grande Rise, South Atlantic), *in* Initial Reports of the Deep Sea Drilling Project, v. 72: Washington, D.C., U.S. Government Printing Office, p. 675–713.

Blow, W. H., 1969, Late Middle Eocene to Recent planktonic foraminiferal biostratigraphy, *in* Bronnimann, R., and Renz, H. H., eds., Proceedings International Conference of Planktonic Microfossils I, Geneva, 1967: Leiden, E. J. Brill, p. 199–421.

Boersma, A., and Shackleton, N. J., 1977, Tertiary oxygen and carbon isotope stratigraphies, Site 357 (mid-latitude South Atlantic), *in* Initial Reports of the Deep Sea Drilling Project, v. 39: Washington, D.C., U.S. Government Printing Office, p. 911–924.

Dean, W. E., Hay, W. W., and Sibuet, J.-C., 1984, Geologic evolution, sedimentation, and paleoenvironments of the Angola Basin and adjacent Walvis Ridge: Synthesis of results of Deep Sea Drilling Project Leg 75, *in* Initial Reports of the Deep Sea Drilling Project, v. 75: Washington, D.C., U.S. Government Printing Office, p. 509–542.

Haq, B. U., 1980, Biogeographic history of Miocene calcareous nannoplankton and paleoceanography of the Atlantic Ocean: Micropaleontology, v. 26, p. 414–443.

Jenkins, D. G., 1978, Neogene planktonic foraminifers from DSDP Leg 40 Sites 360 and 362 in the southeastern Atlantic, *in* Initial Reports of the Deep Sea Drilling Project, v. 40: Washington, D.C., U.S. Government Printing Office, p. 723–732.

Keller, G., 1981a, Planktonic foraminiferal faunas of the equatorial Pacific suggest early Miocene origin of present oceanic circulation: Marine Micropaleontology, v. 6, p. 535–551.

——1981b, Miocene biochronology and paleoceanography of the North Pacific: Marine Micropaleontology, v. 6, p. 269–295.

Keller, G., Barron, J. A., and Burckle, L. H., 1982, North Pacific late Miocene correlations using microfossils, stable isotopes, percent $CaCO_3$, and magnetostratigraphy: Marine Micropaleontology, v. 7, p. 327–357.

Kennett, J. P., 1973, Middle and late Cenozoic planktonic foraminiferal biostratigraphy of the southwest Pacific-DSDP Leg 21, *in* Initial Reports of the Deep Sea Drilling Project, v. 21: Washington, D.C., U.S. Goverment Printing Office, p. 575–639.

——1978, The development of planktonic biogeography in the Southern Ocean during the Cenozoic: Marine Micropaleontology, v. 3, p. 301–345.

Kennett, J. P., Keller, G., Srinivasan, M. A., 1985, Miocene planktonic foraminif-

eral biogeography and paleoceanographic development of the Indo-Pacific region, *in* Kennett, J. P., ed., The Miocene ocean: Paleoceanography and biogeography: Geological Society of America Memoir 163 (this volume).

Kennett, J. P., and Srinivasan, M. S., 1983, Neogene Planktonic Foraminifera: A Phylogenetic Atlas: Stroudsburg, PA, Hutchinson Ross Publishing Company, 265 pp.

Martini, E., 1971, Standard Tertiary and Quaternary calcareous nannoplankton zonation, *in* Farinacci, A., ed., Proceedings of the second Planktonic Conference Roma 1970: Rome, Edizioni Techoscienza, v. 2, p. 739–777.

Matthews, R. K., and Poore, R. Z., 1980, Tertiary $\delta^{18}O$ record and glacio-eustatic sea-level fluctuations: Geology, v. 8, p. 501–504.

Maxwell, A. E., et al., 1970, Initial Reports of the Deep Sea Drilling Project, v. 3: Washington, D.C., U.S. Government Printing Office, p. 71–275.

Prell, W. L., and Curry, W. B., 1981, Faunal and isotopic indices of monsoonal upwelling: Western Arabian Sea: Oceanologica Acta, v. 4, p. 91–98.

Reid, J. L., Nowlin, W. D., and Patzert, W. C., 1976, On the characteristics and circulation of the Southwestern South Atlantic Ocean: Journal of Physical Oceanography, v. 7, p. 62–91.

Savin, S. M., Abel, L., Barrera, E., Hodell, D., Keller, G., Kennett, J. P., Killingley, J., Murphy, M., and Vincent, E., 1985, The evolution of Miocene surface and near-surface marine temperatures: oxygen isotopic evidence, *in* Kennett, J. P., ed., The Miocene ocean: Paleoceanography and biogeography: Geological Society of America Memoir 163 (this volume).

Savin, S. M., Douglas, R. G., and Stehli, F. G., 1975, Tertiary marine paleotemperatures: Geological Society of America Bulletin, v. 86, p. 1499–1510.

Savin, S. M., Douglas, R. G., Keller, G., Killingley, J. S., Shaughnessy, L., Sommer, M. A., Vincent, E., and Woodruff, F., 1981, Miocene benthic foraminiferal isotope records: a synthesis: Marine Micropaleontology, v. 6, p. 432–450.

Sclater, J. G., Meinke, L., Bennett, A., and Murphy, C., 1985, The depth of the ocean through the Neogene, *in* Kennett, J. P., ed., The Miocene ocean: Paleoceanography and biogeography: Geological Society of America Memoir 163 (this volume).

Seisser, W. G., 1980, Late Miocene origin of the Benguela upwelling system off northern Namibia: Science, v. 208, p. 283–285.

Srinivasan, M. S., and Kennett, J. P., 1981a, Neogene planktonic foraminiferal biostratigraphy and evolution: equatorial to subantarctic South Pacific: Marine Micropaleontology, v. 6, p. 499–533.

——1981b, A review of Neogene planktonic foraminiferal biostratigraphy: Applications in the equatorial and South Pacific, *in* The Deep Sea Drilling Project: A Decade of Progress: Tulsa, OK, SEPM Special Publication No. 32, p. 315–432.

Thunell, R. C., and Belyea, P., Neogene planktonic foraminiferal biogeography of the Atlantic Ocean: Micropaleontology, v. 28, no. 4, p. 381–398.

Thunell, R. C., Honjo, S., and Curry, W. B., 1983, Planktonic foraminiferal flux: seasonal variations in the Panama Basin: Earth and Planetary Science Letters, v. 64, no. 1, p. 44–55.

Woodruff, F., Savin, S. M., and Douglas, R. G., 1981, Miocene stable isotope record: detailed deep Pacific Ocean study and its paleoclimatic implications: Marine Micropaleontology, v. 6, p. 617–632.

MANUSCRIPT ACCEPTED BY THE SOCIETY DECEMBER 17, 1984

Typeset by WESType Publishing Services, Inc., Boulder, Colorado
Printed in U.S.A. by Malloy Lithographing, Inc., Ann Arbor, Michigan

Scott W. Starratt
Dept. of Paleontology
U. C. Berkeley
Berkeley, Ca. 94720

Scott W. Starratt
Dept. of Paleontology
U. C. Berkeley
Berkeley, Ca. 94720